MODERN PRACTICE
OF GAS CHROMATOGRAPHY

Modern Practice
of Gas Chromatography

Edited by

ROBERT L. GROB

Professor of Analytical Chemistry
Villanova University
Villanova, Pennsylvania

A WILEY-INTERSCIENCE PUBLICATION

JOHN WILEY & SONS, New York • London • Sydney • Toronto

Library of Congress Cataloging in Publication Data:

Main entry under title:

Modern practice of gas chromatography.

   "A Wiley-Interscience publication."
   Includes index.
   1. Gas chromatography.   I. Grob, Robert Lee.

QD79.C45M63   543'.08   77-779
ISBN 0-471-01564-4

Printed in the United States of America

10 9 8 7 6 5 4 3 2

Contributing Authors

Dr. Frederick J. Debbrecht, Analytical Instrument Development, Avondale, Pa.

Mr. Murrell R. Dobbins, Steroid Laboratory, Hospital of the University of Pennsylvania, Philadelphia, Pa.

Dr. Robert L. Grob, Chemistry Department, Villanova University, Villanova, Pa.

Dr. Mary A. Kaiser, Chemistry Department, University of Georgia, Athens, Ga.

Dr. Eugene J. McGonigle, Merck, Sharp and Dohme Research Laboratories, West Point, Pa.

Dr. Matthew J. O'Brien, Merck, Sharpe and Dohme Research Laboratories, West Point, Pa.

Mr. Harvey L. Pierson, MIS Department, ICI United States, Inc., Wilmington, DE.

Dr. Herbert L. Rothbart, Eastern Utilization Research and Development Division, USDA, Philadelphia, PA.

Mr. Rüder Schill, Hewlett-Packard Company, Avondale, PA.

Dr. Daniel J. Steible, Jr., Stuart Division, ICI United States, Inc., Wilmington, DE.

Mr. James J. Sullivan, Hewlett-Packard Company, Avon-
    dale, PA.

Dr. Walter R. Supina, Supelco, Inc., Supelco Park,
    Bellefonte, PA.

Dr. Joseph C. Touchstone, Steroid Laboratory, Hospit-
    al of the University of Pennsylvania, Philadel-
    phia, PA.

Dr. Gerald R. Umbreit, Greenwood Laboratories, Ken-
    nett Square, PA.

To

Marjorie, Kent, Duane, Michele and Allyson

Foreword

It is appropriate that a book on modern gas chromatography (GC) be written by member-experts from the
Chromatography Forum of the Delaware Valley. This
active scientific body, which lists as a charter member the well-known GC pioneer, the late Stephen Dal
Nogare, has long been active in the development and
dissemination of chromatographic technology. The
Chromatography Forum has held meetings, organized
symposia, sponsored books on liquid and thin-layer
chromatography, add presented awards in chromatography. For several years the Forum has presented a
practical course in GC that inevitably has been over-
subscribed because of its popularity. This textbook
is an outgrowth of this highly successful course.
   The chapters in this book have been prepared by
those personally skilled in the various areas they
have developed. Professor Grob has welded these chapters together into a cohesive treatment, but the in-
dividuality of each expert remains so that the reader
is able to sample the unique experiences and insights
of the writer. It is fortunate that the many con-
tributors to an edited book live and work in the same
area and are members of the same organization. Thus,
the ability to easily communicate and exchange ideas,
develop formats, and closely interact, has resulted
in an edited book with unusual coverage and read-
ability.
   This is a practical book that can be used by workers with a variety of scientific backgrounds. It

contains topics and experiences that have not been developed in other books. Current information, references, and data all are combined to produce an up-to-date treatment.

For several personal reasons I am very gratified that this new book on gas chromatography has been prepared. It will stimulate the development of those interested in using this very powerful method to solve the many chemical problems which still confront the scientific community.

J.J. Kirkland

Wilmington, Delaware
February 1, 1977

Preface

Gas chromatography (GC) is one of the most widely employed analytical techniques today. It is used in
the academic and research laboratories as well as in
industry. GC, in the form of gas-liquid chromatography had its beginning at the end of 1952 in the
accounts of James and Martin {Biochem. J., $\underline{50}$, 679
(1952) and Analyst, $\underline{77}$, 915 (1952)}.
  The wide acceptance and success of this technique
have been due to such features as simplicity, rapidity
of analysis, high sensitivity of detector systems,
efficiency of separations, varied applications and
the use of very small samples (microgram or smaller).
Presently GC is finding use in the concentration of
impurities in the parts per million (ppm) and parts
per billion (ppb) range and in addition to the actual
measurement of impurities at these levels. Without
the use of GC many analytical problems could not be
solved or would involve more intricate and time-consuming techniques.
  This text represents a blending of the basic
theories of chromatography with the experiences of
each author. It is the product of a Gas Chromatography Short Course which has been presented annually
in Philadelphia by the Chromatography Forum of Delaware Valley. The course has been offered each year
with the highest level of participation. It has been
structured for not only beginners but for existing
workers in the field and specialists of other fields
who wish and need to know more about this powerful

technique.

The authors feel that there is a current need for a textbook about gas chromatography. Existing books on the topic are either out of date or do not place the proper emphasis on the technique's modern practices. This situation became painfully apparent when deciding on a text for the short course that subsequently gave birth to this book.

The nomenclature recommended by the IUPAC Committee of Chromatography (1972) and the now fashionable SI units have been utilized as much as possible.

The gas chromatographic technique is explained on the basis of a physical process with correlations to distillation, liquid-liquid extraction, countercurrent distribution, and other separation techniques to give the reader a better appreciation of the basic process of chromatography. Explanation of fundamentals is followed by chapters on columns and column selection, theory and use of detectors, instrumentation necessary for a gas chromatographic system, techniques used for qualitative and quantitative analyses, and data reduction and readout. Subsequent chapters cover specialized areas in which gas chromatographic literature is more scattered and data collection and evaluation are more important.

The format of the book represents what the various authors, collectively and individually, consider to be essential to the student of chromatography. We have tried to be consistent in nomenclature and have furnished representative and comprehensive bibliographies to allow the reader to explore further and in more depth the topics presented. The book is presented in a chapter order that we feel flows smoothly for the explanation of the gas chromatographic process.

Any book, whether a textbook or reference work can not be produced without the help and cooperation of many people. I wish to acknowledge the members of the Chromatography Forum of the Delaware Valley, the contributing authors for their cooperation, professional attitude, and unselfish availability of their time my son Duane for the drawings in several of the chapters and the creation of the cover drawing, Dr. Gerald R. Umbreit for assistance and cooperation beyond those of a contributor, and Drs. Mary A. Kaiser, Matthew J. O'Brien, and John F. Wojcik for helping

with the reading of the chapters. A special acknowl-
edgment is extended to Dr. J.J. Kirkland who reviewed
and offered much assistance with the manuscript. I
also thank him for writing the foreword. Last, but
by no means least, I gratefully acknowledge my wife
Marjorie for her typing and encouragement throughout
this project.

Robert L. Grob

Villanova, Pennsylvania
February 1,1977

Contents

1. Introduction 1
   Robert L. Grob

   PART ONE   THEORY AND BASICS

2. Theory of Gas Chromatography 39
   Robert L. Grob

3. Columns and Column Selection in Gas
   Chromatography 113
   Walter R. Supina

4. Qualitative and Quantitative Analysis by
   Gas Chromatography 151
   Mary A. Kaiser and Frederick J. Debbrecht

   PART TWO   TECHNIQUES AND INSTRUMENTATION

5. Detectors. 213
   James J. Sullivan and Matthew J. O'Brien

6. Instrumentation. 289
   Ruder Schill

7. Trace Analysis by Gas Chromatography 365
   Gerald R. Umbreit

8. Selection of Analytical Data from a Gas
   Chromatographic Laboratory 421
   Harvey L. Pierson and Daniel J. Steible, Jr.

PART THREE   APPLICATIONS

9. Gas Chromatographic Analysis of Food 449
   Herbert L. Rothbart

10. Clinical Applications of Gas Chromatography. . 495
    Joseph C. Touchstone and Murrell R. Dobbins

11. Physicochemical Measurements by Gas
    Chromatography 553
    Mary A. Kaiser

12. Drug Analysis Using Gas Chromatography 591
    Eugene J. McGonigle

    Index 637

MODERN PRACTICE
OF GAS CHROMATOGRAPHY

WHAT IS WRITTEN WITHOUT EFFORT IS IN GENERAL READ
WITHOUT PLEASURE.

Samuel Johnson (1709-1784)
Johnsonian Miscellanies,
Vol. ii, p. 309

CHAPTER 1

Introduction

ROBERT L. GROB

Villanova University

1.1   HISTORY AND DEVELOPMENT OF CHROMATOGRAPHY. 2

1.2   CHROMATOGRAPHIC METHODS. 4

     1.2.1   Classification of Methods. 4

     1.2.2   General Aspects. 6

     1.2.3   Frontal Analysis 6

     1.2.4   Displacement Development 7

     1.2.5   Elution Development. 8
                Summary. 9

     1.2.6   Isotherms. 10

     1.2.7   Process Types in Chromatography. 12

     1.2.8   Linear-Ideal Chromatography. 13

     1.2.9   Linear-Nonideal Chromatography 13

1

      1.2.10 Nonlinear-Ideal Chromatography.   14

      1.2.11 Nonlinear-Nonideal Chromatography   16

1.3   GENERAL ASPECTS OF GAS CHROMATOGRAPHY   16

      1.3.1   Applications of Gas Chromatography.   16

      1.3.2   Types of Detection.   18

      1.3.3   Advantages and Limitations.   19

1.4   DEFINITIONS AND NOMENCLATURE.   21

1.5   LITERATURE OF GAS CHROMATOGRAPHY.   34

1.6   COMMERCIAL INSTRUMENTS.   36

      REFERENCES.   36

1.1   HISTORY AND DEVELOPMENT OF CHROMATOGRAPHY

Many publications have discussed or detailed the history and development of chromatography (1-3). Rather than duplicate these writings, we present in Table 1.1 a chronological listing of events that we feel are the most relevant in the development of the present state of the field. Since the various types of chromatography (liquid, gas, paper, thin-layer, ion exchange) have many features in common, they must all be considered in the development of the field. While the topic of this text, gas chromatography, probably has been the most investigated during the past 25 years, results of these studies have had a great impact on the other types of chromatography, especially modern liquid chromatography.

There will, of course, be those who believe that others should be added to names or events in Table 1.1. We simply wish to show a development of an ever-expanding field and to point out some of the important events that were responsible for the expansion. To attempt an account of contemporary leaders of the field could only result in disagreement with some workers, astonishment by others, and a very long listing which would be very cumbersome to correlate.

TABLE 1.1 Development of the Field of Chromatography

Year (Reference)	Scientist(s)	Comments
1834 (4) 1843 (5)	Runge, F. F.	Used unglazed paper and/or pieces of cloth for spot testing dye mixtures and plant extracts.
1850 (6)	Runge, F. F.	Separation of salt solutions on paper.
1868 (7)	Goppelsroeder, F.	Paper strip(capillary analysis) analysis of dyes, hydrocarbons, alcohols, milk, beer, colloids, drinking and mineral waters, plant and animal pigments.
1878 (8)	Schönbein, C.	Paper strip analysis of liquid solutions.
1897-1903 (9-11)	Day, D. T.	Ascending flow of crude petroleum samples through column packed with finely pulverized fuller's earth.
1906-1907 (12-14)	Tswett, M.	Separated chloroplast pigments on $CaCO_3$ solid phase and petroleum ether liquid phase.
1931 (15)	Kuhn, R. et al.	Liquid-solid chromatography for separating egg yolk xanthophylls.
1940 (16)	Tiselius, A.	Nobel Prize in 1948. Adsorption analyses and electrophoresis.
1940 (17)	Wilson, J. N.	First theoretical paper on chromatography. Assumed complete equilibration and linear sorption isotherms. Qualitatively defined diffusion, rate of adsorption and isotherm nonlinearity.
1941 (18)	Tiselius, A.	Developed liquid chromatography and pointed out frontal analysis, elution analysis, and displacement development.

1941 (19)	Martin, A. J. P. and Synge, R. L.M.	Presented first model that could describe column efficiency. Developed liquid-liquid chromatography. Received Nobel Prize in 1952.
1944 (20)	Consden, R. Gordon, A. H., and Martin, A. J. P.	Development of paper chromatography.
1946 (21)	Claesson, S.	Liquid-solid chromatography with frontal and displacement development analysis. Co-worker A. Tiselius.
1949 (22)	Martin, A. J. P.	Contributed to relationship between retention and thermodynamic equilibrium constant.
1951 (23)	Cremer, E.	Gas-solid chromatography by elution development.
1952 (24)	Phillips, C. S. G.	Liquid-liquid chromatography by frontal technique.
1952 (25)	James, A. T. and Martin, A. J. P.	Introduced gas-liquid chromatography.
1955 (26)	Glueckauf, E.	Derived first comprehensive equation for relationship between HETP and particle size, particle diffusion, and film diffusion in ion exchange.
1956 (27)	Van Deemter, J. J. et al.	Developed rate theory by simplifying work of Lapidus and Ammundson to Gaussian distribution function.
1965 (28)	Giddings, J. C.	Reviewed and extended early theories of chromatography.

1.2 CHROMATOGRAPHIC METHODS

1.2.1 Classification of Methods

In the strictest sense, the term chromatography is a misnomer. Most of the materials chromatographed today are either colorless or, if they were colored, one would not be able to perceive them

in most instances. A number of workers in the field have put
forth contemporary definitions of the term, but not all practi-
tioners of the technique use these terms or even agree with them.
We shall present our own definition, but do not declare it to be
unique or more representative of the process:

Chromatography encompasses a series of techniques having in
common the separation of components of a mixture by a series of
equilibrium operations which result in the entities being separated as a result of their partitioning (differential sorption)
between two different phases; one stationary with a large surface
and the other a moving phase in contact with the first. Chromatography is not restricted to analytical separations. It may be
used in the preparation of pure substances, the study of the kinetics of reactions, structural investigations on the molecular
scale and the determination of physicochemical constants, for example, stability constants of complexes, enthalpy, entropy, and
free energy (see Chapter 11).

Using the definition above (or any other definition of chromatography), one can tabulate numerous variations of the technique
(see Figure 1.1). Our specific concern is the technique of gas
chromatography. For this technique we have available different

ADSORPTION CHROMATOGRAPHY

Liquid-Solid Column Chromatography(LSC)

Paper Chromatography(PC)
Thin-Layer Chromatography(TLC)
Gas-Solid Chromatography(GSC)
    Packed Columns
    Open Tubular Columns

ION EXCHANGE

Liquid-Solid Chromatography(LSC)

Paper Chromatography(PC)
Thin-Layer Chromatography(TLC)

PARTITION CHROMATOGRAPHY

Liquid-Liquid Column Chroma-
            tography(LLC)
Paper Chromatography(PC)
Thin-Layer Chromatography(TLC)
Foam Chromatography(FC)
Emulsion Chromatography(EC)
Gas-Liquid Chromatography(GLC)
    Packed Columns
    Open Tubular Columns

EXCLUSION CHROMATOGRAPHY

Gel Filtration or Permeation
        Chromatography(GPC)
Molecular Sieves

Figure 1.1.

types of columns which may be used to perform the separation.
More details are found in Chapter 3.

1.2.2 General Aspects

The mixture to be separated and analyzed may be either a gas,
liquid, or a solid in some instances. All that is required is
that the materials be stable, have a vapor pressure of ∿0.1 torr
at the operating temperature and interact with the column material (either a solid adsorbent or a liquid stationary phase) and
the mobile phase (carrier gas). The result of this interaction
is the differing distribution of the sample components between
the two phases, resulting in the separation of the sample components into zones or bands. The principle that governs the chromatographic separation is the foundation of most physical methods
of separation, for example, distillation and liquid-liquid
extraction.
   The separation of the sample components may be achieved by one
of three techniques, namely, frontal analysis, displacement development, or elution development.

1.2.3. Frontal Analysis

The liquid or gas mixture is fed into a column containing an adsorbent. The mixture acts as its own mobile phase or carrier and
the separation depends upon the ability of each component, in the
mixture, to become an adsorbate (see Figure 1.2). Once the

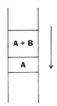

Figure 1.2. Frontal Analysis.   Component B more adsorbed than component A.

column adsorbent has been saturated (i.e., no longer able to adsorb more components) the mixture then flows through with its
original composition. The early use of this technique involved
measuring the change in concentration of the front leaving the
column, hence the name frontal analysis. The least-adsorbed component breaks through first and is the only component to be obtained in a pure form. Figure 1.3 illustrates the integral type
recording for this type of system. In this figure we illustrate

the recording of the fronts obtained from a four-component
sample.

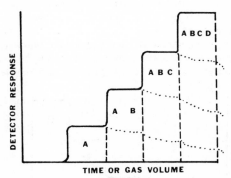

Figure 1.3.   Integral type chromatogram from frontal
analysis.   Component A least absorbed of four components.

Frontal analysis brings with it the requirement of the system
to have convex isotherms (see Section 1.2.6).   This results in
the peaks having sharp fronts and well-formed steps.   An inspec-
tion of Figure 1.3 reflects the problem of analytical frontal
analysis-- it is difficult to calculate initial concentrations
in the sample.   One can, however, determine the number of components present in the sample.   If the isotherms are linear, the
zones may be diffuse.   This may be caused by three important processes: inhomogeneity of the packing, large diffusion effects,
and nonattainment of sorption equilibrium.

1.2.4 Displacement Development

In this technique the developer is contained in the moving phase
which may either be a liquid or a gas (Figure 1.4).   One necessary criterion is that this moving phase must be more adsorbed

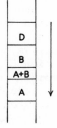

Figure 1.4.   Displacement development.   D is displac-.
er.   D is more absorbed than B which is more absorbed
than A.

than any sample components. One always obtains a single band of
pure component for each component in the sample. In addition,
there is always an overlap zone for each component, which is an
advantage of this technique over frontal analysis. The disadvantage, from the analytical viewpoint, is that the component bands
are not separated by a region of pure mobile phase. The result
of this displacement mechanism shown in Figure 1.5 assumes a
three-component mixture and a developer or displacing agent.

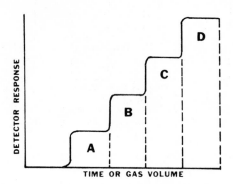

Figure 1.5. Integral type chromatogram from displacement development. Order of adsorption: D> C > B > A.

The step height is utilized for qualitative identification of
components whereas the step length is proportional to the amount
of the component.

As with frontal analysis, displacement analysis requires convex isotherms. Once equilibrium conditions have been attained,
an increase in column length serves no useful purpose in this
technique because the separation is more dependent on equilibrium
conditions than on the size of column.

1.2.5 Elution Development

In this technique, components A and B travel through the column
at rates determined by their retention on the solid packing
(Figure 1.6). If the differences in adsorption are sufficient
and/or the column is long enough, a complete separation of A and
B is possible. Continued addition of eluent causes the emergence
of separated bands or zones from the column. A disadvantage of
this technique is the very long time interval required to remove
a highly adsorbed component. This can be overcome by increasing
the column temperature during the separation process. Figure 1.7
depicts a typical chromatogram for this technique. The position

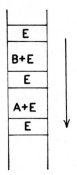

Figure 1.6. Elution development. E = Eluant.
B is more adsorbed than A.

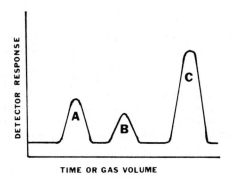

Figure 1.7. Differential chromatogram from elution
development. Order of retention: C > B > A.

of peak maximum on the abcissa qualitatively identifies the component, and the peak area is a measure of the amount of each component.

SUMMARY. The frontal technique does not lend itself to many analytical applications because of the overlap of the bands and the
requirement of a large amount of sample. However, it may be used
to study phase equilibria (isotherms) and for preparative separations. (Many of the industrial chromatographic techniques use
frontal analysis.) Displacement development has applications for
analytical LC (e.g. it may be used as an initial concentrating
step in GC for trace analysis). This technique may also be used
in preparative work. The outstanding disadvantage of both of
these techniques is that the column still contains sample at the
conclusion of the separation. Thus, regeneration of the column
is necessary before it can be used again.
    It is in this regard that elution chromatography offers the

advantage--at the end of a separation, only eluent is left in the column. Thus, the bulk of the discussion in the subsequent chapters, will concern itself with elution GC. Discussion of the isotherms and chromatograms of elution chromatography will be found in Sections 1.2.7, 1.2.11, and in Chapter 2.

1.2.6 Isotherms

An isotherm is a graphical presentation of the interaction of an adsorbent and a sorbate in solution (gas or liquid solvent) at a given temperature. The isotherm is a graphical representation of the partition coefficient or distribution constant, K

$$K = \frac{C_S}{C_G} \tag{1.1}$$

where C_S is the concentration of sorbate in stationary phase or solid surface and C_G is the concentration of sorbate in the gas phase. One plots the concentration of the substance adsorbed per unit mass of adsorbent versus the concentration of the substance in equilibrium with the phase present at the interface. Three types of isotherms are obtainable: one linear and two curved. We will describe the nonlinear isotherms either concave (curved away from abcissa) or convex (curved towards the abcissa). Figure 1.8 depicts these three isotherms.

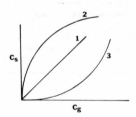

Figure 1.8.   Isotherms. C_s=Conc. at surface or in stationary phase; C_g=conc. in solution at equilibrium; (1) linear isotherm; (2) convex isotherm; (3) concave isotherm.

   The linear isotherm is obtained when the ratio of the concentration of substance adsorbed per unit mass and the concentration of the substance in solution remains constant. This means the partition coefficient or distribution constant, K (see Section 1.4) is a constant over all working concentration ranges. Thus, the front and rear boundaries of the band or zone will be symmetrical.

   The convex isotherm demonstrates that the K value is changing

to a higher ratio as concentration increases. This results in
the component moving through the column at a faster rate thus
causing the front boundary to be self-sharpening and the rear
boundary to be diffuse.

The concave isotherm results from the opposite effect (the K
value is changing to a lower value) and the peak will have a diffuse front boundary and a self-sharpening rear boundary. In
other words, the solute increasingly favors the surface of the
stationary phase as the solution concentration increases. These
effects are depicted in Figure 1.9. When the isotherms curve in
either direction (convex or concave) as concentration is varied,
one obtains complex chromatograms. Changing the sample concentration or physical conditions (temperature, flowrate, pressure,
etc.) can help in converting the rear and front boundaries to
Gaussian shape.

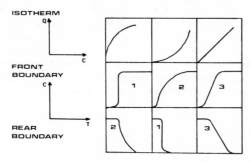

Figure 1.9. The dependence of the boundary profile
on the form of the partition isotherm. (Courtesy of
John Wiley-Interscience.) c=concentration $(cm^3/mole)$
of solute in gas phase; q=concentration in liquid or
adsorbed phase; t=time for band to emerge from
column; (1) self-sharpening profile; (2) diffuse
profile; (3) gaussian profile.

The most frequently applied isotherm equations are those of
Freundlich and Langmuir (29).

1. Freundlich Equation. This equation represents the variation
of adsorption with pressure over a limited range, at constant
temperature.

$$\frac{x}{m} = kp^{1/n} \tag{1.2}$$

where x = mass of adsorbed gas
       m = grams of adsorbing material,
       p = pressure,
       k, n = constants.

The exponent 1/n is usually less than one, indicating amount of adsorbed gas does not increase in proportion to the pressure.   If the exponent, 1/n were unity, the Freundlich equation would be equivalent to the distribution law.   Converting Equation 1.2 to to log form, we obtain,

$$\log\frac{x}{m} = \log k + \frac{1}{n} \log p \tag{1.3}$$

which is an equation of a straight line and thus, the log x/m versus log p relationship is linear (linear isotherm).   If a value of 1/n being unity gives a linear isotherm, then a value of 1/n>1 gives a concave isotherm.   When 1/n<1, a convex isotherm results.

2. Langmuir Equation.   It is probable that adsorbed layers have a thickness of a single molecule because of the rapid decrease in intermolecular forces with distance.   The Langmuir adsorption isotherm equation is

$$\frac{x}{m} = \frac{k_1 k_2 p}{1 + k_1 p} \tag{1.4}$$

where k_1, k_2 = constants for a given system, and
       p = pressure of gas which may be written as

$$\frac{p}{x/m} = \frac{1}{k_1 k_2} + \frac{p}{k_2} \tag{1.5}$$

A plot of p/(x/m) versus p produces a straight line with slope of $1/k_2$ and an intercept of $1/k_1 k_2$.   Deviations from linearity are attributed to nonuniformity leading to various types of adsorption on same surface, that is, nonmonomolecular adsorption on a homogeneous surface.

    1.2.7 Process Types in Chromatography

The process of chromatographic separation can be defined by two conditions:

1. The distribution isotherms (representation of the partition coefficient or distribution constant, K) may be either linear

or nonlinear (see Section 1.2.6).

2. The chromatographic system is either _ideal_ or _nonideal_. _Ideal_ chromatography infers that the exchange between the two phases is thermodynamically reversible. In addition, the equilibrium between the solid granular particles or liquid coated particles and the gas phase is immediate, that is, the mass transfer is very high, and longitudinal and other diffusion processes are small enough to be ignored. In _nonideal_ chromatography these assumptions cannot be made.

Using these two sets of conditions we can then describe four chromatographic systems.

1. Linear-ideal chromatography.
2. Linear-nonideal chromatography.
3. Nonlinear-ideal chromatography.
4. Nonlinear-nonideal chromatography.

1.2.8 Linear-Ideal Chromatography

This is the most direct and simple theory of chromatography. The transport of the solute down the column will depend upon the distribution constant (partition coefficient), K, and the ratio of the amounts of the two phases in the column. Band (zone) shape does not change during this movement through the column. The system could be visualized as illustrated in Figures 1.10, 1.11.

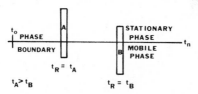

Figure 1.10. Linear ideal chromatography. t_o=start of separation (point of sample injection); t_A=retention time of component A; t_B=retention time of component B; t_n=time for emergence of mobile phase from t_o. $t_A > t_B$.

1.2.9 Linear-Nonideal Chromatography

In this system the bands (zones) broaden because of diffusion effects and nonequilibrium. This broadening mechanism is fairly symmetrical and the resulting elution bands approach the shape of a Gaussian curve. This system best explains liquid or gas partition chromatography. The system may be viewed in two ways:

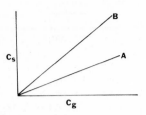

Figure 1.11. Isotherms for linear ideal chromatography. C_s = conc. at surface or in stationary phase;
C_g = conc. in solution at equilibrium.

1. Plate Theory. Envision the chromatographic system as a discontinuous process functioning the same as a distillation or extraction system, that, is comprised of a large number of equivalent plates.

2. Rate Theory. Consider the chromatographic system as a continuous medium where one accounts for mass transfer and diffusion phenomena.

It is these two points of view which are used in the Chapter 2
to discuss the theory of gas chromatography. Linear, nonideal
chromatography may be visualized by the relationships shown in
Figures 1.12 and 1.13.

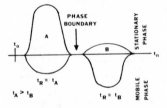

Figure 1.12. Linear nonideal chromatography. t_o= time
at start of separation (point of sample injection);
t_A = retention time of component A; t_B = retention
time of component B; t_n = time for emergence of mobile
phase from t_o. $t_A > t_B$.

1.2.10 Nonlinear-Ideal Chromatography

Liquid-solid chromatography is representative of this theory because nonlinearity effects are usually appreciable. Mass transfer is fast and longitudinal diffusion effects may be ignored in
describing the system. The net result is that the bands (zones)
develop self-sharpening fronts and diffuse rear boundaries. It
is because of this tailing that this technique is to be regarded

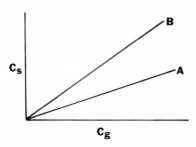

Figure 1.13.   Isotherms for linear nonideal chromatography. C_s = conc. at surface or in stationary
phase; C_g = conc. in solution at equilibrium.

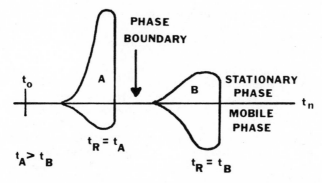

Figure 1.14.   Nonlinear ideal chromatography. t_o =
start of separation (point of sample injection); t_A =
retention time of component A; t_B = retention time
of component B; t_n = time of emergence of mobile
phase from t_o. $t_A > t_B$.

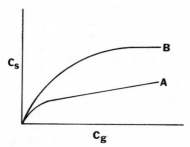

Figure 1.15.   Isotherms for nonlinear ideal chromatography. C_s = conc. at surface or in stationary
phase; C_g = conc. in solution at equilibrium.

as less than suitable for elution analysis. This system may be
represented by Figures 1.14 and 1.15.

1.2.11 Nonlinear-Nonideal Chromatography

Gas-solid chromatography is best described by this theory. Here
one finds diffuse front and rear boundaries with definite tailing
of the rear boundary. Mathematical descriptions of systems of
this type can become very complex; however, with proper assumptions mathematical treatments do fairly represent the experimental data. The bands (zones) are similar to those shown in Figures
1.16 and 1.17.

1.3   GENERAL ASPECTS OF GAS CHROMATOGRAPHY

1.3.1 Applications of Gas Chromatography

Gas chromatography is a very unique and versatile technique. In
its initial stages of development, it was applied to the analysis
of gases and of vapors from very volatile components. The work
of Martin and Synge (36) and then James and Martin (54) in gas-
liquid chromatography (GLC) opened the door for an analytical
technique which has revolutionized chemical separations and analyses. As an analytical tool GC can be used for the direct separation and analysis of gaseous samples, liquid solutions, and
volatile solids.

    If the sample to be analyzed is nonvolatile then one can call
upon the technique of pyrolysis gas chromatography. This is a
modification wherein a nonvolatile sample is pyrolyzed prior to

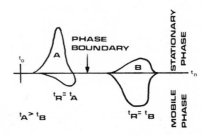

Figure 1.16.   Nonlinear nonideal chromatography.
t_o = start of separation (point of sample injection);
t_A = retention time of component A; t_B = retention
time of component B; t_n = time of emergence of mobile
phase from t_o. $t_A > t_B$.

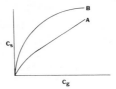

Figure 1.17. Isotherms for nonlinear-nonideal chromatography. C_s = conc. at surface or in stationary
phase; C_g = conc. in solution at equilibrium.

its entering the column. Decomposition products are separated
in the gas chromatographic column after which they are qualitatively and quantitatively determined. Analytical results are obtained from the pyrogram (a chromatogram resulting from the detection of pyrolysis products). This technique can be compared
to mass spectrometry, a technique in which analysis is based on
the nature and distribution of molecular fragments which results
from the bombardment of the sample component with high-speed
electrons. In the pyrolysis GC, the fragments result from chemical decomposition by heat. If the component to be pyrolyzed is
very complex, complete identification of all the fragments may
not be possible. In a case of this type the resulting pyrogram
may be used as a set of "fingerprints" for subsequent study.
     In addition to analysis GC may be used to study structure of
chemical compounds, determine the mechanisms and kinetics of
chemical reactions, and measure isotherms, heats of solution,
heats of adsorption, free energy of solution and/or adsorption,
activity coefficients, and diffusion constants (see Chapter 11).
     Another significant application of GC is in the area of the
preparation of pure substances or narrow fractions as standards
for further investigations. GC also is utilized on an industrial
scale for process monitoring. In adsorption studies, it can be
used to determine specific surface areas (30,31). A novel use is
its utilization to carry out elemental analyses of organic components (32). Distillation curves may also be plotted from gas
chromatographic data.
     GC can be applied to the solution of many problems in various
fields. To enumerate a few:

1. Drugs and Pharmaceuticals (see Chapter 12). Not only is GC
used in the quality control of products of this field, but it has
been applied to the analysis of new products as well as the monitoring of metabolites in biological systems.

2. Environmental Studies. A review of the contemporary field of
air pollution analyses by GC will be appearing very shortly (33).

Many of our chronic diseases (asthma, lung cancer, emphysema, and bronchitis) could result from air pollution or be directly influenced by air pollution. Air samples can be very complex mixtures and GC is easily adapted to the separation and analyses of such mixtures. Two publications concerned with the adaptation of cryogenic gas chromatography to analyses of air samples illustrate this adaptation (34,35).

3. Petroleum Industry. The petroleum companies were among the first to make widespread use of GC. The technique was very successfully used to separate and determine the many components in petroleum products. One of the earlier publications concerning the response of thermal conductivity detectors to concentration resulted from research in the petroleum field (36).

4. Clinical Chemistry (see Chapter 10). Gas chromatography is adaptable to such samples as blood, urine, and/or biological fluids. Compounds such as proteins, carbohydrates, amino acids, fatty acids, steroids, triglycerides, vitamins, and barbiturates are handled by this technique, directly or after preparation of appropriate volatile derivatives.

5. Pesticides and Their Residues. Gas chromatography in combination with selective detectors such as electron capture, phosphorus and electrolytic conductivity detectors (see Chapter 5) have made the detection of such components and their measurement relatively simple. Detailed information in this area may be found in a monograph by Grob (37).

6. Foods. The determination of antioxidants and food preservatives is a very active part of the gas chromatography field. Adaptations and sample types are almost limitless; for example, analysis of fruit juices, wines, beers, syrups, cheeses, beverages, food aromas, oils, dairy products, decomposition products, contaminants, and adulterants. A detailed discussion of this field may be found in Chapter 9.

1.3.2 Types of Detection

A discussion of the various detectors employed in GC is covered in Chapter 5. Our purpose here is only to categorize the detection system according to whether they are an integral type system or a differential type system. This classification is an old one; any detection system can be made integral or differential simply by a modification of the detector electronics. A more modern categorization would be instantaneous (differential) and

<u>cumulative</u> (integral). Chromatograms which result from this classification of detectors are shown in Figure 1.18.

1.3.3 Advantages and Limitations

From the limited discussion so far one can visualize the versatility of this technique. There are many reasons for this and we shall enumerate some of the advantages. It should be stressed that what one person considers a disadvantage may, to someone else be an advantage. Additionally, what may be a current disadvantage, may not be several years from now.

A few broad comments regarding GC would include:

1. An Analytical Technique. It is used not only for qualitative identification of components in a sample, but also quantitative

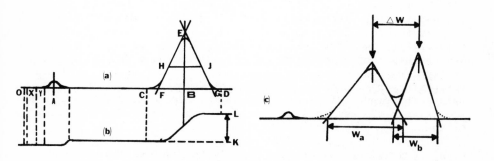

Figure 1.18. Types of chromatograms. (a) Differential chromatogram; (b) integral chromatogram; (c) peak resolution. O, injection point; OX, injector volume; OY, detector volume; OA, holdup volume, V_M; OB, total retention volume, V_R; AB, adjusted retention volume, $V_R - V_M$; CD, peak base; FG, peak width, W; HJ, peak width at half-height, $W_{0.5}$; BE, peak height; E, peak maximum; CHEJD, peak area (space incorporated within these letters); KL, step height of integral chromatogram.

measurements.

2. A Physical Research Technique. It may be used to investigate various parameters of a system, for example, determination of partition coefficients, thermodynamic functions, and adsorption isotherms.

3. A Preparative Technique. Once the analytical conditions have been determined, the system may be scaled up to separate and collect gram amounts of components. This approach is discussed in more detail in Section 2.7.
   Some overall advantages of GC that should be stressed are:

1. Resolution. The technique is applicable to systems containing components with very similar boiling points. By choosing a selective liquid phase or the proper adsorbent one can separate molecules that are very similar physically and chemically. Components that form azeotropic mixtures in ordinary distillation techniques may be separated by GC.

2. Sensitivity. This property of the gas chromatographic system largely accounts for its extensive use. The simplest thermal conductivity detector cells can detect 100 ppm or less. Utilizing a flame ionization detector one can detect a few parts per million; with an electron capture detector or phosphorous detector parts per billion or picograms of solute can easily be measured. This level of sensitivity is more impressive when one considers that the sample size used is of the order of a microliter or less.

3. Analysis Time. Separation of all the components in a sample may take from several seconds up to 30 min. Analyses that routinely take an hour or more may be reduced to a matter of minutes. This fast analysis results from the high diffusion rate in the gas phase and the rapid equilibrium between the moving and the stationary phases.

4. Convenience. The operation of a gas chromatograph is a relatively straightforward operation. It is not difficult to train nontechnical personnel to carry out routine separations.

5. Cost. Compared with many analytical instruments available today, gas chromatographs represent an excellent value.

6. Versatility. GC is easily adapted to analyze samples of permanent gases as well as high boiling liquids or volatile solids.

7. High Separating Power. Since the mobile phase has a low viscosity, very long columns with excellent separating power can be
employed.

8. Assortment of Sensitive Detecting Systems. GC detectors (see
Chapter 5) are relatively simple, highly sensitive, and possess
rapid response.

9. Ease of Recording Data. Detector output from gas chromatographs can be conveniently interfaced with recording potentiometers, integrating systems, computers, and a wide variety of
automatic data storing modules (see Chapter 8).

10. Automation. Gas chromatographs may be used to monitor automatically various chemical processes in which samples may be per-
iodically taken and injected onto a column for separation and detection.

1.4   DEFINITIONS AND NOMENCLATURE

The definitions given in this section are a combination of those
used widely and those recommended by the International Union of
Pure and Applied Chemistry (IUPAC) (38-40). The widely used
symbol, if different from the recommended IUPAC symbol, will be
in parenthesis.

Adjusted retention volume. V'_R The solute retention volume
minus the retention volume for an unretained peak.

$$V'_R = V_R - V_M$$

Adsorbent. An active granular solid used as the column packing in gas-solid chromatography which retains sample components
by adsorptive forces.

Adsorption chromatography. Synonymous with gas-solid chromatography.

Adsorption column. The column used in gas-solid chromatography, consisting of an active granular solid.

Air peak. The peak resulting from a sample component nonre-
tained by the column. This peak can be used to measure the time
necessary for the carrier gas to travel from the point of injection to the detector.

Analysis time. t_{ne} The minimum time required for separation.

$$t_{ne} = 16\ R_s^2\ \frac{H}{\bar{u}}\left(\frac{\alpha}{\alpha-1}\right)^2\ \frac{(1 + k')^3}{k'^2}$$

Attenuator. An electrical component made up of a series of resistances which is used to reduce the input voltage to the recorder by a particular ratio.

Band. Synonymous with zone.

Band area. Synonymous with peak area.

Baseline. That portion of a detector record resulting from only eluent or carrier gas emerging from the column.

Bed volume. Synonymous term to the volume of a packed column.

Capacity factor. (k'). See mass distribution ratio. (In GSC, $V_A > V_L$; thus smaller β values and higher k' values occur).

Capillary column. Synonymous with open tubular column. Small diameter tubing (0.25-1.0 mm i.d.) in which the inner walls are used to support the stationary phase.

Carrier gas. Synonymous with mobile or moving phase. The phase that transports the sample through column.

Chromatogram. A plot of the detector response (which uses effluent concentration or other quantity used to measure the sample component) versus effluent volume or time.

Chromatograph (verb). To separate sample components by chromatography.

Chromatograph (noun). The specific instrument employed to carry out a chromatographic separation.

Chromatography. A technique used for the separation of sample components in which these components distribute themselves between two phases, one stationary and the other mobile. The stationary phase may be a solid or a liquid supported on a solid.

Column. A metal, plastic or glass tube packed or internally coated with the column material through which the sample components and mobile phase (carrier gas) flow and in which the chromatographic separation takes place.

Column bleed. The loss of liquid phase that coats the support within the column.

Column material. The material in the column used to effect the separation. An adsorbent is used in adsorption chromatography; in partition chromatography, the material is a stationary phase distributed over an inert support or coated on the inner walls of the column.

Column volume. X. The total volume of the column which contains the stationary phase. (IUPAC recommends the column dimensions be given as the inner diameter and the height or length of the column occupied by the stationary phase under the specific chromatographic conditions. Dimensions should be given in millimeters or centimeters.)

Component. A compound in the sample mixture.

Concentration distribution ratio. D_C. Ratio of the analytical concentration of a component in the stationary phase to its analytical concentration in the mobile phase.

$$D_C = \frac{\text{amt. Component/cm}^3 \text{ Stationary phase}}{\text{amt. Component/cm}^3 \text{ Mobile phase}}$$

Corrected retention volume. (V_R^o). Total retention volume corrected for pressure gradient across column. $V_R^o = jV_R$.

Dead volume. See hold-up volume.

Detection. A process by which a chromatographic band is recognized.

Detector. A device that signals the presence of a component eluted from a chromatographic column.

Detector volume. The volume of carrier gas (mobile phase) required to fill the detector at the operating temperature.

Differential detector. A detector that responds to the instantaneous difference in composition between the column effluent and the carrier gas (mobile phase).

Displacement chromatography. An elution procedure in which

the eluent contains a compound more effectively retained than the
components of the sample under examination.

Distribution coefficients. The amount of a component in a
specified amount of stationary phase, or in an amount of stationary phase of specified surface area, divided by the analytical
concentration in the mobile phase.

Distribution coefficient in adsorption chromatography with adsorbents of unknown surface area:

$$D_g = \frac{\text{amt. Component/g Dry stationary phase}}{\text{amt. Component/cm}^3 \text{ Mobile phase}}$$

Distribution coefficient in adsorption chromatography with well-
characterized adsorbent of known surface area:

$$D_s = \frac{\text{amt. Component/m}^2 \text{ of Surface}}{\text{amt. Component/cm}^3 \text{ Mobile phase}}$$

Distribution coefficient when it is not practicable to determine
the weight of the solid phase:

$$D_v = \frac{\text{amt. Component in Stationary phase/cm}^3 \text{ of Bed volume}}{\text{amt. Component/cm}^3 \text{ Mobile phase}}$$

Distribution constant. K_D(K). The ratio of the concentration
of a sample component in a single definite form in the stationary
phase to its concentration in the mobile phase at equilibrium.
Both concentrations should be calculated per unit volume of the
phase. IUPAC recommends this term rather than partition coefficient.

$$K_D = \frac{c_S}{c_G}$$

Efficiency of column. Usually measured by column theoretical
plate number. Relates to peak sharpness or column performance.

Effective theoretical plate number. $N(N_{eff})$. A number relating to column performance when resolution is taken into account.

$$N = \frac{4R_S}{(\alpha-1)}$$

Eluent.   The gas (mobile phase) used to effect a separation by elution.

Elution.   The process of transporting a sample component through and out of the column by use of the carrier gas (mobile phase).

Elution chromatography.   A chromatographic separation in which an eluent is passed through the column after injection of the sample.

Filament element.   A fine tungsten or similar wire which is used as the variable resistance sensing element in the thermal conductivity cell chamber.

Flow controller.   A device to regulate the flow of mobile phase (carrier gas) through column.

Flow programming.   A procedure in which the rate of flow of the mobile phase (carrier gas) is systematically increased during a part or the whole of the separation of higher boiling components.

Flowrate. F_c.   The volumetric flow rate of the mobile phase (carrier gas), in cm^3/min, measured at the column temperature and outlet pressure.

Frontal chromatography.   A chromatographic separation in which the sample is fed continuously into the column.

Fronting.   Asymmetry of a peak such that, relative to the baseline, the front of the peak is less sharp than the rear.

Gas chromatography.   Collective noun for those chromatographic methods in which the moving phase is a gas.

Gas-liquid chromatography.   A chromatographic method in which the stationary phase is a liquid distributed on an inert solid support or coated on the column wall and the mobile phase is a gas.   The separation occurs by the partitioning of the sample components between the two phases.

Gas-solid chromatography.   A chromatographic method in which the stationary phase is an active granular solid (adsorbent). The separation is performed by selective adsorption.

Height equivalent to an effective plate. H. The number obtained by dividing the column length by the effective plate number.

$$H = \frac{L}{N_{eff}}$$

Height equivalent to a theoretical plate. h{h (reduced plate height)}. The number obtained by dividing the column length by the theoretical plate number.

$$h = \frac{L}{n}$$

$$= \frac{H}{d} \quad \text{(particle diameter in packed column or tube diameter in capillary column)}$$

Hold-up volume. V_M. Volume of mobile phase from the point of injection of the sample and the point of detection. In GC it is measured at the column outlet temperature and pressure and is a measure of the volume of carrier gas required to elute an unretained component (including injector and detector volumes).

Initial and final temperatures. The temperature range used for a separation in temperature-programmed chromatography.

Injection point. The starting of the chromatogram, which corresponds to the point in time when the sample was introduced into the chromatographic system.

Injection port. Closure column on one side and a septum inlet on the other through which the sample is introduced into system.

Injection temperature. The temperature of the chromatographic system at the injection point.

Injector volume. The volume of carrier gas (mobile phase) required to fill the injection port of the chromatograph.

Integral detector. A detector dependent on the total amount of a sample component passing through it.

Integrator. An electrical or mechanical device employed for a continuous summation of the detector output with respect to time.

The result is a measure of the area of a chromatographic peak (band).

   Internal standard. A pure compound added to a sample in known concentration for the purpose of eliminating the need to measure the sample size in quantitative analysis and for correction of instrument variation.

   Interstitial fraction. ε_I The interstitial volume per unit of a packed column.

$$\varepsilon_I = V_I/X$$

   Interstitial velocity of carrier gas. μ. The linear velocity of the carrier gas inside a packed column calculated as the average over the entire cross section. Under idealized conditions it can be calculated as:

$$\mu = \frac{F}{\varepsilon_I}$$

   Interstitial volume. V_I (V_G). The volume occupied by the mobile phase (carrier gas) in a packed column. This volume does not include the volumes external to the packed section, i.e., volume of sample injector and volume of the detector. In GC it corresponds to the volume which would be occupied by the carrier gas at atmospheric pressure and zero flowrate in the packed section of the column.

   Ionization detector. A chromatographic detector in which the sample measurement is derived from the current produced by the ionization of sample molecules. This ionization may be induced by thermal, radioactive, or other excitation sources.

   Katharometer. Synonymous term used for a thermal conductivity cell. Sometimes spelled catharometer.

   Linear flowrate. $F(\nu)$. The volumetric flowrate of the carrier gas (mobile phase) divided by the area of the cross section of the column. It is expressed as cm^3cm^{-2}/min or cm/min.

   Liquid phase. Synonymous with stationary phase or substrate. A relatively nonvolatile liquid (at operating temperature) which is either sorbed on the solid support or coated on the walls of open tubular columns where it acts as a solvent of the sample. The separation results from differences in solubility of the various sample components.

   Marker. A reference component which is chromatographed with
the sample to aid in the identification of sample components.

   Mass Distribution ratio. $D_m(k')$. The fraction (1-R) of a component in the stationary phase divided by the fraction (R) in the
mobile phase. IUPAC recommends this term in preference to capacity factor.

$$D_m(k') = \frac{1-R}{R} = \frac{K}{\beta} = \frac{c^L v^L}{c^G v^G} = \frac{K(v_L)}{(v_G)}$$

   Mean interstitial velocity of the carrier gas. $\bar{u}$. Interstitial velocity of the carrier gas multiplied by the pressure-
gradient correction factor.

$$\mu = \frac{Fj}{\varepsilon_I}$$

   Moving phase (mobile phase). See carrier gas.

   Net retention volume. V_N. The adjusted retention volume multiplied by the pressure gradient correction factor.

$$V_N = jV'_R$$

   Open tubular columns. See capillary columns.

   Packing material. Active granular solid or stationary phase
plus solid support which is in the column. The term refers to
the conditions existing when the chromatographic separation is
started, whereas the stationary phase refers to the conditions
during the chromatographic separation.

   Partition chromatography. See gas-liquid chromatography.

   Peak. That portion of a differential chromatogram recording
the detector response or eluate concentration when a component
emerges from the column. If the separation is incomplete, two or
more components may appear as one peak (unresolved peak).

   Peak area. Synonymous with band area.

   Peak base. In differential chromatography it is the baseline between the extremities of the peak.

Peak height. The distance between the peak (band) maximum and the peak base, measured in a direction parallel to the detector response axis and perpendicular to the time axis.

Peak maximum. The point of maximum detector response when a sample component elutes from the chromatographic column.

Peak resolution. R_S. The separation of two peaks in terms of their average peak width. See Figure 1.18

$$R_S = \frac{2 \, \Delta W}{W_a + W_b}$$

Peak width. W. The segment of the peak base intercepted by tangents to the inflection points on either side of the peak and projected on to the axis representing time and volume.

Peak width at half-height. $W_{0.5}$. The length of the line parallel to the peak base, which bisects the peak height and terminates at the intersections with the two limbs of the peak, projected on to the axis representing time or volume.

Performance index. PI. Used with open tubular columns, a number (in poise) which provides a relationship between elution time of a component and pressure drop.

$$PI = 30.7H^2 \left(\frac{\mu}{K}\right) \left[\frac{1 + k'}{k' + \dfrac{1}{16}}\right]$$

Phase ratio. β. The ratio of the volume of the mobile phase to the stationary phase in a partition column.

$$\beta = \frac{V_I}{V_S} = \frac{V_G}{V_A}$$

PLOT. Porous Layer Open Tubular column.

Potentiometric recorder. A continuously recording device whose deflection is proportional to the voltage output of the chromatographic detector.

Pressure gradient correction factor. j. The factor that

corrects for the compressibility of the mobile phase in a homogeneously filled column of uniform diameter.

$$j = \frac{3}{2} \left[\frac{(p_i/p_o)^2 - 1}{(p_i/p_o)^3 - 1} \right]$$

Pyrogram. The resulting chromatogram from the sensing of the
fragments of a pyrolyzed sample.

Pyrolysis. A technique by which nonvolatile samples are decomposed in the inlet system and the volatile products are separated in the chromatographic column.

Pyrolysis gas chromatography. The induction of molecular
fragmentation to a chromatographic sample by means of heat.

Pyrometer. An instrument for measuring temperature by the
change in electrical current.

Relative retention. $r_{a/b}$. The adjusted retention volume of a
substance related to that of a reference compound obtained under
identical conditions.

$$r_{a/b} = \frac{(V_g)_a}{(V_g)_b} = \frac{(V_N)_a}{(V_N)_b} = \frac{(V_R')_a}{(V_R')_b}$$

Required plate numbers. n_{ne}. The number of plates necessary
for the separation of two components based upon resolution, R_S, of
1.5.

$$n_{ne} = 16 R_S^2 \left(\frac{\alpha}{\alpha - 1} \right)^2 \left(\frac{1+k'}{k'} \right)^2$$

Resolution. The degree of separation between two peaks.

Retention index. I. A number relating the adjusted retention
volume of a component A to the adjusted retention volumes of normal paraffins. Each n-paraffin is arbitrarily allotted, by definition, an index of one hundred times its carbon number. The
index number of component A is obtained by logarithmic interpolation.

$$I = 100N + 100 \left[\frac{\log V_R'(A) - \log V_R'(N)}{\log V_R'(n) - \log V_R'(N)} \right]$$

N and n are the smaller and larger
n-paraffin, respectively, which
bracket substance A.

Retention time. t_R. The time that has elapsed from injection
of the sample to the recording the peak maximum of the component
band (peak).

Retention volume. V_R. The product of the retention time of a
sample component and the volumetric flowrate of the carrier gas
(mobile phase). IUPAC recommends it be called the Total Retention Volume because it is a term used when the sample is injected
into a flowing stream of the mobile phase. Thus, it includes any
volume contributed by the sample injector and the detector.

Sample. The gas or liquid mixture injected into the chromatographic system for separation and analysis.

Sample injector. A device for introducing liquid or gas
samples into the chromatograph. The sample is introduced directly into the carrier gas stream (e.g., by syringe) or into a
chamber temporarily isolated from the system by valves which can
be changed so as to instantaneously switch the gas stream through
the chamber (gas sampling valve).

Separation. The time elapsed between elution of two successive components, measured on the chromatogram as distance between
the recorded bands.

Separation factor. $\alpha_{a/b}$. The ratio of the distribution ratios or coefficients for two substances (A) and (B) measured under
identical conditions. By convention the separation factor is
usually greater than unity.

$$\alpha_{a/b} = \frac{K_{D_a}}{K_{D_b}} = \frac{D_a}{D_b} = \frac{K_a}{K_b}$$

Separation number. n_{sep}. Possible number of peaks between
two n-paraffin peaks resulting from components of consecutive
carbon number.

$$n_{sep} = \left[\frac{(t_1 - t_2)}{(W_{o.5})_1 + (W_{o.5})_2} \right] - 1$$

Separation temperature. The temperature of the chromatographic column.

SCOT. Support Coated Open Tubular column.

Solid Support. The solid packing material which holds the
liquid phase.

Solute. A synonymous term for the components in a sample.

Solvent. Synonymous with the liquid phase (stationary phase
or substrate).

Solvent efficiency. Synonymous with separation factor.

Span of the recorder. The number of millivolts required to
produce a change in the deflection of the recorder pen from 0 to
100% on the chart scale.

Specific retention volume. V_g. The net retention volume per
gram of stationary phase corrected to 0^0C.

$$V_g = \frac{273}{Tw_L} \frac{V_N}{} = \frac{j}{Tw_L} \frac{V'_R}{}$$

Specific surface area. Area of a solid granular adsorbent expressed as m^2/unit weight (gram) or m^2/cm^3.

Splitter. A fitting attached to the column to divert a portion of the flow. It is used on the inlet side to permit the introduction of very small samples to a capillary column, on the
outlet side to permit introduction of a very small sample of the
effluent to the detector or to permit introduction of effluent to
two detectors simultaneously.

Stationary phase. Synonymous with liquid phase.

Stationary phase fraction. ε_s. The volume of the stationary
phase per unit volume of the packed column.

Stationary phase volume. $V_S (V_L)$. Total volume of stationary
phase liquid on the support material in partition columns.

Substrate. Synonymous with liquid phase.

Surface area. Area of a solid granular adsorbent, A.

Tailing. Asymmetry of a peak such that, relative to the baseline, the front is steeper than the rear.

Temperature programming. The procedure in which the temperature of the column is changed systematically during part or the
whole of the separation.

Theoretical plate number. n. Number defining the efficiency
of the column or sharpness of peaks.

$$n = 16 \ \frac{\text{Peak retention volume}}{\text{Peak width}}^2$$

$$= 16\left(\frac{V_R}{w}\right)^2$$

Thermal conductivity. A physical property of a substance,
measure of its ability to conduct heat from a warmer to a cooler
object.

Thermal conductivity cell. A chamber in which an electrically
heated element will reflect changes in thermal conductivity within the chamber atmosphere. The measurement is possible because
of the change in resistance of the element.

Thermistor bead element. A thermal conductivity detection device in which a small glass-coated semiconductor sphere is used
as the variable resistive element in the cell chamber.

True adsorbent volume. V_A. The weight of the adsorbent packing divided by adsorbent density.

$$V_A = \frac{W_A}{D_A}$$

WCOT. Wall Coated Open Tubular column.

Weight of stationary liquid phase. W_L. Weight of liquid
phase in the column.

Zone. The position and spread of solute within the column.
See band.

1.5   LITERATURE OF GAS CHROMATOGRAPHY

We will not attempt to present an all-inclusive listing of the
literature available in which one may find chromatographic information.   Rather, we will list categories of source material and a
few examples of each category.   The expansion of the chromatographic field has made the reading of all the literature an impossible task.   One must choose a particular aspect of the field
and follow it closely while, at the same time, scanning the
topics and abstracts of the other aspects of chromatography.
   The novice to the field of chromatography would best become
aware of the basic texts which are available.   Once familiar with
these general sources he can then go to specific areas of chromatography and then sample types and/or topics within an area.
With this background he can then pursue the scanning of journal
contents and abstracts of the chromatographic literature.
   Gas chromatography is one of the most active areas of analytical chemistry, but many references in GC will be found in
sources other than just chromatography or analytical chemistry.
Thus, literature searches should take one to the journals on topics where GC may be utilized, for example, journals of biochemistry, organic chemistry, physical chemistry, catalysis,
environmental studies, drug analysis, forensic chemistry, petroleum chemistry, inorganic chemistry.
   A compilation of the journals may be found in the Chemical
Abstracts listing of periodicals abstracted by the Chemical Abstracts of the American Chemical Society, Washington, D. C. ,
20036.   Familiarity with this journal is a must for anyone doing
any type of chemical research.   In addition to these abstracts,
there are other abstract services that are concerned solely with
chromatography references.   Typical of these would be:

1. Gas and Liquid Chromatography Abstracts.   These are published
under the auspices of the Chromatography Discussion Group of
London, England.   They are published quarterly and may be obtained from the Applied Science Publishers, Ltd., Ripple Road,
Barking, Essex, England.   The fourth quarter volume contains an
author and a detailed subject index.   These abstracts are a
worthwhile investment for those working in GC or LC.

2. Gas-Liquid Chromatography Abstracts.   These are published in
separate volumes;   GC abstracts come out monthly and LC abstracts
come out bimonthly.   They are published by the Preston Technical
Abstracts Services, Niles, Illinois 60648.   The last issue of the
year contains an author and detailed subject index.

There are several journals that are dedicated solely to chromatography articles. Among these are Journal of Chromatography, Journal of Chromatographic Science (formerly Journal Gas Chromatography), Chromatographia, and Separation Science. Other prominent journals where one finds a large fraction of the remaining literature are Analytical Chemistry, Nature, Zeitschrift für analytische Chemie, Analyst, Journal of the American Chemical Society, Journal of the Chemical Society, Angewandte Chemie, Brennstoff-Chemie, Bulletin de la Societe Chimique de France, and Revista Italiana delle Sostanze Grasse.

Two excellent publications that review the various techniques and their applications are:

1. Annual Fundamental Reviews in Analytical Chemistry. This is published biannually, in the even years, by the Analytical Chemistry Journal of the American Chemical Society, Washington, D. C. 20036.

2. Annual Applied Reviews in Analytical Chemistry. This is published biannually, in the odd-numbered years, by the Analytical Chemistry Journal of the American Chemical Society, Washington, D. C. 20036.

Additional abstract sources are Chemisches Zentralblatt, Analytical Abstracts, and the Bibliography Section of the Journal of Chromatography. As pointed out previously, any research journal may have an article in which GC was used. In this case, the researcher should go to the journal for the specific area and/or topic, for example, Analytical Biochemistry, Food Technology, Journal of Agricultural and Food Chemistry, Journal of the American Oil Chemist's Society, Clinical Chemistry, Journal of Chemical Physics.

Lastly, there are many books which deal with gas chromatography specifically or chromatography in general:

SUGGESTED READINGS

A. J. M. Keulemans, "Gas Chromatography," Reinhold Corp., New York, 1957.
E. Lederer and M. Lederer (Eds.), "Chromatography," Elsevier, New York, 2nd edit., 1957.
G. Schay, "Theoretische Grundlagen der Gaschromatographie," VEB Verlag der Wissenschaften, Berlin, 1961.
H. A. Szymanski (Ed.), "Progress in Industrial Gas Chromatography," Plenum Press, New York, 1961.

A. A. Zhuhovitskii and N. M. Turkeltaub, "Gazovaya Khromato-
grafiya," Gostroptyehizdat, Moscow, 1962.

S. Dal Nogare and R. S. Juvet, "Gas-Liquid Chromatography.
Theory and Practice," Interscience, New York, 1962.

J. C. Giddings, "Dynamics of Chromatography," Part I. Principles
and Theory, Marcel Dekker, New York, 1965.

W. E. Harris, "Programmed Temperature Gas Chromatography,"
J. Wiley, New York, 1966.

L. S. Ettre and A. Zlatkis (Eds.), "The Practice of Gas Chromatography," Interscience, New York, 1967.

J. Tranchant (Ed.), "Practical Manual of Gas Chromatography,"
Elsevier Scientific Publishing Co., Amsterdam and New York,
1969.

J. Sevcik, "Detectors in Gas Chromatography," Elsevier Scientific
Publishing Co., Amsterdam and New York, 1975.

R. L. Grob (Ed.), "Chromatographic Analysis of the Environment,"
Marcel Dekker Inc., New York, 1975.

1.6   COMMERCIAL INSTRUMENTS

All the leading instrument manufacturers produce and market gas
chromatographs. In addition, there are many smaller speciality
companies who also manufacture and market GC units. Which instrument should be considered depends upon the use to which they
are to be utilized, and this ultimately sets the criteria for
purchase. GC units come in a variety of makes and models, from
simple student instructional types, (e.g., Gow-Mac Instrument
Co.) up to deluxe multicolumn, interchangeable detector types
(e.g., Varian Instrument Division). So as not to infer recommendation for one particular company, the reader is referred to the
Lab Guide issue of the Journal of Analytical Chemistry (41) for a
listing of the instrument manufacturers.

REFERENCES

1. V. Heines, Chem. Technol., $\underline{1}$, 280-285 (1971).

2. L. S. Ettre, Anal. Chem., $\underline{43}$ (14), 20A-31A (1971).

3. G. Zweig and J. Sherma, J. Chromatog. Sci., $\underline{11}$, 279-283
   (1973).

4. F. F. Runge, Farbenchemie, I and II, 1834, 1843.

5. F. F. Runge, Annal. Phys. Chem., XVII, $\underline{31}$, 65 (1834); XVIII,
   $\underline{32}$, 78, (1834).

6. F. F. Runge, Farbenchemie, III, 1850.

7. F. Goppelsroeder, Zeit. Anal. Chem., $\underline{7}$, 195 (1868).

8. C. Schönbein, J. Chem. Soc., 33, 304-306 (1878).

9. D. T. Day, Proc. Am. Phil. Soc., 36, 112 (1897).

10. D. T. Day, Congr. Intern. Pétrole Paris, 1, 53 (1900).

11. D. T. Day, Science, 17, 1007 (1903).

12. M. Twsett, Ber. Deut. Bot. Ges. XXIV, 316 (1906).

13. M. Twsett, Ber. Deut. Bot. Ges. XXIV, 384 (1906).

14. M. Twsett, Ber. Deut. Bot. Ges. XXV, 71-74 (1907).

15. R. Kuhn, A. Winterstein, and E. Lederer, Hoppe-Seyler's Z. Physiol. Chem., 197, 141-160 (1931).

16. A. Tiselius, Ark. Kemi, Mineral. Geol., 14B, (22) (1940).

17. J. N. Wilson, J. Am. Chem. Soc., 62, 1583-1591 (1940).

18. A. Tiselius, Ark. Kemi, Mineral. Geol., 15B(6) (1941).

19. A. J. P. Martin and R. L. M. Synge, Biochem. J. (London), 35, 1358 (1941).

20. R. Consden, A. H. Gordon, and A. J. P. Martin, Biochem. J., 38, 224-232 (1944).

21. S. Claesson, Arkiv. Kemi. Mineral. Geol., 23A(1) (1946).

22. A. J. P. Martin, Biochem. Soc. Sym., 3, 4-15 (1949).

23. E. Cremer and F. Prior, Z. Elektrochem., 55, 66 (1951); E. Cremer and R. Muller, ibid., 55, 217 (1951).

24. C. S. G. Phillips, J. Griffiths, and D. H. Jones, Analyst, 77, 897 (1952).

25. A. T. James and A. J. P. Martin, Biochem. J., 50, 679-690 (1952).

26. E. Glueckauf, in "Ion Exchange and Its Applications," pp.34-36, Soc. Chem. Ind., London, England, 1955.

27. J. J. van Deemter, F. J. Zuiderweg, and A. Klinkenberg, Chem. Eng. Sci., 5, 271-289 (1956).

28. J. C. Giddings, Dynamics of Chromatography, Part I, "Principles and Theory," pp. 13-26, Dekker, New York, 1965.

29. S. Brunauer, The Adsorption of Gases and Vapors, Princeton University Press, Princeton, Vol. I, 1945.

30. R. L. Grob, M. A. Kaiser, and M. J. O'Brien, Am. Lab., 7(6), 13-25 (1975).

31. R. L. Grob, M. A. Kaiser, and M. J. O'Brien, Am. Lab., 7(8), 33-41 (1975).

32. P. W. Rulon, "Organic Microanalysis by Gas Chromatography," Ph.D. Dissertation, Villanova University, May 1976.

33. R. L. Grob, Analysis of the Atmosphere, in Contemporary Topics in Analytical and Clinical Chemistry, D. M. Hercules, M. A. Evenson, G. M. Hieftje, L. R. Snyder (Eds.), Plenum Publishing Co., Vol. I, 1977.

34. J. A. Giannovario, R. J. Gondek, and R. L. Grob, J. Chromatog., 89, 1 (1974).

35. J. A. Giannovario, R. L. Grob, and P. W. Rulon, J. Chromatog., 121, 285-294(1976).

36. D. M. Rosie and R. L. Grob, Anal. Chem., 29, 1263 (1957).

37. R. L. Grob (Ed.), Chromatographic Analysis of the Environment, Dekker, New York, 1975.

38. D. Ambrose, A. T. James, A. I. M. Keulemans, E. Kovats, H. Rock, C. Rovit, and F. H. Stross, in Gas Chromatography 1960, 3rd Sym., Edinburgh, June, 1960, R. P. W. Scott (Ed.), Butterworths, London, 1960, pp. 423-432.

39. D. Ambrose, A. T. James, A. I. M. Keulemans, E. Kovats, H. Rock, C. Rovit, and F. H. Stross, Pure Appl. Chem., 8, 553-562 (1964).

40. Recommendations on Nomenclature for Chromatography, Appendices on Tentative Nomenclature, Symbols, Units and Standards -- No. 15, Information Bulletin, International Union of Pure and Applied Chemistry, February 1972.

41. 1975-1976 Lab Guide, Anal. Chem., 47(10), 1975.

PART ONE

THEORY AND BASICS

Science moves, but slowly slowly, creeping on from point to point.

Alfred, Lord Tennyson (1809–1892)
Locksley Hall, line 134

The usefulness of an analytical technique is directly related to an understanding of its theory. This part of the book will explain the basic concepts of gas chromatography (GC), the purpose and selection of columns, and the interpretation of the chromatographic record, that is, the chromatogram. The information in this part will give the reader sufficient information to understand what GC is, how the separation occurs, and how to relate data to sample composition.

CHAPTER 2

Theory of Gas Chromatography

ROBERT L. GROB

Villanova University

2.1   SEPARATION TECHNIQUES.   40

     2.1.1   Phase Equilibria   43
              Adsorption   47
              Diffusion.   49

     2.1.2   Distillation   52

     2.1.3   Liquid-Liquid Extraction   53
              Equilibrium of Separation.   54
              Maximum Efficiency in Extraction   58

     2.1.4   Countercurrent Extraction   59

2.2   GAS CHROMATOGRAPHY   60

     2.2.1   Plate Theory   61

     2.2.2   Rate Theory   65

           Modifications of the van Deemter Equation . . . 75
           Flow. 77
2.2.3   The Solid Support 81
2.2.4   Mobile Phase. 85
2.2.5   Stationary Phase. 88
2.2.6   Evaluation of Column Operation. 90
           Column Efficiency 92
           Effective Number of Plates. 92
           Resolution. 93
           Required Plate Number 94
           Separation Factor 96
           Separation Number 97
           Analysis Time 97
2.2.7   Preparative Gas Chromatography. 99
           Essential Components of a Preparative GC System 101
           Detectors 103
2.2.8   Gas-Solid Chromatography. 103
           Adsorbent Properties. 104
           Adsorption of Gases at the Solid Surface. . . . 105
           Thermodynamics of the Processes at the Gas Solid
           Interface 109

           REFERENCES 110

2.1   SEPARATION TECHNIQUES

Most separation techniques involve the formation of at least two
phases with the object being to separate and measure the various
constituents. There are various ways of describing a phase, that
is, gas, liquid, and solid. By proper choice of conditions
(temperature and pressure) one is able to convert a solid to a
liquid (melting) or a gas (sublimation), a liquid to a solid
(freezing) or a gas (distillation), and a gas to a liquid (condensation) or solid (condensation). When the phase transition(s)
are completed, one phase, hopefully, will contain the material of
interest and the other(s), materials not of interest. The phases
can then be mechanically or physically separated and the phase
containing the component of interest is retained.
     Since the component(s) of interest can be in one of three
states of matter and these in turn can be converted into one of
three phase types, the scientist has many types of separation at
his disposal. The major classifications are shown in Figure 2.1.
Chromatography is used in four of the nine major types shown.
     In the discussion of separations not only do we include homo-

Figure 2.1 Classification of Separations

GAS			LIQUID			SOLID		
Gas	Liquid	Solid	Gas	Liquid	Solid	Gas	Liquid	Solid
Thermal Diffusion	GIC Condensation Sorption	GSC Sorption	Volatili-zation	LLC Distilla-tion Extraction	LSC Ppt'n Electro-deposition Crystal-ization Ring Oven	Subli-mation	Soln. Zone Refining	Sieving Magnetic Techniques

Figure 2.2 Processes of Separation

CHEMICAL	MECHANICAL	PHYSICAL
Precipitation	Filtration	GLC, GSC, LLC, LSC
Electrodeposition	Centrifugation	Liquid-Liquid Extraction
Masking	Exclusion Chromatography	Distillation
Ion Exchange	Dialysis	Sublimation
		Zone Electrophoresis
		Zone Refining

geneous equilibria, but also heterogeneous equilibria and the
rates at which these equilibria are obtained. If both the equilibrium point and the rate of attainment of said equilibrium are
favorable, the separation can usually be attained in one step.
Less favorable systems utilized multistage operations. Multistage separations are both feasible and attractive.

The actual process of performing the separation for the
classifications in Figure 2.1 can be categorized as to mechanical,
physical, or chemical processes. This is illustrated in Figure
2.2 The measurements of the separated components can be made by
physical, chemical, or biological means. Within each of these
three types of analyses several techniques are used. In the majority of analysis studies most of the discussion relates to an
examination of the theoretical background, the experimental limitations, and the applications of the various techniques for making useful measurements. Methods of analysis are usually defined
in terms of the final measurement made, and thus many give the
impression that this stage constitutes the entire subject of analytical chemistry. A more realistic view of analytical chemistry
is deciding what information is needed from a system, deciding
how to obtain that information, utilizing one or more separations and measurements, collecting and evaluating the experimental data, and finally drawing some conclusion(s) from the
data.

When analyzing materials, any one of the above categories
(sample, separation, measurement) may assume more importance than
another. Obtaining a representative sample may be more difficult
than the separation and/or measurement or the separation may be
more difficult than the sampling or measurement. Two objectives
should be paramount for any analysis: the data must have the required accuracy and precision and be produced in the minimum
time.

The feature that sets chromatography into a special category,
that is, distinguishes it from other separation techniques, is
that one of the phases moves, whereas the other phase is stationary. Combining the states of matter into different pairs we are
able to arrive at a number of different chromatographic techniques (Table 2.1). Viewing chromatography in simple terms, one
would reasonably expect these several types of chromatography
have features in common, and this is the case. The principles by
which separation is achieved are the same even though different
equipment is utilized. Experimentally, chromatography is a relatively simple separation technique. Four components are necessary to perform a chromatographic separation, namely, a column,
a mobile phase, a sample injector, and a detector.

In this text we are concerned with gas chromatography (gas-
liquid and gas-solid) and so we will discuss only this type.

TABLE 2.1   TYPES OF CHROMATOGRAPHY

Mobile Phase	Stationary Phase	Type
Gas	Liquid	GLC
	Solid	GSC
Liquid	Liquid	LLC[1], PC
	Solid	LSC[1], TLC, ion exchange

[1]Includes HPLC and HSLC.

Before discussing the theory of GC per se, let us look at some basic separations and some of the theoretical fundamentals which underlie the technique.

2.1.1   Phase Equilibria

Gas chromatography involves chemical equilibria between phases to bring about a particular separation. Thus, a brief discussion of phase equilibria is pertinent at this point. Phase equilibria separations can be understood with the use of the second law of thermodynamics. The phase rule states that if we have a system of $\underline{C}$ components which are distributed between $\underline{P}$ phases, the composition of each of these phases will be completely defined by $C-1$ concentration terms. Thus, to have the compositions of P phases defined it is necessary to have $P(C-1)$ concentration terms. The temperature and pressure also are variables and are the same for all the phases. Assuming no other forces influence the equilibria it follows that,

$$\text{Total degrees of freedom (variables)} = P(C-1) + 2 \qquad (2.1)$$

Thus, for P phases of C components we have $C(P-1)$ independent equations. It follows then that $C(P-1)$ variables are fixed, which leaves,

$$\{P(C-1) + 2\} - \{C(P-1)\} = C - P + 2 \qquad (2.2)$$

unknown. Thus, this number of variables, $C - P + 2$, must be fixed. Stated another way, this is the value of our limiting degrees of freedom. These degrees of freedom may be represented by F. This gives us the recognizable form of the phase rule.

$$F = C - P + 2 \qquad (2.3)$$

In its strictest sense the phase rule assumes that the equilibrium between phases is not influenced by gravity, electrical or magnetic forces, or by surface action. Thus, the only variables are temperature, pressure, and concentration; if two are fixed, then the third is easily determined (another reason for the constant 2 in Equation 2.3).

We need to define exactly the terms in Equation 2.3.

1. By a phase we mean any homogeneous or physically distinct part of the system which is separated from other parts of the system by definite bounding surfaces.
2. By a component we mean the smallest number of independently variable components from which the composition of each phase can be expressed (directly or in the form of a chemical equation).
3. By a degree of freedom we mean the number of variable factors (e.g., temperature, pressure, and concentration) which must be fixed to completely define a system at equilibrium.

We may treat this system two ways: Hold temperature constant (isothermal conditions) and vary pressure and concentration, or hold pressure constant (isobaric conditions) and vary concentration and temperature. Both situations are shown in Figure 2.3. The plots in Figures 2.3a and 2.3b represent conditions for isothermal and isobaric treatments, respectively.

In this system, $C = 2$. If we choose a point which does not fall on the vapor-liquid equilibrium line, then all three variables must be known to describe the system. However, by choosing a point on the vapor-liquid line phases, $P=2$ and thus, degrees of freedom $F = 2-2+2 = 2$. In other words, only two of the three degrees of freedom (variables) must be known. Referring to Figure 2.3b, if we have a 50/50 mole fraction solution of A and B, the mixture boils at $92^{\circ}C$ and the vapor contains 78 mole % of B. In Figure 2.3a the dotted lines indicate the partial pressure of each of the components, that is, the equation of each line defines Raoult's law:

$$P_A = X_A P_A^o \qquad (2.4)$$

$$P_B = X_B P_B^o \qquad (2.5)$$

where P = partial pressure of A or B
    X = mole fraction of A or B in liquid (by mole fraction we
        mean the number of moles of one component divided by
        the sum of all the moles of all components present in
        the mixture)
    p^0= vapor pressure of pure A or B.
The dotted lines in Figure 2.3 describe the condition that as the
mole percent of either component decreases so does the partial
pressure (since there are less molecules at the surface to exert

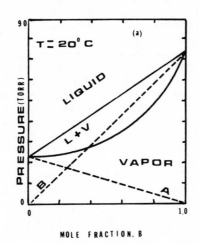

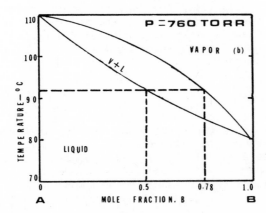

Figure 2.3.   Phase diagram for two-component system.
(a) Isothermal conditions; (b) isobaric conditions.

vapor pressure). The upper solid line (boundary between liquid
and liquid plus vapor) represents the sum of the two dotted
lines. The equation for this line defines Dalton's law:

$$P_A = Y_A p \qquad (2.6)$$

$$P_B = Y_B p \qquad (2.7)$$

where P = partial pressure of A or B,
   Y = mole fraction of A or B in the vapor,
   p = total pressure of the system

   In all the above discussions regarding liquid-vapor equilibria
we have assumed that our representative systems were ideal, that
is, there are no differences in attractions between molecules of
different types. Few systems are ideal and most show some deviation from ideality and do not follow Raoult's law. Deviations
from Raoult's law may be positive or negative. Positive deviations (for binary mixtures) occur when the attraction of like
molecules, A-A or B-B, are stronger than unlike molecules, A-B
(total pressure greater than that computed for ideality). Negative deviations result from the opposite effects (total pressure
lower than that computed for ideality). A mixture of two liquids
can exhibit nonideal behavior by forming an azeotropic mixture
(a constant boiling mixture).
   Raoult's law assumes that the liquid phase is ideal, that is
the partial pressure of component A is equal to the mole fraction
of A, in the liquid, times the vapor pressure of pure A. The
same could be said of component B, and so on. Mathematically we
write Raoult's law as shown in Equations 2.4 and 2.5; therefore
the sum of the partial pressures equals the total pressure of the
system, viz., Dalton's law:

$$P = P_A + P_B \qquad (2.8)$$

This relationship is represented by the plot in Figure 2.4.
Raoult's law is usually followed when X_A is a large value. In
some systems it does hold for low values of X_A. When X_A is a
small value, the system is said to follow Henry's law, which
states that the partial pressure of a component is equal to the
mole fraction in the liquid, multiplied by a constant.

$$P_A = H_A X_A \qquad (2.9)$$

where H_A = Henry's constant. If the system being studied is ideal,

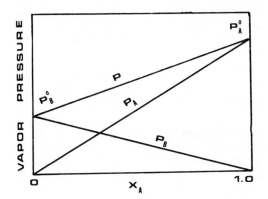

Figure 2.4.   Plot of Raoult's law for two-component system.

Raoult's and Henry's law are identical, that is, the H_A term is the vapor pressure of the pure component, p_A^o.

A short discussion of thermodynamics is necessary to place the topic of equilibrium into proper perspective. From the viewpoint of thermodynamics a system is in equilibrium when the free energy G is equal to zero. Free energy is the energy available to do work. The free energy of a system depends upon the enthalpy (heat content), H and the entropy(disorder or randomness of the molecules), S.

$$G \ = \ H \ - \ TS \qquad\qquad (2.10)$$

and for an isothermal process the free energy change is dependent upon the changes both in enthalpy and entropy.

$$\Delta G \ = \ \Delta H \ - \ T\Delta S \qquad\qquad (2.11)$$

ADSORPTION. Solids have a residual surface force field and there is a tendency for the free energy of this surface to decrease. It is this phenomenon that is responsible for adsorption. Temperature and pressure are the two main variables that affect the process of adsorption. Decreasing the temperature or increasing pressure increases adsorption. At low temperatures ($<^oC$), adsorption of gases increases very rapidly with small changes in pressure. Increasing the temperature causes a decrease in adsorption, which implies evolution of heat in the adsorption process. Curves that show the variation in pressure (p) with temperature (T) are referred to as isosteres. Plotting the log p versus the reciprocal of temperature (1/T) gives a straight line, indicating ΔH is independent of temperature. van't Hoff's equation represents the variation of the equilibrium constant (K) with temperature (T) for a reaction involving gases in terms of the change in heat content (heat of reaction, H, at constant pressure). It is represented as

$$\frac{d \ln K}{dT} = \frac{\Delta H}{RT^2} \tag{2.12}$$

The adsorption capacity of various solids for a specific gas depends primarily upon the effective area of the solids. For a series of gases, the order of increasing adsorption is the same for all solid adsorbents. These similarities hold as long as there are no chemical bonding factors intervening in the adsorption process. Gases that are the most easily liquified are the most readily adsorbed at a solid surface. Table 2.2 depicts data for the adsorption of a number of gases on 1 g of activated charcoal at $15^{\circ}C(1)$.

TABLE 2.2 Correlation of Adsorption of Gases with Critical
Temperature

Gas	Critical Temperature[1] ($^{\circ}K$)	Volume Adsorbed (cm^3)
H_2	33	4.7
N_2	126	8.0
O_2	154	8.2
CO	134	9.3
CH_4	190	16.2
CO_2	304	48.0
N_2O	310	54.0
HC	324	72.0
H_2S	373	99.0
NH_3	406	181.0
Cl_2	417	235.0
SO_2	430	380.0

[1]Critical temperature (maximum temperature at which a gas can be liquified) is related to boiling point.

Adsorption phenomena are divided into two main categories, physical (van der Waals or dispersion forces) and chemical (analogous to valence-bonding). The former type results in small

heats of adsorption (same order of magnitude as heats of vaporization), with the equilibrium of the gas being reversible and
easily attained with changes in temperature and/or pressure.   All
gases exhibit van der Waals adsorption and additionally some exhibit chemisorption.   This second type shows heats of adsorption
of the order of chemical combinations.   Adsorption equilibria are
usually presented as an adsorption isotherms (quantity adsorbed
as a function of pressure with temperature constant)   (See
Chapter 1).

Diffusion.   From the gas laws we know that M/V is a measure of
the density (ρ) of a gas.   Therefore, we can arrive at an equation for the speed of gaseous molecules,

$$(\bar{C}^2)^{\frac{1}{2}} = (\frac{3P}{\rho})^{\frac{1}{2}} \tag{2.13}$$

$\bar{C}^2$ is the mean square velocity for all the molecules in a gas.
Equation 2.13 tells us that the velocity of gaseous molecules is
inversely proportional to the square root of the density.   Thus,
it follows that the rate of diffusion is inversely proportional
to the square root of the molecular weight (M) of a gas.
     Diffusion processes occur in all systems where concentration
differences exist.   Diffusion is the main mechanism which aids in
the elimination of concentration gradients.   Fick's first law of
diffusion defines this phenomenon by correlating mass flow and
concentration gradient.   This law may be shown as

$$\frac{\partial N}{\partial t} = -D\frac{\partial n}{\partial \ell} \tag{2.14}$$

where N = number of molecules passing through a unit surface,
      t = time for molecules to pass unit surface,
      D = diffusion coefficient (weight diffusing across a plane
          1 cm^2 in unit time under a concentration gradient of
          unity),
      n = concentration of gas molecules,
      ℓ = distance molecules diffuse,
      $\partial n/\partial \ell$ = concentration gradient.

The right side of Equation 2.14 carries a negative sign to indicate that diffusion is taking place in the direction of lower
concentration.   Stated otherwise, Equation 2.14 illustrates that
the amount of material diffusing through the unit surface in unit
time is proportional to the concentration gradient in direction
of the diffusion.
     If diffusion is taking place in a system there must be a

conservation of mass in the process. In a gas chromatographic column we are primarily concerned with the longitudinal diffusion, and this can be described by Fick's second law:

$$\frac{\partial n}{\partial t} = D\frac{\partial^2 n}{\partial \ell^2} \qquad (2.15)$$

These diffusion processes are random and one may express these in terms of a statistical distribution. Equation 2.15 can be stated,

$$n = \frac{A}{(t)^{\frac{1}{2}}} e^{-\ell^2/4Dt} \qquad (2.16)$$

where A is a constant. If the system can be given in terms of the total quantity of diffusing material, m, we can solve for the constant A:

$$A = \frac{m}{2(\pi D)^{\frac{1}{2}}} \qquad (2.17)$$

Solving the differential Equation 2.16 and substitution of Equation 2.17 for A gives

$$n = \frac{m}{2(\pi Dt)^{\frac{1}{2}}} e^{-\ell^2/4Dt} \qquad (2.18)$$

Equation 2.18 is an equation of a Gaussian curve. This curve is described by its maximum and width. We can represent the Gaussian curve in terms of Equation 2.18 as shown in Figure 2.5. The base width of the curve is $2(2Dt)^{\frac{1}{2}}$.

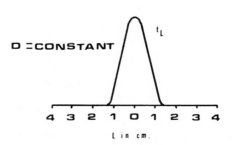

Figure 2.5. Gaussian curve in terms of Fick's second law.

One may study zone broadening in gas chromatography by observing the shape of the elution peak which is Gaussian in ideal systems. The base width of the Gaussian curve is measured in standard deviation units, therefore

$$\sigma = (2Dt)^{\frac{1}{2}} \qquad\qquad (2.19)$$

Squaring of the standard deviation term gives the variance, a term which is used in rate theory description for the gas chromatographic process (see Section 2.2.2).

$$\sigma^2 = 2Dt \qquad\qquad (2.20)$$

As Equations 2.19 and 2.20 illustrate, the diffusion curve is determined by time (time elapsed from beginning to end of separation) and the value of D (diffusion coefficient which is different for each gas). One may obtain the value of D from the kinetic theory of gases.

Up to this point we have been discussing diffusion in terms of molecular or free diffusion where the diffusion rate is determined by molecular collisions and the particle voids which are larger than the mean free path. In packed gas chromatographic columns the diffusion process follows other laws. Under these conditions we can encounter four types of diffusion.

1. Surface or Volmer Diffusion. When pores are small the adsorbed molecules diffuse from the pore walls towards the less densely coated areas. D_g values are $<10^{-3} cm^2/sec$.

2. Knudsen Diffusion. When the pores have diameters <0.1 μm, collisions with the walls are more frequent than intermolecular collisions (also referred to as capillary diffusion). D_g values are approximately $10^{-2} cm^2/sec$.

3. Free or Molecular Diffusion. The type used to explain the above model. Takes place between particles and in pores of diameter >0.1 μm. Here we encounter largest values for D_g, that is, $10^{-1} - 1.0$ cm^2/sec.

The three above types occur in gas-solid chromatography on microporous adsorbents (activated charcoal, molecular sieves). Knudsen diffusion may occur in gas-liquid chromatography supports.

Solid diffusion takes place in pore diameters of about 0.001 $\mu m (10A)$ (D_g values of order of $10^{-5} cm^2/sec$). Since most diffusion types (Volmer, Knudsen, and solid) are orders of magnitude smaller than molecular diffusion, they contribute little to the

longitudinal diffusion process in gas chromatographic columns.
However, they can greatly affect mass transfer rate and this
effect is evident in the broadening of the zones.

Diffusion in liquids is four to five orders of magnitude less
than that found in gases. For this reason we may neglect longitudinal diffusion effects in the liquid phase in zone broadening.
However, longitudinal diffusion must be considered in equilibrium
effects because it may determine the rate of mass transfer.

2.1.2 Distillation

Separations that take place in a chromatographic column are very
similar to other types of separation, and it is for this reason
that we will cover several of the more important separation techniques. A chromatographic column may be likened to a distillation column in many respects; in fact, some of the terminology in
chromatography is taken from distillation theory.

The distillation technique is not used to separate complex
mixtures, but finds its acceptance more for the preparation of
large quantities of pure substances or the separation of complex
mixtures into fractions. The technique depends on the distribution of constituents between the liquid mixture and component
vapors in equilibrium with the mixture; two phases exist because
of the partial evaporation of the liquids. How effective the
distillation becomes depends upon the type equipment employed,
the method of distillation, and the properties of the mixture
components. The distinguishing aspects of distillation and evaporation are that in the former all components are volatile,
whereas in the latter technique volatile components are separated
from nonvolatile components. An example of distillation would be
the separation of ethyl alcohol and benzene. An evaporative
separation would be the separation of water from an aqueous solution of some inorganic salt, for example, sodium sulfate.

We can construct a distillation column that has separate steps
or plates (i.e., a bubble-cap type distillation column). Each
plate would correspond to an evaporation-condensation step. The
phase diagram for such a system is illustrated in Figure 2.3b.
This type of a column is a useful example for describing the concept of plates.

Efficient laboratory distillation systems use a fractionating
column which is a packed column rather than a column with separate plates. Here we cannot refer to the plate or step where the
evaporation-condensation step occurs. Thus, we refer to a
"theoretical plate" that will produce a liquid with a particular
composition. The number of these theoretical plates in a
fractionation column is given the symbol, n. This number of

theoretical plates may be determined experimentally by distilling
a binary mixture and comparing the data obtained with the phase
diagram.

Efficiency of the distillation column is measured by the
height equivalent to a theoretical plate, abbreviated HETP or
simply h. The length of the column is L, thus: h=L/n. The h
value is independent of L, whereas the n value is dependent on L.

This height equivalent to a theoretical plate has been applied
to chromatographic separations. The reader should keep in mind
that actual plates do not exist in a chromatographic column
either.* The n value is a measure of efficiency for the column.
The number of theoretical plates for a chromatographic column is
a relative measure of the zone broadening that occurred during
the passage of the sample through the column. A direct comparison of efficiencies between the distillation technique and the
chromatographic technique is not possible. A given separation
would require more theoretical plates by the chromatographic
technique than it would by the distillation technique.

2.1.3 Liquid-Liquid Extraction

In the previous section (2.1.2) we were concerned with phase
transitions between liquid and vapor and discussed the various
techniques for effecting such changes. In this section we will
look at transferring solute components from one liquid phase to a
second liquid phase. This technique is referred to as liquid-
liquid extraction (LLE). The main restriction on this separation
technique is that the two phases must be immiscible. By immiscible liquids we mean two liquids which are completely insoluble
in each other. A little reflection will reveal it is very difficult to have two liquids that are mutually insoluble. If such a
system were achievable, then the total pressure, P, of the system
would be defined by,

$$P = p_A^o + p_B^o \qquad (2.21)$$

where p_A^o and p_B^o are the vapor pressures of liquids A and B in the
pure state. The composition of the vapor, in terms of moles of

*We visualize the chromatographic column as if it were divided
into a number of regions called theoretical plates. We further
assume that equilibrium exists between the solute in the mobile
and the stationary phases.

each liquid (n_A, n_B) would be,

$$\frac{n_A}{n_B} = \frac{p_A^o}{p_B^o} \tag{2.22}$$

As with any liquid system, this system would boil when the total vapor pressure, P, equals atmospheric pressure. However, the boiling point of a mixture of two immiscible liquids is lower than that of either constituent because the pressure of the mixture is higher at all temperatures. This total vapor pressure is independent of the amounts of the two phases, and thus the boiling point remains constant as long as the two layers are present.

Once a quantity of a third substance (solute) is added to a system of two immiscible liquids, it will distribute or divide between the layers in definite proportions. Applying the phase rule to such a system reveals that we have a system of three components (C) and two phases (P). Thus, the system has three degrees of freedom (F), that is, pressure, temperature, and concentration.

The distribution of the third substance (solute) between the two phases is governed, to a first approximation, by its solubility in each of the two phases. Thus, at a definite temperature, the ratio of the concentration in each phase is a constant. This is the basis of the distribution law first stated by Berthelot and later extended by Nernst. Simply stated it is

$$K = \frac{C_{II}}{C_I} \tag{2.23}$$

where C_I and C_{II} are the concentrations of the solute in liquids I and II.

EQUILIBRIUM OF SEPARATION. If we allow a system of two immiscible liquids, containing a solute (i) to come to equilibrium, we can express our equilibrium distribution coefficient, $\tilde{K}_c$, as

$$(i)_1 \rightleftharpoons (i)_2 \tag{2.24}$$

$$\tilde{K}_c = \frac{\{i\}_2}{\{i\}_1}$$

$$K^o = \frac{(a_i)_2}{(a_i)_1} = \frac{\{i\}_2 \, (\gamma)_2}{\{i\}_1 \, (\gamma)_1}$$

where γ= activity coefficient. Our distribution coefficient can also be expressed in terms of weight and volume:

$$\tilde{K}_C = \frac{(w_i)_2/MW_i}{V_2} \div \frac{(w_i)_1/MW_i}{V_1} \qquad (2.25)$$

where w_i = weight of solute (g),
MW_i = molecular weight of solute,
V_1 and V_2 = volumes of the two phases.
Equation 2.25 becomes

$$\tilde{K}_C = \frac{(w_i)_2}{(w_i)_1} \cdot \frac{V_1}{V_2} \qquad (2.26)$$

The ratio of the total amount of solute in phase 2 to the total amount of solute in phase 1 is known as the capacity factor and is given the symbol, k', thus

$$k' = \frac{(w_i)_2}{(w_i)_1} = \frac{(C_i)_2 V_2}{(C_i)_1 V_1} \qquad (2.27)$$

We can introduce another term, which is used a great many times in chromatography, that of phase ratio, β. It is the ratio of the volumes of the two phases:

$$\beta = \frac{V_1}{V_2} \qquad (2.28)$$

Incorporating Equations 2.27 and 2.28 into Equation 2.26 gives us

$$\tilde{K}_C = k'\beta \qquad (2.29)$$

Another way to consider the efficiency of an extraction system is to present the data in terms of the fraction of solute extracted and the fraction unextracted. The fraction of total solute in a given phase, ϕ, can be written

$$\phi_2 = \frac{C_2 V_2}{C_1 V_1 + C_2 V_2} \qquad (2.30)$$

where C_1 and C_2 = concentrations of solute in two phases,
V_1 and V_2 = volumes of the two phases.
Since $\tilde{K}_c = C_2/C_1$, then

$$\phi_2 = \frac{\tilde{K}_c V_2}{K_c V_2 + V_1} = \frac{\tilde{K}_c (V_2/V_1)}{K_c (V_2/V_1) + 1} \qquad (2.31)$$

Equation 2.27 may be written as

$$\frac{V_1}{V_2} k' = \frac{C_2}{C_1} = \tilde{K}_c \qquad (2.32)$$

Combining Equations 2.31 and 2.32

$$\phi_2 = \frac{(V_1/V_2)k'(V_2/V_1)}{(V_1/V_2)k'(V_2/V_1) + 1} = \frac{k'}{(k'+1)} \qquad (2.33)$$

By a similar substitution manipulation we obtain

$$\phi_2 = \frac{1}{(1 + k')} \qquad (2.34)$$

We may now derive equations for the fraction of solute extracted, ϕ_2, and the fraction unextracted, $(1 - \phi_2)$. Therefore

$$\phi_2 + (1 - \phi_2) = 1 \qquad (2.35)$$

$$\phi_2 = 1 - (1 - \phi_2) \qquad (2.36)$$

Also, the ratio of fraction unextracted to that of the fraction extracted would be same as the capacity factor, k'

$$k' = \frac{\phi_2}{1 - \phi_2} \qquad (2.37)$$

From Equation 2.29, we would have

$$\tilde{K}_c = \frac{\phi_2}{1 - \phi_2} \cdot \frac{V_1}{V_2} \qquad (2.38)$$

giving

$$\phi_2 = \frac{\tilde{K}_c V_1}{V_2 + \tilde{K}_c V_1} \tag{2.39}$$

Dividing by V_2 gives

$$\phi_2 = \frac{\tilde{K}_c(1/\beta)}{1 + \tilde{K}_c(1/\beta)} = \frac{k'}{(1 + k')} \tag{2.40}$$

Thus, Equations 2.33 and 2.40 are similar.
  For fraction unextracted,

$$1 - \phi_2 = \frac{(1 - \tilde{K}_c V_1)}{(V_2 + \tilde{K}_c V_1)} = \frac{(V_2 + \tilde{K}_c V_1 - \tilde{K}_c V_1)}{(V_2 + \tilde{K}_c V_1)} \tag{2.41}$$

$$= \frac{V_2}{(V_2 + \tilde{K}_c V_1)}$$

Dividing by V_2 gives

$$1 - \phi_2 = \frac{1}{(1 + \tilde{K}_c(1/\beta))} = \frac{1}{(1 + k')} \tag{2.42}$$

If $V_1 = V_2$, the $\beta = 1$ and Equations 2.39 and 2.41 become

$$\phi_2 = \frac{\tilde{K}_c}{(1 + \tilde{K}_c)} \tag{2.43}$$

and

$$1 - \phi_2 = \frac{1}{(1 + \tilde{K}_c)} \tag{2.44}$$

Carrying this one step further, if we are performing multiple
extractions, after n extractions we would have

$$(1 - \phi_2)_n = \frac{V_o}{(\tilde{K}_c V_o + V_w)^n} = \left[\frac{V_o}{\tilde{K}_c V_o + V_w}\right]^n \tag{2.45}$$

and

$$(\phi_2)_n = 1-(1-\phi_2)_n = 1 - \left[\frac{V_o}{\widetilde{K}_c V_o + V_w}\right]^n \qquad (2.46)$$

MAXIMUM EFFICIENCY in EXTRACTION. One may determine the maximum possible efficiency of an extraction by considering the fraction of solute that remains in the original phase (raffinate) after equilibrations. ϕ_R is the symbol for this fraction, thus

$$\phi_R = \frac{V_1}{(\widetilde{K}_c V_2 + V_1)} \qquad (2.47)$$

where V_1 = volume of original phase containing all the solute before extraction,
$\quad\ \ V_2$ = volume of extracting phase.
If n extractions are carried out on V_1, we have

$$\phi_R^n = \left[\frac{V_1}{\widetilde{K}_c (\Sigma V_2/n) + V_1}\right]^n \qquad (2.48)$$

where ΣV_2 = fixed total volume of extractant.
The limit of this expression as $n \to \infty$ would be

$$\phi_R^\infty = e^{-\widetilde{K}_c \Sigma V_2/V_1} \qquad (2.49)$$

By plotting percent efficiency of extraction $\{100(\phi_R^n)\}$ versus number of extractions, n, one would obtain a plot like Figure 2.6. The plot shows that a limiting value is approached when $V_2 = 5V_1$ (assuming $\widetilde{K}_c = 1$).

When the value of k' is near to 0.5 or when equal amounts of solute are present in each phase, we evidence the most sensitive changes in $\widetilde{K}_c$ or V_2/V_1. Also, the fractional amount of total solute, ϕ, in a given phase asymptotically approaches one or zero for large or small values, respectively of the capacity factor, k'. What this says is that when k' is large or small, little effect is noted for ϕ when a large change occurs in k'. Under the above conditions it becomes very difficult to remove the last traces of a component from either phase. It is for this reason that more than five extractions accomplishes little in regard to quantitative separation. Thus, when $(1 - \phi) < 0.01$, we assume complete separation of solute from one phase to another phase.

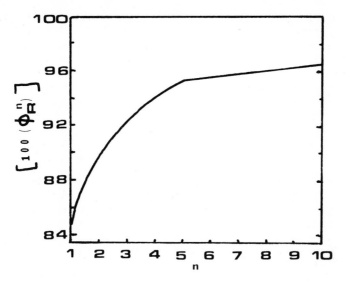

Figure 2.6.   Plot of efficiency versus number of extractions.

2.1.4 Countercurrent Extraction

In Section 2.1.3 we discussed extractions from the viewpoint of one substance being transferred from one phase to another or the separation of two solutes by selective extraction. When we have a system in which the distribution constants (K) or distribution coefficients ($\tilde{K}_C$) differ by 10^3, we can only recover the extracted solute in about 97% purity. Continued extractions will increase yield but not purity. Good separations, with high purity, of two or more solutes can be achieved when there is a difference in the thermodynamic behavior of the various solutes, that is, a difference in the distribution constants (K) or coefficients ($\tilde{K}_C$). A measure of this degree of separation is the separation factor, α, for pairs of solutes which is defined as

$$\alpha = \frac{K_A}{K_B} = \frac{\tilde{K}_{C_A}}{\tilde{K}_{C_B}} = \frac{(C_A)_2/(C_A)_1}{(C_B)_2/(C_B)_1} \tag{2.50}$$

$$= \frac{\{(C_A)_2(C_B)_1\}}{\{(C_A)_1(C_B)_2\}}$$

If there are other equilibria (intraphase) involved then the K
values are replaced by the distribution ratio, D. For extractions, it makes little difference which solute appears in the
numerator because the α term is usually defined in such a manner
that it will have a numerical value >1. Separation factor
applied to chromatographic systems is expressed as a ratio of
retention data (see Section 1.4):

$$\alpha = \frac{(V_R')_A}{(V_R')_B} \qquad (2.51)$$

V_R' = adjusted retention volume. The component which is more retained (greater value of V_R') is placed in the numerator.

Before proceeding, the author wishes to point out that the
separation factor, α, as defined above is the term used by most
people working in chromatography and the term recommended by the
IUPAC. Readers will encounter statements to the contrary in
some references (5,7).

2.2   GAS CHROMATOGRAPHY

It was first pointed out in Chapter 1 that chromatographic separations can be evaluated by the shape of the peaks from a particular system. The shapes of the peaks depend upon the isotherms
that describe the relationship between concentration of solute in
stationary phase to the solute concentration in the carrier gas.
If the isotherms are linear the peaks are Gaussian in shape and
separations proceed with little or no problems. If the isotherms
are nonlinear the peaks become asymmetric. Some isotherms are
linear over a limited range and as long as we work in this
limited range few problems are encountered. If the isotherm is
concave to the concentration axis (so that the distribution ratio
decreases with the increase in solute concentration in the mobile
phase) the band will have a sharp front and a long tail. If, on
the other hand, the isotherm is convex to the concentration axis
(so that the distribution ratio increases with the increase in
solute in mobile phase) the band will have a leading front and a
sharp rear edge.

When explaining chromatography theory on the basis of a discontinuous model we make three assumptions:

1. Equilibrium between solute concentration in the two phases is
   reached instantaneously.

2. Diffusion of solute, in mobile phase, along column axis is
   negligible.
3. Column is packed uniformly.

We do not have the conditions present during all chromatographic
separations.

   If the rate constants for the sorption-desorption processes
are small equilibrium between phases need not be achieved instantaneously. This effect is often called resistance-to-mass
transfer, and thus transport of solute from one phase to another
can be assumed diffusional in nature. As the solute migrates
through the column it is sorbed from the mobile phase into the
stationary phase. Flow is through the void volume of the solid
particles with the result that the solute molecules diffuse
through the interstices to reach surface of stationary phase.
Likewise, the solute has to diffuse from the interior of the
stationary phase to get back into the mobile phase.

   When the term longitudinal diffusion is applied to a chromatographic band we include true longitudinal molecular diffusion
(Section 2.1.1) and apparent longitudinal diffusion or eddy diffusion. True longitudinal diffusion occurs because of concentration gradients within the mobile phase, but eddy diffusion results from uneven velocity profiles because of the large number
of zigzag paths having unequal lengths and widths. As a result
of these diffusion effects some solute molecules move ahead,
while others lag behind. The widening of the band as it moves
down the column is of paramount importance in gas chromatography.
How much the band spreads (peak sharpness) determines the column
efficiency.

2.2.1 Plate Theory

From the equalities shown in Equation 2.50 it follows that

$$\alpha = \frac{k'_A}{k'_B} \qquad (2.52)$$

and that optimum separation occurs when $k'_A k'_B = 1$. If $k'_B = 100$
and $k'_B = 0.01$, then $\alpha = 10^4$ and $k'_A k'_B = 1$. This would indicate a
good separation because 1% of A would remain unextracted and 1%
of B would be extracted. However, if $k'_A = 1.0$ and $k'_B = 10^{-4}$ we
still obtain an α of 10^4, but $k'_A k'_B = 10^{-4}$ meaning 50% of A is in
each phase and 0.01% of B is extracted. B has been significantly
unextracted but not separated from A. It is this second example

of a separation which lends itself to countercurrent distribution
(extraction). In this extraction technique <u>both</u> phases (layers)
are contacted with fresh solvent after each equilibration rather
than only the original phase (raffinate). L. C. Craig (8-11) can
be credited for the refinement of this technique. It is this
technique of extraction that can be used as one of the explanations for what occurs in a chromatographic column. It also is
very illustrative for explaining zone broadening in multistage
processes. What one assumes is that the system is made of individual, discontinuous steps (theoretical plates) and that the
system comes to equilibrium as solute passes from one step
(plate) to the next. Thus, it is referred to as the "plate"
model. It is this model and the "rate" model (discussed in
Section 2.2.2) which may be used to describe the theory of chromatography. Both models arrive at same basic conclusion that the
zone broadening is proportional to the square root of the column
length and that the zone shape follows the normal distribution
law.

Figure 2.7 illustrates the similarity between the countercurrent extraction (CCE) process and the chromatographic process.

The concept of plate theory was originally proposed for the
performance of distillation columns (12). However, Martin and
Synge (13) first applied the plate theory to partition chromatography. The theory assumes that the column is divided into a
number of zones called theoretical plates. One determines the
zone thickness or height equivalent to a theoretical plate (HETP)
by assuming that there is perfect equilibrium between the gas and
liquid phases within each plate. The resulting behavior of the
plate column is calculated on the assumption that the distribution coefficient remains unaffected by the presence of other

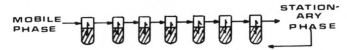

COUNTERCURRENT SYSTEM

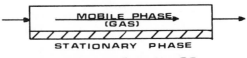

IDEALIZED COLUMN-GC

Figure 2.7. Comparison of countercurrent extraction
and the chromatographic process.

solutes, and that the distribution isotherm is linear. The diffusion of solute in the mobile phase from one plate to another is neglected.

Martin and Synge derived an expression for the total quantity, q_n, of solute in plate n and the volume of mobile phase (carrier gas) that passes through the column.

$$q_n = \frac{1}{(2\pi n)^{\frac{1}{2}}} \exp\left[\frac{-((V/v_R) - n)^2}{2n}\right] \tag{2.53}$$

where v_R = retention volume per plate,
V = total retention volume.

If $q_n = v_R C_n$, where C_n is solute concentration in mobile phase of plate n, then

$$C_n = \frac{1}{v_R(2\pi n)^{\frac{1}{2}}} \exp\left[\frac{-((V/v_R) - n)^2}{2n}\right] \tag{2.54}$$

or

$$C_n = \frac{1}{(v_R\sqrt{n})\sqrt{2\pi}} \exp\left[\frac{-(V - nv_R)^2}{2(v_R\sqrt{n})^{\frac{1}{2}}}\right]$$

Equation 2.54 has the form of the normal error curve and from the geometrical properties of the curve we can show that

$$n = 16(V_R/W)^2 \tag{2.55}$$

where W is the base width of the peak. Equation 2.55 is a measure of the efficiency of a gas chromatographic column. Sometimes the number of plates are measured at the bandwidth at half-height, $W_{0.5}$. From statistics

$$\tag{2.56}$$

$$W = (2/\ln 2)^{\frac{1}{2}} W_{0.5}$$

Equation 2.55 may be expressed as

$$n = 8 \ln 2 (V_R/W_{0.5})^2 \qquad\qquad (2.57)$$

$$= 8 (2.30 \log_{10} 2) (V_R/W_{0.5})^2$$

$$= 5.55 (V_R/W_{0.5})^2$$

As long as the sample occupies less than $0.5(n)^{\frac{1}{2}}$ theoretical plates, there will be no band broadening because of sample size. As the total number of theoretical plates increases, within a column, the maximum space (in terms of theoretical plates) that should be occupied by the sample will also increase. However, the percentage of column length available for sample will decrease (because number of theoretical plates per column length increases), as shown in Table 2.3

TABLE 2.3   SAMPLE SPACE IN TERMS OF THEORETICAL PLATES IN COLUMN

Number of Plates in Column	Maximum Space Available for Sample	
	In Terms of Theoretical Plates[1]	In Terms of % Column Length[2]
4	1	25
100	5	5
400	10	2.5
10,000	50	0.5

[1] $0.5 \sqrt{n}$.   [2] $(0.5 \sqrt{n}/n) 100$.

Having calculated the number of theoretical plates and knowing the length of the column, one may determine the HETP.

$$h = HETP = L/n = (L/16)(W/V_R)^2 = (L/16)(W/t_R)^2 \qquad (2.58)$$

Plate theory disregards the kinetics of mass transfer. Thus, it
reveals little about the factors influencing HETP values. Plate
theory tells us that HETP becomes smaller with decreasing flowrate; however, experimental evidence shows that a plot of HETP
versus flowrate always goes through a minimum.

The "theoretical plate" defined in gas chromatography is not
the same concept as in distillation or other countercurrent mass
transfer operations. In the latter, the number of theoretical
plates represents the number of equilibrium stages on the equilibrium curve of a binary mixture which causes a given concentration change. In other words HETP is the length of column producing a concentration change which corresponds to one equilibrium
stage. In gas chromatography, the number of theoretical plates
is a measure of peak broadening for a single component during its
lifetime in the column. Thus, for a given column of constant
length, the HETP represents the peak broadening as a function of
retention time. In a gas chromatographic column, each component
will yield different n and HETP values. Those solutes with high
retention (high K values) will result in greater numbers of theoretical plates and thus lower HETP values. It is generally found
that the necessary number of theoretical plates for packed gas
chromatographic columns is 10 times greater than in distillation
for a similar separation.

2.2.2 Rate Theory

Although HETP is a useful concept, it is an empirical factor.
Since plate theory does not explain the mechanism that determines
these factors, we must use a more sophisticated approach, the
rate theory, to explain chromatographic behavior. Rate theory is
based on such parameters as rate of mass transfer between stationary and mobile phases, diffusion rate of solute along the
column, carrier gas flowrate, and the hydrodynamics of the mobile
phase.

Glueckauf (14) studied the effect of four factors on the chromatographic process:

1. Diffusion in the mobile phase normal to the direction of flow.
2. Longitudinal diffusion in the mobile phase.
3. Diffusion into the particles.
4. Size of the particle.

The interpretation of the resulting chromatogram will tell how
well a separation has been performed. This interpretation can be

viewed from two points: (a) how well the centers of the solute
zones have been disengaged and (b) how compact are the resulting
zones. Many chromatographic separations accomplish the first
point but not the second, which results in the two zones spreading into each other.

We will look at the three variables that may cause zone
spreading, that is, ordinary diffusion, eddy diffusion, and local
nonequilibrium. Our approach to this discussion will be from the
random walk theory, since the progress of solute molecules
through a column may be viewed as a random process.

First we define three factors:

Ordinary diffusion. This process results when there exists a
region of high concentration and a region of low concentration.
The migration is from the high to the lower concentration region
in the axial direction of the column. This type of diffusion
occurs on the molecular level resulting from the movement of
molecules after collision.

Eddy diffusion. Visualize a column packed with marbles of
equal diameter. The void space along the column is essentially
uniform. As the size of the marbles (particles) becomes smaller
it becomes increasingly more difficult to control uniformity in
size and to prevent crushing or fractionation of the particles.
With the particle size used in analytical gas chromatographic
columns, 60/80 mesh (0.25-0.17mm) it is very difficult to have
all the particles the same diameter and some of these particles
might fit into void spaces between particles. The overall effect
is that the spaces along the column are not uniform. Thus, when
a sample migrates down the column each molecule sees different
paths and each path is of a different length. Some molecules
take long paths others short paths. In addition to these different pathlengths there are variations in the velocities of the
mobile phase. The overall result of these phenomena is that some
molecules lag behind the center of the zone while others move
ahead of the zone. Therefore, the eddy diffusion process results
from flow through randomly spaced particles in the column.

Local nonequilibrium. As the zone of solute molecules
migrates through the column (approximating a Gaussian curve),
there exists a variable concentration profile from leading edge
through the center to the trailing edge. As this zone continues
to migrate down the column, it is constantly bringing an ever-

changing concentration profile in contact with the next part of
the column. This effect results in different rates of achieving
equilibrium along the column. Thus, each section (theoretical
plate) in the column is constantly attempting to equilibrate with
a variable concentration zone in the mobile phase. At one time
the zone attempts to equilibrate with a low concentration in the
mobile phase, and then at another time with a high concentration.
If no flow were present, equilibration proceeds; however, we are
in a dynamic system and there is always flow. These overall processes result in unequilibrium at each theoretical plate. The
overall process is determined by kinetic rate processes that
account for transfer of the solute molecules between the two
phases in the column.

Viewing the zone migration as discussed above, we can conclude
that increasing the mobile phase velocity will increase the nonequilibrium effect. Providing for more rapid exchange of solute
molecules between the mobile and stationary phases decreases the
nonequilibrium effect. Theory tells us that horizontal displacement (perpendicular to flow) is constant throughout the zone,
proportional to velocity of flow but inversely proportional to
rate of restoring equilibrium. On the other hand, vertical displacement (parallel to flow) is proportional to concentration
gradient.

Since the three processes discussed above are all random diffusion processes, we can evaluate the zone broadening from the
viewpoint of a random walk. If a process results from the random
back and forth motion of solute molecules, we then have a concentration profile which is Gaussian in shape (i.e., there is an
equal number of molecules preceeding the zone center as there are
trailing the zone center). The extent of spreading for normal
Gaussian distributed molecules is described by the standard deviation, σ. This band spreading (σ) is defined in the random walk
model by the number of steps taken (n) and the length of each
step (ℓ):

$$\sigma = \ell(n)^{\frac{1}{2}} \tag{2.59}$$

Equation 2.59 states that zone spreading is proportional to step
length but not to the number of steps. For instance, movement is
random; it takes 16 steps to give a displacement 4 times the
average length of each step.

We know from statistical treatment that standard deviations
are not additive. However, variances, the square of the standard
deviation, are additive. In terms of the chromatographic process
three diffusive process variables contribute to zone spreading.
Thus, we can sum these variables in terms of variances to give

the overall contribution of zone spreading. The combined effect
may be shown as:

$$\sigma^2 = \Sigma\sigma_i^2 \qquad (2.60)$$

where the $\Sigma\sigma_i$ term is a sum of each of the three processes:
σ_D for ordinary diffusion, σ_E for eddy diffusion and
σ_C for nonequilibrium diffusion effects.

The ordinary diffusion process term is defined by the Einstein
diffusion equation

$$\sigma_D^2 = 2Dt \qquad (2.61)$$

Here D is the coefficient of diffusion and t represents the time
molecules spend in the mobile phase from the start of the random
process. The term t also can be expressed in terms of the distance the zone has moved (L) and the velocity of the mobile phase
(v); thus,

$$t = \frac{L}{v} \qquad (2.62)$$

and Equation 2.61 becomes

$$\sigma_D^2 = \frac{2DL}{v} \qquad (2.63)$$

The reader should keep in mind when developing a theory of zone
spreading that we must have a point of reference to show how the
spreading develops . This point of reference is the zone center.

The eddy diffusion term, σ_E, describes the change in pathway
and velocity of solute molecules in reference to the zone center.
If the molecules are in a "fast" channel they can migrate ahead
of the zone center, if in a "slow" channel they can lag behind
the zone center. To quantify the eddy diffusion term, we must
describe the step length and the number of steps taken in a
specified period of time. The void or channel between particles
would be expected to be in the order of one particle diameter, d.
As molecules move from one channel to another, their velocity
will be of the order of +d or -d (in respect to the zone center).
So on the average, the molecules will take an equivalent step of
d.

As to the number of steps, we can determine these in terms of
the total column length, L, and the equivalent length of the step
Therefore,

$$n = L/d \qquad (2.64)$$

On reflection it is apparent that channels cannot be regarded as either "fast" or "slow." Rather, they will be a range of veloci-ties with some average value for the entire column length. Also, the column voids or channels will not all be exactly equal to d, but will vary from larger than d to smaller than d, with an over-all average of d. In light of the above description we can equate d for length of step (ℓ) and L/d for number of steps. Substitution into Equation 2.59 gives:

$$\sigma_E = d(L/d)^{\frac{1}{2}} = (Ld)^{\frac{1}{2}} \tag{2.65}$$

This equation states that eddy diffusional effects on zone spreading increase with the square root of zone displacement and particle size.

Equations 2.61 and 2.65 account for the effect of ordinary and eddy diffusion in the zone broadening process. Now we need to express nonequilibrium effects which are concerned with the time the solute molecules spend in the two phases. Let us define a few more terms in order to set up some mathematical relation ships:

k_1 = transition rate of the molecule from mobile phase to sta-tionary phase.

$1/k_1$ = average time required for one sorption to occur.

k_2 = transition rate of molecule from stationary phase to mobile phase.

$1/k_2$ = time required for one desorption to occur.

A molecule in the mobile phase is moving faster than the center of the zone. The velocity of the zone is Rv, where R is the fraction of solute molecules in mobile phase and v is mobile phase velocity. Therefore, 1-R is the fraction of solute mole-cules in the stationary phase with a velocity of zero. Now, molecules move back and forth with respect to the zone center as each phase transfer occurs. In terms of random walk, n is the number of transfers our molecules take between the two phases. In terms of sorptions-desorptions, n is twice the number of de-sorptions (one desorption occurs for each sorption) and the time needed for the solute zone to move through the column (distance= L) at its velocity Rv is

$$t = \frac{L}{R_v} \tag{2.66}$$

During this time, t, the molecules will spend the fraction R in mobile phase and the fraction (1-R) in the stationary phase. So the time the fraction of molecules, (1-R), spend in the station-ary phase will be,

$$t = \frac{(1-R)L}{R_v} \qquad (2.67)$$

The number of desorptions is the time the molecules spend in the stationary phase (Equation 2.67) divided by $1/k_2$,

$$n_{des} = \frac{(1-R)L/Rv}{1/k_2} = \frac{k_2(1-R)L}{Rv} \qquad (2.68)$$

Since there are twice the phase transfers as there are desorption processes the number of steps (n) is equal to two times Equation 2.68, or

$$n = \frac{2k_2(1-R)L}{Rv} \qquad (2.69)$$

To obtain a value for the distance a molecule moves back with respect to the zone center, ℓ, we need to consider $1/k_2$, the lifetime of a molecule in the stationary phase. The center of the zone moves forward $\{Rv \times (1/k_2)\}$ or (Rv/k_2), during the time the molecule is in the stationary phase. Thus, our step length also is Rv/k_2. By similar reasoning we arrive at the same value for the forward movement of molecules ahead of the zone center.

We now can describe an equation for the effect of nonequilibrium on zone spreading viewed as a random walk. Substituting Rv/k_2 for ℓ and $2k_2(1-R)L/Rv$ for n in Equation 2.59, we have

$$\sigma_c = \frac{Rv}{k_2} \left[\frac{2k_2(1-R)L}{Rv} \right]^{\frac{1}{2}} \qquad (2.70)$$

$$= \left[\frac{2R(1-R)Lv}{k_2} \right]^{\frac{1}{2}}$$

Equation 2.70 indicates that an increase in flow velocity causes an increase in nonequilibrium effects. Providing for rapid exchange of solute molecules between phases decreases these effects.

We may now return to Equation 2.60 and substitute equations 2.61, 2.65, and 2.70.

$$\sigma^2 = 2Dt + Ld + \frac{2R(1-R)Lv}{k_2} \qquad (2.71)$$

$t = L/v$, so,

$$\sigma^2 = \frac{2DL}{v} + Ld + \frac{2R(1-R)Lv}{k_2} \qquad (2.72)$$

$$\sigma^2 = L\left[2D/v + d + \frac{2R(1-R)Lv}{k_2}\right] \qquad (2.72)$$

Martin and Synge (15) introduced height equivalent to a theoretical plate, h, as a measure of zone spreading,

$$h = \sigma^2/L \qquad (2.73)$$

So Equation 2.72 may be written as

$$h = \frac{2D}{v} + d + \frac{2R(1-R)v}{k_2} \qquad (2.74)$$

Rearrangement of Equation 2.74 we have

$$h = d + \frac{2D}{v} + \frac{2R(1-R)v}{k_2} \qquad (2.75)$$

To find the correct flow velocity, v, which gives the minimum plate height, we take the first derivative of Equation 2.74 and set dh/dv equal to 0. This results in

$$v = \left[\frac{k_2 D}{R(1-R)}\right]^{\frac{1}{2}} \qquad (2.76)$$

The van Deemter Equation (16) is used for describing the process of GC. This equation was evolved from the earlier work (17) and then was extended with Glueckauf's theory. The equation was derived from consideration of the resistance to mass transfer between the two phases as arising from diffusion.

$$h = 2D_c/u + (8/\pi^2)(k'/(1+k')^2)(d_f^2/D_\ell)u \qquad (2.77)$$

where D_c = overall longitudinal diffusivity of solute in gas phase,
   k' = capacity factor,
   d_f = effective film thickness of liquid phase,
   D_ℓ = diffusivity of solute in liquid phase,
   u = apparent linear flowrate of gas phase.
The first term in Equation 2.77 is the contribution due to overall longitudinal diffusion and the second term is contribution due to resistance to mass transfer in liquid phase.
   The overall longitudinal diffusivity (D_c) is the sum of apparent longitudinal diffusivity (D_a) and true molecular diffusivity (D_g).

$$D_c = D_a + \gamma D_g \qquad\qquad (2.78)$$

The factor γ is used to account for irregular diffusion patterns and usually is less than unity because molecular diffusivity is smaller in packed columns than in open tubes.

Klinkenberg and Sjenitjer (18) statistically showed that

$$D_a = \lambda u dp \qquad\qquad (2.79)$$

where λ = a dimensionless constant characteristic of packing. This equation is indicative of how good or bad the packing homogeneity is in the column; for regular packings, $\lambda < 1$; for non-uniform packed columns with channels, $\lambda > 1$. The term dp is the particle diameter in centimeters.

Uneven distribution of the stationary phase liquid on the solid support particles causes band dispersion. This may be rationalized if one considers that molecules entering a thin part of the liquid film permeate faster than in a thick part of the liquid film. This effect causes some molecules to spend more time in the liquid phase than other molecules with their slow movement through the column and resulting widening of the band.

Considering all the effects discussed above and combining Equations 2.77 to 2.79 we have the expression:

$$h = 2\lambda dp + 2\lambda D_g/u + (8/\pi^2)(k'/(1+k')^2)(d_f^2/D\ell)u \qquad (2.80)$$

This equation predicts that for maximum column performance we must minimize the contribution of each term while still maintaining a constant linear flowrate. The first term accounts for the geometry of the packing, the second for longitudinal diffusion in gas phase, and the third for resistance to mass transfer process.

The general form for the van Deemter equation is

$$h = A + B/u + Cu \qquad\qquad (2.81)$$

A representation of this equation is in Figure 2.8, which shows the effect of h with changes in linear gas velocity. Equation 2.81 is that of a hyperbola having a minimum at velocity $u = (B/C)^{\frac{1}{2}}$ and a minimum h value (h_{min}) at $A + 2(BC)^{\frac{1}{2}}$. The constants may be graphically calculated from an experimental plot of h versus linear gas velocity as shown in Figure 2.8.

It is also useful to plot h versus $1/u$ (Figure 2.9). In both presentations (Figures 2.8, 2.9) the intercept of the linear portion of the h plot will equal $2\lambda dp$. Thus, if the particle size

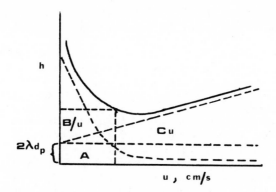

Figure 2.8. van Deemter plot. Change in h versus linear gas velocity, u. h_{min} = A + $(2BC)^{1/2}$. u_{opt} = $(BC)^{1/2}$.

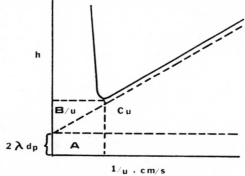

Figure 2.9. Rate theory equation plotted as h versus 1/u. h_{min} = A + $(2BC)^{1/2}$. u_{opt} = $(B/C)^{1/2}$.

(dp) is known, λ can be calculated and a measure of packing regularity is obtained. From the slope of the linear part of the h-u curve one also can estimate the film thickness, d_f, if D_1 and k' are known (resistance to mass transfer in liquid phase term).

The constants, A, B, C can also be determined by the method of least squares. A gradual approximation of B may be calculated from a plot of h-Cu versus 1/u, and C can be approximated from a plot of h-Bu versus u.

Let us take a better look at the effect of the terms of Equation 2.80 on plate height. The contribution of the $2\lambda dp$ term can be decreased by reducing the particle size. However, as the particle size becomes very small, the pressure drop through the column increases. The value of λ usually increases as dp decreases. Of the three terms in Equation 2.80, only this first one is independent of linear flowrate.

The second term, $2\gamma D_g/u$, is a measure of the effect of molecular diffusion on zone spreading. This term may be decreased by reducing the molecular diffusivity D_g. We know from the kinetic theory of gases that the D_g value depends upon nature of the vapor, and the temperature and pressure of the system. Diffusion in low molecular weight gases (H_2 and He) is high compared to that in higher molecular weight gases (N_2 or CO_2). If this was the only criterion for choice of carrier gas one would choose N_2 or CO_2, rather than He. This is evidenced by the fact that optimum gas velocity is governed by $(B/C)^{\frac{1}{2}}$. One obtains a value for $(B/C)^{\frac{1}{2}}$ by differentiating Equation 2.81 with respect to u and then setting dh/du=0. $u_{opt}(h_{min})$ is then equal to $(B/C)^{\frac{1}{2}}$. However, other factors affect the choice of a carrier gas, such as the effect of the sensitivity of the detector employed. If in a particular system the C term is small and high flowrates are allowable (this reduces term $2\gamma D_g/u$), the nature of the carrier is not too important. For columns of low permeability a low molecular weight (less viscous) gas might be the best choice (e.g., He). If the C term is large and low flowrates are used, the term $2\gamma D_g/u$ becomes important and the carrier gas can exert influence on HETP. In this case, high molecular weight gases (e.g., N_2 or CO_2) would be preferred because solute diffusion coefficients would be small.

The third term of Equation 2.80 accounts for resistance to mass transfer in liquid phase. An obvious way of reducing this term is to reduce the liquid film thickness d_f. This causes a reduction in k' and an increase in the term $k'/(1 + k')^2$. However, using thinly coated column packings increases the probability of adsorption of solute molecules on the surface of support material, which might result in peaks tailing.

The k' term is temperature dependent, so we increase k' and decrease $k'/(1 + k')^2$ by lowering of temperature. Lowering of temperature increases viscosity and thus decreases D_1. Therefore, the effects of the factors $k'/(1+k')^2$ and $1/D_1$ counteract each other.

Thus it can be seen that the observed HETP is not only a function of column packing but depends upon operating conditions and the properties of the solute. This is why different values of HETP (or different numbers of theoretical plates per unit column length) are obtained for various solutes.

Many modifications to the original van Deemter plate height equation have appeared in the literature (19-23). Some account for mass transfer in the gas phase (19,20) and other modifications were made for velocity distribution because of retarded

flow of interfacial resistance (21,22). Improvements were
attempted, usually stochastic theories based on random walk
theory (23). However, we elaborated upon the work of Giddings
(24), who described plate height contributions as a function of
the diffusional character of zone broadening by accounting for
local nonequilibrium.

MODIFICATIONS OF THE van DEEMTER EQUATION. If one accounts for
the fact that resistance to mass transfer can occur in the stationary phase as well as in the mobile phase, Equation 2.81 may
be written as

$$h = A + B/u + C_\ell u + C_g U \qquad (2.82)$$

The last term accounts for the resistance to mass transfer in the
gas phase. Low-loaded liquid coatings cause the C_g term to be
significant. Equation 2.82 was further extended to account for
velocity distributions due to retarded gas flow in the layers
(C_1) and the interaction of the two types of gas resistance (C_2):

$$h = A + B/u + C_\ell u + C_g u + C_1 u + C_2 u \qquad (2.83)$$

The term $C_g u$ may be defined as,

$$C_g u = c_a (k'^2/(1 + k')^2)(d_g^2/D_g) u \qquad (2.84)$$

c_a is a proportionality constant, d_g the gas diffusional path
length, and D_g the diffusion coefficient of solute molecules in
gas phase.

   The C_1 term becomes significant with rapidly eluted but poorly
sorbed components. The value of C_1 depends upon the particle
size of the packing.

$$C_1 u = (c_b d_p^2/D_g) u \qquad (2.85)$$

where c_b is a proportionality factor approximately equal to unity.
Giddings (25) realized that the processes in the gas phase cannot
be considered independent with respect to their effect on h.
Thus, he stated that the term A (flow characteristic) and the
effect of resistance to mass transfer in gas phase must be
treated dependently. So Equation 2.82 becomes

$$h = 1/(\frac{1}{A} + \frac{1}{C_1 u}) + \frac{B}{u} + C_\ell u + C_g u + H_e \qquad (2.86)$$

The term

$$\{ 1/(\frac{1}{A} + \frac{1}{C_1 u}) \}$$

results from the merging of the eddy diffusion term and the velocity distribution term ($C_1 u$) of Equations 2.83 and 2.85. The term H_e is introduced to account for the characteristics of the equipment used in the system. The first term (Equation 2.86) is not simple (26,27). Depending upon the nature of the packing and of the flow, five possible mechanisms can take place; so our term becomes a summation term.

$$h = \sum_{i=1}^{5} \frac{1}{1/A + 1/C_1 u} + \frac{B}{u} + C_\ell u + C_g u + H_e \qquad (2.87)$$

The five possible mechanisms of band broadening occur because of:

1. Flow through channels between particles.
2. Flow through particles.
3. Flow because of uneven flow channels.
4. Flow between inhomogeneous regions.
5. Flow throughout the entire column length.

All the preceding discussion has assumed no compressibility of the gas stream. With columns where the pressure drop is large, the change in gas velocity should be considered (gas expansion also causes zone spreading). DeFord et al. (28) demonstrated the importance of a pressure correction and after considering the A term to be negligible developed the following equation:

$$h = \frac{B^o}{p_o u_o} + (C_g^o + C_1^o) p_o u_o f + C_\ell^o u_o j \qquad (2.88)$$

where B^o, C_g^o, C_1^o and C_ℓ^o = coefficients determined by measuring h
for various outlet pressures and outlet velocities,

p^o = outlet pressure,

u_o = outlet gas velocity,

j^o = James and Martin pressure correction factor

$$= \frac{3}{2} \frac{(p_i/p_o)^2 - 1}{(p_i/p_o)^3 - 1} ,$$

f = pressure correction

$$= p_i (p_o + 1) j^2 / 2.$$

f is usually unity and can be neglected except in accurate theoretical work.

FLOW.   The rate at which zones migrate down the column is dependent upon equilibrium conditions and mobile phase velocity; on the other hand, how the zone broadens depends upon flow conditions in the column, longitudinal diffusion, and the rate of mass transfer.   Since there are various types of columns used in gas chromatography, namely, open tubular columns, support coated open tubular columns, packed capillary columns, and analytical packed columns, we should look at the conditions of flow in a gas chromatographic column.   Our discussion of flow will be restricted to Newtonian fluids, that is, those in which the viscosity remains constant at a given temperature.

   Flow through an open tube is characterized by the dimensionless Reynolds' number,

$$Re = \frac{\rho dv}{\eta} = \frac{dG}{\mu} \tag{2.89}$$

where ρ = fluid density (g/cm^3),
       d = tube diameter (cm),
       v = fluid velocity (cm/sec),
       η = fluid viscosity (poise),
       G = mass velocity $(g/cm^2 \ sec)$,
       μ = absolute viscosity (g/cm sec).
Inertial forces of the fluid increase with density and the square of velocity (ρv^2) while viscous forces decrease with increasing diameter of tube $(\eta v/d)$ and increase with viscosity and velocity. High Reynolds' numbers (Re>4000) result in turbulent flow; with low Reynolds' number (Re<2000) the flow is laminar.   Laminar flow results from formation of layers of fluid with different velocities after a certain flow distance, as illustrated in Figure 2.10A.   Flow at the walls is zero and increases approaching the center of the tubes.   The laminar flow pattern results from layers of mobile phase with different velocities travelling parallel to each other.   The maximum flow at the center is twice the average flow velocity of the fluid.   Molecules in the fluid can exchange between fluid layers by molecular diffusion.   Most open tubular columns operate under laminar flow conditions.

   Turbulent flow results because of the radial mixing of layers to equalize flowrates.   The mixing of the layers is due to the increased eddies and mass transfer occurs by eddy diffusivity. Turbulent diffusivity increases in proportion to mean flow velocity as depicted in Figure 2.10B.   Figure 2.10C represents plug flow which is unattainable in practice but does suggest a model

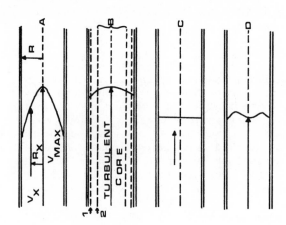

Figure 2.10.   Flow profiles in tubes and packed columns.   (A) Laminar flow.   r = tube radius, V_x =stream
path velocity at radial position r_x. V_{max} = maximum
flow velocity at tube center.   (B) Turbulent flow.
1 = laminar sublayer. 2=buffer layer. (C) Plug flow.
(D) Flow in a packed column. Effect is more pronounced the smaller the ratio of tube diameter: particle
size.

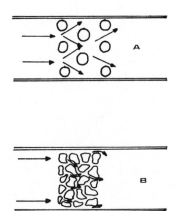

Figure 2.11.   Representative flow through a packed
column.   (A) Simplified diagram of column with "uniform" particles.   (B) Representative diagram of column with experimental particles.   How tortuous the
path becomes depends on the particle packing structure.   Plug flow usually results.

from which other flows may be considered. The flow usually attained in packed columns is illustrated in Figure 2.10D.

A considerable difference exists between flow through an open column and packed column as illustrated by Figure 2.11. Darcy's law, which governs flow through packed columns, states that flow velocity is proportional to the pressure gradient,

$$v_o = \frac{B^o}{\eta} \frac{(p_o - p_i)}{L}$$
(2.90)

True average fluid velocity may be expressed as

$$v = \frac{B^o}{\varepsilon\eta} \frac{(p_o - p_i)}{L}$$
(2.91)

and mean velocity of a fluid is represented by

$$v \text{ (mean velocity)} = \frac{(p_i - p_o)r^2}{8\eta L}$$
(2.92)

Combining Equations 2.90 and 2.92 gives the specific permeability coefficient, B^o:

$$B^o = \frac{r^2}{8}$$
(2.93)

If we express the free cross-section of the column bed by the interparticle porosity (knowing total porosity of packed beds with porous particles is larger because of intraparticle space), we can obtain the true average fluid velocity, v,

$$v = \frac{B^o (p_o - p_i)}{\varepsilon\eta L}$$
(2.94)

where v_o = superficial velocity (average velocity without packing)
  B^o = specific permeability coefficient (1 darcy = $10^{-8} cm^2$),
  ε = interparticle porosity (0.4±0.03),
  η = mobile phase velocity,
  L = column length,
  p_o = outlet pressure,
  p_i = inlet pressure.

Thus, by combining the cross-sectional area of the tube, $r^2\pi$, and the mean velocity (Equation 2.92) we can come up with an expression for the volume flowrate F:

$$F = \frac{(p_i - p_o)r^4\pi}{8\eta L}$$
(2.95)

Flow in packed columns may be expressed in terms of modified Reynolds' numbers (Re)m, which take into account a geometric factor for the diameter of the particle rather than the diameter of the column (see Equation 2.89).

$$(Re)m = \frac{\rho v d_p}{\eta} = \frac{d_p G}{\mu} \qquad (2.96)$$

For laminar flow Reynolds' (Re)m values are less than and with turbulent flow (Re)m>200. Packed gas chromatographic columns normally operate with a (Re)m of <10, so they may be considered operating with laminar flow.

Assuming a column of diameter (d) 0.3 cm and particles of diameter (d_p) 0.02 cm, Table 2.4 shows the change in Reynolds' number with flowrate.

TABLE 2.4 Δ(Re)m WITH MOBILE PHASE FLOWRATE

Flowrate	(Re)m for H_2	(Re)m for N_2
25 cm^3/min	0.11	0.78
50 cm^3/min	0.22	1.56

In gas chromatographic procedures the carrier gas flow usually is measured after the column via a soap bubble flowmeter. To obtain the average flowrate, $\bar{F}$, in the column, one must account for three factors:

1. Compressibility correction, j.
2. Correction for flow being measured at room temperature, T_o, rather than column temperature, T, that is, T/T_o.
3. Correction factor for vapor pressure of water, p_w, when using flowmeter, p_o-p_w/p_o, where p_o is atmospheric pressure.

The initial flow, F_o into the column in terms of the measured flow, F_m is

$$F_o = F_m(T/T_o) \ \{(p_o-p_w)/p_o\} \qquad (2.97)$$

The average flowrate in the column then is determined by

$$\bar{F} = jF_m(T/T_o) \ \{(p_o-p_w)/p_o\} \qquad (2.98)$$

This average flowrate term should be used to measure precise retention volumes.

2.2.3   The Solid Support

One should keep two things in mind when choosing a support: (1) structure and (2) surface characteristics. Structure contributes to the efficiency of the support, whereas the surface characteristics govern the support's participation in the resulting separations. The perfect column material would be chemically inert towards all types of samples. It would have a large surface area so that liquid phase could be spread in a thin film and structure of the surface would be such that it would properly retain the liquid film. However, large surface area is not a guarantee of an efficient column.

The most commonly used column support materials are made from diatomite. Other materials include sand, Teflon, inorganic salts, glass beads, porous layer beads, porous polymers, carbon blacks, etc. We will discuss the diatomite supports in some detail and additional information may be obtained in Chapter 3.

Basically, two types of supports are made from diatomite. One is pink and derived from firebrick and the other is white and derived from filter-aid. German diatomite firebrick is known as Sterchmal. Diatomite itself is a diatomaceous earth, as is the German kieselguhr. Diatomite is composed of diatom skeletons or single-celled algae which has accumulated in very large beds in numerous parts of the world. The skeletons consist of a hydrated microamorphorous silica with some minor impurities, (e.g. metallic oxides). The various species of diatoms number well over 10,000 from both fresh water and salt water sources. Many levels of pore structure, in the diatoms cell wall, cause these diatomites to have large surface areas (20 m^2/g). The basic chemical difference between the pink and white diatomite may be summarized as follows:

1. The white diatomite or filter-aid is prepared by mixing with a small amount of flux (e.g., sodium carbonate), and calcining (burning) at temperatures greater than 900°C. This process converts the original light-gray diatomite to white diatomite. The change in color is believed to be the result of converting the iron oxide to a colorless sodium iron silicate.
2. The pink or brick diatomite has been crushed, blended and pressed into bricks, which are calcined (burned) at temperatures greater than 900°C. During the process the mineral impurities form complex oxides and/or silicates.

It is the oxide of iron which is credited for the pink color.

It has been well established that the surface of the diatomites
are covered with silanol (Si-OH) and siloxane (Si-O-Si) groups.
The pink diatomite is more adsorptive than the white, this difference being due to the greater surface area/unit volume rather
than in any fundamental surface characteristic. The pink diatomite is slightly acidic (pH of 6-7) whereas the white diatomite
is slightly basic (pH of 8-10). Both types of diatomites have
two sites for adsorption: (1) van der Waals sites and (2) hydrogen-bonding sites. Hydrogen-bonding sites are more important
and there are two different types for hydrogen bonding, that is,
silanol groups, which act as a proton donor and the siloxane
group where the group acts as a proton acceptor. Thus, samples
to be analyzed by GC containing compounds which form strong
hydrogen-bonds, (e.g., water, alcohol, and amines) may show considerable tailing; whereas those compounds which hydrogen-bond to
a lesser degree (e.g., ketones, esters) do not tail as much.
    The elimination of adsorption sites (i.e, deactivation of sur-
face) can be performed several ways:

1. Removal by acid washing.
2. Removal by reaction of silanol groups.
3. Saturation of surface with a liquid substrate phase.
4. Coating with solid material.

It is not entirely clear what is accomplished by acid treatment.
It is generally believed that some species, perhaps iron, is removed from the support. Regarding reaction at silanol groups, it
has been suggested (29) that when treated with DMCS (dimethyldichlorosilane) there is a reaction between the surface hydroxy
groups:

$$
\begin{array}{c}
\overset{\displaystyle OH \quad OH}{\underset{\displaystyle |\qquad|}{-Si-O-Si-}} + (CH_3)_2-Si-Cl_2 = \quad
\overset{\displaystyle H_3C \diagdown \quad \diagup CH_3}{\underset{\displaystyle \underset{\displaystyle \underset{\displaystyle -Si - O - Si-}{|\qquad\qquad|}}{O \diagup \qquad \diagdown O}}{Si}} + 2HCl
\end{array}
$$

silyl ethers

If only one hydroxy group is available the reaction may be:

$$
\begin{array}{c}
CH_3 \\
| \\
H_3C-Si-Cl \\
| \\
\end{array}
$$

$$
\begin{array}{ccccc}
OH & & & & O \\
| \quad | & & & & | \qquad | \\
-Si-O-Si- & + & (CH_3)_2-Si-Cl_2 & = & -Si - O - Si- \; + \; HCl \\
| \quad | & & & & | \qquad | \\
\end{array}
$$

silyl ethers

In the case of HMDS (hexamethyldisilazane) the following reaction
has been proposed:

$$
\begin{array}{ccccc}
OH \quad OH & & H & & \\
| \quad\; | & & | & & | \qquad | \\
-Si-O-Si- & + & (CH_3)_3-Si-N-Si-(CH_3)_3 & = & -Si -O - Si- \; + \; NH_3 \\
| \quad\; | & & & & | \qquad | \\
& & & & O \qquad\quad O \\
\end{array}
$$

$$
H_3C- \;\; Si-CH_3 \; CH_3-Si-CH_3
$$

$$
\qquad\qquad | \qquad\qquad\quad | \\
\qquad\qquad CH_3 \qquad\qquad CH_3
$$

Silyl Ethers

It should be noted that silanization reduces the surface area of
the support. Thus, one generally should not use more than 10%
stationary phase loading.

Small particles should be used in gas chromatographic columns
since the HETP is directly proportional to particle diameter.
However, column permeability is proportional (and pressure drop
is inversely proportional) to the square of the particle diameter. Therefore, if particles are too small pressure requirements increase tremendously.

It has become the practice to refer to chromatographic
supports in terms of the mesh range. When sieving particles for
chromatographic columns both the Tyler Standard Screens and the
United States Standard Series are frequently used. Tyler screens
are identified by the actual number of meshes per linear inch.
U. S. Sieves are identified by either micrometer (micron) designations or arbitrary numbers. Thus, a material referred to as
60/80 means particles which will pass a 60 mesh screen but not an
80 mesh screen. You may also see this written as -60+80 mesh.
Particle size is much better expressed in micrometers (microns),
therefore 60/80 mesh would correspond to 250-177 micrometers
(micron) particle size range. Table 2.5 shows the conversion of
column packing particle sizes. The table shows the relation

among mesh size, micrometers, millimeters and inches. Table 2.6
shows the relationship between particle size and sieve size.

TABLE 2.5 CONVERSION TABLE OF COLUMN PACKING PARTICLES

Mesh Size	Micrometers	Millimeters	Inches
4	4760 μm	4.76 mm	0.185 in
6	3360	3.36	0.131
8	2380	2.38	0.093
12	1680	1.68	0.065
16	1190	1.19	0.046
20	840	0.84	0.0328
30	590	0.59	0.0232
40	420	0.42	0.0164
50	297	0.29	0.0116
60	250	0.25	0.0097
70	210	0.21	0.0082
80	177	0.17	0.0069
100	149	0.14	0.0058
140	105	0.10	0.0041
200	74	0.07	0.0029
230	62	0.06	0.0024
270	53	0.05	0.0021
325	44	0.04	0.0017
400	37	0.03	0.0015
625	20	0.02	0.0008
1250	10	0.01	0.0004
2500	5	0.005	0.0002

TABLE 2.6 RELATION BETWEEN PARTICLE SIZE AND SCREEN OPENINGS

Sieve Size	Top Screen Openings (μm)	Bottom Screen Openings (μm)	Micrometer Spread
10/30	2000	590	1410
30/60	590	250	340
35/80	500	177	323
45/60	350	250	100
60/80	250	177	73
80/100	177	149	28
100/120	149	125	24
120/140	125	105	20
100/140	147	105	42

Lack of the proper amount of packing in a gas chromatographic column often is the source of a poor separation. How can one tell when a column is properly packed? The answer is twofold: by column performance (efficiency) and peak symmetry. Many factors affect column performance, but one of the easiest ways to check is the amount of packing per foot of column length (g/ft). If the amount of packing varies greatly from its optimum value, poor separations can result. Knowledge of the number of grams per foot will allow one to predict column performance.

Loosely packed columns generally are inefficient. A column that is too tightly packed gives excessive pressure drop or may even become completely plugged because the support particles have been broken and fines are present. Applied Science Laboratories, Inc. has prepared a used column packing guide which is reproduced in Table 2.7. A column packed within ±10% of the values shown in the table will provide satisfactory efficiency.

2.2.4   Mobile Phase

Sample components are transported through the column by means of a proper carrier gas. Use of a gas as the mobile phase enables rapid equilibrations between moving and stationary phases, resulting in the high performance of the GC technique. Gaseous mobile phases have low flow resistance, allowing long columns with high separating power. Detection of the emerging gaseous sample is simple with the use of highly sensitive detectors (see Chapter 5).

The mobile phase is an important component of overall gas chromatographic system. Constant and reproducible flow conditions should be maintained at all times. To achieve this the mobile phase must have the proper auxiliary components.

1. Gas Cylinder or Generator. In most cases the supply of mobile phase is from commercial cylinders. These are connected to the carrier gas system of the chromatograph by means of a reducing valve. Hydrogen necessary for a flame-ionization detector may be supplied from a commercial tank or electrolytically generated.

2. Purifier. Commercially prepared tank gases are usually technical grade which can contain oil vapors, oxygen, and water. The impurities present in these gases reduce column activity and life as well as interfere with the proper functioning of the detector(s). Oxygen may be removed by passing carrier gas through

TABLE 2.7 COLUMN PACKING GUIDE INFORMATION (g/ft)

Mesh Size of Support	Type of Metal Tubing	O.D. (inches) of Tubing	Supports Used (grams packing/ft column length)				
			GAS CHROM S,A,P,Z,Q	GAS CHROM R,RA,RP,Rz Chromosorb P	Chromosorb W	Chromosorb G	Porous Polymer Beads
45/60	SS	1/8	0.45^2	0.55	0.5	–	–
		3/16	–	2.1	–	–	–
		1/4	–	4.0^3	2.8^2	4.6	–
60/80	SS	1/8	0.4	0.6	0.4	0.75	0.5
		1/4	2.4	3.5	2.8	4.6	2.5
80/100	SS	1/8	0.45	0.6	0.45	0.8	0.5
		1/4	2.7	3.7	2.9	4.7	2.6
100/120	SS	1/8	0.5	0.6	0.5	0.8	0.5
		1/4	2.7	3.7	2.9	4.7	2.8^2
45/60	Al	1/8	–	0.4^3	0.3^3	–	–
60/80	Al	1/8	0.25^2	0.4	0.25^2	–	–
		1/4	1.4	2.8^2	–	3.2	–
		3/8	5.1	6.8^3	–	10.3	–
80/100	Al	1/8	0.3	0.4	0.3^2	–	–
		1/4	1.7	–	–	–	–
		3/8	5.3^2	–	–	10.3	–
100/120	Al	1/8	0.3	0.45^2	–	–	–
		1/4	1.7^2	–	–	–	1.6^3
		3/8	5.3^2	–	–	–	5.0
45/60	Cu	1/4	–	3.4	–	–	–
60/80	Cu	1/8	0.3	0.35^2	–	–	–
		1/4	1.6	3.2^3	–	–	–
80/100	Cu	1/8	0.3	0.45^2	0.3	–	–

[1]Porous Polymer Beads must be packed tightly; figures are thought to be minimum acceptable value.   [2]Estimated.   [3]Limited data

(Reproduced from Applied Science Laboratories, Inc. GAS-CHROM Newsletter, Jan/Feb.1970, Technical Bulletin No. 7, p.4).

86

a catalyst (Cu or Ni) at 100-105°C.   To remove CO and hydrocarbon
impurities, the same catalysts may be operated in excess of
600°C.   Palladium is very efficient for removing oxygen from
hydrogen streams.   Carbon dioxide may be removed by passing the
gas through solid soda-lime or soda-asbestos.   Oil vapors and
other heavier contaminants are best removed with activated charcoal.   Molecular sieves serve well for removing water.   Purification of air for flame ionization detector is combusted over
quartz at 800°C.

3. Pressure and/or Flow Controls.   Gas valve regulators supplied
by most gas suppliers are adequate to control pressure and flow.
Maintaining constant flowrate is accomplished with a large pressure outside of column.

4. Pressure-Measuring Device.   Accurate pressure measurements are
made with a manometer if needed.   In most instruments the carrier
gas, hydrogen, and air (for flame ionization detector) are provided with separate gauges.

5. Flowmeter.   To monitor flow of carrier gas, a variety of devices are available, such as differential capillary, thermal conductivity, ionization, rotameters, and calibrated soap-film tubes.
Measurement of the flow may either be continuous or intermittent,
and the flowmeter may be placed either in front of the column or
at the carrier gas outlet.   The soap-film type is most commonly
used because of its economy and ease of operation.

6. Preheater.   Using an optional piece of equipment, the carrier
gas is heated before it enters the sample injector.   This prevents condensation of high boiling components and subsequent
blocking of outlets.   When using a thermal conductivity detector
another advantage of using a preheater is that it ensures that
identical thermal conditions are in effect with the reference and
measuring sides of the cell.

   Any gas may be used as the carrier as long as it does not react with the sample and/or stationary phase.   However, other properties must be considered depending upon the type detector employed.   With a thermal conductivity detector one uses a gas with
high heat conductivity because thermal conductivity of a gas is
inversely proportional to the square root of the molecular weight.
Thus, very low molecular weight gases are optimum.   Helium is

generally preferred, since hydrogen is reactive and inflammable.
Argon is used with β-radiation ionization detectors and at times
with the gas density balance detector.

As was shown in Section 2.2.2, the efficiency of a column is
not only a function of linear flow velocity but also carrier gas
type. Helium generally is the best compromise as a carrier gas.
However, Loyd (20) has shown that nitrogen permits the highest
column efficiency, but not the greatest separation. Nitrogen is
the choice for highly loaded columns whereas hydrogen is best for
lower loaded columns. Optimum flowrate and retention time are
both dependent on the cross-sectional area of the column, and
therefore a narrower bore column gives a more efficient separation in the same time.

2.2.5 Stationary Phase

One of the reasons for the wide acceptance of gas-liquid chromatography is that there exists such a variety of liquid phases
with different properties. Because of this large number of liquid phases there has been a great amount of work to clarify the
interaction between liquid phase and the solute molecules. There
is hope that some theoretical basis can be found for choosing a
liquid phase to accomplish a particular separation and lately
there has been an effort to decrease the number of liquid phases
which are used. We now wish to discuss in general terms the role
of the liquid phase and describe some of the criteria needed to
discuss its role in a chromatographic separation.

In choosing a liquid phase some fundamental criteria must be
considered:

1. Is the liquid phase selective toward the components to be
   separated?
2. Will there be any irreversible reactions between the liquid
   phase and the components of the mixture to be separated?
3. Does the liquid phase have a low vapor pressure at the operating temperature and is it thermally stable?

Let us look at some of the information available to answer the
above questions, although we would like to point out that we are
not now attempting to develop a pattern by which one would choose
the proper liquid substrate as this is discussed in Chapter 3.
The vapor pressure of the liquid phase should be less than 0.1
torr at the operating temperature of the column. This value can
change depending upon the detector used (see Chapter 5), as bleed
from the liquid phase will cause noise and elevated background

signal and thus decrease sensitivity.   Information from a plot of
vapor pressure versus temperature is not always completely infor-
mative because adsorption of the liquid phase on the solid support decreases the actual vapor pressure of the liquid phase.
Other than its effect on the detector noise, liquid phase bleed
may interfere with analytical results and determine the life of
the column.   Also some supports may have a catalytic effect to
decompose the liquid phase, thus reducing its life in the column.
Contaminants in the carrier gas (e.g., O_2) also may interfere
with the stability of a liquid phase.

Two other properties of the liquid phase to be considered are
viscosity and wetting ability.   Ideally liquid phases should have
low viscosity and high wetting ability (ability to form a uniform
film on the solid support).

It is uninformative to refer to liquid phase as being selective, since all liquid phases are selective to varying degrees.
Selectivity refers to the relative retention of two components
and gives no information regarding the mechanism of separation.
Most separations depend upon boiling point differences, variations in molecular weights of the components, and/or the structure of the components.

The relative volatility or separation factor (α) depends upon
the interactions of the solute and the liquid phase, that is,
van der Waals cohesive forces.   These cohesive forces may be divided into three types:

1. London Dispersion Forces.   These are due to the attraction of
dipoles which arise from the arrangement of the elementary
charges.   Dispersion forces act between all molecular types and
especially in the separation of nonpolar substances (e.g., saturated hydrocarbons).

2. Debye Induction Forces.   These forces result from interaction
between permanent and induced dipoles.

3. Keesom Orientation Forces.   These forces result from the interaction of two permanent dipoles, with the hydrogen bond being
the most important.   Hydrogen bonds are stronger than dispersion
or inductive forces.

If two components have the same vapor pressure, separation can
be achieved on the basis of several properties.   These properties
are (in the order of their ease of separation): (a) difference in
the functional groups, (b) isomers with polar functional groups,

(c) isomers with no functional groups.

Polarity is another property that has been used to tabulate
liquid phases. By polarity we mean the electrical field effect
in the immediate vicinity of the molecule which depends on the
number, nature and arrangement of the atoms and on the type of
bond and the groups. Rohrschneider (30) introduced a polarity
scale, P*, which ranks solvents according to their polarity.

$$P* = a \{ \log(_2V_{R,p}/_1V_{R,p}) - \log(_2V_{R,u}/_1V_{R,u}) \} \qquad (2.99)$$

where a = constant,
        subscripts 1 and 2 = butane and butadiene, respectively,
        p = a phase with polarity P*,
        u = a nonpolar phase, squalane (standard).
        On the scale, P* = 100 for β, β'-oxydipropionitrile (polar
        liquid, and P* = 0 for squalane (nonpolar liquid). All
        other liquids fall between these two limits. This scale
has two very good features. First, it permits a rapid selection
of a liquid by minimizing the number of different liquid phases
since many have equivalent P* values. Second, it allows us to
pick solvents on a general basis of polarity when we have many
components to separate. Details of this polarity scale are given
in Chapter 3.

2.2.6 Evaluation of Column Operation

There are several parameters which can be used to evaluate the
operation of a column and to obtain infomation about a specific
system. If we plot the concentration of solute (in %) versus
volume of mobile phase or number of plate volumes for the tenth,
twentieth, and fiftieth plate in the column, we will obtain a
plot as shown in Figure 2.12. Improved separation of component
peaks is possible for columns that have a larger plate number.
Similar information is obtained if we plot concentration of solute (% of total) versus plate number. Figure 2.13 shows the band
positions after 50, 100, and 200 equilibrations with the mobile
phase.

A good gas chromatographic column is considered to have high
separation power, high speed of operation, and high capacity.
One of these factors can be improved usually at the expense of
another. Thus, a number of column parameters need to be discussed so that we can arrive at an efficient operation of a
column. We now look at several of these and illustrate with
appropriate relationships.

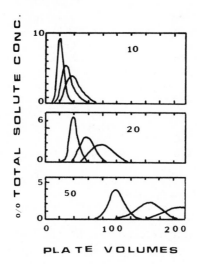

Figure 2.12.   Elution peaks for three solutes from various plate columns.   Top: 10 plates.   Middle: 20 plates.   Bottom: 50 plates.

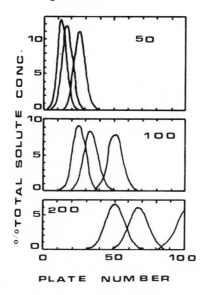

Figure 2.13.   Plate position of components after variable equilibrations.   Top: 50 equilibrations. Middle: 100 equilibrations.   Bottom: 200 equilibrations.

COLUMN EFFICIENCY.   Two methods are available for expressing the
efficiency of a column in terms of HETP: one is to measure the
peak (Figure 1.18) width at the baseline (Equation 2.55); the
other is to measure the peak width at half-height (Equation 2.57).

In determining n, we assume that the detector signal changes
linearly with concentration.   If it does not, then n cannot measure column efficiency precisely.   If Equations 2.55 or 2.57 are
used to evaluate peaks that are not symmetrical, positive deviations of 10-20% may result.   Since n depends on column operating
conditions, these should be stated when efficiency is determined.

There are several ways by which one may calculate column
efficiency other than the two equations shown (Equations 2.55,
2.57).   Figure 2.14 and Table 2.8 illustrate other ways in which
the information may be obtained.

TABLE 2.8   CALCULATION OF COLUMN EFFICIENCY FROM CHROMATOGRAMS

Standard Deviation Terms	Measurements	Plate Number $n=$
$A/h(2\pi)^{\frac{1}{2}}$	t_R and Band area A and Height h	$2\pi(t_R h/A)^2$
$W_i/2$	t_R and width at inflection points $(0.607h)W_i$	$4(t_R/W_i)^2$
$W_{0.5}/(8\ln2)^{\frac{1}{2}}$	t_R and width at half-height $W_{0.5}$	$5.55(t_R/W_{0.5})^2$
$W_b/4$	t_R and baseline width W_b	$16(t_R/W_b)^2$

EFFECTIVE NUMBER OF PLATES.   Desty et al. (31) introduced the
term effective number of plates, N, to characterize open tubular
columns.   In this relationship adjusted retention volume, V_R', in
lieu of total retention volume V_R, is used to determine the number of plates.   This equation is identical to Purnell's separation factor discussed below.

$$N = 16(V_R'/W)^2 = 16(t_R' /W)^2 \qquad (2.100)$$

This N value is useful for comparing a packed and an open tubular
column when both are used for the same separation.   Open tubular

columns generally have larger number of theoretical plates. One
can translate regular number of plates, n, to effective number of
plates by the expression:

$$N = n \left(\frac{k'}{1 + k'}\right)^2 \qquad (2.101)$$

as well as the plate height to the effective plate height:

$$H = h \left(\frac{1 + k'}{k'}\right)^2 \qquad (2.102)$$

Similarly, the number of theoretical plates per unit time can be
calculated:

$$n/t_R = \bar{u}(k')^2/t_R(1 + k')^2 \qquad (2.103)$$

where $\bar{u}$ is the average linear gas velocity. The above relationship accounts for characteristic column parameters, thus offering a way to compare different type columns.

RESOLUTION.   The separation of two components as the peaks appear
on the chromatogram (see Figure 1.18) is characterized by

$$R_S = \frac{2\Delta W}{W_1 + W_2} \qquad (2.104)$$

where $\Delta W = t_{R2} - t_{R1}$. If the peak widths are equal, that is
$W_1 = W_2$, then Equation 2.104 may be rewritten

$$R_S = \frac{\Delta W}{W} \qquad (2.105)$$

The two peaks will touch at the baseline when ΔW is equal to 4σ.

$$t_{R_2} - t_{R_1} = W \qquad (2.106)$$

Therefore, if the two peaks are separated by a distance 4σ, then
$R_S = 1$.   If the peaks are separated by a 6σ, then $R_S = 1.5$.
  Resolution also may be expressed in terms of Kovát's retention
index (see Chapter 3.).

$$R_S = \frac{I_2 - I_1}{W_{0.5}f} \qquad (2.107)$$

where f = correction factor (1.699) because $4\sigma = W = 1.699W_{0.5}$.
A more general expression for resolution is

$$R_s = \sqrt{n/4} \; \left(\frac{\alpha-1}{\alpha}\right) \; \left(\frac{k'}{1+k'}\right) \qquad (2.108)$$

where, n and k' refer to the later eluted compound of the pair.
Since α and k' are constant for a given column (under isothermal
conditions), resolution will be dependent on the number of theoretical plates, n. The k' term generally increases with a temperature decrease as does α, but to a lesser extent. The result
is that at low temperatures one finds that fewer theoretical
plates or a shorter column are required for the same separation.

REQUIRED PLATE NUMBER. Knowing the capacity factor, k', and the
separation factor, α, one is able to calculate the required number of plates (n_{ne}) for the separation of two components (the k'
value refers to the more readily sorbed component).

$$n_{ne} = 16R_s^2 \; \left(\frac{\alpha}{\alpha-1}\right)^2 \; \left(\frac{1+k'}{k'}\right)^2 \qquad (2.109)$$

The R_s value is set at the 6 α level or 1.5. In terms of the required effective number of plates Equation 2.109 would be

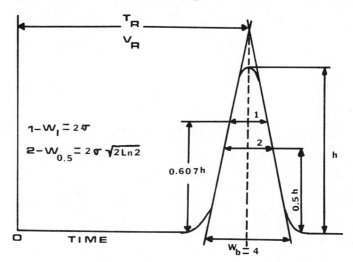

Figure 2.14. Characteristic data of the peak for
calculation of column efficiency.

$$N = 16R_s^2 \; (\frac{\alpha}{\alpha-1})^2 \tag{2.110}$$

Taking into account the phase ratio, β we can write Equation 2.109 as

$$n_{ne} = 16R_s^2 \; (\frac{\alpha}{\alpha-1})^2 \; (\frac{\beta}{k_2'} +1)^2 \tag{2.111}$$

Equations 2.109 and 2.111 illustrate that the required number of plates will depend on the partition characteristics of the column and the relative volatility of the two components, that is on K and β. Table 2.9 gives the values of the last term of Equation 2.109 for various values of k'. These data point up a few interesting conclusions: If k' < 5 the plate numbers are controlled mainly by column parameters; if k' > 5 the plate numbers are controlled by relative volatility of components. The data also illustrate that k' values greater than 20 cause theoretical number of plates, n, and effective number of plates, N, to be of the same order of magnitude, that is,

$$n \simeq N \tag{2.112}$$

TABLE 2.9   VALUES FOR LAST TERM OF EQUATION 2.109

k'	0.25	0.5	1.0	5.0	10	20	50	100
$(1 + k'/k')^2$	25	9	4	1.44	1.21	1.11	1.04	1.02

The relationship in Equation 2.109 also can be used to determine the length of column necessary for a separation L_{ne}. We know that

$$n = L/h$$

thus,

$$L_{ne} = 16R_s^2 h \; (\alpha/\alpha-1)^2 (1+k_1'/k_1')^2 \tag{2.113}$$

Unfortunately, Equation 2.113 is of little practical importance because the h value for the more readily sorbed component must be known and this is not readily available from independent data.

Let us give some examples from the use of Equation 2.109.
Table 2.10 gives the number of theoretical plates for various
values of α and k', assuming R_S is at 6 α(1.5). Using data in
Table 2.10 and Equation 2.109 we can make an approximate comparison between packed and open tubular columns. As a first approximation, β values of packed columns are from 5 - 30 and for open
tubular columns from 100 - 1000; thus, 10 - 100-fold difference
in k'. Examination of the data in Table 2.10 shows that when α
= 1.05 and k' = 5.0 we would need 22,861 plates in a packed
column, which would correspond to an open tubular column with k'
= 0.5 having 142,884 plates. Although a greater number of plates
is predicted for the open tubular column this is relatively easy
to attain because longer columns of this type have high permeability and smaller pressure drop than the packed columns.

TABLE 2.10   NUMBER OF THEORETICAL PLATES FOR VALUES OF α AND k'
(R_S at 6 σ = 1.5)

α	1.05	1.10	1.50	2.00	3.00
k'					
0.1	1,920,996	527,076	39,204	17,424	9,801
0.2	571,536	156,816	11,664	5,184	2,916
0.5	142,884	39,204	2,916	1,296	729
1.0	63,504	17,424	1,296	576	324
2.0	35,519	9,801	729	324	182
5.0	22,861	6,273	467	207	117
8.0	20,004	5,489	408	181	102
10.0	19,210	5,271	392	173	97

SEPARATION FACTOR. The reader will recall that the separation
factor, α, in Section 2.1.4, is the same as the relative volatility term used in distillation theory. In 1959, Purnell
(32,33) introduced another separation factor (S) term to describe
the efficiency of a column. It can be used very conveniently to
describe efficiency of open tubular columns:

$$S = 16(V_R'/W)^2 = 16(t_R'/W)^2 \qquad (2.114)$$

where V_R', t_R' = adj.retention volume and adj. retention time respectively. Equation 2.114 may be written as a thermodynamic
quantity which is characteristic of the separation but independent of the column. In this form we assume resolution (R_S) at the

6 σ level or having a value of 1.5. Therefore, from Equation
2.110

$$S = 36 \left(\frac{\alpha}{\alpha-1}\right)^2 \qquad (2.115)$$

SEPARATION NUMBER. As an extension of the term separation factor
(S) discussed above, we also can calculate a separation number
(n_{sep}) as another way of describing column efficiency (34). By
separation number we mean the number of possible peaks which
appear between two n-paraffin peaks with consecutive carbon
numbers. It may be calculated by

$$n_{sep} = \left[\frac{(t_1-t_2)}{(W_{0.5})_1 + (W_{0.5})_2} \right] - 1 \qquad (2.116)$$

This equation may be used to characterize capillary columns or
when employing programmed pressure or temperature conditions for
packed columns.

ANALYSIS TIME. If possible we like to perform the chromatographic separation in a minimum time. Time is important in analysis but it is particularly important in process chromatography.
Analysis time is based upon the solute component which is more
readily sorbed. Using the equation for determination of retention time,

$$t = \frac{L(1+k')}{\bar{u}} = \frac{nh}{\bar{u}} (1+k') \qquad (2.117)$$

and substituting the value for the required number of plates, n_{ne}
for n (Equation 2.109), we arrive at an equation for the minimum
analysis time, t_{ne}:

$$t_{ne} = 16R_s^2 \frac{h}{\bar{u}} \left(\frac{\alpha}{\alpha-1}\right)^2 \frac{(1 + k')^3}{(k')^2} \qquad (2.118)$$

The term $h/\bar{u}$ can be expressed in terms of the modified van

Deemter equation (Section 2.2.2, Equation 2.81).

$$h/\bar{u} = A/\bar{u} + B/\bar{u}^2 + C_\ell + C_g \tag{2.119}$$

For minimum analysis time high linear gas velocities are used so the first two terms on right side of Equation 2.119 may be neglected. Therefore,

$$h/\bar{u} = C_\ell + C_g \tag{2.120}$$

Substituting of Equations 2.109 and 2.120 into Equation 2.118 we obtain

$$t_{ne} = n_{ne}(C_\ell + C_g)(1+k') \tag{2.121}$$

This equation indicates that minimal separation time depends on plate numbers, capacity factor, and resistance to mass transfer. It should be pointed out that the analysis times calculated from Equation 2.121 also depend on the desired resolution. Our example calculations were made on the basis of resolution, $R_s = 1.5$. For a resolution of 1.00, even shorter analysis times can be achieved.

Figure 2.15 shows a representation of an idealized separation of component zones and the corresponding chromatographic peaks for a three-component system. With columns of increasing number of plates, we see better resolution as column efficiency increases.

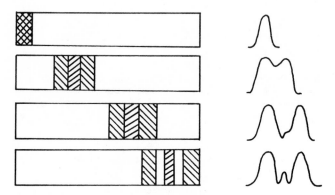

Figure 2.15. Idealized separation process in a gas chromatograph, 3 components; 2 major and 1 minor.

2.2.7   Preparative Gas Chromatography

Gas chromatography can be utilized for preparative-scale separations as well as for analysis. For the qualitative and quantitative analyses of a small sample (see Chapter 4), microliter or microgram sample sizes are used. Once having worked out the conditions for such an analytical separation, one may wish to isolate larger amounts of one or more components of a mixture. Use of GC for this purpose is referred to as preparative gas chromatography.

Preparation of milligram quantities of substances can readily be performed with an analytical gas chromatographic column by repetitive injection and collection. Larger sample quantities require modifications to an analytical apparatus, but are more easily obtained with the use of a special preparative unit. It has been postulated that the sample size approximately increases with the fourth power of the column diameter (up to 1 g).

Samples larger than 1 g necessitate different gas chromatographic equipment than for analysis.

Preparative GC involves the consideration of several possibilities:

1. Automatic Sample Injection and Fraction Collection. The manual injection of repetitive analytical size samples is one of the simplest and most obvious approaches to preparative GC. This may be improved by automatically injecting the samples and collecting fractions with a mechanical system.

2. Rotating Columns and Moving Bed Equipment. Preparative GC can be carried out where the sample is continuously introduced into a set of columns which may move in a transverse direction to the mobile phase flow and sample injection point. Using such an arrangement, the elution point on the cylindrical base, x_i, is

$$x_i = 2\pi r a t_{R_i}$$ (2.122)

where i = component and t_{R_i} = retention time of component i,
        a = number of revolutions of cylinder/unit time,
        r = cylinder radius.
Maximum efficiency is realized when the nth component is eluted adjacent to the first component. The frequency of revolutions must be

$$a = 1/(t_{R_n} - t_{R_i}) \qquad\qquad (2.123)$$

where t_{R_n} = retention time of nth component.

One expects this system to have the efficiency of a single column arrangement. However, optimum efficiencies are not realized because of the column-to-column variations and the tendency to use excess feed volumes.

3. Large-diameter Columns (5-20 cm). This has been one of the primary approaches to preparative GC, the assumption being that sample size can be increased in proportion to cross-sectional area of the column. Actually this is not strictly the case. The basic problem with large-diameter columns is that lateral mixing becomes inefficient with increased column diameter (35). There are several ways to overcome this deficiency: (a) Placing rings in the column at various intervals (e.g., 10 cm for a 100-cm column). These act as baffles, thus causing the mobile phase at the wall to be forced back to the column center (36), (b) Constructing the preparative columns from two concentric tubes with the column packing between the walls. This eliminates the center of the column packing and is reputed to provide a more uniform flow pattern (37).

4. Small-diameter Columns in Parallel (15-20 mm). This is an alternative to increased column diameter. Such a system maintains the inherent efficiencies of small diameter columns, but with increased total cross-sectional area. The disadvantage of this system is the development of an efficient manifold system for disbursing the sample uniformly into matched columns to obtain identical retention characteristics (38).

5. Long Small-diameter Columns (8-12 mm). This setup does permit useful separations; however, separation efficiency deteriorates quickly since the column is overloaded with sample. The justification for such a system is that preparative-size sample capacity is proportional to the amount of solid support, mesh size and column dimensions (39).

6. Sample Recirculation. This technique involves the recirculation of the solute mixture between two columns until a satisfactory separation is obtained. One is able to provide a

large number of theoretical plates from two short columns; however, this system is restricted to two neighboring components.
If the mixture contains other components with longer retention
times, they complicate the separation unless removed from the
system (40).

In preparative GC one is not primarily concerned with resolution or theoretical plates. The basic reasons for the technique is its high separation efficiency and high speed.
Partitioning is much faster in gas phase systems than liquid-
liquid or liquid-solid systems. This results from the high surface area relative to the liquid volume which provides shorter
equilibrations than one could achieve in, say, preparative distillation. Distillations also are limited in terms of high-
quality separations of two-component systems and the large sample
volume required for efficient columns.
In preparative GC long columns can be more of a disadvantage
than an aid. Longer length gives increased retention times and
more time for unwanted diffusion effects to become pronounced.
Twenty feet is a practical maximum for column length in preparative GC. If good resolution cannot be obtained on an analytical
column with proper adjustment of flow, temperature, and sample
size, increasing column length generally will not improve the
situation.
Verzle (41) studied variables in preparative GC including
column length, particle size, liquid substrate loading, and
sample size. Studies showed that column length can be increased
if mesh size of particles are decreased (larger particles). This
permits reasonable flows of carrier gas with modest inlet
pressures.

ESSENTIAL COMPONENTS OF A PREPARATIVE GC SYSTEM. The most important components for the system are: (a) the sample inlet system,
(b) the column, and (c) the collection system.
It is just as important in preparative GC as in analytical GC
to introduce the sample as a "plug" to maintain the resolution.
As noted previously, the longer the component spends in the
column the more it diffuses with resulting peak broadening. Unfortunately it is not easy to inject a "plug" of sample in preparative GC. Injection of a large liquid sample results in a
large amount of vapor and relatively large pressures in the injection port. As a result there exists a large back-pressure
against the syringe, causing a longer time for sample injection
and possibly a safety hazard. One of the better ways of circum-
venting this problem is to inject sample as a liquid (low temperature), and then begin to temperature program (see Section 1.5

and Chapter 6). Another trick is to inject liquid at a rate so
as to create vapor at the same rate as the carrier gas flowrate.
Purnell (42) has calculated that if the sample bandwidth is less
than 25% of the average peak bandwidth, then the effects of injection have a minimal influence on the final peak (band) width.

We now will focus our attention on the various component parts
of a preparative GC column: the column tube, the solid support,
the liquid substrate, the carrier gas, and column length and
diameter.

In preparative GC it is particularly important that solvent
efficiency (α) should be maximized as in analytical GC. A liquid
substrate should be used that will give the maximum separation of
peaks (high selectivity). The higher the α term the fewer theoretical plates (n) required and the larger samples that can be
injected.

The number of required theoretical plates may be calculated by
Equation 2.109. As the value of k' increases beyond k' = 5, the
second term of Equation 2.109 becomes less effective and can be
ignored. Thus, at k'>5 the number of theoretical plates is especially dependent on the α term. Generally, the maximum sample
size increases in proportion to column diameter (d^4), and if
column diameter remains constant the volume of sample that can be
injected is in proportion to column length. There is also a relationship between maximum sample size and liquid loading. However, this effect is only significant when using small sample
sizes (39). Horvath (43) states a range of 10 to 300 mm as the
optimum inside diameter and a liquid loading greater than 20% w/w
for a preparative column.

Bayer et al. (36) studied the efficiency of preparative columns and showed that, when using the van Deemter equation (Section 2.2.2), an additional term had to be introduced to account
for the nonuniform flow distribution. This can be represented by

$$H_p = \frac{E \, d^{0.58}}{u^{1.806}} \qquad (2.124)$$

where H_p = preparative term to be added to van Deemter equation,
      E = correlation factor,
      d = column diameter,
      u = linear gas velocity.

Increased column diameter may cause a reduction in the homogeneity of the packing, resulting in flowrate distortion of the
component zones across the column cross-section and higher h
(HETP) values. Diminution of column efficiency also can be related to (1) the nonuniformity of liquid coating which in turn
affects mass transfer, (2) the radial temperature gradient that

exists within the large diameter column, and (3) variation in the
carrier gas between the wall and center of column (this is
especially true in large coiled columns). The above mentioned
variables generally increase in effect as the column diameter
increases.

One of the greatest problems with packing preparative columns
results from the radial separation of particles; that is, large
particles tend towards the column walls whereas fines remain more
to the center. Since the use of larger diameter columns tend to
introduce more problems, the chromatographer is advised to purchase prepacked columns (see Chapter 3).

DETECTORS. Most of the work in preparative GC has utilized
either thermal conductivity detectors or flame ionization detectors. With thermal conductivity detectors, one normally
passes the entire sample through the detector. However, by-pass
designs are necessary to work with the more sensitive thermal
conductivity detectors now on the market. When using flame ionization detectors one must use a by-pass arrangement, allowing
about 0.1-1.0% of the column effluent to be burned for signal
purposes. Detectors that have slow response time and a wide
linear range are recommended in preparative GC. The reader is
referred to Chapter 5 for details concerning detectors.

2.2.8 Gas-Solid Chromatography

All previous discussion regarding the theory of GC has concerned
itself with gas-liquid chromatography. By and large the majority
of GC investigations have dealt with gas-liquid chromatography
rather than gas-solid chromatography. In recent years, however,
more attention has been given to gas-solid chromatography, a
unique and versatile technique.

Some of the reasons for this interest are:

1. Higher temperatures are possible ($500^\circ C$).
2. Availability of solid surfaces which are stable and do not
   undergo chemical reactions (e.g., oxidation).
3. High column efficiencies are possible due to no liquid phase
   contribution to band spreading.
4. Great specificity of surfaces for solute molecule configurations.
5. Elimination of liquid substrate bleed effects.
6. The adaptability of this technique to study the physico-chem-
   ical measurements possible at solid surfaces (e.g., isotherms

and heats and entropies of adsorption).

7. Evaluation of catalysts, kinetics of catalytic processes, and catalyst reactions in general.

There are some modifications of the definitions and terms used in gas-liquid chromatography compared to those used in gas-solid chromatography. First, the phase ratio is described as

$$\beta = V_G/V_A \qquad (2.125)$$

where V_A = true adsorbent volume (wt.$_{adsorbent}$ ÷ density = V_A).

In the rate theory of gas-solid chromatography, the equation for h has essentially the same terms except that C_k replaces C_ℓ. C_k is a term characteristic of adsorption kinetics. Equation 2.87 then becomes

$$h = \sum_{i=1}^{5} \frac{1}{1/A + 1/C_1 u} + \frac{B}{u} + C_k u + C_g + H_e \qquad (2.126)$$

Theoretical considerations indicate that, on homogeneous surfaces, C_k is smaller than C_ℓ, and this implies highly efficient applications of gas-solid chromatography.

One of the big obstacles still existing in gas-solid chromatography is the lack of adequate adsorbent structure descriptions, as well as the distribution and dimensions of the pores. Another is the lack of reproducibility of adsorbents, not only among manufacturers (different products, presumably the same) but within the same manufacturer (different lots).

ADSORBENT PROPERTIES. In the use of a solid adsorbent for gas-solid chromatography, its properties must be considered from several points of view. First, the specific surface area (m^2/g) is is important. The greater the surface area the higher probability of some sorption process occurring. Also, the more active sites per unit area, the more reactive will be the sorbate molecules with the sorbent surface. Secondly, the chemical composition of surface layer and its crystal structure is of interest. Information of this type allows one to speculate more correctly regarding wanted or unwanted reactions which may take place at the gas-solid interface. Lastly, a knowledge of the porous structure is helpful in identifying molecules that may selectively be trapped or sorbed on the surface. In spite of the fact that much work has been done to determine the surface area of solids, main information regarding adsorption phenomena comes from the analysis of adsorption isotherms.

ADSORPTION OF GASES AT THE SOLID SURFACE. No matter what the
process at the gas-solid surface, physical adsorption of the gas
or vapor on the solid surface is part of the mechanism. The distance between sorbed molecules is shorter than the distances
found between molecules of a real gas but these distances are
larger than those encountered in chemical interactions. The interaction energy between molecules and the surface of a sorbent
may be estimated initially by assuming that the molecules are
spherical in shape and located in a field of infinite sorbent and
secondly, considering the temperature to be great enough to reduce the effect of molecular interactions to a negligible value.
Thus, by introducing N gas molecules at a temperature T and a
pressure P, into a container holding a sorbent with a uniform
surface, one can calculate the "apparent volume," V_a:

$$V_a = \frac{NkT}{P} \tag{2.127}$$

If the "dead" volume, V_g, represents the sorbent uniform surface
we can designate V_o as the limiting value of V_a as P goes to
zero, thus

$$\frac{1}{V_a} = \frac{1}{V_o} = \frac{PV_a C}{kTV_o^3} \tag{2.128}$$

and

$$V_o = V_g + \int V_g \exp(-E/kT) \, dV \tag{2.129}$$

where E = potential energy of gas molecules in sorbent field.
$1/V_o$ is the intercept of the linear plot of $1/V_a$ versus PV_a. E
may then be calculated from the integration of Equation 2.129.
     The carrier gas may have a significant effect on the separa-
tion process. Adsorbents with high specific surfaces adsorb the
carrier gas thus decreasing some of the active sites (adsorption
centers). This results in a change of component adsorption,
which is in proportion to the adsorption capacity of the carrier
gas. A change in carrier gas from hydrogen or helium to nitrogen
may produce sharper peaks because of the higher adsorption capacity for nitrogen (44).
     Sorbates (solute molecules) may be grouped according to their
intermolecular interactions. These groupings are based on electronic configurations, electron density, and functional groups in
the molecule.

Group A molecules. Molecules having spherically symmetrical electron shells, for example, noble gases and the saturated hydrocarbons having only sigma bonds between the carbon atoms. Molecules of this type will interact nonspecifically, through dispersion forces resulting from concordant electronic motion in the interacting molecules.

Group B molecules. Molecules that have a concentrated electron density (negative charge)(e.g., unsaturated and aromatic hydrocarbons) and molecules having pi electron bonds (e.g., N_2, H_2O, ROH, ROR, RCOR, NH_3, NH_2R, NHR_2, NR_3, RSH, RCN).

Group C molecules. These include molecules with locally concentrated positive charges within small radius linkages but these should not be adjacent groups with concentrated electron densities (e.g., -OH or =NH groups). Organometallic compounds exemplify this group. This type of compound interacts specifically with Group B molecules but nonspecifically with Group A molecules.

Group D molecules. These are molecules with adjacent bonds one with positive charge and one with electron density on the periphery of the other. Molecules with -OH and =NH functional groups, for example, H_2O, ROH, and 1^0 and 2^0 amines constitute this group. Group D molecules may interact specifically with Groups B and C molecules and with each other; however, they interact nonspecifically with Group A molecules. Adsorbents are classified either specific or nonspecific. The specificity results from the type molecules or functional groups attached to the adsorbent surface. Three classifications result:

Nonspecific adsorbents. No functional groups or exchange ions on the surface. Adsorbents of this type are carbon black, boron nitride and polymeric saturated hydrocarbons (e.g., polyethylene).

Specific adsorbents with positive surface charges. Acidic hydroxyl groups (hydroxylated acid oxides such as silica), aprotic acid centers, or small radius cations (zeolites) on the surface. Adsorbents of this type will interact with molecules which have locally concentrated electron densities, that is, Group B and Group D molecules.

Specific adsorbents with electron densities on surface.
Graphitized carbon blacks with dense monolayers of Group B molecules or macromolecules deposited on surface. Adsorbents with a
functional group, for example, cyano, nitrile, or carbonyl, would
also be included in this category.

The adsorbents may also be classified according to their
structure. These classifications are summarized in Table 2.11

TABLE 2.11 ADSORBENTS, CLASSIFIED BY TYPE

Type	Description and Classification
I-Nonporous	Nonporous mono- and polycrystalline sorbents, (e.g., graphitized carbon black, NaCl). Porous amorphous sorbents (e.g., Aerosil and thermal blacks).
II-Uniform wide pores	Large-pore glasses, wide-pore Xerogels and com- pressed powders made from nonporous particles (>100A in size and specific surface areas <300 m^2/g).
III-Uniform fine pores	Amorphous fine-pore Xerogels, fine-pore glasses, many activated charcoals, and porous crystals (type A and X Zeolites).
IV-Nonuniform pores	Chalklike silica gels obtained from hydrolysis of salts from strong acids in a silicate solu- tion.

In gas-solid chromatography, retention of sorbate components
is determined by:

1. The chemical nature and geometric pore structure of the
   sorbent.
2. Molecular weight of sorbate molecules and their geometric and
   electronic structures.
3. Temperature of the column.

Column separating power depends upon selectivity of the sorbent and diffuseness or spreading of chromatographic bands moving
through the sorbent. Thus, all things being equal, a chromatography column will be most effective when the bands are less diffuse. Band spreading is caused by thermodynamic, kinetic, injection, and diffusion effects. These may be summarized as:

1. Nonsymmetrical band spreading may be attributed to nonlinearity of the equilibrium adsorption isotherm (i.e., deviation of isotherm from Henry's Law). This causes the sorbate to move through the column at different rates (rate dependent upon sorbate concentration).

2. Peak diffuseness can be attributed to the various processes occuring during the transport of the sorbate through the column. The diffusion processes are very complex, this complexity being due to: (a) ordinary diffusion in the gas phase, (b) band movement through particle layers of different size and shape and packed in various ways, causing diffuseness related to a nonuniform distribution of gas flowrates over each cross-sectional area, (c) difference in local flow velocities from the average velocity through the column. Columns can exhibit what is referred to as "wall effect," which means that flow at the walls is higher than the average of the column because resistance at the wall is less. The great effectiveness of capillary columns is due mainly to the absence of specific diffusion processes caused by particles. However, one does observe diffuseness in capillary or unpacked columns resulting from the parabolic velocity distribution over the column cross-section. The gas velocity is higher at the center and lower near the walls than the average velocity of the band.

3. Peak diffuseness can be a result of the kinetics of the sorption-desorption process (i.e., slow mass transfer or exchange at sorbent surfaces). Peak diffusion in this case is usually nonsymmetrical because the rates of sorption and desorption are not the same. The band spreading due to the final rate of mass exchange is closely related to the diffusion phenomena. Physical adsorption, for all practical purposes, is instantaneous. The overall process of sorption however, consists of several parts: (a) movement of sorbate molecule toward sorbent surface, resulting from intergrain diffusion (outer diffusion), (b) movement of sorbate molecules to inside of pores (i.e., internal diffusion of the sorbate molecules in the pores and surface diffusion in the pores), (c) the sorption process in general.

4. Apparent diffusion may be influenced by the time it takes for sample injection.

Listed below are some of the more common adsorbents that have been used for gas-solid chromatography:

1. Carbon. These are nonspecific type adsorbents because of the lack of functional groups, ions, or unsaturated bonds. Most interactions are due to dispersion forces.

2. Metal Oxides. (a) <u>Silica gel</u>. This is a specific type adsorbent because of the free hydroxyl groups on surface. Polar molecules are easily separated. Wide-pore silicas with homogeneous
surfaces are used for analytical gas-solid chromatography, (b)
<u>Alumina</u>. This is used more in liquid-solid chromatography than
in gas-solid chromatography. In gas-solid chromatography most
peaks are symmetrical in shape. It is sometimes coated for use
in gas-solid chromatography.

3. Zeolites. Porous structure and good adsorption properties
make them usable in gas-solid chromatography. They are a specific type adsorbent, with cavities allowing sieving action for
molecules able to enter "holes" (windows).

4. Inorganic Adsorbents. These have two general classifications:
(a) inorganic salts (e.g., alkali metal nitrates and halides (45),
alkaline earth halides (46), vanadium, manganese, and cobalt
chlorides (47), and barium sulfate (48). (b) inorganic salts
coated on surfaces of silica, alumina, carbon, and so on.

5. Organic Adsorbents. (a) Organic crystal compounds (e.g.,
benzophenone on firebrick, anthraquinone on graphitized carbon
black, phthalic anhydride and/or phthalic acid isomers on Chromosorb G). (b) Liquid phases below their melting point (e.g.,
SE-30 on Chromosorb and Carbowax 20M on Chromosorb). (c) Organic
clay derivatives, (e.g., Bentone 34). (d) Porous polymers (.e.g.,
Porapaks, Chromosorb 101 and 102, Polysorb, and Synachrom).

THERMODYNAMICS OF THE PROCESSES AT THE GAS-SOLID INTERFACE. The
processes taking place at the gas-solid interface may be inter-
preted by the use of adsorption thermodynamics. A term much used
in chromatography is the retention volume, that is, the volume of
carrier gas needed to transport the sample molecule through the
column and into the detector system. In gas-solid chromatography
we refer to the retention volume in the same way but the sample
molecule is more specifically referred to as the sorbate. Nomenclature between gas-solid and gas-liquid chromatography does not
necessarily agree. In gas-liquid chromatography, retention volume is represented by V_R (meaning the total retention volume).
V_R has been used in gas-solid chromatography to represent
corrected retention volume (49), whereas corrected retention volume in gas-liquid chromatography is given the symbol V_R^o. Specific retention volume in gas-solid chromatography is V_s^T, but in

gas-liquid chromatography it is V_g.

The determination of the adsorption parameter (V_g) at three temperatures permits the calculation of the free energy, enthalpy, and entropy of adsorption. A plot of log V_g versus 1/T has a slope of $-\Delta H_{ads}/2.3R$. ΔG_{ads} is obtained by

$$-\Delta G_{ads} = RT \ln K = RT \ln V_g \qquad (2.130)$$

ΔS_{ads} is then obtained with proper substitution in Equation 2.131.

$$\log V_g = \frac{-\Delta H_{ads}}{2.3RT} + \frac{\Delta S_{ads}}{2.3R} \qquad (2.131)$$

These calculations may be performed manually or by use of an appropriate computer program. The reader may obtain more details in reference (50) and/or Chapter 11.

REFERENCES

1. S. Glasstone, Textbook of Physical Chemistry, Van Nostrand, New York, 1951, p. 1197.
2. E. R. Gilliland, Ind. Eng. Chem., 26, 681 (1934).
3. J. K. Clarke and A. R. Ubbelohde, J. Chem. Soc., 2050 (1957).
4. G. H. Morrison and H. Freiser, Solvent Extraction in Analytical Chemistry, John Wiley and Sons, New York, 1957.
5. J. M. Miller, Separation Methods in Chemical Analysis, John Wiley and Sons, New York, 1975.
6. B. L. Karger, L. R. Snyder, and C. Horvath, An Introduction to Separation Science, Wiley-Interscience, New York, 1973, p. 12.
7. E. B. Sandell, Anal. Chem., 40, 834 (1968).
8. L. C. Craig, J. Biol. Chem., 155, 519 (1944).
9. B. Williamson and L. C. Craig, J. Biol. Chem., 168, 687 (1947).
10. L. C. Craig, O. Post, Anal. Chem., 21, 500 (1949).
11. L. C. Craig, Anal. Chem., 22, 1346 (1950).
12. W. A. Peters, Ind. Eng. Chem., 14, 476 (1922).
13. A. J. P. Martin and R. L. M. Synge, Biochem. J., 35, 1359 (1941).
14. E. Glueckauf, Disc. Faraday Soc., No.7, 199-213 (1949).
15. A. J. P. Martin and R. L. M. Synge, Biochem. J., 35, 1358 (1941).
16. J. J. van Deemter, F. J. Ziuderweg, and A. Klinkenberg, Chem. Eng. Sci., 5, 271 (1956).

17. L. Lapidus and N. R. Amundson, J. Phys. Chem., _56_, 984 (1952).
18. A. Klinkenberg and F. Sjenitzer, Chem. Eng. Sci., _5_, 258
    (1956)
19. E. Glueckauf, M. J. E. Goley, and J. H. Purnell, Ann. N.Y.
    Acad. Sci., _72_, 612 (1959).
20. R. J. Loyd, B. O. Ayers, and F. W. Karasek, Anal. Chem., _32_
    698 (1960).
21. W. L. Jones, Anal. Chem., _33_, 829 (1961)
22. R. Kieselbach, Anal. Chem., _33_, 806 (1961)
23. J. H. Beynon, S. Clough, D. A. Crooks, and G. R. Lester,
    Trans. Farad. Soc., _54_,705 (1958).
24. J. C. Giddings, Dynamics of Chromatography, Part I, Vol. I,
    Chromatographic Science Series, Marcel Dekker, New York,
    1965.
25. J. C. Giddings and R. A. Robinson, Anal. Chem., _34_, 885
    (1962).
26. J. C. Giddings, Anal. Chem., _35_, 2215 (1963).
27. J. C. Giddings, in "Gas Chromatography, 1964," A. Goldup
    (Ed.), Butterworths, London, 1964, p.3.
28. D. D. DeFord, R. J. Loyd and B. O. Ayers, Anal. Chem., _35_,
    426 (1963).
29. J. Bohemen, S. H. Langer, R. H. Perrett, and J. H. Purnell,
    J. Chem. Soc., 2444 (1960).
30. L. Rohrschneider, in Advances in Chromatography, Vol. IV,
    J. C. Giddings and R. A. Keller (Eds.), Marcel Dekker, New
    York, 1968.
31. D. H. Desty, A. Goldup and W. T. Swanton, in Lectures on Gas
    Chromatography -- 1962, H. A. Szymanski (Ed.), Plenum Press,
    New York, 1963, p. 105.
32. J. H. Purnell, Nature, _184_, 2009 (1959).
33. J. H. Purnell, J. Chem. Soc., _54_, 1268 (1960).
34. R. Kaiser, Z. Anal. Chem., _189_, 11 (1962).
35. J. H. Purnell, Ann. N. Y. Acad. Sci., _72_, 614 (1959).
36. E. Bayer, K. P. Hupe, and H. Mack, Anal. Chem., _35_, 492
    (1963).
37. J. C. Giddings, Anal. Chem., _34_, 37 (1962).
38. J. H. Purnell, J. Roy. Inst. Chem., _82_, 586 (1958).
39. M. Verzele, J. Bouche, A. DeBruyne, and M. Verstappe, J.
    Chromatog., _18_, 253 (1965).
40. A. J. P. Martin, Gas Chromatography, V. J. Coates, H. Noebels
    and I. Fagerson (Eds.), Academic Press, New York 1958, p.237.
41. M. Verzele, J. Gas Chromatog., _3_, 186 (1965).
42. J. H. Purnell, Gas Chromatography, John Wiley and Sons, New
    York, 1962, pp. 113-116.
43. C. S. Horvath, in The Practaice of Gas Chromatography, L. S.
    Ettre and A. Zlatkis (Eds.), Interscience, New York, 1967.

44. A. C. Locke, and W. W. Brandt, in Gas Chromatography, 1962,
    4th Symposium, Hamburg, M. van Swaay (Ed.), Butterworths,
    London, 1962, p. 66.
45. R. L. Grob, G. W. Weinert, and J. W. Drelich, J. Chromatog.,
    30, 305-324 (1967).
46. R. L. Grob, R. J. Gondek, and T. A. Scales, J. Chromatog.,
    53, 477-486 (1970).
47. R. L. Grob and E. J. McGonigle, J. Chromatog., 59, 13-20
    (1971); 101, 39-50 (1974).
48. L. D. Belyakova, A. V. Kiselev, and G. A. Soloyan, Chroma-
    tographia, 3, 254-259 (1970).
49. D. T. Sawyer and D. J. Brookman, Anal. Chem., 40, 1847 (1968).
50. R. L. Grob, in Progress in Analytical Chemistry, Vol. 8,
    I. V. Simmons and G. W. Ewing (Eds.), Plenum Press, New York,
    1976, pp. 151-194.

CHAPTER 3

Columns and
Column Selection
in Gas Chromatography

WALTER R. SUPINA

Supelco, Inc.

3.1 FACTORS AFFECTING PEAK SEPARATION 114
     3.1.1. Column Separating Power 116
     3.1.2 Efficiency. 117
     3.1.3 Selectivity 119
     3.1.4 Peak Tailing. 119
3.2 SOLID SUPPORTS 120
     3.2.1 Supports for Gas-Liquid Chromatography 120
     3.2.2 Adsorbents for Gas-Solid Chromatography 123
3.3 STATIONARY PHASE 127
     3.3.1 Concentration of Stationary Phase 132
     3.3.2 Temperature Effects 133
3.4 PROBLEMS OF GAS CHROMATOGRAPHY 134
     3.4.1 Tubing 134
     3.4.2 Glass Wool. 135
     3.4.3 Carrier Gas 135
     3.4.4 No Peaks 136
     3.4.5 Selecting a Column for a Specific Application . 137
     3.4.6 Preparation of Column Packings. 138

     3.4.7 Filling the Column 143
     3.4.8 Conditioning the Column 144
3.5   OPEN TUBULAR (CAPILLARY) COLUMNS 145
     3.5.1 Description and Capabilities 145
     3.5.2 Column Performance 147
     3.5.3 Summary 148
References 148

The column is the heart of the chromatograph and provides versatility in types of analyses that can be obtained with a single instrument. This versatility is due to the wide variety of column materials available, but can also be a problem when it comes time to choose appropriate materials. The purpose of this chapter is to provide the reader with a basic understanding in the composition and use of columns.

The column system consists of (a) tubing, (b) a packing material which is either an adsorbent or a solid support or adsorbent coated with a stationary phase, and (c) packing retainers such as glass wool inserted into the tubing ends. A carrier gas is used to move the sample components through the column during the analysis. The degree of separation of various components and speed of analysis is affected by column temperature.

Proper consideration of all of these variables is important if successful results are to be obtained. Shortcuts taken to save time in selection, preparation, or operation of the column may only result in inadequate results or delays. The reader is encouraged to study the few references cited to obtain a better understanding of the workings of a chromatographic column.

3.1   FACTORS AFFECTING PEAK SEPARATION

When two peaks are not completely separated the problem could be due to improper choice of a column or the manner in which the column is operated. Figure 3.1a illustrates a typical case of two unresolved peaks which are quite broad and tail excessively. To obtain an effective solution, it is important to first identify the cause of the problem. If the problem is due to the wrong choice of a column, then variations in operating conditions may do little to improve the separation. Therefore, a little time spent in analyzing the problem may go a long way in helping to analyze the sample. Figure 3.1b shows the improvement which can be obtained by increasing the separating power with the use of a longer column of the same material. Note here that both compon-

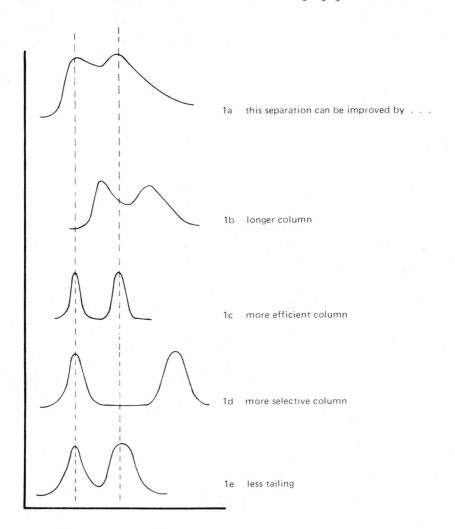

1a   this separation can be improved by . . .

1b   longer column

1c   more efficient column

1d   more selective column

1e   less tailing

Figure 3.1

ents are eluted somewhat later as a result of the increased column length.   A more <u>efficient</u> column as shown in Figure 3.1c also
provides the necessary separation.   This column may differ from
the one used for Figure 3.1a only in the manner in which it was
prepared or operated.   A more <u>selective</u> column will move the two
peaks farther apart as in Figure 3.1d.   The selectivity is determined largely by the type of column material and to a lesser extent by column temperature.   Reduction of <u>peak tailing</u> is

important not only for the obvious purpose of separating the two components, but also to assure good quantitative analysis. Tailing can be minimized by use of proper column materials and in some cases by derivitization of the sample. Discussion of derivitization techniques may be found in Chapters 7 and 12.

3.1.1 Column Separating Power

The separation can be improved by increasing the column length as this increases the number of theoretical plates which can be generated by the column. The theoretical plate concept is similar to that used in distillation, but in chromatography no actual "plates" are present. However, the concept is a useful one in that it provides a convenient means of expressing the separating power of a column. If

$$n = 16 \left[\frac{V_{R_1}}{W_1} \right]^2 = \text{Total theoretical plates} \qquad (3.1)$$

where V_{R_1} is the retention volume of a peak and W_1 is the width of the tangents to the peak at the baseline intercept as in Figure 3.2. (Retention time instead of retention volume may be used for these calculations.) The ability of a column to separate two

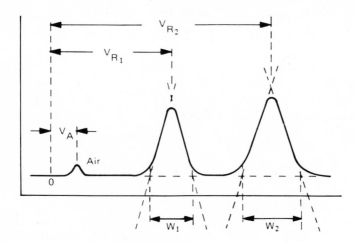

Figure 3.2. Column efficiency parameters used in Equation 1. See text for details.

components is a function of n, and not necessarily the length of
the column. A poorly prepared 6-foot column would give the same
degree of separation as a good 3-foot column if both generated
the same number of theoretical plates (but analysis time would be
doubled).

3.1.2 Efficiency

The relative "goodness" of a column is expressed in terms of efficiency, n/L, as plates per foot. A 6-foot column having 2000
plates would only be half as efficient as a 3-foot column with
the same number of plates. Although the total number of plates,
n, influences the degree to which peaks will be resolved, column
efficiency is a measure of how well the column has been prepared
and operated. A performance of 1000 plates per foot can be obtained but 500 is reasonable; anything less is indicative of a
problem. Column efficiency is also expressed as h, which is the
length of column (expressed in millimeters) equivalent to one
theoretical plate. This efficiency is related to column variables by the van Deemter equation:

$$h = \text{HETP} = \underbrace{2\lambda d_p}_{\substack{\text{Eddy} \\ \text{diffusion}}} + \underbrace{\frac{2\gamma D_{gas}}{\bar{\mu}}}_{\substack{\text{Molecular} \\ \text{diffusion}}} + \underbrace{\frac{8}{\pi^2} \frac{k'}{(k'+1)^2} \cdot \frac{d_f^2 \bar{\mu}}{D_{liq.}}}_{\substack{\text{Mass transfer} \\ \text{in liquid}}} \qquad (3.2)$$

where d_p = diameter of support particle,
λ = "packing term,"
γ = tortuosity factor,
D_{gas} = diffusivity of sample in gas,
$D_{liq.}$ = diffusivity of sample in liquid,
$\bar{\mu}$ = average linear velocity of carrier gas,
d_f = effective film thickness of liquid,
k' = capacity factor = ratio of amount of sample in sta-
             tionary phase to amount in carrier gas in a given
             segment of column = K/β.

Understanding the van Deemter equation is important if one is
to obtain the best possible performance of chromatographic columns, as indicated by a low value of h. A more thorough discussion of the equation has been given by Schupp (1). Each of
three major terms are affected by either the choice of column
materials or the manner in which the column is prepared or operated.

The first term is independent of carrier gas flowrate and shows that h decreases as particle size decreases. This means that 100/120 mesh would give narrower peaks than 80/100. However, these smaller particles are more difficult to pack uniformly, and this is reflected in λ which shows that the better a column is packed, the lower the value of h. If the reader prepares his own columns and has been using 60/80 mesh, it would be unwise to change immediately to 100/120 as the columns would probably be too tightly packed. Instead, he should develop experience with 80/100 mesh and later change to 100/120. But if he purchases columns, it is advisable to use 100/120 mesh supports when highest possible column efficiency is desired. The larger diameter particles should be used only to avoid excessive pressure drop in very long columns. The second term of the van Deemter equation reflects the effect on efficiency of diffusion of the sample in carrier gas. The geometry of the column packing is represented by γ, the tortuosity factor; the more uniform the particle size and shape, the less tortuous path the molecules must follow. If support particles are fractured during preparation of the packing or filling of the column, the resulting smaller pieces will reduce the efficiency of the column.

The last term of the van Deemter equation shows that h increases as the square of the film thickness; in other words, the lower the percentage of stationary phase the more efficient the column.

Since carrier gas velocity enters into both the second and third terms, its effect is more complex, but there is a certain velocity at which column efficiency is best, as shown in Figure 3.3. Below the optimum velocity the change is quite drastic, but above this point a more gradual decrease in efficiency takes place. Although somewhat higher efficiencies can be obtained

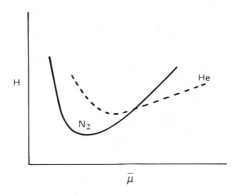

Figure 3.3

with nitrogen, helium provides a more usable range of flowrates without serious sacrifice in efficiency. It is not necessary to operate columns at the optimum flowrate unless the separation is an extremely difficult one, but one should be aware of the effect of flowrate on column efficiency.

3.1.3 Selectivity

The selectivity of a column determines the degree by which the maxima of the two peaks are separated as in Figure 3.1c. The following factors affect the selectivity of a column:

1. Solid support (or adsorbent).
2. Solid support (or adsorbent) treatment.
3. Stationary phase.
4. Stationary phase concentration.
5. Column temperature.

   If two components are almost separated, adjustment of column temperature, phase concentration, or solid support might be suf-ficient to effect a complete separation. If the problem is more difficult, selection of a new stationary phase might be in order. This will be discussed in more detail later.

3.1.4 Peak Tailing

This phenomena can be an indication of adsorption of the sample by the solid support. By using the same stationary phase but a different support one might eliminate the tailing as in Figure 3.1c. Tailing can be responsible for errors in identification as well as errors in quantitation. Scholz (2) has shown that the retention time of a compound changes with sample size if the com-pound is adsorbed by the column, as shown in Figure 3.4. Note that the tails of each peak merge into a single tail, but the maxima of the peaks shift. The smaller the sample, the later the peak. A good illustration would be analysis of methanol on a nonpolar column such as SE-30, OV-1, or SP-2100. Kirkland (3) has shown that the use of Teflon as a support reduces the tailing of acids. However, not many stationary phases can be effectively coated on Teflon, and this type of support also has temperature limitations. Therefore, chemical treatment such as silanization is used to render the diatomite supports more inert. The tubing is often the cause of adsorption and in many cases glass is

required.   Whenever glass is used, it should always be silanized;
this also applies to glass wool.

3.2   SOLID SUPPORTS

   3.2.1 Supports for Gas-Liquid Chromatography

Two excellent reviews on solid supports have been published by
Ottenstein (4,5).   Those users of gas chromatography (GC) should
read these articles to gain a better understanding of the role
played by this important component of the chromatographic column.
In addition, Supina (6) has covered the topic in more detail than
can be done here.   Many of the problems associated with poor
separations can be attributed to inadequacies of the support for
a particular application.   The purpose of the support is to provide an inert surface onto which the stationary phase can be deposited.   Many materials have been investigated, but the only
ones that are used to any extent are the diatomites and Teflon.
The diatomite supports can be further divided into three main
types.   (a)   The firebrick-derived materials such as Chromosorb P
and Gas Chrom R are used primarily for analysis of nonpolar compounds such as hydrocarbons; alcohols, amines, acids, and other
polar substances are adsorbed and resulting chromatographic peaks
tail severely.   (b)   The white diatomite supports derived from
filteraids are the most widely used and include Chromosorb W,
Supelcoport, Gas Chrom Q, and Anakrom ABS.   (c)   Chromosorb G is
the most significant of the many attempts to discover or develop

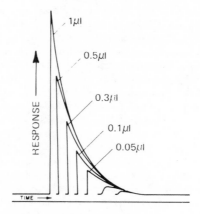

Figure 3.4.

a better support. It is almost as inert as Chromosorb W, but has
very good particle strength and is less likely to be damaged during preparations of packings or filling of columns. However, for
analysis of very sensitive compounds such as pesticides, the
greater inertness of supports such as Chromosorb W is preferred.
A good rule of thumb is that Chromosorb G is worth considering if
stainless steel columns are satisfactory for the analysis.

Deactivation of all diatomite supports is necessary for most
applications since the surface contains mineral impurities which
can cause decomposition of the sample or stationary phase. These
impurities can be removed by thoroughly washing the support with
hydrochloric acid and then washing to neutrality. Most commercial supports are available in an acid-washed grade.

Surface silanol groups present on the diatomite supports also
are reactive and can cause peak tailing by hydrogen bonding with
polar samples. The degree of peak tailing depends on the type of
sample and increases in this order: hydrocarbons, ethers, esters,
amines, alcohols. As concentration of such polar compounds is
decreased, the tailing is more pronounced and the analysis becomes more difficult. This tailing can be minimized by treatment
of the support with dimethyldichlorosilane and then neutralizing
with methanol. This procedure is a very critical one and it is
inadvisable for the user to do it as results will probably be
less satisfactory than with commercially available materials. It
is equally important to purchase the best grade of any brand of
support. Some manufacturers offer two or three grades of silane-
treated supports, but only the best should be used. For example,
it is false economy to use Chromosorb W HMDS or Chromosorb W-AW
DMCS instead of Chromosorb WHP. The more commonly used supports
are listed in Table 3.1 according to type and chemical treatment.

When the chemical treatment of the support proves to be inadequate, it may be necessary to use priming and/or tail reducers.
Priming is the repeated injection of compounds similar to the
sample being analyzed (or even the sample itself) until the peak
height for a given sample size remains constant, the purpose being to saturate the active sites on the column and to then perform the analysis. This technique must be used with caution as
its effectiveness can gradually wear off, and it is difficult to
tell when the column begins to adsorb the sample. Calibrations
with mixtures of known composition should be made before and
after each analysis. If this is done only at the beginning and
end of each day, and the results do not agree, the dependability
of all analyses performed during the day must be questioned.

Tail reducers are polar substances added in small quantities
along with the stationary phase, and are used most frequently
with nonpolar phases. Phosphoric acid is used with polyesters(7)
for analysis of fatty acids and has been found effective for

TABLE 3.1 DIATOMITE SUPPORTS -- BRAND EQUIVALENTS[1]

Prepared From	Mfgr.[2]	Nonacid Washed	Acid Washed	Acid Washed DMCS Treated	Other Treatment	Designated
Celite	A	Chromosorb W NAW	Chromosorb W AW	Chromosorb W AW-DMCS	HMDS	Chromosorb W-HMDS
Filteraid	A			Chromosorb W HP*		
	B			Supelcoport		
	C	Gas Chrom CL	Gas Chrom CLA	Gas Chrom CLZ	Acid + Base	Gas Chrom CLP
	D			Diatoport S		
	E			Varaport 30		
Other	C	Gas Chrom S	Gas Chrom A	Gas Chrom Z	Acid, Base & DMCS	Gas Chrom Q*
Filteraid	C				Acid, Base & DMCS	Gas Chrom P
						Anakrom ABS
	F	Anakrom U	Anakrom A		Acid, Base & DMCS	Anakrom Q
						Anakrom SD*
C-22 Firebrick	A	Chromosorb P NAW	Chromosorb P AW	Chromosorb P AW-DMCS	HMDS	Chromosorb P HMDS
	C	Gas Chrom R	Gas Chrom RA	Gas Chrom RZ		

[1]Where a manufacturer has several grades of the same type, the best grade of that brand is marked by an asterisk (*).

[2]A: Johns-Manville, B: Supelco,Inc., C: Applied Science, D: Hewlett-Packard, E: Varian, F: Analabs.

122

other acidic compounds such as phenols (8). Basic compounds such
as amines can be more readily chromatographed if substances such
as THEED (5) or potassium hydroxide are used as tail reducers.
The upper temperature limit of some columns may be lowered by the
use of certain tail reducers because these additives are often
less stable than the stationary phase itself. If columns are
operated at too high a temperature there may not be any noticeable increase in bleed, but the effectiveness of the tail reducer
may be lost. Because of this, tail reducers were used to a lesser degree as new column materials such as porous polymers became
available. Increasing interest in trace analysis has necessitated
development of columns in which adsorption is minimized, and tail
reducers are used in many of these applications.

Although diatomite supports are the most widely used, there
are situations where an even more inert material is required.
The analysis of very polar or corrosive substances are often performed with halocarbon supports such as Fluoropak-80, Kel-F, and
Teflon; for corrosive substances the support must also be coated
with a halocarbon stationary phase. Chromosorb T, a sieved form
of Teflon 6, is now recognized as the best available with regard
to column efficiency and inertness. Special precautions are
necessary when preparing packings and columns if reasonable performance is to be obtained (3,6,9). Another limitation is the
type of stationary phase that can be used; Kirkland (3) has shown
that phases such as Carbowax and squalane give much more efficient columns than do those prepared with diglycerol. The most
common reason for failure when working with soft Teflon supports
is improper handling of the material; thorough study of the references cited above will help to minimize the problem.

3.2.2 Adsorbents for Gas-Solid Chromatography

This type of support is usually used without any stationary phase,
but in special cases small amounts are added to obtain special
effects. In gas-liquid chromatography, every effort is made to
eliminate adsorption, whereas this is used to advantage in gas-
solid chromatography. The technique is used mostly for analysis
of gases and low-boiling compounds; some adsorbents are particularly suited for analysis of aqueous solutions in that water can
be easily quantitated. Kieselev and Yashin (10) have presented
an excellent review of adsorption chromatography and stress the
relationship between performance of a column and the structure of
the adsorbent. Jeffrey and Kipping (11) emphasize separations
rather than theory, providing considerable detail for each
example.

Porous polymers, such as Porapak and the Chromosorb Century Series, are the most widely used adsorbents. Porapak Q and Chromosorb 102 are both styrenedivinylbenzene polymers, have similar separation characteristics and are the most widely used of the porous polymers. The other members of the respective series differ in composition and consequently in selectivity. Dave (12) and Supina (6) have compared the properties of the two series; these references are useful in selection of the appropriate porous polymer for a specific separation. The porous polymers are useful for analysis of permanent gases and for short chain polar compounds such as glycols, acids, amines. Water is eluted as a symmetrical peak with these materials. It is important to note that for most polar compounds such as acids or amines, it is desirable to use glass columns rather than metal.

Whereas, most of the varius porous polymers differ from each other only in selectivity, there are two major exceptions. Chromosorb 103 was developed specifically for analysis of amines and therefore is not suitable for acidic compounds. Tenax, because of its excellent thermal stability, can be used at much higher temperatures than the other porous polymers. Because of the minimal bleed from this material it has found use as a trapping medium for concentration of trace components (13) in air. These compounds are then desorbed from the Tenax and subsequently analyzed, permitting detection at much lower levels than by direct analysis of the air.

The zeolites, commonly referred to as molecular sieve 5A, 13X, and so on, are best known for their ability to separate oxygen and nitrogen, but can also be used for other permanent gases. A major limitation is that CO_2 is permanently adsorbed at usual operating temperatures; even though it is eventually released at higher temperatures, the results are not quantitative. This problem has been solved by the use of a precolumn which retards the CO_2 and allows all the other components to pass on to the molecular sieve column as shown in Figure 3.5. Molecular sieve columns can vary significantly in separation characteristics. This appears to be due to inadequate control of activation procedures and to variations in the molecular sieve itself (14). If the molecular sieves are not properly activated, all of the permanent gases are eluted quickly with little or no separation. Activation decreases the water content of the molecular sieve; therefore, it is imperative that the carrier gas be thoroughly dry or columns will deteriorate rapidly with use.

Silica gel has been used for the analysis of carbon dioxide and other permanent gases, but porous polymers now are used for many of these applications. One drawback of the use of silica gel is that its characteristics vary depending on the type used

(10) and many literature references do not supply sufficient information. New grades such as Chromosil developed specifically
for analysis of sulfur compounds at trace levels as shown in Figure 3.6 are more standardized in chromatographic properties. A
major advancement in standardization of silica materials was the
development of Spherosil and Porasil (15). These are porous,
spherical silica beads available in six types, each with a different surface area and pore size.

Carbon has been used in several forms. Activated charcoal
found limited usage because of significant batch-to-batch variations. Kaiser (16) developed a carbonaceous material with pore
structure similar to zeolites and referred to it as carbon molecular sieve. It is useful only for gases and very short chain
compounds. It is unique in that it can be used for the analysis
of oxygen, nitrogen, and carbon dioxide as shown in Figure 3.7.

Graphitized carbons provide a means for separation of a variety of polar compounds at trace levels. However, they must be
coated with a stationary phase, and also in some cases with a
tail reducer. The use of these materials for gas chromatography
has been greatly advanced by the work of Bruner, Di Corcia,
Liberti, and associates (17,18). Examples of separations possible with this type of material are shown in Figures 3.8 and 3.9.
The graphitized carbons are extremely fragile and considerable
care must be used in preparing packings and filling columns.

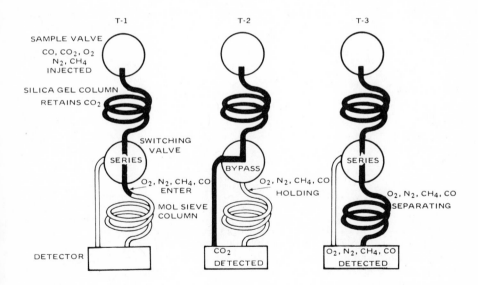

Figure 3.5.

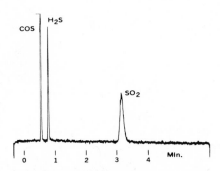

6 ft. x 1/8" O.D. Teflon (FEP) Chromosil 310; Col. Temp.: 40°C: Flow rate: 20 ml/min. (2b)

Figure 3.6.

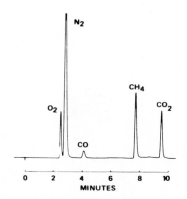

Column: 9 ft. x 1/8" S.S. with Carbosieve-B 120/140 mesh: Col. Temp.: 35°C to 175°C, 4 min. hold, prog. 30°C/min; Flow Rate: 50 ml/min. helium. (2a)

Figure 3.7.

3.3   STATIONARY PHASE

The primary requirements of a stationary phase are to provide separation of the sample with reasonable column life.   Therefore, in addition to having suitable selectivity, the phase should have reasonable chemical and thermal stability.   Many catalogs list upper temperature limits for stationary phases, but these should be used only as approximations because the true limit depends upon the type of detector used and the amount of column bleed one can tolerate to get the job done.   Even if a phase is stable to 250°C, the column will last much longer if the temperature is limited to 200°C.   Excessive temperatures result not only in shorter column life, but also in more rapid fouling of the detector.

The selection of the appropriate stationary phase for a specific separation can be done in a logical manner.   The first step

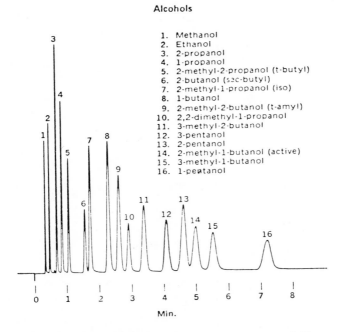

Alcohols

1. Methanol
2. Ethanol
3. 2-propanol
4. 1-propanol
5. 2-methyl-2-propanol (t-butyl)
6. 2-butanol (sec-butyl)
7. 2-methyl-1-propanol (iso)
8. 1-butanol
9. 2-methyl-2-butanol (t-amyl)
10. 2,2-dimethyl-1-propanol
11. 3-methyl-2-butanol
12. 3-pentanol
13. 2-pentanol
14. 2-methyl-1-butanol (active)
15. 3-methyl-1-butanol
16. 1-pentanol

Min.

80/100 Carbopack C/0.2% Carbowax 1500, 6 ft. x 2 mm ID Glass, Col. Temp.: 125°C, Flow Rate: 20 ml/min. N₂ @ 31 psi. Sample Size: 0.02 μl, Det.: FID.

Figure 3.8. Alcohols.

should be to attempt an analysis with one column. If the sample is a gas, one of the porous polymers would be a suitable choice. If it is a liquid, a nonpolar column such as OV-101 or SP-2100 could be used. Even if this first column does not provide the necessary separation, it can provide information that will be helpful in selection of the next column. The next step would be to determine whether or not the same compounds or similar ones have been separated by someone else. The easiest thing to do is to ask someone who is working with similar materials. If this is not possible, then a thorough search of the literature is in order. Fortunately, there are two major abstracting services which simplify the job appreciably. The Preston Technical Abstracts (19) are issued on a monthly basis and provide a convenient means for keeping abreast of current publications. The publisher offers a computer search program and will do a literature search for a fee.

One of the most useful publications is the series of abstracts published by the Chromatography Discussion Group of Great Britain (20). These have been published annually since 1958 and two five-year cumulative indices are available. The major feature of the abstracts is the comprehensive index at the back of each volume. This series of abstracts is so valuable that it should be readily available to anyone working in GC. The cost of a complete set is less than the cost spent by trying a few columns that do not work.

Attendance at meetings of local chromatography groups (such as

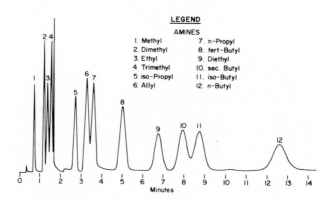

4% Carbowax 20 M + 0.8% KOH on Carbopack B 1.4 meters x 4mm I.D. Glass Column. Col. Temp.: 91.5°C. Carrier Gas Flow: 50 ml/min. helium.

Figure 3.9.   Amines.

the Chromatography Forum) provides an opportunity for exchange of
information on an informal basis. At these meetings it is possible to meet outstanding speakers to explore in depth many problems. The local sections of the American Chemical Society
could advise whether or not such a group is active within their
area. If not, why not start one?

The suitability of a stationary phase for a specific separation depends upon the selectivity of the phase. This is a measure of the degree to which polar compounds are retarded relative
to their elution on a nonpolar phase. A systematic method for
expressing the retention data is based on retention indices. For
this sytem, the retention indices of the n-paraffins are by definition equal to 100 times the number of carbon atoms in the
molecule. For example, the retention index for n-hexane is 600
and for n-octane 800. These values are defined and apply regardless of the column used and regardless of the temperature.
For all other compounds except the n-paraffins, the column and
operating conditions must be specified or the data will be meaningless. The retention indices for all compounds other than the
normal paraffins are determined as follows:

First inject a mixture of benzene, hexane, heptane and octane
onto the column and then determine the retention times for the
four components. In the next step subtract from each of the retention times the retention time for an air peak to correct for
the dead volume in the system. The result is referred to as the
justed retention time (t'R) and is used for calculating relative retention times. The logarithm of the t'_R is plotted
against the retention times as a function of retention index is
for all practical purposes a straight line over a range of several carbon numbers. Assume that on this column, hexane has an adjusted retention time of 15 min, heptane has 19 min, and octane
has approximately 25 min. Under the same conditions, benzene
would have a adjusted retention time of 17 min. Construct the
semi-log plot for the three normal paraffins (Figure 3.10). Fif-
teen minutes on the log retention scale and 600 on the retention
index scale would give one point; the 19 min versus 700, and 25
min versus 800 would give the other two points. Connecting the
three points by a straight line would give the plot which is the
basis for determining retention indices on this 20% squalane
column at 100°C. This curve would be accurate for compounds
which are eluted on this column between 15 and 25 min. The retention index 649 for benzene can now be determined from the
curve at a value of 17 min which is the adjusted retention time.
for benzene. Retention indices can be estimated by extrapolation
but this should be done only as a last resort; the most accuracy
is obtained by bracketing the compound with n-paraffins. What
does this number 649 mean? It indicates that benzene elutes

approximately halfway between hexane and heptane on a logarithmic
time scale.

The selectivity of a stationary phase for a particular compound can be measured in terms of the degree to which the retention index differs from the retention index obtained with a nonpolar column. As shown in Figure 3.10, the retention index for
benzene was 649 on the squalane column. If the experiment is repeated in exactly the same manner but using dinonylphthalate as
a stationary phase, the retention index is 733. This increase,
ΔI, of 84 units of retention index indicates that the dinonylphthalate column will retard the benzene slightly more than will
the squalane column. Under the same conditions, a highly polar
phase such as SP-2340 would give a retention index of 1169 which
would be a ΔI of 520 units. An excellent review of the retention
index system has been presented by Ettre (22).

The apparent polarity of the dinonylphthalate column, expressed by a ΔI of 84, is a measure of retardation of aromatic
and olefinic substances. Since the chromatographer is interested
in the selectivity of a column for a variety of functional
groups, it is important to classify each of the stationary phases
by their ability to retard compounds other than benzene. This
has been done by Rohrschneider (22) and further developed by
McReynolds (23). The system is discussed in detail with many
examples by Supina (6). The Rohrschneider/McReynolds (R/M) system involves the measurement of the retention indices for several
compounds on a given column to determine the degree to which each
is retarded. In each case, the retention indices are compared to

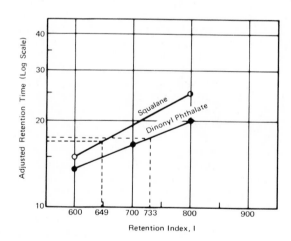

Figure 3.10.

those obtained for the same compounds on a squalane column.
These ΔI values are listed as in Table 3.2, where the x', y', z',
u', and s' values indicate the polarity of the phases for each
type of compound. Most chromatography supply catalogs now contain these "McReynolds constants" for the stationary phases
offered for sale.

TABLE 3.2[1]

Stationary Phase	x'	y'	z'	u'	s'
SE-30	15	53	44	64	41
DC-200	16	57	45	66	43
Dinonylphthalate	83	183	147	231	159
OV-17	119	158	162	243	202
SP-2250	119	158	162	243	202
SP-2401	146	238	358	468	310
Carbowax 20M	322	536	368	572	510
SP-2340	523	757	659	942	801

[1] x' = ΔI for benzene, y' = ΔI for n-butanol, z' = ΔI for 2-
pentanone, u' = ΔI for nitropropane, s' =ΔI for pyridine.

One of the most valuable uses for these constants is in identifying stationary phases which are similar such as (a) SE-30 and
DC-200, (b) OV-17 and SP-2250. In some cases, the choice between
similar materials would be based on other factors, such as availability or thermal stability. For example, high-temperature
applications would benefit by the use of SE-30 rather than DC-
200. The data also indicate that OV-17 and SP-2250 are similar
and either would be suitable for most applications. This type of
comparison can be very helpful in avoiding unnecessary duplication of columns. When new stationary phases are introduced, the
suppliers usually announce the McReynolds constants. This makes
it possible to compare the characteristics of the new phase with
phases already in use. An extensive tabulation of McReynolds
constants has been published by Supina (6).
    Another use of this system is to select columns which have
unique characteristics. The fluorosilicones, such as SP-2401,
are the only phases for which the z' value is greater than the y'
value, which indicates that ketones would be eluted after alcohols. With polyglycols such as Carbowax 20M, the reverse is

true. The most important point to remember in using the R/M system is that it is used to select columns for separation of compounds differing in functionality such as alcohols from ketones, aromatics from aliphatics, saturates from unsaturates. It is of no value for separation of a homologous series or isomers. There have been many papers published on the R/M system and related topics in recent years, primarily in the Journal of Chromatography and the Journal of Chromatographic Science.

In analysis of polar compounds, tailing is minimized if the stationary phase is similar in composition to the types of sample being analyzed. For example, alcohols with carbowax, esters with polyesters, and amines with tetraethylenepentamine. If these phases do not give the desired separation, it may be advisable to add them to another phase as a tail reducer.

In selection of an appropriate column for a specific separation, it is important to understand why columns which may have been tried do not provide the desired results. This type of information may then lead to a better choice for the next attempt. Points to consider are:

1. Do the peaks tail?
2. Are all the components eluted?
3. What is the difference in structure of the compounds to be separated?
4. What is the boiling point difference of the compounds?

If there is a significant difference in boiling points of two components having different functional groups, it is usually easier to separate them if the higher boiling compound is eluted last. For analysis of a trace contaminant, it is better to have this material elute before the major component if possible.

3.3.1 Concentration of Stationary Phase

The analysis time depends upon the length of column, concentration of the phase, and column temperature. A 6-foot column is a commonly used length, but shorter columns will often provide adequate separation and in less time. The concentration of stationary phase and column temperature are related in that increasing the amount of stationary phase will require a higher column temperature to obtain the analysis within the same period of time. However, even though two such analyses might be done within the same time, the column with the lower concentration of stationary phase operated at the lower temperature would give a chromatogram in which the early peaks would be eluted closer together than for a column with more phase operated at a higher temperature. Note that in Figure 3.11 the C_{24} compound is eluted in 30 min on both the 20% column at 200°C and the 5% column at 180°C. With the latter column however the early peaks are not well resolved and half the analysis time is required for the last peak.

One might be led to believe that high concentrations are desirable, but limitations on thermal stability of stationary phase generally dictates the operating temperature of most columns. Also as the concentration of stationary phase increases, the column efficiency decreases with a resulting broadening of peaks. Decreasing the concentration of stationary phase also tends to decrease the effective polarity of the phase; is is more noticeable with the higher polarity phases and concentrations below 15%. Therefore a separation which is readily accomplished with a column containing 15% stationary phase may be more difficult with 5%.

3.3.2 Temperature Effects

Column temperature is another variable that is useful in improving separations since one class of compounds may be affected more than another by change in column temperature. Compounds from two such classes which are not separable at one temperature may be completely resolved if the temperature is changed 20°C. This effect is also helpful in tentative identification of compounds since change of retention index with temperature for various classes of compounds has been the subject of many papers. A more detailed discussion of the effect of stationary phase concentration and temperature on separations has been published (6).

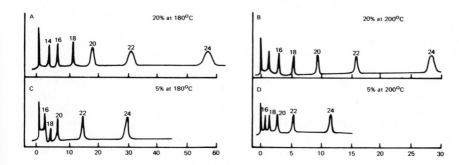

Figure 3.11.   Effect of temperature and percentage of liquid phase on GC column separation.

3.4   PROBLEMS OF GAS CHROMATOGRAPHY

     3.4.1 Tubing

The tubing used to hold the column packing often contributes to
many of the problems attributed to the column packing. In many
cases the type of tubing to be used is selected primarily on the
basis of convenience or cost, without adequate consideration being given to the difficulties that might be encountered. Metal
is not more convenient than glass nor is copper less expensive
than stainless steel if the results are not satisfactory. If one
is in doubt, glass should be used unless sufficient effort is
made to determine the suitability of metal for the specific
application. In a quality control situation where the same types
of compounds are analyzed repeatedly, it might be worthwhile to
determine whether or not there are problems associated with the
use of metal. However, in a laboratory where sample type varies
from day to day, it would be more practical to use glass.
     Glass tubing is the best choice for analysis of most com-
pounds. The only exception might be for corrosive substances
such as halogens and for trace analysis of reactive sulfur compounds. In all cases where glass is desirable, it is equally important to thoroughly clean this inside surface to remove any
residual materials and then to silane treat this surface. This
is best done by filling the empty column with a 5% solution of
dimethyldichlorosilane in toluene and allowing to stand for 5-10
min. The solution is first displaced with toluene, then with
methanol; the column is then purged with dry nitrogen or air and
is ready for filling. Reagents such as BSA, Silyl-8, and Rejuv-8
should not be used for this treatment as they are not reactive
enough. Inadequate treatment can cause the same tailing and loss
of compound which is typical of inadequately prepared column
packings.
     Stainless steel can be used for analysis of hydrocarbons and
moderately polar compounds such as esters. Cleanliness and proper treatment of the inside surface is very important particularly
if the tubing is to be used for analysis of more polar substances.
Rinsing with solvents may be adequate preparation for many analyses but certainly not for general usage. A given tubing may
appear to have a shiny, smooth inside surface, but microscopic
impurities may lead to unsatisfactory results. Therefore, it is
advisable to obtain tubing from a reputable supplier of chromatographic materials rather than from a tubing vendor.
     Copper and aluminum are also used but to a much lesser extent,
and for good reason. Oxides of these metals are good adsorbents

and can cause serious tailing and even reactions with certain
compounds. Also, column efficiencies are usually lower and this
may be due to the difficulty of properly filling the columns.
The economy of using these metals is questionable.

3.4.2 Glass Wool

Glass wool is used to hold the packing in place within the column
and often causes either tailing or ghosting. If glass wool is
suspected as the cause of the problem, it should be removed from
the inlet side of the column to see if this improves the analysis.
The surface of the glass wool is similar to that of glass tubing
and, therefore, it should be similarly treated. For most applications, it is advisable to use pyrex glass wool which has been
silane treated in a manner similar to the glass tubing treatment.
With analysis of acidic compounds such as barbiturates, fatty
acids, and phenols, it is helpful to use pyrex glass wool that
has been soaked in a dilute solution of phosphoric acid and air
dried. Even with these treatments, it is best to use as little
glass wool as possible. Substitutes such as stainless steel
screens or fritted discs are convenient and eliminate these problems and are available from most suppliers of chromatographic
materials.

3.4.3 Carrier Gas

One of the most common problems associated with deterioration of
columns is the presence of water and/or oxygen in the carrier gas.
This is more common with nitrogen than with helium. Stationary
phases such as polyesters, FFAP, SP-1000, and carbowax depolymer-
ize rapidly with 100 ppm of water in the carrier gas; columns may
be satisfactory the first day and destroyed the next. If this is
occurring, peaks for polar compounds will begin to tail and this
will become progressively worse. It is not necessary to have a
fancy device to remove the water, but it should have the capacity
to remove large quantities of water.
   A pipe approximately 3 feet long, 1 inch in diameter filled
with 1/2 lb of molecular sieve is preferable to a shorter, fatter
device as the contact time will be greater in the longer tube.
It is equally important to change the molecular sieve regularly.
If this is done each time a tank of gas is changed, there will be
little doubt as to the effectiveness of the drying tube. The in-
dicators that change color in presence of moisture are not satis-

factory because they are not sensitive enough. By the time they
change color, the moisture in the carrier gas may have already
destroyed the column.

Oxygen in the carrier gas can rapidly destroy column materials
prepared from carbon; all peaks will tail severely even after
a few hours of column use. Oxygen will change the separating
characteristics of stationary phases such as polyesters and
carbowax. The symptoms are more subtle here in that the column
may still seem to be functioning properly, but the polarity may
be gradually decreasing, with the rate of change being dependent
on the concentration of oxygen. The presence of oxygen in the
carrier gas can cause serious problems if an electron capture detector is being used, and this will be indicated by a rapid decrease in sensitivity. Removal of the oxygen from the carrier
gas can be accomplished by means of catalytic systems. Some are
available in disposable cartridge units while others utilize a
heated system with larger capacity. Whenever the oxygen or water
removal systems are installed or changed, it is important to
purge the entire chromatographic system for at least 12 hours before the column is heated in order to remove residual moisture
and oxygen.

3.4.4 No Peaks

When no peaks are observed upon injection of a sample, it is
common to blame the column but there are other factors that could
cause this. Is the sample actually getting into the column? The
syringe needle may become clogged with parts of the septum or insoluble material from the sample. Loose plungers can permit the
sample to be blown back by the pressure of the carrier gas. The
syringe is a precision device and should be handled with care.
Those with burred or bent needles, or bent plungers should be re-
paired before use.

A defective detector can often be the source of problems.
Most manuals supplied by chromatograph manufacturers provide instructions for proper maintenance and troubleshooting. When
checking the detector, it is advisable to use a sample that will
not be adsorbed by the column, such as a hydrocarbon. For electron capture detectors a suitable test substance would be a halocarbon, rather than a sensitive chlorinated pesticide.

Leaks in the flow system are one of the most common sources of
problems. The fact that a rotameter indicates that the flowrate
is satisfactory does not assure that there is gas flowing through
the column, it merely means that there is gas going through the
rotameter. A leaking septum or connection at the head of the

column would not be detected by the rotameter. Therefore, the
best place to check flowrate is at the end of the column before
connecting it to the detector.

Flow controllers should maintain a constant carrier gas flowrate during temperature programming. In order to function properly, the carrier gas pressure to the flow controller usually
must be 15-20 psig greater than the desired pressure at the column at the maximum temperature. If this is not done, or if the
controller is defective, flow rates will decrease as the column
temperature rises.

3.4.5 Selecting A Column for a Specific Application

When selecting a column for a particular analysis, one should
consider the entire system, not just the stationary phase. This
should include the support, stationary phase, type of tubing,
packing retainers such as glass wool. Equally important are instrument capabilities such as temperature programming, on-column
injection, special valving arrangements for back flush or for use
of two columns, special detectors. While there can be no specific procedure to follow for determining the conditions for each
application, general guidelines can be given. First, one should
determine if the analysis has been done successfully by others.
This involves a search of the literature or by discussing the
problem with someone who may have experience with the same or
similar analyses. It is important to consider improving upon the
technique; previous work may have been done with obsolescent column materials.

If no information on this type of analysis is available, the
next step is to obtain a preliminary analysis for the mixture.
In this analysis all of the components may not be separated, but
information can be obtained which will be useful in selecting a
more appropriate column. This first chromatogram should be examined for peak symmetry, analysis time, and degree of separation. If the peaks tail severely, the column is not compatible
with the sample and a different tubing and/or packing should be
selected. When the sample is strongly acidic, the packing should
be acidic in character; conversely, strongly basic samples require basic packings. If the sample contains hydroxyl groups as
in the case of alcohols, it is helpful to use either a stationary phase or a tail reducer which also contains hydroxyl groups
such as Carbowax. One would not use a packing containing acidic
materials such as FFAP, SP-1000, or H_3PO_4 for analysis of amines,
as there would be a reaction between sample and column.

One method of reducing tailing is to prepare a derivative of
the substance being analyzed. Compounds such as acids, alcohols,
and amines are eluted as symmetrical peaks when chromatographed
as methyl esters, trimethylsilyl ethers, and so on. In many
cases the same column can be used for both the free form and the
derivatized version. However, if the column has been used to
analyze the derivative without removing the excess derivatizing
reagent, the column may no longer be suitable for analysis of the
underivatized form. An example would be the use of FFAP for analysis of methyl esters. If excess methylating agent is injected
onto this column, the acidic groups of the phase will be esterified and would not be effective for analysis of free acids.
Highly corrosive or reactive substances require column materials
which will not react with the samples. One extreme case is HF
for which a halogenated oil on Chromosorb T packed into Teflon
tubing must be used.

After determining the type of column packing which is suitable
for a particular separation, it is necessary to select the concentration of stationary phase. While 10% may be better than 3%
for a given case, if one already has a 3% column, it is worth
trying it first rather than to prepare another column. If the
chromatographer has neither, the choice depends largely on the
molecular weight range of the sample components. In isothermal
analysis, for mixtures up to approximately C_6, a 15% concentration is useful, but higher concentrations may be used. From C_6
to C_{20} it is convenient to use 10% and above this the concentration should be reduced to 3-5% depending on the molecular weight
and the polarity of the compounds. With temperature-programmed
analysis the concentration of stationary phase is kept as low as
possible to minimize baseline drift but the effect of concentration on effective polarity must be taken into consideration.

There are many factors to consider in selecting a column for
a specific separation. Table 3.3 lists a variety of applications
along with columns that have been found suitable for each analysis. While each of these columns has been used successfully for
the class of compounds shown, they will not separate all compounds of that class. Separation of isomers can be difficult and
special columns are often required. However, the columns listed
are useful for the first step in the analytical problem.

3.4.6 Preparation of Column Packings

The economics of your laboratory may be such that it is advisable
to purchase packings already prepared. Nevertheless, it is advisable to be familiar with the techniques and to have on hand

TABLE 3.3

Compound	Recommended Phase or Packing	Suitable Equivalent	Equivalent Not Recommended	Support[1] Treatment	Tubing[2]	Remarks
Acids						
C_1-C_6	*Porapak Q	*Chromosorb 102		*	R	Isomer interference
	SP-1200/H_3PO_4			AW		Isomers separated
C_{6+}	DEGS-PS	EGS/H_3PO_4		AW		Will separate saturates and unsaturates
Alcohols						
C_1-C_5	*Chromosorb 102	*Porapak Q		*	R	Will not separate all isomers
	*Carbopack C/0.2% Carbowax 1500			*		Isomers separated
C_{6+}	SP-2100	OV-101, SE-30, OV-1	DC-200, SF-96 JXR, UC-W98	S	R	
C_{6+}	SP-2300	Silar-SC		S	U	Saturates and unsaturates separated
	10% Carbowax 20M			S	U	Good peak symmetry, not thermally stable
Amines						
C_1-C_5	*Chromosorb 103			*	N	Not for trace analysis
C_{6+}	*Carbopack B/4% Carbowax/KOH			*	N	Trace, isomers
	10% Carbowax 20M + 2% KOH			AW	N	
	10% Apiezon L + 2% KOH			AW	N	
Amino Acids	0.65% EGA			AW	N	As n-butyl-TFA derivatives

Carbohydrates TMS derivatives	15% SP-2330	15% Silar 9C	15% EGS	AW	U	
Alditol Acetates	3% SP-2340	3% Silar 10C		S	R	
Fatty Acid Esters C_2-C_6	Most columns					
C_{6+}	10% SP-2330	Silar 9C	DEGS, EGS, EGSS-X	AW	U	Better thermal stability than DEGS or other polyesters
Gases O_2-N_2	*Mol Sieve 5A			*	U	Will not elute CO_2
	*Carbosieve B	*Carbon Molecular Sieve		*	U	Will also separate CO_2, but requires long column for O_2/N_2
Co, CO_2, N_2, CH_4	*Porapak Q	*Chromosorb 102		*	U	
Hydrocarbons C_1-C_3	*Carbosieve B	*Carbon Molecular Sieve		*	U	
C_1-C_4 unsaturates	*Carbopack C/0.19% Picric Acid			*	U	
Xylenes	5% SP-1200/5% Bentone 34	5% DIDP/5% Bentone		S	U	
Aliphatics from Aromatics	10% tris (2-cyanoethoxy propane)			AW	U	Retards aromatics
Medium boiling	10% SP-2100	OV-101, SE-30 OV-1	DC-200, SF-96 UC W98, JXR	S	U	

Application	Column 1	Column 2	Column 3			Comments
High boiling	3% SP-2100	OV-101, SE-30	DC-200, SF-96 UC W98, JXR	S	U	
Very high boiling	1% Dexsil 300	OV-1		S	U	
Pesticides	1.5% SP-2250/ 1.95% SP-2401	1.5% OV-17/ 1.95% OV-210	1.5% OV-17/ 1.95% QF-1	S	N	
Phenols	10% SP-2100	OV-1, SE-30, OV-101	DC-200, SF-96 UC W98, JXR	S	N	Will not separate M/P isomers
	*Carbopack C/ 0.1% SP-1000			*	N	Slow, will separate M/P
Steroids	3% SP-2100	OV-1, OV-101, SE-30	UC W98, DC-200, JXR, SF96	S	N	
	3% SP-2250	OV-17		S	N	
	3% SP-2401	OV-210	QF-1	S	N	
	3% SP-2340	Silar-10C		S	N	
Sulfur Gases						
@% levels	*Porapak QS	*Chromosorb 102		*	T	
@ ppb levels	*Supelpak S	Porapak QS/ H_3PO_4		*	T	

*These packings are adsorbents, or coated adsorbents, and do not utilize conventional supports.

[1]AW: acid-washed diatomite such as Chromosorb W; S: silane-treated diatomite such as Chromosorb W HP, Gas-Chrom Q, or Supelcoport.

[2]N: glass tubing necessary; R: glass tubing recommended; U: glass tubing unnecessary; T: Teflon tubing.

the necessary ingredients in case an emergency situation arises. As in any manufacturing operation, a certain amount of skill must be developed and the more critical the application, the more skill required. Packings prepared with low concentrations of silicones, such as 3% SE-30, which are to be used for analysis of sensitive steroids are difficult to prepare. In contrast it is relatively simple to prepare a good batch of 15% Carbowax 20M on Chromosorb W. There are many techniques described in the literature for preparation of packings but a "cookbook procedure" will not assure good results. Instead, a few simple rules must be observed:

1. The stationary phase must be coated uniformly over the surface of the support.
2. The support particles must not be damaged.
3. The stationary phase must not be oxidized, hydrolyzed, or evaporated during the preparation.

As long as these are observed, one should obtain a good packing material. Problems usually occur where in trying to obtain a uniform coating by thoroughly mixing the packing, support particles are fractured exposing untreated, uncoated surfaces. The two most widely used techniques for coating the support are the evaporative and the solution coating methods, which are described here but are covered in more detail elsewhere (6).

In the evaporative technique the stationary phase is dissolved in an appropriate solvent, the support added and the solvent is then slowly evaporated. All of the stationary phase is thereby deposited on the support. However, unless the support is mixed thoroughly during the evaporation process, the stationary phase may be deposited more heavily on some particles than on others. Rotary evaporators have been used in order to obtain a more uniform coating, but the tumbling action can cause breakage of particles. A more suitable method is to pour the slurry into an open tray and to stir this until the solvent is evaporated. Only when there is no liquid remaining in the bottom of the tray should heat be applied, and then only slightly. This is done conveniently with an infrared lamp. If the packing is one which is subject to oxidation or reaction with CO_2 in the atmosphere (such as KOH), the evaporation step should be done under an inert atmosphere.

The solution coating technique involves preparing an excess of solution of known concentration of the stationary phase dissolved in appropriate solvent. A known weight of support is then added slowly to a known volume of solution while the solution is stirred until all of the support is in suspension. This slurry

is then poured into a Buchner funnel and excess solvent is removed. Two critical steps must be observed: (a) the slurry must
be poured rapidly so that the support settles into a smooth cake
rather than an inverted cone, and (b) suction to the filter flask
must be discontinued as soon as the solvent ceases to flow and
begins to drip.

At this point the volume of filtrate is measured and the
amount held up by the support is calculated. Since the concentration of stationary phase in the solution is known, the amount
of phase deposited on the support can easily be calculated. For
the novice, a little experience is necessary in selecting the
appropriate concentration of solution to obtain the desired concentration of phase in the packing. As an approximation, if the
solution of stationary phase has a viscosity similar to that of
water, a 10% solution will result in a packing with a 10% concentration of stationary phase. Some phases such as Carbowax, SE-
30, and OV-1 produce highly viscous solutions which will give a
10% packing from a 5% solution.

After the suction to the filter flask is disconnected, the
packing is then transferred to a tray for drying: This is accomplished by inverting the funnel and tapping gently; the packing
should not be removed by scraping. The damp packing should be
air dried until all traces of solvent are gone and then heated
gently with an infrared lamp.

The solution coating technique in most cases is preferred to
the evaporative method because there is more uniform deposition
of the phase and because there is less chance for damage of
support particles. The evaporative procedure is more convenient
for preparing high concentration packings (15-20%) of the viscous
phases such as Carbowax, and for preparing packings with two or
more stationary phases which are not soluble in a common solvent.

3.4.7 Filling the Column

Before the tubing is filled it is imperative that the inner surface be properly cleaned and treated. Often one obtains metal
tubing that has not been manufactured specifically for chromatographic use, and for certain analyses no amount of cleaning
with solvents will suffice. Even many chromatographic grades of
tubing contain residual oils and greases and these must be removed by thorough washing with polar and nonpolar solvents.
Glass columns may contain residual caustic materials and should
be thoroughly cleaned with diluted acid. They should then be silane treated since glass columns are usually used for analysis of
sensitive compounds.

The object in packing the column is to fill it as tightly as possible without fracturing or deforming the particles. Many techniques have been described but their effectiveness depends primarily on the skill and experience of the person filling the column. For metal columns 6 feet and shorter it is advisable to fill the column after it has been coiled in the same manner as a glass column. First, a large wad of glass wool is placed partially into the end of the column which will connect to the detector. The excess glass wool is bent back around the outside of the tubing and a piece of rubber tubing is connected to the column. This method assures that the glass wool will not be pulled out of the column during filling. The other end of the rubber tubing is then connected to a vacuum source such as an aspirator or pump. A funnel is then connected to the other end of the column and the packing added slowly while vacuum is applied at the other end. During the entire filling operation it is advisable to provide a slight agitation to the packing by vibrating or tapping the column at approximately the point at which the packing is settling. When the column is filled, a small piece of silane-treated glass wool or fritted disk is placed in the inlet end; the vacuum source is then disconnected from the outlet end. The large glass wool plug is then removed from the outlet and replaced with a small piece of silane-treated glass wool or a fritted disk.

Long metal columns up to 20 feet are packed in straight sections and then carefully coiled with a minimum of flexing of the tubing. For still longer columns it is advisable to pack shorter lengths and to connect these with a Swagelok union which has been bored out to permit a butt to butt seal of the column ends.

Glass U-tube columns are usually filled by gravity without the need for vacuum. It is best to add packing to both arms simultaneously, or in small segments to each arm. Silane-treated glass wool should be used sparingly and is not really needed or recommended in the inlet side.

3.4.8 Conditioning the Column

Most columns must be conditioned prior to connection to the detector in order to purge volatile components that would foul the detector and cause unsteady baselines. The conditioning procedure depends upon the column material used; excessive conditioning temperatures result in short column life. Normal carrier gas flow should be maintained during conditioning in most cases.

Adsorbents such as silica gel and molecular sieve are heated to approximately 300°C overnight. This releases any substances that may have been adsorbed during preparation of the column and

also activates the adsorbent. The separating characteristics of
this type of column may be affected if it is not adequately activated.

Columns prepared with polymeric stationary phases such as
Carbowax, polyesters, or polyphenyl ethers, should be conditioned
at a temperature at which the column will be used. These stationary phases contain polymers of varying molecular weight and
the conditioning helps to remove the more volatile portions.
While conditioning overnight will usually suffice, the level of
column bleed will gradually decrease and will be indicated by
drifting baseline.

Silicones require special conditioning. Technical grade materials such as UC-W98, DC-200, SF-96, QF-1, and JXR should be
treated as any other polymeric phases, and conditioned overnight
at slightly above the planned operating temperature. The thermal
stability of these silicones varies depending upon the molecular
weight and chemical structure. These materials are not recommended for high temperature use even though the literature contains
many references to use at temperature up to 300°C.

Chromatographic grade silicones such as materials sold under
the designations OV, SE-30, Silar, and SP have superior thermal
stability characteristics and require a special conditioning
procedure. These stationary phases usually consist of a very
narrow molecular weight range of polymers. Conditioning is used
only to rid the column of residual solvent and not for the purpose of removing portions of the stationary phase. Columns
should be heated slowly to slightly above the operating temperature, with an hour or two allowed for this to take place. The
column is then ready for use. Conditioning such columns at 300°C
if the columns are to be used at 230°C will only shorten the
column life.

Columns should never be removed from the chromatograph while
the packing material is still hot. The recommended procedure is
to turn off oven and injector heaters, open the oven and allow to
cool for approximately 30 min. The carrier gas is then turned
off, the exit end uncoupled, and when no flow can be measured at
the exit of the column, the inlet is disconnected. The two ends
of the column should then be sealed with a suitable cap to prevent contamination of the packing material.

3.5   OPEN TUBULAR (CAPILLARY) COLUMNS

3.5.1 Description and Capabilities

The possible use of open tubular columns (OTC) was first suggested by Martin in 1956 but it wasn't until two years later that Golay published the theoretical and practical results for their use. Over the years two excellent books have been published on their preparation and application (24,25).

Several types of open tubular columns are possible:

1. WCOT Columns. Wall-coated open tubular columns wherein the internal wall is coated with a thin layer of liquid (liquid substrate).

2. PLOT Columns. Porous-layer open tubular columns wherein the internal wall is coated with a layer of adsorbent support. If the support is then coated with a liquid phase it is referred to as a SCOT Column (i.e., support-coated open tubular column).

The inside diameter of these columns is usually <1 mm. The SCOT columns contain a larger amount of stationary phase (liquid coating) per column segment, thus sample capacity is higher than the standard open tubular column (WCOT).

Column gas velocity is dependent upon the column length, carrier gas viscosity, pressure drop through the column and the column permeability. Column permeability is expressed by the specific permeability coefficient, B_o, which may be calculated by the following expression.

$$B_o = 2\eta\epsilon \; Lp_o u_o / (p_i^2 - p_o^2) = cm^2 \tag{3.3}$$

where p_i = inlet pressure (dyne/cm^2),
p_o = outlet pressure (dyne/cm^2),
η = viscosity of carrier gas (dyne-sec/cm^2),
L = length of column (cm),
$u_o = F_c L/V_c = \bar{u}a/j$,
V_c = tube volume (cm^3),
$\bar{u}$ = average linear velocity of air peak,
a = packing porosity,
ϵ = interparticle porosity.

If $\underline{a}$ is known, B_o can be calculated from the retention time of the air peak.

The permeability of open tubular columns can be calculated by use of the Hagen-Poiseuille equation:

$$B_o = \frac{d^2}{32} \tag{3.4}$$

where d is the tube diameter.

Permeability of open tubular columns is 10-100 times greater than the permeability of packed columns. As a result open tubular columns can be longer than packed columns by the same factors.

Since open tubular columns have low sample capacity this necessitates smaller sample sizes which in turn requires the use of sensitive detectors. The PLOT or SCOT columns, however, have a larger sample capacity than the WCOT columns, thus permitting larger sample sizes. These columns (PLOT, SCOT) can be operated without splitters and therefore, less sensitive detectors. One potential use for the SCOT columns is in the area of trace analysis.

Increased versatility of these types of columns is possible by modification of the internal walls. Several modifications which are applicable:

1. Tube Wall Deactivation. This may be accomplished with the use of small quantities of highly polar liquid, coating the wall with a fine layer of Hg, Pt, Ag, or Au, or if glass columns are used treating the glass tube with trimethyl chlorsilane.
2. Coating with Porous Solid Layer. The internal wall may be coated (1-100 μ layers) with graphitized carbon black, silica gel, alumina, or iron oxide. These columns may be used in gas-solid chromatographic studies. Coating the solid layer with a liquid substrate permits their use in gas-liquid chromatography. The solid layer does not effect the column permeability but does allow a surface for making SCOT columns with high sample capacities.
3. Increased Surface Area of Walls. If glass columns are used the surface area may be increased by treatment with hot ammonium or sodium hydroxides. If aluminum columns are used a similar treatment can be employed. In the former case a layer of porous silica is formed, in the latter a layer of porous alumina.

3.5.2 Column Performance

The high efficiency of open tubular columns makes the choice of stationary phases less complex than with packed columns. A separation that is incomplete on a packed column would have better resolution using an open tubular column. The separation of the xylene isomers on a packed column using 7,8-benzoquinoline requires 20 min and 14,600 theoretical plates; whereas the same separation can be performed on an open tubular column in less time but using squalane and utilizing 68,000 theoretical plates.

Another advantage of open tubular columns is that most separations can be performed at a lower temperature than with packed
columns.

Precision of retention data with open tubular columns is
better than packed columns and one is able to see changes in the
separation factors when a change is made in the carrier gas
(e.g., changing from helium to hydrogen).

Comparison of column performance among the different types of
columns is not easy. Resolution and analysis time at a minimum
temperature seem to be the best criteria for comparison. Golay
(26) suggests the calculation of performance index (P.I.) for
intercomparison of columns.

$$
P.I. = \frac{(w_{0.5})^4}{t_R^3 (t_R - 0.94t_o)} t_o (p_i - p_o) \tag{3.5}
$$

where t_o = retention time of air peak,
t_R = retention time of solute peak.

An ideal open tubular column would have a P.I. = 0.1 poise.
Older packed columns would have P.I. = 1000 poise (four powers of
ten worse). Golay assigned P.I. values of 0.2-1.0 poise for open
tubular columns and P.I. values of 10-20 poise for highly efficient packed columns.

3.5.3 Summary

Open tubular columns enable the chromatographer to effect separations impossible by other techniques, for example, separation of
isomers. Their use permits a smaller inventory of columns and a
smaller selection of stationary phases. Their main advantage is
their high efficiency and their disadvantage is their small load
capacity.

REFERENCES

1. O. E. Schupp, Gas Chromatography, John Wiley and Sons, New
   York, 1968.
2. R. A. Scholz and W. W. Brandt, Third Int. Gas Chromatog. Symposium, N. Brenner, J. E. Calvin, M. D. Weiss (Eds.),
   Academic Press, New York, 1962. p. 7.
3. J. J. Kirkland, Anal. Chem., 35, 2003 (1963).
4. D. M. Ottenstein, J. Chromatog. Sci., 11, 136 (1973).

5. D. M. Ottenstein, Advances in Chromatography, Vol. 3, J. C. Giddings and R. A. Keller (Eds.), Marcel Dekker, New York, 1966, p. 137.
6. W. R. Supina, The Packed Column in Gas Chromatography, Supelco, Bellefonte, PA, 1974.
7. L. D. Metcalfe, J. Gas Chromatog., 1, 7 (1963).
8. A. R. Paterson, Gas Chromatography, Instrum. Soc. America, June 1959, H. J. Noebels, R. F. Wall, and N. Brenner (Eds.), Academic Press, New York, 1961, p. 323.
9. Bulletin FF-124, Johns-Manville Corp., Celite Div., Denver, CO.
10. A. V. Kiselev and Y. I. Yashin, Gas-Adsorption Chromatography, Plenum, New York, 1969.
11. P. G. Jeffrey and P. J. Kipping, Gas Analysis by Gas Chromatography, 2nd edit. Pergamon Press, London, 1972.
12. S. B. Dave, J. Chromatog. Sci., 7, 389 (1969).
13. A. Zlatkis, H. A. Lichtenstein, and A. Tishbee, Chromatographia, 6, (2) 67-70 (1973).
14. D. M. Ottenstein, personal communication.
15. C. L. Guillemin, M. Lepage, R. Bean, and A. J. De Vries, Anal. Chem., 39, 940-945 (1967).
16. R. Kaiser, Chromatographia, 1, 199 (1968).
17. F. Bruner, A. Liberti, M. Possanzini, and I Allegrini, Anal. Chem., 44, 2070 (1972).
18. A. Di Corcia, Anal. Chem., 45, 492 (1973)
19. Preston Technical Abstracts Co., P. O. Box 312, Niles, IL.
20. Chromatography Discussion Group, c/o Trent Polytechnic, Burton St., Nottingham, NG1 4 BU England.
21. L. Ettre, Anal. Chem., 36, No. 8, 31A-41A (1964).
22. L. Rohrschneider, Advances in Chromatography, Vol. IV, Marcel Dekker, New York, 1967.
23. W. O. McReynolds, J. Chromatog. Sci., 8, 685 (1970).
24. R. Kaiser, Chromatographie in der Gasphase, II. Teil. Kapillar-Chromatographie, Bibliograph. Institut, Mannheim, 1962.
25. L. S. Ettre, Open Tubular Columns in Gas Chromatography, Plenum Press, New York, 1965.
26. M. J. E. Golay, Nature, 180, 435 (1957).

CHAPTER 4

Qualitative and Quantitative Analysis by Gas Chromatography

I   Qualitative Analysis

MARY A. KAISER

University of Georgia

4.1   IDENTIFICATION FROM GAS CHROMATOGRAPHIC DATA ONLY . . . 153
      4.1.1 Retention Data. 153
      4.1.2 Plots of Log Retention versus Carbon Number . . . 154
      4.1.3 Kovat's Retention Index 155
      4.1.4 Multiple Columns. 157
      4.1.5 Relative Detector Response. 158
            Selective Detectors 158
            Molecular Weight Chromatography 158
      4.1.6 Simple Pretreatment 160
            Extractions 160
            Beroza's "p" Value. 160
            Water-Air Equilibration 161
      4.1.7 Tandem Gas Chromatographic Operation 161
            Subtractive Precolumns. 161

151

        Carbon Skeleton 162
        Controlled Pyrolysis 162
4.2   IDENTIFICATION BY GAS CHROMATOGRAPHIC AND OTHER DATA. . 162
     4.2.1 Elemental and Functional Analysis 163
     4.2.2 Other Instrumental Techniques 164
     4.2.3 Trapping. 165
4.3   LOGIC OF QUALITATIVE ANALYSIS 166

II Quantitative Analysis

FREDERICK J. DEBBRECHT

Analytical Instrument Development

4.4   GENERAL DISCUSSION. 166
4.5   PEAK MEASUREMENT 167
     4.5.1 Peak Height 167
     4.5.2 Height and Width at Half-Height 170
     4.5.3 Triangulation 171
     4.5.4 Cut and Weigh 172
     4.5.5 Planimeter 172
     4.5.6 Disc Integrator 172
     4.5.7 Electronic Integrators and Computers. 175
     4.5.8 Comparison of Peak Size Measurements. 176
4.6   STANDARDIZATION 179
     4.6.1 General 179
     4.6.2 Internal Normalization 181
     4.6.3 External Standardization 184
         Static Gas Standards. 187
         Dynamic Gas Standards 190
     4.6.4 Internal Standardization 199
     4.6.5 Standardization Summary 201
4.7   ERROR DISCUSSION. 203
     4.7.1 Sampling Techniques 203
     4.7.2 Sample Introduction 203
         Syringe Injection 204
         Gas Sample Valve 207
         Miscellaneous Sampling Devices 208

      4.7.3 Gas Chromatographic System Errors • • • • • • • • 209
REFERENCES • 210

PART I

4.1   IDENTIFICATION FROM GAS CHROMATOGRAPHIC DATA ONLY

     4.1.1 Retention Data

Qualitative analysis by gas chromatography (GC) in the classical
sense involves the comparison of retention data of an unknown
sample with that of a known sample. The alternate approach involves combination and comparison of gas chromatographic data
with data from other instrumental and chemical methods.

     A chromatographic peak provides valuable information, namely,
the elapsed time from the injection point or the difference in
elution times of two peaks (qualitative information), the peak
shape (qualitative or quantitative information), and the peak
size (quantitative information). The simplest qualitative tool is
simply the comparison of retention data from known and unknown
samples. A chromatogram illustrating the commonly used retention
nomenclature is given in Figure 4.1. The retention time (t_R) is
the time elapsed from injection of the sample component to the
recording of the peak maximum. The retention volume (V_R) is the
product of the retention time and the flow rate of the carrier
gas. Generally, the adjusted retention time (t_R') or adjusted retention volume (V_R') and the relative retention ($r_{A/B}$) are used
for qualitative analysis. Adjusted retention time (volume) is
the difference between the retention time (volume) of the sample
and an inert component (usually air). The relative retention is
the ratio of the adjusted retention time (or volume) of a standard to the adjusted retention time (or volume) of the unknown.
(see Chapter 2).

     It must be emphasized at this point that, just because the re-
tention times of an unknown and a known component are the same,
the identity of each is not necessarily the same. In many fields
of analysis the mixtures frequently encountered contain a number
of similar components that will possess either the same, or nearly the same, retention characteristics. However, if the retention time of an unknown and a known in a well-designed system do
not agree, then it can be said that the unknown and the known are
absolutely not the same.

     There are many factors that must be considered when comparing

retention measurements. The precision of the data generally depends upon the ability of the instrument to control temperature
and flowrate. A change in the temperature of approximately 30°C
changes the retention time by a factor of two. Thus, to maintain
a 1% repeatability in the temperature measurement, the column
temperature must be held within ± 0.1°C. A 1% change in the
flowrate affects the retention time by 1%. Sample size also
plays an important role (Figure 4.2). If too much sample is introduced onto the column for its diameter and length, "leading"
peaks appear. Since the retention time is measured from the peak
maximum, the retention time will appear to increase with large
samples. This problem can be remedied by simply decreasing the
sample size.

An attempt to compare retention times on two different columns
of the same type can be difficult, at best. Differences in packing density, liquid loading, activity of the support, age and
previous use of the packing, and variations in the composition of
the column wall can lead to large differences in retention measurements between the two columns. If one must use two
separate columns of the same type, then relative retention data
is preferred since this measurement is reasonably constant for
columns of the same type, it is not as subject to temperature and
flow changes, and it is easy to obtain.

4.1.2 Plots of Log Retention Versus Carbon Number

A linear dependence exists between the logarithm of the retention

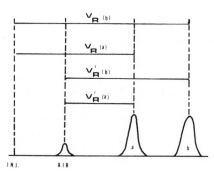

Figure 4.1. Chromatogram illustrating retention nomenclature. V_R = retention volume, V_R' = adjusted
retention volume, $r_{a/b}$ = relative retention =
$V_R'(a)/V_R'(b)$.

volume of fatty acids and the number of carbon atoms in the molecule. This relationship has been shown to hold for many classes
of compounds, such as alkanes, olefins, aldehydes, ketones, alcohols, acetates, acetals, esters, sulfoxides, nitro-derivatives,
aliphatic amines, pyridine homologs, aromatic hydrocarbons, di-
alkyl ethers, thiols, alkylnitrates, substituted tetrahydrofuran,
and furan. A typical series of plots of the logarithm of the retention time versus carbon number are given in Figure 4.3.

It must be emphasized that this method of identification is
valid only for members of a homologous or pseudo-homologous series. Thus, it must be known what type of compound is being analyzed before this technique may be applied. Generally, the lower
members of the series cannot be included since the relationship
is not strictly linear and can deviate significantly at lower
carbon numbers.

4.1.3 Kovat's Retention Index

Wehrli and Kovats (1) introduced the concept of the retention index to help confirm the structure of organic molecules. This
method utilizes a series of normal alkanes as a reference base
instead of one compound as in the relative retention method.
Identification can be assisted with the use of the retention

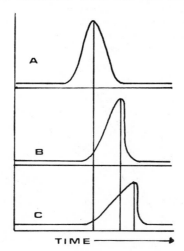

Figure 4.2. Effect of sample size on retention time.
(A) Column not overloaded; (B) column slightly overloaded; (C) column severely overloaded.

index, I,

$$I = 100N + 100 \left[\frac{\log V_R'(A) - \log V_R'(N)}{\log V_R'(n) - \log V_R'(N)} \right] \tag{4.1}$$

where N and n are the smaller and larger n-paraffin, respectively, which bracket substance A, and V_R' is the adjusted retention volume. The retention indices for even-numbered n-alkanes are defined as 100 times the number of carbon atoms in the molecule for every temperature and for every liquid phase (e.g., octane = 800, decane = 1000).

In practice, the retention index is simply derived from a plot of the logarithm of the adjusted retention time versus carbon number times 100 (Figure 4.4). To obtain a retention index, the compound of interest and at least three hydrocarbon standards are injected onto the column. At least one of the hydrocarbons must elute before the compound of interest and at least one must elute after it. A plot of the logarithm of the adjusted retention time versus the Kovats index is constructed from the hydrocarbon data. The logarithm of the adjusted retention time of the unknown is calculated and the Kovats index determined from the curve (Figure 4.4).

Many factors can influence the Kovats index which can make it unreliable at times for the characterization of gas chromatographic behavior, although it generally varies less than the relative retention with temperature, flow, and column variation. However, for many it is the preferred method of reporting retention data.

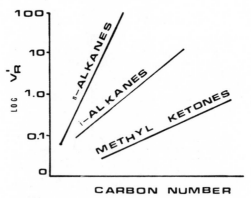

Figure 4.3. Logarithm of adjusted retention time versus carbon number.

4.1.4 Multiple Columns

The use of two or more columns improves the probability that the
identity of an unknown compound is the same as that of a compound
with identical retention times; however, these data alone are not
conclusive proof. The reliability of the identification depends
on the efficiency and polarity of the columns used. With efficient columns, the probability of having two or more components
under one peak diminishes and the peaks are generally well resolved. Care must be taken in selecting columns to be certain the
columns have different selectivities and not just different
names. Thus, the McReynold's constants (Chapter 3) must be compared and should be quite different for each column. For example, in the analysis of pesticides, four different liquid
phases might be chosen arbitrarily (e.g., OV-1, UCW-98, SE-30,
and DC-200). However, the relative retention in this case will
be the same for all four columns since they are all methyl silicones and have essentially the same McReynold's constants (2).

An unequivocal identification of an unknown compound is unlikely by chromatographic processes alone. Not the least of the
reasons for this is the need for comparison to standards thereby
assuming reasonable prior assurance of the possible identity of
the unknown. It should be noted that in addition to retention
time measurements obtained on two or more column systems, if
reasonable care has been exercised, quantitative measures of the
suspect compound should also correspond, thus providing an additional secondary identification. In other words, whatever the
unknown compound may be, it cannot be a mixture of two components
on one column and a single component on the second column without

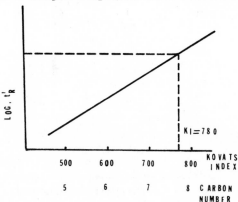

Figure 4.4. Plot of logarithm of adjusted retention time versus Kovats index.

quantitative measure detecting this fact. The value of this particular observation is commonly ignored.

Information on the structure of an unknown peak can be obtained from the difference in the retention indices on polar and nonpolar stationary phases:

$$\Delta I = I(polar) - I(nonpolar) \qquad (4.2)$$

For a particular homologous series, ΔI is a specific value which
is determined by the character of the functional group(s) of the
molecule.

Takacs and co-workers (3,4) calculated the Kovats index for
paraffins, olefins, cyclic hydrocarbons, and homologs of benzene
on the basis of molecular structures. The index was divided into
three additive portions: atomic index, bond index, and sample-
stationary phase index components.

4.1.5 Relative Detector Response

SELECTIVE DETECTORS. Comparison of the relative detector response from two or more detectors can aid in the identification
or classification of an unknown component. Generally, the component is chromatographed on one column and the effluent split
and fed to two or more detectors. Commonly used pairs of detectors are the phosphorous and electron capture, flame ionization and radioactivity, and flame ionization and phosphorous detectors. The electron capture detector allows the identification
of substances containing atoms of phosphorous, oxygen, nitrogen,
and halogens in a complex mixture while remaining insensitive to
other substances. Flame photometric detectors are useful with
phosphorous or sulfur containing compounds. The flame ionization
detector is especially sensitive to hydrocarbons. (For a complete discussion of specific and nonspecific detectors, see
Chapter 5).

MOLECULAR WEIGHT CHROMATOGRAPHY. The molecular weight of a component can be obtained through mass chromatography. This relies
on two gas density detectors, two columns, and two carrier gases.
A diagram of a typical mass chromatographic system is given in
Figure 4.5. The sample is introduced into the injection chamber
by syringe, gas sampling valve, pyrolysis unit, or reaction
chamber and trapped on two separate trapping columns. After the
sample has been trapped, it is displaced from the traps by back-
flushing and heating and swept onto two matched gas chromato-

graphic columns using two different carrier gases. The carrier
gases are chosen on the basis of significant difference in molecular weight (for example, CO_2 (44 g/mole) and SF_6 (146 g/mole)).
The sample is then separated on the column and the eluate passed
through each gas density balance detector (Figure 5.22). Thus,
two peaks are recorded for each component (Figure 4.6). The molecular weight of a component is obtained from the ratio of the
two peak heights or areas by use of the following equation:

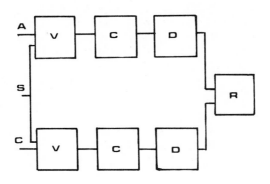

Figure 4.5. Mass chromatograph. (A) Carrier A inlet;
(S) sample inlet; (B) carrier B inlet; (V) valve/trap
system; (C) chromatographic column; (D) gas density
detector; (R) recorder.

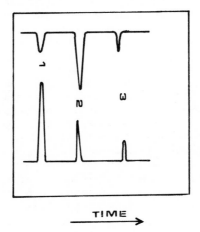

Figure 4.6.   Mass chromatogram.

$$MW = \frac{K\,(A_1/A_2)\,\,(MW_{CG_2} - MW_{CG_1})}{(K(A_1/A_2)) - 1} \qquad (4.3)$$

where K is an instrumental constant, A_1 and A_2 are the area responses for the unknown component from detector 1 and detector 2, respectively, and MW_{CG_1} and MW_{CG_2} are the molecular weights of carrier gas 1 and carrier gas 2, respectively. In practice, A_1 and A_2 are measured for known compounds and K is determined for the experimental conditions. Then the molecular weight of the unknown is determined by obtaining its area ratio and using the K previously obtained for known compounds.

The molecular weight and the Kovats retention index can then be combined to aid in the identification of the component. A linear relationship exists between the molecular weight and retention index for a homologous series of compounds. The relationship varies for each class of compound; thus, a clue can be obtained regarding the type of compound present which can be verified by some other technique.

4.1.6 Simple Pretreatment

A few minutes devoted to simple pretreatment of the sample can save many hours of interpretation of complex data. Procedures as filtration, extraction, or distillation can be readily accomplished and will simplify the identification of the separated components.

EXTRACTIONS. Simple partition between two phases can add another valuable piece of information about the sample. A gas chromatographic analysis before and after extraction indicates the character of the components present. For example, carboxylic acids are readily separated from phenolic compounds by extracting a nonaqueous solution of the sample with dilute aqueous sodium bicarbonate. The carboxylic acids are almost completely transferred to the aqueous phase, whereas the phenolic constituents remain in the organic layer. Additional information on extractants for specific classes can be found in most organic analysis textbooks or inferred from solubility tables.

BEROZA'S "p" VALUE. Beroza and co-workers (5) devised a simple

extraction method to aid in the characterization of pesticides.
The method consists of determining the partition coefficient ("p"
value) of a pesticide between two immiscible solvents. A detailed description of this technique is given in Chapter 7. The
general application of this simple preliminary extraction to the
qualitative identification of a wide variety of organic materials
should not be overlooked.

WATER-AIR EQUILIBRATION. McAuliffe (6) introduced a multiple
phase equilibrium procedure for the qualitative separation of
hydrocarbons from water-soluble organic compounds. For n-alkanes,
more than 99% were found to partition in the gas phase after two
equilibrations with equal volumes of gas and aqueous solution.
Cycloalkanes require three equilibrations to be essentially completely removed, and oxygen-containing organic compounds (e.g.,
alcohols, aldehydes, ketones, and acids) remain in the aqueous
layer. Thus, after equilibration with equal volumes of gas, an
immediate clue is given regarding the identification of the compound. More details of this technique can be found in Chapter 7.
   This technique also provides two additional pieces of information, the distribution coefficient(D_s or D_g) and the initial
concentration of the unknown.

4.1.7 Tandem Gas Chromatographic Operation

Tandem gas chromatographic operation involves the use of chemical
reactions in conjunction with the chromatographic analysis.
These reactions can occur ahead of the injection port, in a precolumn reactor, within the column, or in a postcolumn reactor.
The most simple application of tandem operation is the separation
of components followed by reaction with a single peak. These reaction products can then be analyzed chromatographically.

SUBTRACTIVE PRECOLUMNS. For many applications the mixture to be
analyzed is so complex that the only reasonable method of analysis requires the removal of certain classes of compounds. This
process can be easily implemented by the use of a reactive precolumn. For example, a precolumn of potassium hydroxide can be
used to remove acid vapors. The mixture could then be chromatographed with and without the precolumn to identify which peaks
had acid character. A discussion of precolumn reagents is given
by Littlewood (7). Potential packing materials for precolumns
may also be found in the trace analysis literature. (see Chapter

7).

CARBON SKELETON. The technique of precolumn catalytic hydrogenation can be applied to reduce certain unsaturated compounds to
their parent hydrocarbons. Compounds analyzed by this technique
include esters, ketones, aldehydes, amines, epoxides, nitriles,
halides, sulfides, and fatty acids. Fatty acids usually give a
hydrocarbon that is the next lower homolog than the parent acid.
For most systems utilizing hydrogenation, hydrogen is also used
as the carrier gas. Usually 1% palladium or platinum on a non-
adsorptive porous support such as AW-Chromosorb P is used as the
catalytic packing material.

CONTROLLED PYROLYSIS. The principle of controlled pyrolysis or
controlled thermolytic dissociation for identification of chromatographic effluents lies with the examination of the pattern
("fingerprint") produced. The peak selected for identification
is transferred with continuous flow from the gas chromatograph
through a gold coil reactor helically wound on a heated stainless
steel core, then through a second gas chromatograph for identification of the pyrolysis products. The products are identified
by comparing the Kovats retention indices to those of standard
compounds and by enhancing the peaks with selected standard compounds. The "fingerprint" can also be obtained with increased
certainty by coupling a mass spectrometer to the second chroma-
trograph. The controlled pyrolysis technique can be especially
useful in forensic and toxicological applications where direct
comparison of samples is necessary.
    Information concerning functional groups absent or present in
a molecule can be obtained by determining the concentration
ratios of the small molecules produced on pyrolysis (CO, CO_2,
CH_4, C_2H_4, C_2H_6, C_3H_6, NH_3, H_2S, and H_2O). "Large molecule" pyrograms (C_4H_8 and larger) in combination with "small molecule"
pyrograms can give additional information regarding the functional groups present. Neither technique can yield a priori identification of molecules by pyrolysis at this time.

4.2   IDENTIFICATION BY GAS CHROMATOGRAPHIC AND OTHER DATA

Certainly the more discrete pieces of information obtainable concerning an unknown compound, the easier it will be to obtain a
confident identification. Microchemical tests such as functional

group classification, boiling point, elemental analysis, and derivative formation as well as infrared spectroscopy, nuclear magnetic resonance spectrometry, coulometry, flame photometry, and
ultraviolet-visible spectroscopy are also helpful aids when used
in conjunction with gas chromatographic analysis.

4.2.1 Elemental and Functional Analysis

The major reason that GC is not generally used for qualitative
analysis is that it cannot differentiate or identify indisputably
the structure of the molecule. Therefore, it is necessary to
perform additional tests on the separated peak to ascertain its
functionality and elemental composition.

There are many books and articles available regarding microanalysis, so it will not be extensively reviewed here. Usually,
it is necessary to trap the peak then perform whatever specific
microanalysis techniques necessary to confirm the peak's identity.
Several commercial instruments are available for elemental analysis (usually carbon, hydrogen, nitrogen, and halogens). These
instruments usually require between 0.1 and 3 mg of sample and
often employ trapping systems for quantitative analysis.

Hoff and Feit (8) reacted samples in a 2-cm^3 hypodermic syringe before injection onto the gas chromatographic column. Reagents were selected either to remove certain functional groups
or to alter them to obtain different peaks. Reagents used included metallic sodium, ozone, hydrogen, sulfuric acid, hydroxylamine, sodium hydroxide (20%), sodium borohydride (15%), and
potassium permanganate (concentrated).

A stream-splitter attached to the exit tube of a thermal conductivity detector can be used to identify the functional groups
of gas chromatographic effluents. Table 4.1 lists functional
group tests and limits of detection. A review of elemental analysis by GC is given by Rezl and Janak (9).

Crippen's book (10) gives an extensive compilation of the
techniques of organic compound identification with the assistance
of GC. It includes a step-by-step account of the preliminary examination, physical property measurements, and functional group
classification tests.

Gas chromatographic methods for the qualitative analysis of
complex systems such as biological materials and bacteria, proteins, steroids and prostaglandins, and triglycerides have been
developed.

TABLE 4.1   FUNCTIONAL GROUP CLASSIFICATION TESTS[1]

Compound Type	Reagent	Type of Positive Text	Minimum Detectable Amount (μg)
Alcohols	$K_2Cr_2O_7$-HNO_3	Blue color	20
	Ceric nitrate	Amber color	100
Aldehydes	2,4-DNP	Yellow ppt.	20
	Schiff's	Pink color	50
Ketones	2,4-DNP	Yellow ppt.	20
Esters	Ferric hydroxamate	Red color	40
Mercaptans	Sodium nitroprusside	Red color	50
	Isatin	Green color	100
	$Pb(OAc)_2$	Yellow ppt.	100
Sulfides	Sodium nitroprusside	Red color	50
Disulfides	Sodium nitroprusside	Red color	50
	Isatin	Green color	100
Amines	Hinsberg	Orange color	100
	Sodium nitroprusside	Red color, 1^0 Blue color, 2^0	50
Nitriles	Ferric hydroxamate-propylene glycol	Red color	40
Aromatics	HCHO-H_2SO_4	Red-wine color	20
Aliphatic unsaturation	HCHO-H_2SO_4	Red-wine color	40
Alkyl halide	Alc. $AgNO_3$	White ppt.	20

[1]Reprinted with permission from Anal. Chem., _32_, 1379(1960).

4.2.2 Other Instrumental Techniques

A technique that is becoming more common for the identification
of compounds is the combination of gas chromatography/mass spectrometry. In part, this is due to the decreasing cost, increasing sensitivity, and decreasing scan time of mass spectrometry
equipment. Where the total mass scan can be obtained from a
single component, this is generally a highly valid technique.
While it does not require a prior knowledge or reasonable suspicion of the identity of the suspect component, the most conclusive indentification will be re-creation of the same mass spectrum from a known standard. The spectrum obtained from an unknown, if not immediately decipherable, will provide a significant number of clues to the probable identity, thus limiting the

the need either for searching reference spectra or for the generation of reference spectra.

The second most used instrumental technique is that of infrared spectrometry. In general the first instrumental method to
consider is the one most readily available. In a few cases, notably mass spectrometry, the technique may be used in tandem with
the gas chromatograph, but most techniques require trapping of
the peaks as discussed in Section 4.2.3.

Consideration must be given to the quantity of sample needed
for the minimum detection limits of the instrumental technique
used. A number of techniques have been ranked in order of increasing amounts of material needed as follows: mass spectroscopy
(1 - 10 µg), chemical spot tests (1 - 100 µg), infrared and
ultraviolet spectroscopy (10 - 200 µg), melting point (0.1 -1 mg),
elemental analysis (0.5 - 5 mg), boiling point (1 - 10 mg), functional group analysis (1 - 20 mg), and nuclear magnetic resonance
spectroscopy (1 - 25 mg).

4.2.3 Trapping

Trapping a sample as it elutes from the column followed by some
other identification or classification technique is often utilized with gas chromatographic analysis. The most common trapping
devices are the cold trap, the gas scrubber (gas washing bottle),
the evacuated bulb, and the adsorbent postcolumn.

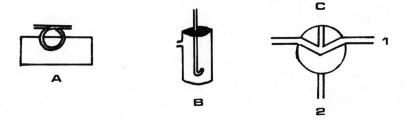

Figure 4.7   Traps (A)Simple coil cold trap; (B) Gas
scrubbing trap; (C) Evacuated bulb trap; (1) to outlet
of gas chromatograph; (2) to evacuated gas-sampling
bulb.

A simple cold trap can be constructed from small-diameter
glass tubing and connected with some flexible inert tubing to the
outlet port of the chromatograph (Figure 4.7A). Part of the coil
should be immersed in a liquid coolant such as liquid nitrogen
(-196°C), dry ice - acetone (-86°C), sodium chloride - ice (1:2)
(-21°C), or ice - water slush (0°C). (One should not use liquid
nitrogen when air or oxygen is being used as a carrier gas because of the explosion hazard as liquid oxygen accumulates.) The

upper part of the coil should be above the coolant liquid so that loss of sample due to too rapid cooling (fogging) can be avoided.

A gas-washing bottle (Figure 4.7B) may also be used for trapping. This technique is especially useful in conjunction with infrared analysis. The sample is simply bubbled through the anhydrous solvent as it exits the chromatographic column. The solution is then placed in a liquid sample infrared cell. A matching cell containing only the solvent is placed in the reference beam. An infrared spectrum of the sample may then be recorded.

Evacuated bulbs (Figure 4.7C) are generally used for trapping volatile components. Since this technique does not concentrate the sample, additional sample preparation may be required. For substances with high infrared absorptivity, the sample may be trapped directly in an evacuated infrared gas cell and analyzed directly. For nonvolatile samples that may condense on the inside walls, the cell must be heated before analysis.

An adsorbent postcolumn can also be used to trap eluting peaks. Packing materials such as Tenax-GC (Enka N.V., The Netherlands), Porapak N and Porapak R (Waters Associates), Carbosieve B, and 20% DC-200 have been tested as sampling tubes for concentrating organic compounds in air. Tenax-GC and Porapak N seemed to have the widest general applicability. Tenax-GC was more suitable for higher-boiling compounds, and Porapak N was more suitable for lower-boiling organics (20-100°C).

4.3   LOGIC OF QUALITATIVE ANALYSIS

The most important factor in qualitative analysis by GC is the collection of as much information as possible about the sample. The place to begin gathering this information is people involved in the collection of the sample. The sampling location, the person taking the sample, the method of sampling and sample handling should be known. The sample matrix (solvent, etc.) should be investigated to determine the source of the chromatographic peaks. A "pure" material should be utilized to compare with the unknown sample.

Furthermore, the chemist should always be alert to unknown peaks originating from simple decomposition in storage or decomposition or isomerization under chromatographic conditions. One should keep in mind that the identification of an unknown by GC can easily turn into a major research project.

PART II

QUANTITATIVE ANALYSIS

4.4   GENERAL DISCUSSION

The technique of GC is at best a mediocre tool for qualitative analysis. As has been shown previously, it is best used with other techniques to answer the question of what is present in the

sample. The rapid growth of GC over the last two decades cannot
be explained by ease of operation, the simplicity of the technique, the relative low cost of the instrument, or the wide range
of the types of samples capable of being handled. That growth
comes from the fact that it has all of the above attributes and
provides an answer to the question of "how much?" Its reason for
existence is that it is an excellent quantitative analytical tool
whether one is quantifying micrograms of heptachlor in a liter of
water or one volume carbon monoxide in a million volumes of air.

Sometimes we get carried away with the latest advancement in
instrumentation or with the perfectly symmetrical peaks obtained
with a certain system. These are only means to an end, perhaps
very necessary means, but they are not the end. The end is a
number that tells us how much of a component is in a sample.
Without the ability of GC to supply that number with reasonable
accuracy this entire book would not be written. The tremendous
advances in instrumentation, theory, columns, applications, and
technique are all justified because they provide more accurate
and precise analyses or analyses for materials not previously
handled.

The remainder of this Chapter will deal with the techniques
used to obtain the answer to the question of "how much?" from the
information given by the chromatograph. The quantitative principle of GC depends on the fact that the size of a chromatographic
peak is proportional to the amount of material. The first aspect
to be considered is the technique of determining peak size.
Next, the factors that influence peak size and thus introduce
errors will be considered. Finally the problem of relating peak
size to quantity of material will be discussed.

4.5   PEAK MEASUREMENT

The size of a chromatographic peak is proportional to the amount
of material contributing to that peak and the size of this peak
can be measured by a number of ways. Each of these will be considered individually.

Two basic concepts can be used for peak size. The first is
simply the measurement of the height of the peak. The second
involves the measurement of area with a wide variety of methods
available.

4.5.1 Peak Height

Peak height is the simplest and easiest of the measurement techniques. As shown in Figure 4.8, the baseline is drawn in connecting the baseline segments both before and after the peak (line
AB in the figure). This line is the best estimate of the detector output had there been no detectable amount of material present that contributed to the peak. The height of the peak is
then measured from this baseline vertically to the peak maximum.
(line CD in the figure). This height is proportional to the
amount of material contributing to the peak if nothing in the
system changes that could cause a change in the width of the peak
between sample and standard.

Factors that can influence the peak width are generally instrumental or technical in nature. The temperature of the column
changes the retention time of the material thus changing the
width of the peak. To a first approximation the ratio of retention time to peak width will stay constant for a given component
on a given column. Temperature can influence retention time
approximately 3%/$^{\circ}$C. A 1°C change in column temperature between
the standard and the unknown chromatograms can cause a 3% change
in retention time and thus a 3% change in peak width. This
change in width will be accompanied by a compensating change in
peak height such that height times width remains constant. The
height then will change 3%/$^{\circ}$C. This means that to maintain

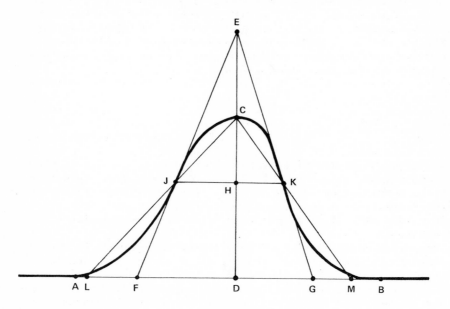

Figure 4.8.   Constructions for peak size measurements.

analysis at 1% accuracy using peak height measurement the temperature of the column must be controlled within $\pm$ 0.3°C and preferably to better than $\pm$0.1°C, assuming the temperature change of the column to be the only factor producing error.

The carrier gas flow also produces a change of retention time and thus peak width and peak height. To a first approximation, a 1% change in flow will change retention time 1%; thus peak width and peak height are changed 1%. Therefore, using the peak height measurement requires that the flowrate between the standard and the sample chromatograms be constant within 1% to maintain an accuracy of 1%. The above consideration regarding the effect of flow on the error of peak height measurement is independent of the major error consideration regarding constant flow. Several detectors, notably thermal conductivity, are flow sensitive; that is, the sensitivity or electrical output for a given amount of material varies with flow. This flow effect affects any method of peak measurement and is really not an error of size measurement. It simply says that flow control is needed regardless of the method of peak measurement.

Reproducibility of peak height is also quite dependent on the reproducibility of the sample injection. This is especially important on early and thus normally quite sharp, narrow peaks. On such early peaks the width of the peak is controlled more by the injection time rather than the chromatographic process. A fraction of a second increase in injection time can double the width of these peaks and reduce peak height 50%. The peaks most subject to error in peak height measurement from injection problems are those with retention volumes between one and two times the hold-up volume of the column. Peaks beyond five to ten times the hold-up volume are negligibly affected by injection technique.

When there is column adsorption of a particular component in the system, the peak will show some tailing. This may not be evident at high concentrations of a component but with low concentrations a significant portion of the component may be in the tail. This means that at low concentrations the relationship of peak height to amount of material may not be linear due to the amount of material in the tail. For quantitative analysis· it is best to avoid adsorption by a better choice of column regardless of the technique of peak measurement. However, with adsorption, peak height may give a significant error with low amounts of material.

The final consideration of peak height measurement is the phenomenon of column overload. When a large amount of a component is injected onto a chromatographic column, the liquid or adsorptive phase becomes saturated with the material, causing a broadening of the peak. This then reduces the height contributing to a nonlinear relationship between peak height and amount of

material with high amounts of material. This is independent of
any detector nonlinearity at high concentrations. Overloading
can be observed by careful observation of the peak shape. There
is a sloping front edge with a sharp tail or, in some cases, a
sharp front with a sloping tail. The peak maximum also moves
with this distortion to longer times with the sloping front and
shorter times with the sloping tail. This overload distortion is
a function of the amount of liquid phase per unit length of column. It occurs more readily then on small diameter columns and
on packings with low percentage of liquid phase.

   4.5.2 Height and Width at Half-Height

Contrary to peak height measurement, there are a number of techniques used for peak area measurement. Some of these are manual
techniques and others make use of instrumental accessories to
provide an area measurement. The discussion that follows will
consider all of these techniques from the manual through the instrumental, in that order.
   In the height-width measurement the area is determined by multiplying the height of the peak by the width of the peak at one-
half the height. This technique requires the construction of the
baseline AB in Figure 4.8 and the measurement of the height of
the peak CD as in the peak height technique. Point H is then determined as being halfway between points C and D such that distance DH is one-half the height CD. Line JK is then measured and
is thus the peak width at half height. The product of CD and JK
is the exact area of the triangle CLM. It is a close approximation of the true area of the chromatographic peak. It includes
an area below the line JL not a part of the peak but it excludes
some peak area above the line JC that is a part of the peak. To
the extent these areas compensate each other, the area of the
triangle CLM is equal to the area of the chromatographic peak.
   If the baseline is sloping for any reason the measurement becomes a bit more complicated. Figure 4.9 is constructed as Figure 4.8 with the same parts of the peak and construction labeled
with the same letters. The baseline, AB, is constructed as the
best extension of the baseline before and after the peak. The
peak height, CD is constructed vertically from the peak maximum
to the baseline. The midpoint, H, is located as before. Line JK
is then drawn through point H and parallel to the baseline, AB.
The desired peak height is the distance CD. However, the width
at half height is not the distance JK; it is only the horizontal
segment of this, namely, the distance NP. Thus the verticals JN
and PK must be constructed from points J and K. The true width

at half height is then the distance NP. This would be the width
measured with no slope in the baseline. Note that the distance
JK could be used if it is corrected by the cosine of the angle
JHN or the angle of the baseline to a true horizontal. The important point here is that points J and K are the true points on
the peak one half the height of the peak up from the baseline.
What is wanted for the width measurement then is the real time
separation of points J and K which is given by the horizontal
component only of the distance between them.

    The various errors of this technique will be summarized at the
end of this section for all area techniques.

    4.5.3 Triangulation

Triangulation always involves the construction of the baseline
AB, then tangents to the peak are drawn at the inflection points
of the peak. These tangents are lines EF and EG in Figure 4.8,

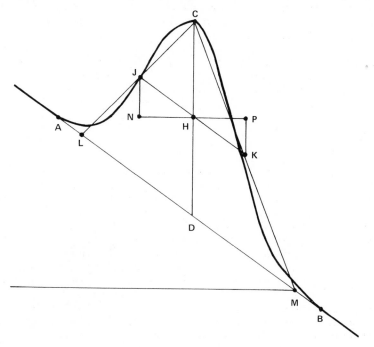

Figure 4.9. Constructions for peak size measurements
on sloping base line.

and along with the baseline, form the triangle EFG. The area of
this triangle is the height, ED, times one-half the base, FG.
This area closely approximates the area of the peak.

The comparison of the various area techniques presented later
will discuss the problem of this technique. For this reason a
more detailed discussion here regarding sloping baselines is not
warranted.

4.5.4 Cut and Weigh

The technique sometimes referred to as "paper dolls" involves
drawing the baseline, AB, as before. Then the peak is carefully
cut out of the chart paper and weighed on an analytical balance.
This weight is then converted to an area by weighing a known area
cut from the same chart near the peak. The major advantage of
the technique is that it can accommodate distorted and tailing
peaks giving a true measure of the area. The major problem is
the inhomogeneity of the paper and the destruction of the chromatogram. Certainly copies can be made prior to cutting or the
copy itself could be cut. Ball, Harris, and Habgood in their
studies (11) indicate the chart paper itself was much better than
bond used for copies as far as homogeneity was concerned.

The general errors associated with this technique are reserved
for comparison at the end of this section.

4.5.5 Planimeter

As the cut and weigh method above, the planimeter method is a
perimeter method that makes use of a surveyor's or draftsman's
instrument called a planimeter. Using this technique the baseline is drawn as usual. The perimeter of the peak is then traced
using the eyepiece containing cross-hairs of the planimeter.
When the starting point is reached the dial reads a number proportional to the area. On some planimeters the number is proportional to area via a settable scale factor. On other instruments
the factor must be determined by measuring a known area. The
major advantage of the technique is the ability to handle distorted and tailing peaks to produce a true area. The major problems
are the painstaking nature of tracing the peak and the use of a
tool not normally found in a laboratory.

4.5.6 Disc Integrator

The simple ball disc integrator attaches to the recorder as illustrated in Figure 4.10. The integrator pen draws a trace on
about 10% of the chart leaving 90% for the chromatogram as drawn
by the recorder pen. The integrator pen is linked mechanically
to the ball through the cam and roller and the ball rides on the
disc that rotates at a fixed speed. When the recorder pen de-
flects the ball (which is linked to the recorder pen drive), it
moves away from the center of the disc and begins to rotate at a
speed proportional to its distance from the center. The roller
begins rotating at the speed of rotation of the ball. The cam
then causes the integrator pen to oscillate on the chart at a
rate proportional to the ball rotation. Thus the number of integrator pen excursions between the beginning and end of the
chromatographic peak is directly proportional to the peak area.
A single excursion is assigned a value of 100 counts. A partial
excursion generally can be estimated to ±1 count.

A typical tracing of the integrator pen is illustrated in Figure 4.11. The 10-sec interval AB prior to the peak shows a trace
having a positive count rate of 1.8/sec. This means that the
baseline is simply not at the electrical zero of the integrator
and counts are being generated for no peak. This is not uncommon
because of the sensitivity of the integrator. A correction can
be applied to the peak area for this small offset by determining
the counts per second, 1.8 in this case, and multiplying by the
peak width in seconds. In this case the correction is subtracted
since the drift is in the same direction as the peak. The

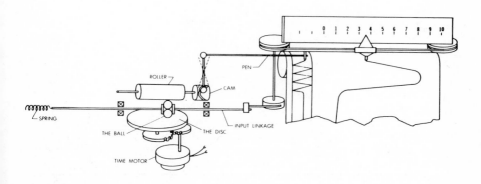

Figure 4.10. Mechanical schematic of disc integrator
(courtesy Disc Instruments).

vertical lines at points B and F are extended down from the chromatographic peak at its start and finish and in this time frame
represent 29 sec. From point B to point C, 54 counts are recorded. From C to D (two full excursions) 200 counts have been recorded. Note that at points D and E the integrator pen produces
a "blip" above the line level. The pen does this every 600
counts to ease the counting effort and to eliminate the need to
speed up the chart speed simply for counting convenience on high
peaks. Thus between C and E 800 counts have been received (eight
full excursions of the integrator pen). From E to F 70 counts
were recorded making the total for the peak (from B to F) 924
counts. In the 10-sec interval from F to G after the peak 7
counts were obtained or an offset rate of 0.7 counts/second.
Since this rate is different than the 1.8 counts/sec at the front
of the pen, we are told that the baseline is drifting slightly
negative. In this case it is now closer to the electrical zero.
If the drift is assumed constant during the peak the average offset is then 1.25 counts/sec. Since the peak width was 29 sec the

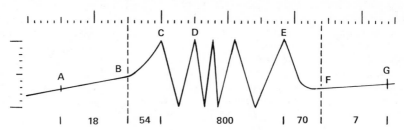

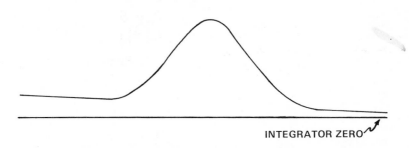

Figure 4.11.   Interpretation of disc integrator readout.

correction would be 36.25 counts to be subtracted from the 924
counts for the peak. Thus the correct integrated peak area is
888 counts.

The mechanical ball and disc integrator is rapid, precise, and
accurate. Reproducibility between instruments, individuals, and
laboratories is really quite good. The integrator itself is
quite economical (less than $1000) but it does require a recorder
compatible with the integrator. The limitation in the accuracy
appears to be in the recorder itself since it is the recorder pen
drive itself that actuates the integrator. Thus any dead band or
damping in the response of the recorder will be reflected in the
performance of the integrator.

4.5.7 Electronic Integrators and Computers

Electronic integrators are fed the detector signal directly without attenuation. This voltage signal first goes to a voltage to
frequency converter which generates an output pulse rate proportional to the voltage. These pulses are then counted over the
duration of the chromatographic peak. A slope detector senses
when the peak begins and ends. During this interval the pulses
are accumulated and then printed out as a measure of the peak
area. Most integrators will also print the time of the peak maximum at the time the area is printed.

The major advantages of the electronic integrator are the
speed and accuracy with which the area is obtained. These devices operate on the detector signal only and are thus limited
only by the detector. Their wide dynamic range permits the integration of both trace and major components without attenuation.
The high count rate and sensitive voltage detection insure accuracy well beyond any other mode of peak measurement. The only
drawback of electronic integrators, if indeed it can be called a
drawback, is their cost. They start at slightly under $2000 and
go up from there. In general the cost is proportional to the
features and capability of the instrument.

The output of electronic integrators can be readily fed into a
computer system. Here the raw area data can be compared to calibration data, calculations made, and complete analysis printed
out. An alternate is to feed the detector output voltage directly into a computer using the proper interface. The capability of
the computer can be used to integrate and calculate results.

The technique of electronic data handling is somewhat of a
speciality in itself and is reserved for a special chapter (Chapter 8) in this book. Due to the wide variety of electronic integrators designed especially for GC presently available, it is

difficult to discuss them in much more detail without considering
each individually. The individual manufacturers provide detailed discussions of the features of their hardware and manuals
on their proper use. This is the best source of any further detailed information. Additional information may be found in Chapter 6.

4.5.8 Comparison of Peak Size Measurements

In the fall of 1966 a survey of over 1600 practicing gas chromatographers in the United States was made and reported by Gill and
Habgood (12) on measurement techniques then in use. These are
reported in Table 4.2. Even though 10 years have passed since
this survey the technique of GC had been fairly well established
by then. The large number of respondents would also tend to support the validity of the data. In this survey the technique of
triangulation would have to be assumed to include the technique
of height times width at half height. If one were to speculate
how this has changed over the decade since the survey, one would
have to conclude that use of the direct integration techniques of
electronic integrators, computers, and tape systems has certainly
increased. This is probably at the expense of planimeter, ball
and disc integrator, cut and weigh and triangulation, in that
order.

TABLE 4.2   PEAK SIZE TECHNIQUES IN USE IN 1966[1]

Technique	Relative Usage(%)
Peak height	28.0
Triangulation	16.9
Planimeter	15.5
Cut and weigh	6.4
Disc integrators	20.8
Digital electronic integrators	8.5
Computers	2.4
Tape systems	1.4

[1]Data of Gill and Habgood (12).

Ball, Harris, and Habgood (11, 13, 14) in a series of papers
have considered the various manual techniques for the peak size
measurements. The treatment was both theoretical and experimental in that standard peaks were given to a sizeable group for manual measurement. Five techniques were studied: peak height,
height and width at half height, triangulation, planimetry, and
cut and weigh. The summary of actual measurement errors are
listed in Table 4.3. Peak shape is defined as the peak height
divided by the peak width at half height. In manual methods of
peak area measurement accuracy and precision of measurement degrade considerably as the peak shape becomes extreme. This
occurs with very sharp peaks (peak shape greater than 10) or very
flat peaks (peak shape less than 0.5). Thus there is an optimum
peak shape that the chromatographer should strive to achieve.
The attenuator of the chromatograph can be used to adjust peak
height and the chart speed can adjust peak width. In all cases
of measurement the bigger the area the better the precision of
measurement. In the case of peak height the narrower the peak
for a given area the better the precision. In all cases it is
best to record the peak at the minimum attenuation (maximum sensitivity)and still maintain the peak on scale. Since most chromatographic attenuators work in steps of two this places the peak
height between 50 and 100% of the chart width. For manual area
measurements the chart speed should then be selected to give an
optimum peak shape between 1.0 and 10. Note that increasing the
chart speed for peak height measurements accomplishes nothing.
    The point should be well made here that the errors mentioned
in Table 4.3 are for the measurement technique only. They do not
represent the precision expected of the analysis.

TABLE 4.3   CONDITIONS FOR LEAST ERROR IN PEAK SIZE MEASUREMENT.

| Measurement Technique | Relative Least Error(%) | | Peak Shape[1] for Least Error |
	1.5 cm^{2}Area	15 cm^{2}Area	
Peak height	<1	<0.5	>1
Height x width @ 1/2 H	2	0.5	5
Triangulation	3.5	1.5	1
Planimeter	4	0.6	1-10
Cut and weigh	3.2	2	1-10

[1]Peak shape defined as peak height/width at half-height.

Condal–Busch (15) has pointed out that triangulation gives 97% of the true area of a Gaussian peak while height and width at half height gives only 90% of the true area. However, since chromatographic peaks are not truly Gaussian the error is less. Condal–Bush also concludes that since standards and unknowns are measured the same way this error becomes insignificant compared to the actual measurement error itself.

McNair and Bonelli (16) report on a study made comparing a number of techniques for area measurement wherein the entire chromatographic system was analyzed. An eight-component sample was used. The relative standard deviation of 10 replicate analyses by the different techniques is recorded in Table 4.4. In general the data in Tables 4.3 and 4.4 are consistent if one remembers Table 4.3 is measurement technique precision only and Table 4.4 is entire system precision.

TABLE 4.4   PRECISION OF AREA MEASUREMENT

Measurement Technique	Relative Std. Dev. (%)
Planimeter	4.06
Triangulation	4.06
Height x width @ 1/2 height	2.58
Cut and weigh	1.74
Ball and disc	1.29
Electronic	0.44

The problem of peak measurement on a sloping baseline has to be considered. On peak height, planimeter, and cut and weigh techniques this is only a problem of drawing the proper baseline under the peak. As discussed previously, it is more complex for height-width and ball and disc techniques. In the case of electronic integration, severe error may be introduced or accurate correction may be made simply depending on the features and capability of the particular integrator. Errors of this type and their solutions have been discussed in detail (17–21).

A final point of consideration is the measurement time required. Certainly electronic integration is by far the fastest and the ball and disc integrator would also be considered fast. The manual methods in increasing slowness would be peak height, height and width, triangulation, planimeter, and finally cut and weigh technique.

In evaluating all of the above, references (11-21) especially
(11) and (14) and current practices the following conclusions
have to be made regarding peak size measurement.

1. Where the time saved and accuracy obtained can justify the expense, electronic or computer integration is the preferred
   approach. In general an integrator capable of handling drifting baseline and fused peaks accurately, though more expensive, is much preferred to the less expensive integrators
   without these capabilities.
2. The ball and disc integrator is capable of excellent results
   on all but the most exacting analyses. The recorder used with
   it should be of top quality and in excellent working order to
   obtain full capability of the integrator.
3. Peak height due to its simplicity, speed, and inherent measurement precision is the preferred manual method. Chromatographic conditions are much more critical here than in any
   other measurement technique. Today's instrumentation helps in
   this regard. However more frequent standardization is the
   real solution.
4. The time required and the difficulty of accurate tangent construction makes triangulation a method that cannot be recommended under any circumstance.
5. Height-width is the preferred manual area technique assuming
   reasonable peak shapes.
6. Perimeter methods should be used on irregularly shaped peaks.
7. The cut and weigh method is quite time consuming. However
   with adequate control of variable paper density it has real
   value for irregularly shaped peaks.

4.6   STANDARDIZATION

   4.6.1 General

With techniques of peak measurement in hand the next important
step in quantitative analysis is to convert the size of the peak
into some measure of quantity of the particular material of interest. In some fashion this involves chromatographing known
amounts of the materials to be analyzed and measuring their peak
sizes, then, depending on the technique to be used, relating the
unknown peaks to the known amounts through peak size.
   There is always the question of standards (known amounts of
material generally in a matrix) regarding their preparation in

the laboratory versus the purchase of ready-made standards. In
general standards should be as close to the unknown samples as
possible not only in the amounts of the materials to be analyzed
but also in the matrix of the sample itself. In all cases this
requirement would dictate the preparation of standards in the
laboratory. There is also the question of stability of the standards. With elapsed time loss of either the matrix (such as hexane evaporation from a solution of pesticides in hexane) or the
components of interest (such as adsorption of xylene on container walls of a 50 ppm standard of xylene in air) cause the standard to be unreliable. In general without prior knowledge this
dictates that standards be prepared, used, and then discarded all
within a short period of time.

Generally it is much easier to buy gas standards already prepared and analyzed. Experience here would indicate that these
standards be viewed skeptically until credibility has been established for a given source for these standards. Certainly
rather specialized equipment is needed to prepare a gas mixture
with known concentrations of components, but in some cases this
is the only reliable way to obtain standards. This discussion is
amplified as follows.

The question of purity arises regarding materials used to prepare standards. Two problems occur here, the purity of the component of interest and the purity of the matrix. Fortunately, GC
can be used to check purity of chemicals in a reasonable fashion.
If a small (1 µl) sample of a "pure" liquid is injected into a
chromatograph and the detector system operated a reasonably high
sensitivity, impurities will be observed. Without even identifying these impurities it is generally possible to make some
comment on the purity of the chemical relative to its use in a
standard. This does require the use of a general type detector
(such as thermal conductivity) rather than any of the specific
detectors. If no impurities are observed where one might be expected to see approximately 0.05% of most materials, then it is
not too unreasonable to assume that the purity is better than
99.5%. This could certainly be used to build a standard well
within ±1% accuracy assuming no other problems. There are a number of loopholes in this approach. Certainly the column system
is overloaded for 1 µl of essentially a pure component. This
causes the major peak to broaden and to obscure possibly an impurity very close to the major peak. In general, suspected impurities will be close to the major constituent due to similarities in chemical properties and boiling points. The advantage of
checking compounds to be used as standards by chromatography is
that contamination can be detected. This contamination may have
been introduced by previous users of the chemical not using good
analytical technique, by the inadvertent use of unclean contain-

ers, and possibly by mislabeling.

In general one cannot be too critical regarding standard purity. However a realistic approach must be taken. If an analysis
is required at the 10 ppm level to ±1 ppm (10% relative) it is
not reasonable to spend time and money obtaining standards with
reliability to better than 1% and perhaps even 5% would be sufficient.

A reasonable approach to any standard preparation is to obtain
the best accuracy in the standard that one can get quickly, and
then see if this accuracy will be the limiting factor in the final analysis. If, however, it may be limiting then further work
is needed to improve the accuracy of the standard. In this light
the separating power of GC should not be overlooked. The trapping techniques discussed for qualitative analysis by other techniques can be used to isolate small amounts of pure material for
standard preparation.

With this introduction to standardization, three techniques
will be discussed and the use of standards will be covered in
each technique.

4.6.2 Internal Normalization

To use the technique of internal normalization a sample is injected into the chromatograph and peaks obtained for <u>all</u> of the
sample components. Generally, area measurements are used for all
peaks though peak heights can be used. The basic calculations
are shown in Table 4.5 for an assumed sample containing four components, A, B, C, and D. If the peak areas are simply added one
can calculate the area percent. This, however, does not take into account the fact that different materials will have different
responses in the detector for a given weight. These different
responses may be determined either absolutely as area per unit
weight or may simply be determined relative to each other for a
given analysis. The latter approach is indicated in Table 4.5.
Here the standard is prepared by adding known weights of the pure
components to each other and calculating the weight percent as
shown. The standard is then chromatographed and the areas of the
four peaks are measured. Area percents are listed to show their
relation to weight percents for this mixture. The areas are then
divided by weight percent to give a response per unit weight (relative). Component A was chosen as a reference and assigned a
response factor of 1.000. The other weight responses are thus divided by 194.6, the weight response of component A, to determine
the response factors relative to component A. Generally speaking
these response factors should be constant as long as the operat-

ing conditions of the detector remain constant. The flame ionization detector is relatively insensitive to flow and temperature
changes making it almost ideal for internal normalization. One
should be able to reproduce these response factors over long periods of time with flame detection.

At this time the peak areas in the unknown can be corrected
for individual variation in detector response by dividing each
area by the response factor for that component. The result is
then a corrected peak area. The weight percent is then determined by dividing the corrected area by the total corrected area
and multiplying by 100.

The biggest advantage of this technique becomes obvious when
one notices that the amounts of the standard and the unknown
actually injected did not enter into the calculation nor was it
assumed that these amounts were equal. Aside from some of the
manual methods for peak size measurement, the ability to know the
amount of sample actually injected provides the largest error
source in quantitative analysis by GC. Good technique would dictate that an attempt be made to inject identical amounts of both
standard and unknown. This requires a uniform effort then on the
part of the chromatographic system regarding injection vaporization, sample loading on the column, and response in the detector.
To improve accuracy two standards may be prepared such that the
component levels in the unknown are bracketted in the standards.
Results obtained using this technique on round robin samples was
reported by Emory (22). This paper also provides some excellent
data on various methods of peak measurement.

The major disadvantage of this technique is that the entire
mixture must be separated and detected in the chromatographic
system. All peaks must be standardized via response factors
whether their analysis is needed or not. Internal normalization
also requires that a detector be used that responds somewhat uniformly to all components. This technique cannot be used with
electron capture and flame photometric detectors, for instance.

With care internal normalization can be used where peak size
is measured by height instead of area, though this is rare. The
response factor is now subject to slight variations in column
temperature, injection technique, carrier flow, and the like, all
mentioned under peak height measurement previously. This
approach requires that the standard mix for response factors to
be run as close in time to the unknown as possible and new response factors to be determined each time. Note also that response factors determined from area measurement in no way are the
same for those determined from peak height.

The above example has the four components in approximately the
same concentrations. This is certainly not necessary and in
practice is seldom attained. However a major concern using this

TABLE 4.5   INTERNAL NORMALIZATION

Standard

Component	Wts. Taken	Wt. (%)	Peak Area	Area (%)	Area wt.%	Response Factor, F
A	0.3786 g	21.74	4231	22.41	194.6	1.000
B	0.4692	26.94	5087	26.94	188.8	0.970
C	0.5291	30.38	5691	30.14	187.3	0.962
D	0.3648	20.94	3872	20.51	184.9	0.950
Total	1.7417 g	100.00%	18881	100.00%		

Unknown

Component	Peak Area	Area (%)	Area F	Wt. (%)
A	3962	20.98	3962	20.33
B	5791	30.66	5970	30.65
C	4926	26.08	5120	26.28
D	4206	22.27	4427	22.73
	18885	99.99%	19479	99.99%

technique is that the chromatographic system can handle the absolute amounts injected of all components in a linear fashion.
This means that the detector systems must still be responding
linearly to the absolute amount of each component even if one represents 99% of the sample and is not the component of interest.
Certainly smaller sizes can be used but here again practicality
enters in.

One way to avoid the nonlinear problem is to dilute the standard and unknown with a compatible solvent that is fully resolved
chromatographically from all of the sample components. These dilutions need not be accurately made or need not be the same for
the standard and unknown. Good practice dictates that they be
approximately the same for each. This is merely a technique for
injecting a smaller amount of the standard and sample into the
chromatograph. Since the calculations do not involve sample size,
this dilution is not a factor; the solvent and any solvent impurity peaks are not measured and are not to be considered in the
calculations.

In theory the internal normalization technique may appear ideal. But when analyzing real-life samples which may contain many components, some of which may be unresolved chromatographically and of no interest to the analyst, one of the other two techniques has advantages and is generally employed.

4.6.3 External Standardization

The technique of external standardization involves the preparation of standards at the same levels of concentration as the unknowns in the same matrix as the unknowns. These standards are then run chromatographically under identical conditions as the sample. A direct relationship between peak size and composition of one or more components can be then established. The unknowns are then compared graphically or mathematically to the standards for analysis.

This technique allows the analysis of only one component or several in the same sample. Standards can be prepared with all components of interest in each standard and the range of composition of the standards should cover the entire range expected in the unknowns. The peak size is then plotted against either absolute amounts of each component or its concentration in the matrix, generally the latter.

Figure 4.12 shows a typical calibration curve for four methyl ketones in an air matrix in which peak heights were used as the size measurement. Notice that at some of the higher concentrations the actual chromatograms were obtained at sensitivity (attenuation) settings different than the lower concentrations. These are all related to a fixed attenuator setting by multiplying all sizes by the attenuator setting used for that peak. The peak sizes are then the values (either height or area) that would be obtained if the chromatogram were run at an attenuator setting of 1, the maximum sensitivity, and the recorder chart large enough to keep the peak on scale. This is why chromatographers refer to attenuator settings as "times 32" or "times 512," and why attenuators are marked with increasing numbers for decreasing sensitivity. The peak size times the attenuator setting gives the peak sizes at a constant sensitivity (attenuator set at 1). In some cases it is more convenient to relate the peak size to an attenuation other than 1. This can be done by dividing the size at attenuation 1 by the desired attenuation, as was done in Figure 4.6 for an attenuation of 16.

Five separate standards were used to prepare Figure 4.6 and all four components were present in each standard. Two different calibration scales were used to separate the curves for ease of

identification. Two very important items can be learned from the
calibration curve. In general (and in Figure 4.6), the curves
are straight lines, and secondly they pass through the origin.
These two requirements are most important for they determine that
under the conditions of analysis and over the concentration range
covered, (1) the column has not been overloaded, (2) the detector
has not been overloaded, (3) the electronics are responding linearly, and (4) there is no apparent component adsorption in the
injection port, the column, the detector, and associated plumbing.

   At some point in any system, as the amount of component doubles, the peak size will not quite double. The column may overload distorting peak shape, the detector capacity may be exceeded
or some other phenomena. Where possible one should operate below
this point by using a smaller sample size or by diluting the
sample. While it is possible to do quantitative analysis in a

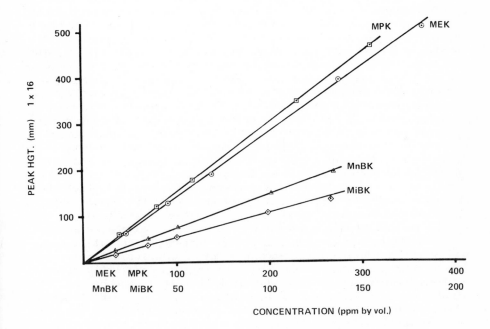

Figure 4.12. Calibration curves for four methyl ketones in an air matrix: 2-butanone (MEK), 2-pentanone
(MPK), 2-hexanone (MnBK), 2-methyl, 4-pentanone (MnBK).

region where the system is nonlinear, this requires that the calibration curve be very well defined in the nonlinear region,
meaning a large number of standards. It also means the calibration curve has to be redefined each time unknowns are analyzed.
This obviously is quite time-consuming and should be avoided if
at all possible.

Adsorption problems and/or sample degradation are generally
the cause of the calibration curve's not extrapolating through
the origin. These often can be avoided by proper sample handling
and by proper choice of columns, both materials of construction
and packing. It is possible to work with a calibration curve
that does not pass through the origin, but this also requires
that the calibration curves be generated quite frequently.

It is generally possible to obtain calibration curves as in
Figure 4.6 where the concentration region of interest is linear
and where the plot extrapolates through the origin. When one is
satisfied that these two conditions are indeed met in a given
analytical system, it is not necessary to regenerate these curves
frequently by running various concentrations of standards.
Slight changes of flowrates and temperatures of the detector and
column may change the sensitivity of the system and perhaps even
the response relationship between various components in the
sample, but they will not change the linearity and origin situation. For day-to-day calibration of the same system one needs to
run only one standard and simply ratio concentration and peak
size for each component of interest between the standard and the
unknown. What one does, in effect, by this approach is to re-
calculate the slope of the calibration curve with the new standard. However, with any new system or any new analysis the two
basic requirements should again be verified by running several
concentrations and plotting the calibration curve.

The major problem with external standardization is that the
sample size of the standards and unknowns must be known accurately. One should attempt to make them equal so that the size of
the standard and size of the unknown divides out of the calculation. If sample size varies slightly the peak size must be corrected to unit sample size for standards before the calibration
curve is plotted and for unknowns before the calculation is made.
Sample size obviously enters into the calculations. As stated
earlier reproducing and measuring sample size is the biggest
single error in quantitative analysis by GC. Considerable attention must be given to this technique of sample injection and will
be done later in the discussion of general errors. It should be
noted that, in the generation of calibration curves, it is
absolutely unacceptable to vary the amount of a component injected by varying the amount of a single standard rather than by using the same amount of different standards having different con-

centrations. There is no doubt that doubling the sample size
doubles the absolute amounts of each component injected into the
chromatograph. But there is no guarantee that the chromatographic system will double the response obtained in the presence of
double the amount of the matrix. Sample sizes for all standards
and unknowns should be kept the same within the errors of size
measurement.

Due to the ease of reproducing injection volumes of gas with a
gas-sampling valve and the difficulty of appyling the technique
of internal standardization discussed below for gas samples, external standardization is the preferred approach to the analysis
of gas samples. For these reasons considerable attention will be
given to the preparation of gas standards and the problems associated with gas analysis. In many cases this touches also on the
area of trace analysis since much of the gas analysis done today
is the analysis of trace components in an air matrix.

STATIC GAS STANDARDS. All static methods involve mixing known
amounts of gases or vapors together in some form of a container.
These amounts may be measured by volumes or pressure depending on
the types of equipment available. Mixes of the permanent gases
in the percentage range are generally reliable. However, they
should not be used as primary standards without verification and
prior experience (23). These mixtures are generally analyzed and
thus become secondary standards.

Difficulties are encountered with these mixtures as the concentration of some of the components gets into the 1-100 ppm
(v/v) range. Reaction and adsorption become problems even for
gases normally considered fairly nonreactive. One report (24) of
two CO standards certified at 26 ppm and 41 ppm by the same supplier gave 51 ppm for the second one (25% error) with the instrument calibrated using the first one. Two conclusions come from
this: (a) at least one "certified" standard is wrong, and (b)
even "certified" standards should not be trusted implicitly without verification.

Recently, introduction of pretreated cylinders with proprietary coatings or treatments have shown some promise of overcoming
reaction and adsorption of even some reactive gases (25). Even
assuming that a mixture stays constant in such cylinders, there
is still the problem of knowing the true concentration. If the
mixture does not stay constant, the situation is impossible.

When using compressed gas standards, extreme care is needed in
the hardware used between the cylinder and the injection of the
standard into the chromatograph. In many cases the cylinder supplier can recommend the proper valving and regulators to use.
The question here is not just, "what is safe?" but in addition

what will not add to or subtract from the standard gas passing
through it. In some cases valves rather than regulators should
be chosen. For safety's sake one should not rely on the cylinder
valve for control.

Standards are available today in small pressurized cans that
are extremely convenient to use with a gas-sampling valve for injection. Again supplier reliability with verification are a
must.

Laboratory preparation of standard mixtures can be made. In
general the static methods are used only for low concentrations
in a matrix gas. Fixed-volume containers, made from inert materials, capable of being sealed, and having a resealable septum
system can be used. One-gallon glass jugs with the lid modified
for a septum are very common. On small containers the volume can
be determined by weighing the container before and after filling
with water and then converting the weight of water to a volume.
In some cases quite large containers are used and here the volume
is generally calculated from measured dimensions. In either case
some means has to be provided to facilitate the mixing of the
mixture to provide homogeneity. Diffusion is not sufficient. In
small containers this can be a piece of heavy-gauge aluminum foil
that can be shaken in the container. In large containers it is
generally a fan blade or blower. The container is thoroughly
flushed with the matrix gas until it is reasonable to assume the
container has matrix gas only. It is then sealed and a small
volume withdrawn through the septum for analysis by GC. This is
to ensure that the matrix in the container is free of the component to be added to within the error of the needed standard. For-
getting this simple check can cause many problems and wasted effort. For gases, a gas-tight syringe is flushed thoroughly with
the component to be added, filled with the needed amount of pure
component, and then emptied into the container through the septum. The concentration is simply a ratio of the volumes as
shown:

$$\% \ A = \frac{\text{vol A added}}{\text{Container vol}} \times 100 \tag{4.4}$$

or,

$$\text{ppm A} = \frac{\text{vol A added}}{\text{Container vol}} \times 10^6 \tag{4.5}$$

The concentrations are volume or mole percent or parts-per-mill-
ion (v/v). This is the usual method of presenting gas concentrations as opposed to those on a weight basis. The container must
be thoroughly mixed to ensure homogeneity. The two major sources
of error of this technique come from lack of good mixing and not

being sure the syringe volume used contained 100% of the desired
component.   One never knows when both conditions have been satisfied; therefore, overcaution is the word.

Known concentrations of vapors can be prepared in the same way
by injecting a known volume of a volatile liquid into the container using a microliter syringe normally used for liquid sample
injection into a gas chromatograph.   The density and molecular
weight of the component are needed for the calculation.

$$\text{ppm A} = \frac{\text{Vol A x den A x 24.45 x } 10^6}{\text{MWA x Cont. vol}} \tag{4.6}$$

where Vol A   = µl of A added,
      den A   = density of A (g/cm^3),
      24.45   = molar volume at $25^\circ C$ and 760 torr,
      MWA   = molecular weight of A (g/mole),
      Cont.vol = volume of container (cm^3).

The liquid syringe must be touched against the side of the container or the foil to get the final amount of injected liquid off
the needle prior to its withdrawal from the container.   The container must again be thoroughly mixed to ensure both complete evaporation of the liquid and a homogeneous mixture.   If the temperature and absolute pressure of the matrix gas in the container
are different than $25^\circ C$ and 760 torr (the conditions of the molar
volume used), either the container volume must be corrected to
these conditions or the molar volume must be corrected to the
conditions of the matrix gas.   Differences of $3^\circ C$ or 7 torr
cause a 1% error.   Generally larger differences should be corrected.   The important point to remember is that the volume of the
vapor (calculated in the equation via liquid volume, density,
molecular weight, and molar volume) must be at the same temperature and pressure as the matrix gas in order to calculate a volume ratio such as volume percent or volume part-per-million.

Several gases or vapors may be added to the container by
either technique to provide standards for a number of components.
A disadvantage of a fixed volume container is that the sample is
depleted as withdrawals are made.   Generally about 10 cm^3 would
be withdrawn to adequately flush a 1 cm^3 volume of a gas sampling
valve.   Two such withdrawals will deplete a two liter container
by 1%.   This depletion will cause a dilution of the standard air
either from small leaks in the container or as the syringe is
withdrawn with the sample under reduced pressure.   Adsorption
with time can be a serious problem especially with vapors.   Generally, the best practice is to prepare the standard using intermittent mixing over a period of 15 - 30 min.   Then the chemist
should use the standard perhaps in duplicate or triplicate and

and discard the standard. Unless experience has indicated a
longer period of stability for a given system, these static standards should be trusted no longer than 1 hour.

Plastic bags have been used to overcome the problem of fixed
volume (26-28). However, other problems are introduced. The
volume of the matrix gas must now be measured accurately each
time a standard is made. Generally this is done by filling the
bag with a constant flowrate for a fixed period of time. Components can be added to the matrix as it is flowing into the bag.
Mixing may be done by gently kneading the bag. Calculations are
the same as those for the fixed volume container. Adsorption
problems can be considerable depending on the components and the
bag material. Bags are also quite susceptible to small leaks
that can cause serious error, especially in the volume of matrix
gas added. The same time frame of standard preparation and use
in fixed volume containers would be reasonable to apply to bags
without other experience.

DYNAMIC GAS STANDARDS. Dynamic methods are basically flow dilution systems providing a continuous flowing calibration gas. In
this approach two or more pure gases flow at a constant, known
flowrate into a mixing junction. Dynamic standards have two maj-
or advantages that make the technique desirable and worth the
effort to set them up. The first is that adsorption problems are
virtually eliminated in the generation and sampling systems because of the constant flowing system. This is extremely important in the preparation of trace standards of reasonably polar and
adsorptive materials. The second advantage is that the flowrate
of one or more of the components can generally be easily changed,
thus providing various concentrations of standards for calibration curves. This becomes important in the initial evaluation of
a system for analysis.

For mixtures in the percentage range, the dynamic mixing technique is reasonably straightforward. Flows can be accurately
controlled and using a technique such as a soap film flow meter
can be measured reasonably accurately. In general, however, continuous in-line flow meters are used, the most common being rotameters. It is a very unusual rotameter that can be read and set
to within 1% accuracy over even 50% of its scale. Too often the
rotameter is read and the value for flowrate is assumed accurate
without fully appreciating the reading error involved. In general, the biggest error in dynamic standards is the lack of accuracy in one or more of the flowrates. Again, one has to be concerned about pressure and temperature of the gases and that these
are the same either by actuality or by calculation. One should
also be concerned with these two variables as they affect the

means used to measure the volumetric flowrate.

As mentioned earlier, the simple form of dynamic dilution works well in the percent range. However, attempts fail to produce a 5 ppm methane in air standard by mixing a 1 cm^3/min methane flow with a 200 ℓ/min air flow simply because of the problems of measuring the high and low flows accurately and conveniently. Generally a double dilution technique works here. First a dynamic standard of 2000 ppm is generated by a flow of 15 cm^3/min methane and 7.5 ℓ/min air. Then 20 cm^3/min of the 2000 ppm standard is mixed with 8 ℓ/min air to produce 5 ppm. Properly, the air flow to produce the 2000 ppm standard should be 7.485 ℓ/min (7.5-0.015). The total flow of the 2000 ppm is then 7.5 ℓ/min. The same goes, of course, for the second dilution. The equation used to keep these concentrations straight in successive dilutions is as follows:

$$F_1 \times C_1 = F_2 \times C_2 \tag{4.7}$$

where F_1 = flowrate of concentration C_1 (cm^3/min),
      F_2 = flowrate of concentration C_2 (cm^3/min).
Thus for the second dilution above

$$20 \ cm^3/min \times 2000 \ ppm = 8000 \ cm^3/min \times 5 \ ppm$$

The accuracy of multiple dilutions fades as more and more dilutions are made because of the added errors of additional flow measurements. In the double dilution above, four flow measurements are needed, two for each dilution. Fortunately, however, multiple dilutions are used to produce low concentrations where perhaps analysis accuracy of ± 10% would be acceptable.

Low flowrates of gases can be delivered to a larger volume flowrate of a diluent gas by the use of small motor-driven syringes (29). This is one way of accurately delivering low volumetric flowrates. Generally periods no longer than an hour are used since the syringe must be refilled. Back-diffusion of the diluent gas into the syringe volume at low delivery rates is a problem here. Also the downstream pressure of the standard thus prepared cannot change since this can cause a pumping action in and out of the syringe volume.

A technique of making known vapor concentrations of reasonably volatile liquids in a diluent gas involves the use of the vapor pressure of the liquid (30). The diluent gas is passed through successive thermostatted bubblers obtaining a mixture determined by the saturation vapor pressure. Thus for ethanol, if the bubblers were maintained at 20°C (ethanol vapor pressure at 20°C is 43.9 torrs and the diluent gas flow maintained low enough to ensure saturation, a dynamic standard is generated with the following concentration

$$C = \frac{Sat.\ V.P.}{Total\ P} \times 100 = \frac{43.9}{760} \times 100 = 5.78\% \qquad (4.8)$$

At this temperature the vapor pressure changes about 5% /^oC requiring bubbler thermostatting to better than $\pm 0.2^oC$ for a 1% standard accuracy. It also is important to know accurately the total pressure at the final bubbler since this is also used in the calculation. This was assumed to be 760 torrs to illustrate the preceding calculation, but must be measured in practice.

Using the vapor pressure technique, two ways can be used to change the concentration. First, the vapor pressure can be changed by changing the temperature. This can be quite time-consuming in that true thermal equilibrium is required for each concentration. Also, the temperature must be kept lower than any subsequent temperature the developed standard will see to prevent condensation and thus a loss of the standard. The second way is to change the total pressure under which the bubbler system is working. Since this can only be done for pressures greater than any subsequent pressure to which the standard will be exposed, it can require a sophisticated experimental setup. In general, for multiple concentrations one standard is prepared in this fashion and is then diluted by a second diluent gas stream. This requires that both the original bubbler flow and the diluent flow be accurately measured, whereas with the single concentration provided by the vapor pressure system, the bubbler flow need not be known accurately as it does not enter into the calculations. One only has to be assured that the gas is saturated.

Another approach to vapor standards is to use the diffusion of vapor through a capillary to add small amounts of vapor to a flowing gas stream (31-33). The theory and practice are reasonably well defined. The concentration is determined by knowing the rate of diffusion and using the following equation:

$$C = \frac{R \times K}{F} \qquad (4.9)$$

where C = concentration (ppm, v/v),
   R = diffusion rate (μg/min),
   F = diluent gas flowrate (ℓ/min),
   K = 24.45/mol. wt. (μl/μg) at 25^oC and 760 torr.
Once again the diluent flowrate must be at the same conditions as the K factor used or vice versa. To ratio gas volumes they must be measured at the same temperature and pressure. The K

factor simply converts the diffusion rate in weight per unit time
to vapor volume per unit time.

Theory predicts the diffusion rate by the following equation
(31).

$$r = 2.303 \frac{DMPA}{RTL} \left(\log \frac{P}{P - p}\right) \tag{4.10}$$

where r = diffusion rate (g/sec),
     D = diffusion coefficient (cm^2/sec),
     M = molecular weight (g/mole),
     P = total air pressure (atm),
     A = diffusion cross-sectional area (cm^2),
     p = partial pressure of sample at T^0 (atm),
     R = gas constant (cm^3-atm/mole-^0K),
     T = temperature (^0K)
     L = length of diffusion path (cm).

By incorporating R into the constant, converting both pressure
(P and p) into torrs from atmospheres, and converting the rate
into µg/min from g/sec this equation is obtained.

$$R = 2.216 \times 10^3 \frac{DMPA}{TL} \log\left(\frac{P}{P - p}\right) \tag{4.11}$$

where R = diffusion rate (µg/min),
     P = total pressure (torr),
     p = partial pressure of sample (torr),
All other terms as above.

Using this equation with vapor pressures and diffusion coefficients from data in the literature and very accurate measurements
of area and length, the diffusion rate generally can be calculated to within about 5%. Thus the only way to build dynamic standards using the diffusion technique is to determine the rate in a
given system. One such system is to use a diffusion tube as
shown in Figure 4.13. The bulb of the tube is loaded with liquid
to about 80% of its capacity (perhaps about 5 cm^3). The capillary length is variable up to about 15 cm and the capillary diameter perhaps 0.5 cm. This capillary is placed in a thermostatted
chamber permitting a dilution gas flow across the tube. The diffusion rate is then determined by weight loss over several hours
or several days using a good analytical balance. Only the gas
flowrate need be measured then to generate a primary standard.
Diffusion rates can be measured during the life of one filling
while the diffusion tube is in use. Different materials can be
filled in the same tube; or it can be refilled with the same material. Only pure materials can be used, not mixtures. Several

tubes, however, can be put in the same gas stream to generate a
multiple standard. The concentration of the standard may be varied over a wide range by variation of the dilution gas flowrate.
This is preferred to a temperature change of the diffusion tube.
Again the temperature control of the diffusion tube is critical.
General practice is to maintain the temperature constant to within ±0.1°C.

A clever means of dynamic generation of standards at the part-
per-million level involves permeation through a polymer. In 1966
O'Keeffe and Ortman (34) described this technique primarily for
air pollution standards. A condensable gas or vapor is sealed as
a liquid in a Teflon tube under its saturation vapor pressure as
shown in Figure 4.14. After an initial equilibration period the
vapor permeates through the tube wall at a constant rate. This
rate is determined by weight loss over a period of time. Temperature must be controlled to within ±0.1°C to maintain 1% accuracy.
In use the tube is thermostatted in a chamber that permits a diluent gas to fully flush the chamber. The concentration is then
determined by the same equation used for diffusion tubes. However, since the rate is generally much less in permeation tubes
it is usually reported in ng/min.

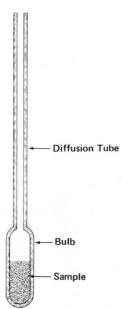

Figure 4.13. Cross-sectional diagram of a diffusion
tube (courtesy Analytical Instrument Development).

$$C = \frac{R \times K}{F} \tag{4.12}$$

where C = concentration (ppm, v/v),
       R = permeation rate (ng/min),
       F = diluent gas flowrate (cm^3/min),
       K = 24.45/MW (nl/ng) at 25°C and 760 torrs.

   Again, to accurately ratio gas volumes they must be at the same temperature and pressure. Either F is corrected to 25°C and 760 torrs or the K factor is adjusted to the conditions under which F was measured.

   Typical materials available in permeation tubes for operation at 30°C are listed in Table 4.6 along with average rates per centimeter length of tube and the K factor. As the length of the tube increases the permeation rate increases in reasonable proportion. The data in Table 4.6 are for tubes 0.25 inches o.d. and a wall thickness of 0.062 inches. A typical example might be an SO$_2$ tube 5 cm in length at a rate of 1350 ng/min. If the dilution gas flow across the tube is 1.35 ℓ/min the concentration would be

$$C = \frac{1350 \times 0.382}{1350} = 0.38 \text{ ppm}$$

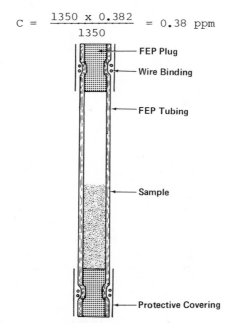

Figure 4.14.   Cross-sectional diagram of a permeation tube (courtesy Analytical Instrument Development).

If the flow is increased to 2.7 ℓ/min the concentration is cut in half. Likewise, at 675 cm^3/min the concentration would be 0.76 ppm. Longer tube length, thinner tube walls, and higher temperature all increase the permeation rate, thus the need to control tube temperature to within 0.1°C. Generally, to prevent the diluent gas from cooling the tube in practice, a low flow is passed across the tube and is then diluted with a higher flow downstream from the tube. The sum of both flows must be used in the calculation.

TABLE 4.6   PERMEATION RATES FOR CHEMICALS IN PERMEATION TUBES AT 30°C

Chemical	K	Permeation Rate (ng/min/cm)	
		0.062-inch wall	0.030-inch wall
Sulfur dioxide	0.382	270	780
Nitrogen dioxide	0.532	1150	
Hydrogen sulfide	0.719	230	
Chlorine	0.345	1350	
Ammonia	1.439	175	
Propane	0.556	35	
Butane	0.422	12	26
Methyl mercaptan	0.509		61
Ethyl chloride	0.379		55
Vinyl chloride	0.391		460
Propylene	0.582	130	

Recently, tubes for higher temperature operation containing some common industrial solvents have been introduced. Some of these are listed in Table 4.7. These permit low concentration standards to be prepared for some industrial hygiene-type analyses.

Permeation tubes are not refilled, have a limited life and cannot be turned off. However their life can be prolonged during periods of non-use by storing them in a refrigerator to reduce the permeation rate. Not many solutes are practical for use in permeation tubes. However, when the technique can be used it is generally preferred as a means of standard preparation.

Significant space has been devoted to gas standards because of the difficulty in preparing known standards. The fact that such a wide variety of techniques are in use attest to the problem.

TABLE 4.7   PERMEATION RATES FOR CHEMICALS IN PERMEATION TUBES
AT 70°C

Chemical	Rate (ng/min/cm)	K
1,1,1-Trichloroethane	113	0.183
Trichloroethylene	1060	0.186
Chloroform	715	0.205
Carbon tetrachloride	265	0.159
Methylene chloride	1690	0.288
Acetone	305	0.422
Methyl ethyl ketone	140	0.340
1,2-Dichloroethane	331	0.247
Benzene	207	0.313
Toluene	120	0.266
o-Xylene	37	0.231
Ethyl benzene	35	0.231
Cyclohexane	20	0.291
n-Hexane	162	0.284
Methanol	245	0.764
Vinyl acetate	450	0.284

On the other hand, liquid standards are quite straightforward and
reasonable analytical technique can insure reliable standards.

In general liquid standards are prepared in a solvent matrix
which should be the same as the matrix of the unknown. In many
cases the liquid may be an extraction solvent or simply a dilution solvent, depending on the type of analysis and the form of
the original sample. If the solvent choice is left to the analyst (as opposed to being prescribed by a procedure), the solvent
has to be chosen such that it does not interfere with any of the
potential sample components. For trace analysis it is important
that the solvent be checked for impurities and that these impurities will not be confused with sample components. Chromatographing the solvent at the maximum sensitivity to be used in the analysis is referred to as "blanking the solvent." It is very important to blank the solvent each time it is used to be sure it
has not been inadvertently contaminated. Also in trace analysis
it is preferred to have a solvent elute from the column following
the sample components of interest rather than ahead of the sample.

Standards are prepared by adding known weights of materials to
a volumetric flask and then diluting to volume with the solvent

or matrix. The approach is best illustrated with an example of
the analysis of benzene in toluene at the 0.01% (weight) level.
A standard is prepared by weighing 100 mg benzene into a 10-cm^3
volumetric flask. This is diluted to the mark with benzene-free
toluene. This can be used as a master standard. Each cubic
centimeter of solution contains 10 mg benzene. The master standard is then used to prepare several additional standards as
follows:

Standard	µl Master Std.	mg Benzene	% Wt/vol	% Wt/wt
1	50	0.50	0.0050	0.00577
2	75	0.75	0.0075	0.00866
3	100	1.00	0.0100	0.01155
4	150	1.50	0.0150	0.01732

Each standard is prepared in a 10-cm^3 volumetric flask and the
proper amount of master standard is then diluted to the mark with
benzene-free toluene giving the concentrations shown above. The
weight to weight percent simply assumes a toluene density of
0.866 g/cm^3. The calibration curve can now be run using the four
standards. Peak size can be plotted against absolute weight of
benzene in the injected sample or against weight percent depending on the final form needed for the unknown. Assuming a 1-µl
injection, standard #3 would provide a benzene peak for 0.1 µg
benzene. This is the convenience of preparing the dilution to
volume and calculating weight percent by density. Assuming the
density of the solution is the same as the pure toluene can introduce an error of no more than 0.01% relative at this level.
It does assume the density of the unknown is also the same as the
toluene. These assumptions should not be made for solutions in
the percent range, for at this level the standards should be prepared by weight of each component.

There are several advantages of the double standard preparation as used above. Significant amounts of toluene are conserved
and the standards are prepared by single volume measurements (except for the one weight measurement). Also, other components can
be added to the second set of standards at the time of preparation from other master standards. These components can then be
varied independently of each other.

Obviously solid samples can be made up by weight in a solvent
as above. This is generally the technique used for such materials as pesticides.

Reliability of liquid standards over a period of time is generally quite good if kept in sealed containers. They should not
be stored for any length of time in volumetric flasks but small

vials are quite convenient. However a word of caution about the caps is in order. Plastic or plastic-coated cardboard liners in vial caps pose serious problems in most cases. Solvents dissolve or leach a number of materials from these caps, generally causing gross interference with the standards. In general, foil-lined caps should be used unless these are known to produce problems. Tightness of the seal is important to prevent selective evaporation of components or solvent from the vial. Homemade inert cap liners may be inert but generally do not adequately seal the vial. Evaporation is generally the major reason why liquid standards become nonstandards. Chemical knowledge of the components should be also considered as far as reactivity and adsorption are concerned regarding the useful life of standards.

4.6.4 Internal Standardization

The technique of internal standardization may best be understood by referring to Table 4.5, which outlines the method of internal normalization. It is assumed in this instance only component C is of interest for analysis and that the unknown contains no component A. If a standard containing known weights of both A and C is prepared and chromatographed, the response factor, F, can still be determined. This is shown in Table 4.8 assuming the same weights and areas as before. In practice several standards should be made and a plot of weight ratio as abscissa and area ratio as ordinate is made. This plot must be linear for the particular system. Once the linearity is established for a given sample type and system only one standard mix need be used to define the slope of that plot. Note that the response factor, F, is the slope of that line. Therefore the standard is actually used to determine the factor, F. Notice that the response factor for C in both techniques is identical whether ratios or percents are used.

   The unknown is now ready to be run. Since no A is present in the unknown, a known weight of A component is added to a known weight of the sample.

   This mixture is then chromatographed and the area ratio of components C to A is measured. Knowing F, the ratio of the area ratio and weight ratio and the area ratio in the unknown, the weight ratio of the unknown can be calculated.

$$\frac{W_C}{W_A} = \frac{A_C}{A_A \times F} \qquad\qquad (4.13)$$

where W_C and W_A = weights of C and A, respectively,
$\quad A_C$ and A_A = areas of C and A, respectively,
$\quad F$ $\qquad$ = response factor.
Since the weight of A added to the sample is known, the weight of C in the sample can be calculated

$$W_C = \frac{A_C \times W_A}{A_A \times F} \qquad (4.14)$$

And since the weight of the sample is known

$$\% \ C = \frac{W_C}{\text{sample wt.}} \times 100 \qquad (4.15)$$

In practice, a master standard of component A and one of component C are prepared on a weight to volume basis in a solvent. Mixing known volumes of each of these two standards can provide a variety weight ratios of the two materials for the initial linearity check. The standard of component A can also be used to add a known amount of A to a known weight of sample.

TABLE 4.8 INTERNAL STANDARDIZATION

Component	Weight	Wt. Ratio C/A	Area	Area Ratio C/A	F = Area Ratio / Wt. Ratio
A	0.3786		4231		
		1.398		1.345	0.962
C	0.5291		5691		

In the preceding example area was used to measure peak size since that was the technique used in the example for the internal normalization. The point must be made that peak height can be used as the size measure just as well as peak area. The same advantages of peak height measurement are present in this method of standardization as in any other. Likewise the same requirement for frequent standardization is present.

In this instance A is referred to as the internal standard. All of the advantages of the internal normalization technique,

such as not knowing the exact sample size and the noncritical aspects of dilution, carry over to this technique. The major disadvantage of internal normalization, namely the necessity of
measuring all components of the sample, does not carry over into
this technique. The cautions under internal normalization regarding system overload apply but they apply only to the components of interest and the internal standard, not to the entire
sample.

With attention to the purity of the standards and to the lack
of interference of any solvent impurities, the precision of the
internal standard method is controlled by the ability to quantitate peak size. That certainly qualifies this technique as the
most precise method of quantitative analysis by GC, and where
precision is paramount, the internal standard technique should be
applied. Its advantages far outweigh the slight increase in
effort required for standard and sample preparation. An excellent, detailed, how-to approach for the internal standardization
technique as applied to a practical problem has been detailed by
Barbato, Umbreit, and Leibrand (35).

The preceding discussion of sample storage of external liquid
standards certainly apply to the standards prepared for the internal standardization technique. There is one further consideration in this regard and that is in the proper selection of the
internal standard for a given analysis. The first step is to
chromatograph a typical sample and identify the component or components to be analyzed. The internal standard is then chosen
such that it must:

1. Elute from the column adequately separated from all sample
   components.
2. Elute as near as possible to the desired component(s) and,
   ideally, before the last sample peak so that analysis time is
   not increased.
3. Be similar in functional group type to the component(s) of interest, or if such a compound is not readily available, an
   appropriate hydrocarbon should be substituted.
4. Be stable under the required analytical conditions and nonreactive with sample components.
5. Be sufficiently nonvolatile to allow for storage of standard
   solutions for significant periods of time.

Several attempts may be necessary to find the best internal
standard for a given analysis but the effort is worthwhile if
highest precision is needed.

4.6.5 Standardization Summary

In all three methods of standardization, standards and samples are chromatographed, the standards being known and the samples unknown. Peak sizes can then be determined for both. The difference in the three methods is in the second piece of information needed to relate the standard to the sample:

- In internal normalization this relationship is that in both the standard and unknown the analyzed peaks total 100%.
- In external standardization this relationship is the accurately known amounts of standard and unknown actually injected into the chromatograph.
- In internal standardization this relationship is the accurately known amount of different material added to an accurately known amount of the standard and unknown.

The errors associated with standardization have been discussed throughout the above section but should be summarized:

- Standard purity and known standards must be checked and not assumed.
- Linearity of response versus absolute amount injected must be confirmed for each different sample type and for each different set of chromatographic operating conditions. This linearity cannot be assumed. Nonlinearity may result from column overload, detector overload, or adsorption problems.
- Proper attention to good analytical practices is important but most especially as it regards proper "blanking" of solvents, syringes, and all sample handling equipment. The high sensitivity for small amounts of material in most detector systems increases the importance of cleanliness.

Attention has already been given to the errors associated with peak size measurement and with standardization. There are many other places in the chromatographic process where errors enter into quantitative analytical GC. Detailed analysis of most of these error sources is not possible, especially in the confines of this chapter, but they should and will be mentioned and briefly discussed. Most of the error sources are generally obvious; it may indeed seem even ridiculous that some have to be mentioned. However, the mere fact that they are obvious tends to slowly place them in the overlooked category. One has to be constantly reminded of these errors until the consideration of them becomes habitual with each problem.

These general errors can be broken into two categories. The first one is in the general area of sampling, the problems of getting the sample from where it is into the GC. The second area is the GC system itself.

4.7   ERROR DISCUSSION

4.7.1 Sampling Techniques

The methods used to obtain samples and physically transport them
to the gas chromatograph is really no different for GC than for
any analytical technique. However, since GC has the inherent
capability to do trace analysis, it becomes even more critical to
observe the best analytical sampling techniques. Some major
areas of concern are obvious.

The sample taken must be the sample that one wants to analyze.
Since very little sample is required for gas chromatographic analysis it is very easy to take a small sample that stands a good
chance of not being representative of the environment to be analyzed. Small differences in homogeneity, or lack thereof, become quite apparent on two small samples supposedly taken from
the same bulk sample.

Problems of adsorption, evaporation, and reaction of samples
following the sampling procedure prior to analysis must be considered. The discussion regarding storage and handling of gas
and liquid standards under external normalization above certainly
apply even more with the unknown samples. Time between sampling
and analysis must be kept to a minimum. In addition, this time
element should be checked with standards to insure that samples
do not change with time or to at least define the extent of the
error if no other solution is possible.

Containers for sampling, and indeed all sampling equipment,
must be checked to determine the contribution to error. This becomes especially important if the sample must undergo some processing prior to the analysis. This processing may be extraction,
preliminary cleanup by column chromatography, and even reaction
such as esterification. All of these steps must be proven in a
given system or known to insure either quantitative sample handling or the reproducibility of the processing. It is not sufficient to assume that "since Joe Blow at Podunk obtained 82.3%
efficiency in the methyl esterification of adipic acid 3 years
ago" the same efficiency is valid for a procedure that attempts
to duplicate Joe Blow's procedure today. Reaction or extraction
efficiencies must be established again.

4.7.2 Sample Introduction

As mentioned above, when a known sample size is required, as in the external standardization technique, the measurement of that sample size will generally be the limiting error in the analysis. However, there are other errors that improper sample injection can introduce into the analysis other than sample size. Thus, it will be beneficial to examine the various methods of sample injection and both types of errors associated with them.

SYRINGE INJECTION. The use of syringe is by far the most common mode of sample introduction into the chromatograph. Today there are a number of excellent syringes on the market designed for gas chromatography. The most common syringe in use today for liquids has 10 μl total volume. With today's greater use of smaller-diameter and lower-loaded columns coupled with better and more sensitive detectors, sample sizes keep getting smaller. Generally liquid samples of about 1 μl size are used. In a sample this size a component of interest should be less than 1% of the injected sample. For concentrated samples this means sample dilution with a compatible solvent. An error can be introduced here if the solvent contains impurities that have the same retention time as any component of interest or if it contains even some of the same material. As with any use of solvents in GC, the solvent has to be "blanked" before it is used.

The use of a 10- μl syringe to deliver a 1-μl volume has a certain error associated with the accuracy to which the syringe markings can be read and the plunger set. This uncertainty alone can contribute 2% to 5% error in a 1-μl volume.

Many using GC are acquainted with the problem of injecting a volatile liquid into a hot injection port of a gas chromatograph. The error associated with this phenomenon overwhelms the reading error without use of the proper technique. The basic problem is this: With the syringe properly loaded to the 1-μl mark, the amount of liquid contained in the syringe is that 1 μl in the barrel plus the amount in the needle. When the liquid is injected, the 1 μl goes into the chromatograph, but also any of the liquid remaining in the needle after injection and prior to withdrawal also evaporates. This may be the entire volume in the needle which will be approximately 0.8 μl. The actual volume in the needle can be determined by loading a syringe with a liquid, running the plunger to zero, wiping the droplet off the needle, slowly drawing the plunger back until the liquid-air interface can be seen in the barrel, and then measuring the liquid slug in volume on the syringe. Knowing this total holdup on a given syringe can permit one to measure the amount actually injected. If the needle volume is 0.8 μl and the plunger is set at 1.0 μl, the total liquid in the syringe is 1.8 μl. Following the injection the plunger is withdrawn and the amount of liquid remaining in

the needle is measured. If this now is 0.3 µl, then an amount of
1.5 µl was injected, 1.0 µl by actual injection and 0.5 µl by
evaporation from the needle. There are two problems here.
First, four syringe readings are needed (plunger and liquid-air
interface, each on initial and final syringe loading), thus giving rise to two reading errors. The second error is worse in
that its magnitude cannot be known with certainty. That is, the
amount that is evaporated from the needle may not (and generally
is not) representative of the true sample concentration due to
selective evaporation of the more volatile components of the
sample.

A technique used to overcome this selective evaporation is to
draw some pure solvent into the syringe (say 1.5 µl), then about
1 µl of air, about 1 µl of sample, and finally about 1.5 µl of
air. The sample slug is then measured in the barrel between the
two liquid-air interfaces (two syringe readings). When this is
injected, only pure solvent is left in the needle and the amount
that evaporates is not important. All of the measured sample
volume will be injected.

Another solution to liquid injection is the use of a 1-µl
total volume syringe. This syringe actually uses the internal
volume of the needle for the sample volume. The plunger is a
fine wire going the full length of the needle. The volume readout is actually accomplished on a glass barrel with an indicator
inside the barrel much the same as any other syringe. However,
the actual liquid held in the syringe is in the needle only.
Initially these syringes were plagued with leak problems. For
the past several years, however, due to design improvement, they
have been performing satisfactorily. The accuracy of a 1-µl injection is generally within 1% using these syringes, but they are
more expensive.

Finally, proper handling technique is very important, especially when it comes to wiping the outside of the needle and the
droplet at the tip of the needle prior to injection. Any residual liquid on the outside of the needle will be caught in the
septum puncture and will slowly enter the column. This produces
broad tailing, especially of the solvent, making separations
difficult as well as introducing an unknown amount of sample. On
the other hand, liquid in the needle cannot be removed via capillary action of the wiping towel.

All of the above points about liquid injection should be considered even when using a standard technique that does not require an accurate volume to be known. Selective evaporation cannot be tolerated even with the internal standard method. The
size measurement errors obvious from the above discussion certainly point to the substantial advantage of the internal standard technique for accurate analysis.

There are reasonably good syringes available today for injection of gas samples. Generally, gas samples are in the range of 1 cm^3 in size. These syringes have a very snug fitting Teflon plunger allowing a gas-tight seal between the plunger and barrel. The gas sample of known volume can be injected into the gas chromatograph using this technique. Even though there is a small (less than 1%) reading error with these syringes, there is a different form of sample injection error. Common pressures at the head of columns of average length at optimum flows will be approximately 30 psig or 3 atm absolute. When the syringe is inserted into the chromatograph through the septum, the first thing that happens is that the carrier gas rushes through the needle into the syringe volume until the pressure in the syringe is 3 atm. Even the smallest leak between barrel and plunger will cause a significant sample loss due to this increased pressure and only erratic results is any indication of this. Next the plunger is depressed injecting a mixture of carrier gas and sample. The residual volume of the syringe, which may be 0.1 to 0.2 cm^3 in a 1 cm^3 syringe, was at atmospheric pressure when the sample was measured but is now at 3 atm. If the carrier had completely and homogeneously mixed with the sample at 3 atm pressure prior to sample injection, no error would be introduced. However, this actually never happens and the extent of mixing and the error introduced by the lack of mixing is not known. Assuming no mixing and a 0.2-cm^3 residual syringe volume, the sample remaining would be 0.2 cm^3 at 3 atm, or 0.6 cm^3 at atmospheric pressure. Since the syringe contained 1.2 cm^3 originally, only 0.6 cm^3 was introduced instead of 1 cm^3. This is a 40% error on sample size!! This is certainly the worst case but the magnitude indicates that the level of error is significant even with a perfectly functioning syringe. Gas work with syringes should be carried out with a syringe and needle having a minimum residual volume.

The leak around the plunger and barrel in "gas-tight" syringes has been mentioned above. A tight, stiff acting plunger is necessary but not sufficient for a gas-tight seal. If 99% of the seal is tight, the entire sample can still be lost out of the 1% of the seal that does leak. A tight plunger can give rise to another error. If the needle plugs due to particulates in the gas sample, septum coring, or whatever, no sample will get in the syringe and the gas chromatograph. Many a "detector malfunction" has been solved by syringe needle replacement. The stiff plunger makes a plugged needle difficult to notice. One final comment on gas-tight syringes. A large number of these have replaceable needles using a standard Luer fitting. This is very convenient for economical needle replacement if the needle becomes burred, bent, broken, or plugged. However, most glass to metal Luer

fittings will leak gas at 3 atm pressure. This is a leak error
not normally considered but should be the first check placed on a
new gas-tight syringe.

GAS SAMPLE VALVE. With all the problems associated with syringe
injection of gas samples, it is not surprising that a more accurate way of injecting gas samples has been found. This system
makes use of a gas sampling valve. There are a number of these
valves on the market using either rotary or push-pull actuation
and interchangeable volumes are standard. A schematic for a
rotary valve is shown in Figure 4.15. In the load position the
volume of the valve is connected to the in and out load ports.
In use the sample is pulled through the valve by a pump, squeeze
bulb, or even a syringe used in the suction mode. If the gas is
under pressure it is allowed to flow through the valve. Sufficient volume of gas is needed to insure that the "loop" or valve
volume contains the sample to be analyzed. For a 1-cm^3 sample
loop volume generally 10 cm^3 of gas is a sufficient flush. The
valve is then rotated to the inject position. This action places
whatever is in the valve loop into the carrier gas flow where it
is carried directly to the column for separation. The biggest

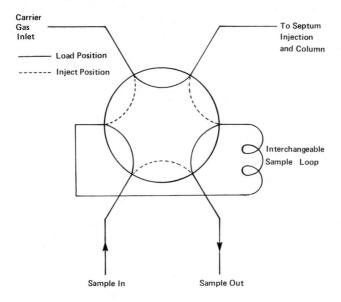

Figure 4.15. Flow schematic of a typical rotary gas
sample valve.

error, namely that of volume, is now fixed. If standards are run
with the same fixed volume as the sample, the actual volume need
not be known with a high degree of accuracy since it is the same
for both standard and unknown and will cancel out of the calculation. However, there are two other parameters that also must
be held constant in the use of a gas-sampling valve to insure
that the same amount of sample is injected of standard and unknown: temperature and pressure. Either a $3^{\circ}C$ difference or a 7-
torr difference will cause a 1% change in the amount of gas
sample. The practical solution to these variables is simply to
run standards and unknowns as close in time as possible such that
these parameters do not vary significantly. If the samples are
hot, such as stack gases, it may be necessary to maintain all
sample lines and the gas-sampling valve at an elevated temperature. Obviously, standards must be sampled at that same elevated
temperature. If the sample is under reduced pressure, it may be
necessary to load the gas-sampling valve under reduced pressure.
The pressure is then usually measured using a manometer to provide proper correction or, preferably, to permit the standard to
be handled at the same pressure.

It may be tempting to increase the loop volume to increase the
amount of sample for trace analysis. Before this is done the
system should be examined. If 1/8-inch diameter columns are being used, a reasonable flowrate is 30 cm^3/min at atmospheric
pressure. But if the pressure at the head of the column is 3 atm
(not unreasonable), the volumetric flow at the head and through
the sample loop is only 10 cm^3/min, or 6 sec/cm^3. If the loop
volume is increased to 5 cm^3 it will take 30 sec to sweep the
sample onto the column. Thus no peak can be any narrower than
30 sec and a large volume loop can completely destroy the separating efficiency of the chromatographic process. Again, as with
any analytical problem, a common-sense, logical examination of
the whole picture will point up problem areas.

MISCELLANEOUS SAMPLING DEVICES. There are a number of specialized ways of introducing the sample into the chromatograph and
these are generally designed for a specific problem. A liquid-
sampling valve is valuable for high-pressure liquid samples and
preserves the sample integrity. These valves work the same in
principle as the gas-sampling valve except that the much smaller
volume is held internally in the valve as opposed to an external
loop.

To increase the precision of the amount of sample injected, a
known weight of sample, solid or liquid, can be introduced via a
sealed indium tube. When the tube is placed in the hot injection
port the tube melts, releasing the total sample into the chroma-

tograph. The injection port is generally modified in such a way
as to allow insertion of the tube through a gas-lock system. The
same approach has also been used with sealed glass tubes which
are crushed in the injection port.

All of these different techniques have some problems of their
own but can be quite precise as far as amounts of sample actually
injected is concerned. Since their use is not general, but used
for rather specific sampling problems, no further discussion is
warranted in this chapter.

4.7.3 Gas Chromatographic System Errors

Most of the problems associated with the processing of the sample
through the column and then its detection are basically covered
in specific chapters of this book. However, some areas deserve
special mention as they relate to quantitative analysis.

The major concern is that the character of the sample is not
changed in the injection port, the column, or the detector prior
to its actual detection. Thermal decomposition, catalyzed or
thermal reaction, and adsorption of part or all of the sample
will contribute to error in the analysis. Problems such as these
may be determined by using the chromatograph itself first to detect possible problems by unexpected results and then confirmation of the problem by variation of the actual operating parameters of the chromatograph.

Adsorption problems are generally indicated by the calibration
curve not psssing through the origin, and in some cases by nonlinearity of the curve. A change of the column may be the
answer. Perhaps increased temperature will reduce the problem to
a workable level. Even though it is not desirable, some adsorption can be tolerated and still give quantitative results but
frequent recalibration is critical.

Sometimes thermal decomposition and reaction can be shown by
variation of injection port temperature, and possibly column
temperature. The only real solution is to operate at as low a
temperature as possible and perhaps use "on-column" injection.
Low-loaded columns sometimes help. Use of glass columns and
glass injection port liners often relieve the problem of unwanted
thermal degradation and may help in some cases. However, in all
cases the precision and accuracy of the quantitative analysis
will be affected until a solution is found or until a decision is
made to "live with it."

Detector errors are basically concerned with the time constant
of the detector and its linearity. The time constant certainly
can affect the peak height on narrow, sharp peaks and this may or

may not show up as nonlinearity. Assuming a good detector system, the basic linearity concern is with overload. This points to the necessity to initially establish a calibration curve and assure its linearity over the entire range of the samples. Extrapolation is dangerous.

All manual methods of quantifying peak size make use of the recorder tracing. Some consideration has already been given to the peak height and width as determined by recorder chart width and chart speed. Both should be maximized for the size measurement technique used. In addition the recorder may have a limiting time constant as far as response to rapid peaks are concerned. This possibility should be considered along with the detector when time constant problems are suspected.

In total summary, GC is an excellent analytical tool for quantitative analysis. However, common sense must be used in handling problems and the entire system should be understood. Best technique should be used to standardize and for sample handling. The "weakest link" concept is no more pronounced than it is in quantitative GC.

REFERENCES

1. A. Wehrli and E. Kovats, Helv. Chim. Acta, 42, 2709(1959).
2. W. O. McReynolds, J. Chromatogr. Sci., 8, 685(1970).
3. J. Takacs, C. Szita, and G. Tarjan, J. Chromatogr., 56, 1 (1971).
4. J. Takacs, Z. Talas, I. Bemath, G. Czako, and A. Fischer, J. Chromatogr., 67, 203(1972).
5. M. Beroza and M. C. Bowman, Anal. Chem., 37, 291 (1965).
6. C. McAuliffe, Chem. Tech., 1, 46 (January 1971).
7. A. B. Littlewood, Chromatographia, 1 (3/4), 133 (1968).
8. J. E. Hoff and E. D. Feit, Anal. Chem., 35, 1298 (1963).
9. V. Rezl and J. Janak, J. Chromatogr., 81, 233 (1973).
10. R. C. Crippen, Identification of Organic Compounds with the Aid of Gas Chromatography, McGraw-Hill, New York, 1973.
11. D. L. Ball, W. E. Harris, and H. W. Habgood, J. Gas Chromatogr., 5, 613 (1967).
12. J. M. Gill and H. W. Habgood, J. Gas Chromatogr., 5, 595 (1967).
13. D. L. Ball, W. E. Harris, and H. W. Habgood, Anal. Chem., 40, 1113 (1968).
14. D. L. Ball, W. E. Harris, and H. W. Habgood, Anal. Chem., 40, 129 (1968).
15. L. Condal-Busch, J. Chem. Ed., 41 A235 (1964).
16. H. M. McNair and E. J. Bonelli, Basic Gas Chromatography, Varian Aerograph, Walnut Creek, CA, 1968, p. 156.

17. L. Mikkelsen and I. Davidson, Paper 260, Pittsburgh Conference on Analytical Chemistry and Applied Spectroscopy, March 4, 1971.
18. L. Mikkelsen, Paper 54, Pittsburgh Conference on Analytical Chemistry and Applied Spectroscopy, March 5, 1973.
19. F. Baumann, F. Tao, J. Gas Chromatogr., 5, 621 (1967).
20. H. A. Hancock, L. A. Dahm, and J. F. Muldoon, J. Chromatogr. Sci., 8, 57 (1970).
21. J. D. Hettinger, J. R. Hubbard, J. M. Gill, and L. A. Miller, J. Chromatogr. Sci., 9, 710 (1971).
22. E. M. Emery, J. Gas Chromatogr., 5, 596 (1967).
23. B. E. Saltzman, The Industrial Environment—Its Evaluation and Control, Chapter 12, U. S. Department of Health, Education, and Welfare, National Institute for Occupational Safety and Health, Washington, D.C., 1973.
24. D. T. Mage, J. Air Poll. Control Assoc., 23, 970 (1973).
25. H. A. Grieco and W. M. Hans, Industr. Res. 39 (March 1974).
26. F. J. Schuette, Atmospher. Environ., 1, 515 (1967).
27. R. A. Baker and R. C. Doerr, Int. J. Air Poll., 2, 142 (1959).
28. W. K. Wilson and H. Buchberg, Ind. Eng. Chem., 50, 1705 (1958).
29. G. O. Nelson and K. S. Griggs, Rev. Sci. Instrum., 39, 927 (1968).
30. R. M. Nash and J. R. Lynch, J. Amer. Ind. Hygiene Assoc., 32, 552 (1971).
31. A. P. Altshuller, I. R. Cohen, Anal. Chem., 32, 802 (1960).
32. J. M. H. Fortuin, Anal. Chem. Acta, 15, 521 (1956).
33. J. M. McKelvey and H. E. Hollscher, Anal. Chem., 29, 123 (1957).
34. A. E. O'Keeffe and G. C. Ortman, Anal. Chem., 38, 760 (1966).
35. P. C. Barbato, G. R. Umbreit, and R. J. Leibrand, Internal Standard Technique for Quantitative Gas Chromatographic Analysis, Applications Lab Report 1005, August 1966, Hewlett-Packard, Avondale, PA 19311.

TECHNIQUES AND INSTRUMENTATION

Books must follow sciences, and not sciences books.

Proposition touching Amendment of Laws

The use of the gas chromatographic technique requires rather simple and inexpensive instrumentation. The basic requirements are a gas source, a flow controller a sample system, a column, a detector, a flowmeter, and the necessary electronic equipment to record the changes in the column effluent composition. Various pieces of equipment are available, the use of each being decided by the sample type. This part of the book contains chapters on detectors, instrumentation requirements for the gas chromatographic system, the coupling of electronic equipment with chromatographic systems and the technique of trace analysis. Complete understanding of this part requires the knowledge of part one.

CHAPTER 5

Detectors

I

JAMES J. SULLIVAN

Hewlett-Packard Company

5.1 INTRODUCTION 216
5.2 GENERAL ASPECTS. 216
    5.2.1 Signal-to-Noise Ratio 217
    5.2.2 Minimum Detectable Level 219
    5.2.3 Sensitivity. 220
    5.2.4 Response Factor. 221
    5.2.5 Universality and Specificity 222
5.3 QUANTITATIVE FUNCTIONS OF DETECTORS 224
    5.3.1 Stability of Response Factors. 224
    5.3.2 Linear Dynamic Range 225
    5.3.3 Nonlinear Quantification 226
    5.3.4 Predictability of Response Factors 227
    5.3.5 Quenching and Enhancement. 227
5.4 THERMAL CONDUCTIVITY DETECTOR 228
    5.4.1 Early History. 228
    5.4.2 Principles of Operation. 229
        Typical Conditions 231
        Other Heat Transfer Mechanisms 231
    5.4.3 Practical Detectors. 234
    5.4.4 Other Variables of TC Detectors. 237

5.4.5 Response Factors   239
5.4.6 Noise and Minimum Detectable Level   240
5.4.7 Practical Operating Hints   241
      Oxidation   241
      Sample Damage   242
      Overnight   242
      Operating Conditions   242
      Carrier Gas   243
      Temperature Programming   243
      Other Plumbing Configurations   243
5.5   FLAME IONIZATION DETECTOR   244
      5.5.1 History   245
      5.5.2 Flame Chemistry   246
      5.5.3 FID Design 248
            Jet   248
            Air Flow 249
            Carrier Gas and Hydrogen Flows 249
            Exhaust Flow 249
            Thermal Control   250
            Ion Collection 250
            Electrometer 252
            Ignitors 252
            Polarity 252
      5.5.4 Response Factors 252
            Effective Carbon Number   253

II

MATTHEW J. O'BRIEN

Merck, Sharp and Dohme Research Laboratories

5.6   ELECTRON CAPTURE DETECTOR   254
      5.6.1 Historical 255
      5.6.2 Principles of Operation   256
            Cell Design   256
            Radiation Source 256
            Flow Requirements   259
            Voltage Requirements 259
      5.6.3 Response Factors 261
      5.6.4 Linearity of Response   264
      5.6.5 Minimum Detectable Level 264

                5.6.6   Practical Operating Hints. 264
                5.6.7   Applications 265
5.7   THE HELIUM/ARGON DETECTOR 265
5.8   CROSS-SECTION DETECTOR. 266
5.9   FLAME PHOTOMETRIC DETECTOR. 266
                5.9.1   Principles of Operation. 267
                        Gas Flow 268
                5.9.2   Response Factors 268
                5.9.3   Minimal Detectable Level 269
                5.9.4   Linearity of Response. 269
                5.9.5   Applications 269
5.10   ALKALI FLAME IONIZATION DETECTOR (AFID)/THERMIONIC. . . 269
        DETECTOR (TID)
                5.10.1 Operating Principles. 270
                5.10.2 Flowrates. 270
                5.10.3 Applied Potential and Polarity of Electrodes . . 271
                5.10.4 Alkali Metal Salts 272
                5.10.5 Response Factors 273
                5.10.6 Minimum Detectable Level 274
                5.10.7 Linearity of Response. 274
5.11   FLAME EMISSION DETECTORS. 274
5.12   MICROWAVE PLASMA DETECTOR (MPD) 275
5.13   THE GAS DENSITY BALANCE 276
5.14   ELECTROMETRIC DETECTORS 278
                5.14.1 Coulometric Detectors 278
                5.14.2 The Reaction Coulometer. 280
                5.14.3 The Electrolytic Conductivity Detector 280
5.15   THE ULTRASONIC DETECTOR 281
5.16   THE PIEZOELECTRIC SORPTION DETECTOR 282
5.17   RADIOACTIVITY DETECTORS 282
5.18   MISCELLANEOUS DETECTORS 283
                5.18.1 Semiconductive Thin-Film Detector. 283
                5.18.2 Dielectric Constant Detector 284
                5.18.3 Brunel Mass Detector 284
                5.18.4 Hydrogen Flame Temperature Detector. 284
                5.18.5 The Titration Detector 284
                5.18.6 Gas Volume/Pressure Detectors. 285
                5.18.7 Flow Impedance Detector. 285
5.19   MASS SPECTROMETRY 285
5.20   ULTRAVIOLET/FLUORESCENCE DETECTORS. 286
5.21   INFRARED DETECTORS. 286
REFERENCES. 286

5.1   INTRODUCTION

The first public reference to the need for better detectors for
gas chromatography (GC) was probably A. J. P. Martin's lecture
(1) when he received the Nobel prize with R. L. M. Synge, for
partition chromatography. Detectors have always received more
attention in GC than they have in the other major analytical
fields, such as mass spectroscopy, infrared spectroscopy, nuclear
magnetic resonance, atomic absorption, or spectrophotometry. In
none of these other techniques is there such a bewildering variety of detectors. The reason for this seems to be that in other
fields the input to the detector is a signal, such as a beam of
radiation. In GC, the input to the detector is a flow of chemicals. The signal is whatever the detector has been designed to
produce from this flow of compounds. Thus, a detector in GC (or
in liquid chromatography) has the same functional description as
many complete analytical instruments: chemical compounds in,
signal voltage out.
      It is useful to consider gas chromatographic detectors in this
way, and to compare them to other analytical instruments:

- Unlike the common spectroscopic instruments, the GC detector
  ordinarily must complete its analysis in a few seconds,
- The GC detector compares favorably with atomic absorption and
  emission spectroscopy in that the sample is completely vola-
  tilized and relatively free of matrix effects,
- The GC detector has better sensitivity than many analytical
  techniques; some detectors can sense 10^{-14} g/sec,
- The GC detector can have a huge dynamic range; better than
  10^7 is achievable,
- Unlike many analytical instruments, detectors in GC occasionally are used to help identify compounds; most applications
  are to quantify known compounds.

5.2   GENERAL ASPECTS

Detectors are used in many different ways. Table 5.1 gives some
of the qualitative and quantitative functions of detectors with
the corresponding requirements of each. No detector performs all
these functions. A particular detector may be used mostly for
one purpose during one analysis, and for others during a later
analysis.

TABLE 5.1 FUNCTIONS OF GC DETECTORS

Functions	Corresponding Requirement
Qualitative	
Sense trace levels	Minimum detectable level, sensitivity
Sense wide variety of compounds	Universality of response
Sense peaks unresolved from background	Specificity of response
Confirm a molecular property of a compound	Predictability of specific response
Identify unknown compounds	Produce spectra, interpretable or comparable with a library of known spectra
Quantitative	
Measure relative amounts of compounds	Stability of response factors
	Linearity over working range
Measure compounds without standards	Predictability of response factors
Measure peaks in the presence of background	Specificity
	Freedom from quenching or enhancement of response

5.2.1 Signal-to-Noise Ratio

Figure 5.1 shows a portion of a chromatogram. Three parameters are illustrated, the peak width, W, the noise, N, and the peak height, S. Peak height is measured in any convenient units from the base of the peak to its maximum. The width of the peak (in seconds) is defined in various ways: (a) the peak width at half the height (illustrated), (b) the base of the triangle which most closely matches the shape of the peak, (c) a multiple of the variance, or second moment, of the peak shape, (d) the ratio (in consistent units) 2A/S where A is the area of the peak. The first method is used here because it is the easiest to measure. It is not as accurate as (c) or (d).

Noise is usually taken as a peak-to-peak measurement over some

time period--usually the peak width. It is the average distance
(in the same units as peak height) between the highest excursion
of the baseline to the lowest excursion during one peak width or
other standard of time. Since it is an average measurement, it
is often convenient to use a section of baseline several peak
widths long. Lines are drawn just below the highest few maxima
and just above the lowest few minima. This method admits some
statistical bias in the measurement, but for measuring several
different chromatograms, an analyst is probably biased equally in
each case. It is important that the gain be set so that the
noise is more than a few percent of full scale. This is because
all recorders have some deadbanding, that is, diminished or no
response for signal changes corresponding to a fraction of a percent of full scale. If an attempt is made to measure a noise
level of 1% or so, the measurement will be too low. The width of
the line drawn by the pen also adds uncertainty under these conditions. It is important to run the chart recorder fast. Low
chart speeds cause the noise excursions to overlap, making a
broad line down the paper. It has been our experience that most
people overestimate the noise. This is probably a psychological
effect, because the center of the noise looks more dense than it
really is.

     The baseline can have other shapes than that shown in Figure
5.1. For instance, Figure 5.2 shows two other baseline perturba-
tions. Drift is a slow, uniform change of the baseline over
times similar to the entire chromatogram. In severe cases, drift
makes it difficult to keep the pen on the recorder until at least
the chromatogram is finished. Electronic integrators can suffer
in performance due to drift. Drift is often due to temperature
change or the column bleed changes on a new column and so is controllable. By contrast, noise is usually due to things within
the detector, so is a more fundamental limit in performance.

     A subcategory is sometimes distinguished between fast noise
(here taken to be much faster than the peak) and wander.

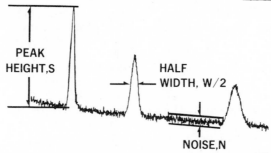

Figure 5.1. Portion of a chromatogram illustrating
peak height, peak half-width, and noise.

When the baseline varies randomly over times similar to the peak
width, it is called <u>wander</u> (Figure 5.2). Wander interferes most with interpreting a chart recording or the operation
of an electronic integrator. It can come either from the detector or from extraneous sources, such as a small leak at a column
fitting.

A detector specification to indicate a peak is measured by the
<u>signal-to-noise</u> ratio, S/N. Here, noise includes both fast noise
as well as wander. Drift is ignored. The signal-to-noise ratio
is useful because it is easy to measure, but it is not good as a
detector specification. Consider what would have happened to
Figure 5.1 if the peak resolution had been worse by a factor of
two. The width would have been twice as large, the peak height
would be down by half, and the signal-to-noise ratio would only
be half as large.

5.2.2 Minimum Detectable Level

It is important to specify detectors independent of column parameters and of sample size. One parameter that does this is <u>mini-
mum detectable level</u>, MDL. It is the "<u>level</u>" of sample in the
detector at the maximum of the peak, when the signal-to-noise
ratio is two. The term <u>detectability</u> is sometimes used for MDL.
Variations of this definition are sometimes given which require
the signal-to-noise ratio to be either one, three, or five. The
parameter is also defined sometimes in terms of root-mean square
(rms) noise. Peak-to-peak noise can be taken as six times rms
noise.

The <u>level</u> is commonly given as g/sec. If the peak in Figure
5.1 corresponds to W of 20 seconds for a component of 570 picograms, the level, in this case a mass flowrate, is 570 pg divided
by the half width, W/2, or 57 pg/sec. If the signal-to-noise

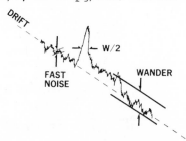

Figure 5.2. Portion of a chromatogram showing distinction between drift, fast noise and wander, relative to the peak half-width, W/2.

ratio, S/N, in Figure 5.1 is 18, then the level corresponding to S/N = 2 is 6.5 pg/sec. This is the minimum detectable level.

There are some detectors (thermal conductivity, for example) which respond to the concentration in the detector, rather than the mass flowrate. Consider such a detector used with a quarter-inch column, where the typical carrier flowrate, F_c, is 1 cm^3/sec. Suppose the MDL is measured to be 16 ng/sec. The MDL can be made a factor of two better (8 ng/sec) by going to an eighth inch column (where the flow is typically 0.5 cm^3/sec). This is because, when the same mass flowrate of compound is going through the detector as before, there is less carrier gas diluting the sample, so the detector performs with high sensitivity.

To make minimum detectable level independent of column flow, it is useful to let level be a concentration. Then MDL has units of g/cm^3. In the example given above, the minimum detectable level is 16 ng/cm^3 for both flowrates. Often the minimum detectable level is expressed in terms of number of molecules, with units of moles/sec, or moles/cm^3. For detectors which respond selectively to certain elements, MDL is expressed in g/sec or g/cm^3, but the weight refers to the weight of the atoms that the detector monitors. Thus, a nitrogen detector which gives a signal-to-noise ratio of two for 10 pg/sec of azobenzene has an MDL of 10 pg/sec for the compound and 1.5 pg/sec for nitrogen, since the compound is 15% nitrogen by weight. In order to distinguish elemental MDL's from those involving weight of compound, for elemental MDL the notation g(X)/sec or g(X)/cm^3 will be used, where X is the element being detected. The MDL's for some common detectors are given in Table 5.2.

TABLE 5.2

Detector	MDL		Sample Compound
Thermal conductivity (TCD)	5×10^{-10}	g/cm^3	Propane
Flame ionization (FID)	10^{-12}	g(C)/sec	Propane
Electron capture (ECD)	10^{-16}	moles/cm^3	Lindane
Flame photometric (FPD)	10^{-10}	g(S)/sec	Thiophene
	2×10^{-12}	g(P)/sec	Tributylphosphate
Alkali flame (AFID)	5×10^{-14}	g(N)/sec	Azobenzene
	5×10^{-15}	g(P)/sec	Tributylphosphate

5.2.3 Sensitivity

The word <u>sensitivity</u> is like beauty, truth, or happiness. It
describes a quality which is always good, but which is not well-
defined, and is used in different ways. The term is supposed to
mean the ratio of signal to sample size. It has also been used
as the ratio of sample size to signal. The term is also used
when minimum detectable level is meant. Consider a gas chromatograph with two detectors, a flame ionization detector (FID) and
a thermal conductivity detector (TCD), which can be alternately
connected to the end of the column. Almost no one will disagree
with the statement that "the flame detector is more sensitive
than the TC detector." The statement will still be accepted even
if the controls on the instrument are adjusted to give a bigger
response on the recorder or integrator for the TCD than for the
flame detector. When different detectors are compared in this
way, sensitivity almost always means minimum detectable level.
If the first definition of sensitivity is used, then "high sensi-
vity" is better; if detectable level is meant, then "low sensitivity" is better. To avoid confusion, it is better to refer to
minimum detectable level or to response factor.

5.2.4 Response Factor

A response factor is a ratio of <u>signal</u>-to-<u>sample</u> size. There are
two kinds of response factors. Some response factors are numbers
used in calculating the quantities required in a chemical analysis. They can be in any convenient form, including the inverse,
that is, sample size divided by signal. They may, if desired,
involve an internal standard, take account of efficiency of
sample workup, or be expressed in arbitrary units. The response
factors considered here, however, are meant to characterize detectors. They should be independent of carrier flow, F_C and the
units of sample size should reflect the way the detector works.
For any particular detector, there are two ways to define response factor, depending on whether peak height, S, and peak width
W, are measured, or whether an integrator makes areas, A, available. They can be shown to be equivalent to the extent that
$SW/2 = A$.

When <u>signal</u> is the peak height, the <u>sample size</u> is the mass
flowrate through the detector at the peak maximum, for the mass
flow sensitive detector. For the concentration-sensitive detect-
or, the sample size is the concentration in the detector at the
peak maximum. Table 5.3 gives the equations for response factor
in terms of weight, M, of compound injected.

Peak heights and areas should be expressed in terms of what
comes out of the detector, not what is presented on the recorder

or printed by the integrator. Expressing a peak height in units
of "58% of full scale, range 100, attenuation 128" is ambiguous.
The instrument could have an electrometer with "100 x 128" corresponding to either 12.8 nanoamps at the input, or to 64 nanoamps.
Similarly, the recorder might be different than the commonly used
1 mV full scale. It is better to say "58% of full scale, recorder sensitivity 12.8 nA full scale," or simply "the peak height is
7.42 nA." The same considerations apply to integrator "counts."
Integrators are usually calibrated in volt-seconds. A TCD, which
has an output directly in volts, with no intervening amplification before the integrator, has an area in volt-seconds. A flame
ionization detector, which has an electrometer to convert a low-
level current (amperes) into volts, has an area in units of
ampere-seconds, or coulombs.

TABLE 5.3   EQUATIONS FOR RESPONSE FACTOR IN TERMS OF WEIGHT (M)
            OF COMPOUND INJECTED, PEAK WIDTH (W), AND FLOW (F).

	Based on Peak Height (S)	Based on Integrator Area (A)
Mass-flow sensitive detectors	SW/2M	A/M
Concentration sensitive detectors	SWF/2M	AF/M

5.2.5 Universality and Specificity

The popularity of GC as an analytical technique in many areas depends on the fact that all of the compounds of interest in an important sample can be detected. For instance, in petroleum and
petrochemical labs, it is the rule that all of the compounds can
be measured at very low levels with the flame ionization detector.
In this case, the detector is "universal." In a natural gas analysis, however, the same detector would not have universal response. This is because several of the important constituents,
such as N_2 and CO, give little or no response on the FID. In
this case a TCD is used, which is "universal" for this analysis.
    Consider the chromatograms in Figure 5.3. This is a mixture
of 10 pesticide standards added to milk extract and run on

several detectors (2). The FID responds, not only to the standards, but also to every organic component which comes out of the
extract. This detector is totally unsuited for this analysis
without a considerably increased amount of wet chemical extraction. What is needed is a family of specific detectors, which
have little or no response for interfering material, and very
good response for some or all of the compounds of interest. In
the example, the specific detectors allow the compounds to be
analyzed, where before they could not be resolved from the background.

Specificity is usually defined as the ratio of the amount of
hydrocarbon to the amount of sample that gives the same response.
The amounts can either be molecular weights or weights of specific elements in the compounds. Most applications of specific detectors are similar to Figure 5.3, that is, small components must
be found within a great mass of overlapping, not well-character-
ized material. Under these circumstances a selectivity less than
1000 is of marginal value. Selectivities are quoted at 10^6 and
higher. For instance, the electron capture detector (ECD) responds a millionfold better to lindane, a chlorinated pesticide,

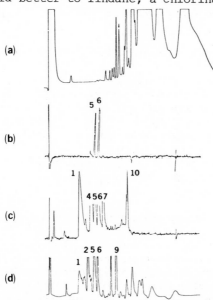

Figure 5.3. Chromatograms of a mixture of 10 pesticide standards, extracted from milk, run on various
detectors (ref. 3). (a) Flame ionization detector,
(b) flame photometric detector, sulfur mode, (c) flame
photometric detector, phosphorus mode, (d) electron
capture detector.

than it does to hexane. This is small consolation when an analysis of a food extract for chlorinated pesticides gives a most prominent (and unexpected) peak that later turns out to be some natural product at the trace level. The problem here is the predictability of specific response. If the ECD was specific only for polychlorinated material (it is not) then the unexpected peak would have to be some chlorinated (and presumably man-made) product, perhaps a polychlorinated biphenyl. Thus the detector cannot be used to confirm polychlorination, even if the detector is found to be highly specific for a compound. In Figure 5.3d peak 6 has very good response on the ECD, but contains no chlorine. The detector can be used to confirm electronegativity, but there are no good theories linking the structure of practical compounds to electronegativity. In contrast to the ECD, the microcoulometric detector can be made quantitatively specific for chlorine, at the expense of sensitivity. In the food extract example, good procedure would be to rerun the extract on a coulometric detector.

5.3   QUANTITATIVE FUNCTIONS OF DETECTORS

Perhaps most gas chromatographs today are used for quantitative analysis. As was shown in Chapter 4, there is a variety of methods of quantifying a compound in a mixture by GC. All of the common techniques involve comparing the chromatogram of the unknown with either a calibration standard or with other peaks in the same chromatogram. The key, then, to good quantification is that two peaks can be produced with dependable relative response factors. There are two ways that the relative response factors can vary. Either the detector can change with time, (instability) or it can change with amount (nonlinearity).

5.3.1 Stability of Response Factors

The stability of detector response varies from changes less than 1 percent per month of operation, to greater than 100 percent per hour. The former case allows a gas chromatograph with an automatic sampler to produce reliably high precision analyses over a weekend of unattended operation. The latter case means that a run of a standard blend, immediately followed by the chromatogram of an unknown, might give some indication of how much is in the unknown mixture. Instabilities may be due to inadequacies in the instrument design, usually poor control of temperature, pressure, or flow. The newer commercial instruments tend to be much better than the earlier ones. The manufacturers have contributed more to improving stability than to other areas of detector design.

Stability is rarely quantified or specified for a detector. This is because it depends on so many parameters. The exceptions are dependence on temperature, pressure and flow. These can usually be found in the literature.

5.3.2 Linear Dynamic Range

There is no accepted definition of linearity. The term <u>linear dynamic range</u> (or simply <u>dynamic range</u>) is the ratio of the largest "linear" level of sample to the smallest "linear" level. Consider the problems in defining this term exactly:

1. Does linearity mean that the response factor is constant for different sample sizes--or that response, when plotted against sample size on log-log paper gives a straight line? The latter definition allows sublinear or superlinear behavior-- where response follows a power law with an exponent different than one.
2. Once linearity is defined, what accuracy is required? The values 1, 5, 20, and 100% have all been used.
3. What is the smallest "linear" level? Is it the minimum detectable level, or is it a level sufficiently higher than the noise to allow a measurement to be made with a precision equivalent to the allowable deviation from linearity? Understandably, instrument manufacturers prefer the former definition, chromatographers with the responsibility of specifying the accuracy of their analyses prefer the latter.
4. What is the largest "linear" level? Is it where peak height deviates from linearity by the specified amount, or where peak area, which is usually more linear, deviates by the specified amount?
5. Should practical or optimal instrumental conditions be used? To produce impressive linearity measurements, it helps to: (a) use easily chromatographed compounds, like propane, (b) use minimum oven temperature, low-loaded columns, low temperature injection port and detector, (c) use fast, very sharp peaks, (d) avoid solvent peaks, either with a gas sample, or with a solvent that is less volatile than the sample, (e) for selective detectors, use a compound with the highest possible response factor, (f) before running, let the instrument equilibrate for as long as possible.

We prefer the following definition of linear dynamic range for purposes of comparing detectors: Linear dynamic range is the range of sample concentration over which the response factor based on area measurement varies by less than plus or minus 20%. Chromatographic conditions must be specified and preferably are close to those commonly used in practical analyses. The sample should be such that it can be used as an internal standard in practical work or is a common analyte. If the linearity appears to extend down to the lowest levels, then the minimum detectable level is taken as the lower limit to the linear dynamic range. The linear dynamic range should be demonstrated experimentally by running one or two different concentrations for each decade of the dynamic range. Figure 5.4 shows a convenient plot used to accent linearity (3). A perfectly linear system (including the column) gives a straight, horizontal line. The arrows indicate the limits of the linear dynamic range. All samples used are from the same dilution series. The same syringe is used to inject the same volume each time.

There have been many linearity plots published of the logarithm of detector response versus logarithm of sample size. This kind of plot gives the <u>appearance</u> of much better linearity.

5.3.3 Nonlinear Quantification

Sometimes it happens that the chromatographic system is not linear enough for the accuracy needed for the analysis but the reproducibility is good enough to justify trying to correct for the nonlinearity. This is most often done with a calibration curve. The peak is measured, either area or peak height, and a calibrated curve is used to calculate graphically the corresponding amount of material in the peak. Another method useful for small

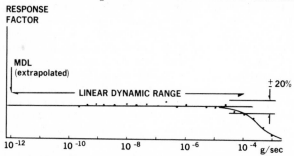

Figure 5.4. Method of plotting response factors to demonstrate linearity. Detector is FID; compound is methane.

amounts of nonlinearity and small ranges of sample size is two
point calibration. Two, instead of just one, standards are run,
one larger than the unknown, the other smaller than the unknown.
Linear interpolation between the responses of the two standards
is used to get an approximate response factor at the concentration of the unknown. These methods can be used for nonlinearities due to the extraction process or due to adsorption on the
column, as well as detector nonlinearities. In almost every case
it is better to modify the chromatographic conditions to achieve
linear operation, if possible, than to correct for nonlinearity
later.

5.3.4 Predictability of Response Factors

The TC detector and the FID are usually within a factor of two of
the same response for an unknown organic compound as for other
peaks in the run. This is a desirable property. Not only does
it allow estimates to be made for the amount of unknown peaks,
but it insures that there are no major volatile impurities if a
chromatogram does not show them. This allows manufacturers of
chemical products to specify the level of organic volatile impurities without identifying them in each case.
    There is no widely used detector that has predictable absolute
response factors. This is why predicted rather than calibrated
response factors are only used when calibration is not practical,
such as for unknowns, or when a closely related compound has been
calibrated, allowing extrapolation to the desired response
factor.

5.3.5 Quenching and Enhancement

When specific detectors are used to quantify a certain class of
compounds in the presence of a background of many nonspecific
compounds, it is not enough that the specificity be sufficient
to get well-resolved peaks for the desired compounds. It is also
important that the presence of the nonspecific background does no
not change the response factors for the compounds of interest.
The flame photometric detector, which is sensitive to sulfur or
phosphorus, shows diminished response to sulfur compounds, if
large amounts of hydrocarbons are eluted simultaneously. This
happens even though there is no significant response on the
chromatogram from the interfering hydrocarbons. The opposite
effect can also occur. In some of the newer types of electron

capture detectors, low-level pesticide responses are improved by
quite small increases in the amount of column bleed. This enhancement effect is especially surprising, since in the older
types of electron capture detectors (constant frequency type),
the presence of bleed quenched the pesticide response. To guard
against this type of problem, it is important to run four kinds
of samples: (a) the standard, a mixture of known concentrations
of the compounds to be determined, (b) the blank, a mixture as
similar as possible to the unknowns, except that it lacks the
compounds to be determined, (c) the unknowns, the mixtures to be
analyzed, and (d) the spiked sample, a blank or previously analyzed unknown, to which a known amount of standard is added. The
standard is used for calibration; the blank checks whether the
chromatographic separation and detector selectivity is adequate;
the spiked sample checks whether the response factor is the same
in the presence of background as it is in the standard mixture.

5.4   THERMAL CONDUCTIVITY DETECTOR

The thermal conductivity (TC) detector is the best known of a
class of detectors known as bulk property detectors. These are
sensitive to some overall property of the effluent. Often the
measured property is a physical parameter, rather than a chemical
one. The distinguishing characteristic of a bulk property detector is that there is a significant response in the absence of
sample. A bulk property detector, when sample comes through,
measures a change of property from the baseline value, A to A+
ΔA. It is characteristic that the baseline value, A, is very
large compared to sample responses, ΔA. The reason a large baseline response is so common is that these detectors are intended
to have almost universal response. When a property is chosen so
that any sample will make it change, it is difficult to avoid
having sensitivity to pure carrier gas as well. Thus the baseline response, A, is due to the "response" of carrier gas, and as
a result, is usually sensitive to variations in things that affect this: for instance, temperature and pressure variations any-
where in the system.

5.4.1 Early History

The historical accident that led to chromatographers standardiz-
ing on TC was that TC had already been developed as a common

laboratory technique. The first devices recognizable as rela-
tives of today's TC detector were described in the 1880s. During
the 1920s and 1930s, many instrument companies both in Europe and
the U. S. sold devices for a variety of applications. Major uses
included gas analysis in the chemical industries, including process control. Power stations used the instrument for fuel and
stack gas analysis. The industries concerned with gases (fuel
gas manufacture, air products, and nitrogen fixation) naturally
were big users of thermal conductivity.

There were some gas-solid chromatographic techniques developed before the development of gas-liquid chromatography (4). In
these early experiments it must have seemed natural to use TC,
since it was such an established technique. About 1955 papers
on TC began to be published regularly. The effects of different
carrier gases on sensitivity were not easily sorted out and it
took a surprisingly long time for the advantages of helium or
hydrogen carrier gas to be appreciated.

By the late 1950s the TC detector was technically mature. But
there was a general dissatisfaction with the problem of sensitivity, in spite of many scientists' efforts to improve this.
When the FID became widespread, it replaced TC in trace analysis.
Since then the TCD remains the easy and inexpensive detector. It
is used for trace analysis only when more sensitive detectors
cannot be used.

5.4.2 Principles of Operation

Thermal conductivity, λ, is the property of a material of trans-
mitting heat when subjected to a temperature difference. It is
defined as the heat flow, Q, through unit thickness of material,
x, of unit cross section, A, when the temperatures on each side
differ by unity. In terms of the parameters shown in Figure 5.5,
the heat flow is

$$Q = \frac{A(T_1-T_2)\,\lambda}{x} \tag{5.1}$$

Thermal conductivity can also be measured in a cylindrical geometry. This arrangement makes it much easier to avoid other
losses of heat besides conduction through the gas. The equation
for heat transfer in a cylindrical geometry is the same as above,
except that the dimensions (A/x) in Equation 5.1 are replaced by
a geometry factor, G:

$$Q = G(T_1-T_2)\lambda \tag{5.2}$$

The TC can be measured by supplying heat at a known rate, Q, and measuring the temperatures of the center conductor and the outside wall. Let the center be a wire of known electrical resistance, R ohms. Then the heat flow can be supplied by an electric current, I amperes,

$$Q = I^2 R/J \tag{5.3}$$

where J is 4.184 W/cal-sec. To measure the temperature of the wire, use a wire whose resistance, R, is temperature dependent.

$$R = R_o (1+\alpha T) \tag{5.4}$$

The wire now doubled as a heating element and as a resistance thermometer. To find the temperature, T_1, use the voltage across the wire, V, to calculate the resistance (at temperature), V/I, and substitute in Equation 5.4.

From the temperatures, heat flow, and geometry, it is possible to calculate TC. In a TC detector for GC, however, it is not important to measure the absolute value of λ for the column effluent. What is important is that when samples are present in the carrier gas, there is a small change in λ. This changes T_1, which in turn can be detected by the change in the resistance:

$$\Delta R = \alpha \cdot R_o \cdot \Delta T \tag{5.5}$$

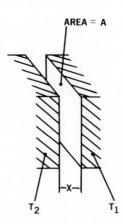

Figure 5.5. Thermal conductivity between two planar surfaces, of temperatures T_1 and T_2, of cross-sectional area A, separated by a distance x. See Equation 5.1.

and

$$\Delta T = - \frac{Q}{G\lambda^2} \cdot \Delta\lambda = \frac{T_1 - T_2}{\lambda} \Delta\lambda \qquad (5.6)$$

These equations can be combined to give:

$$\Delta R = - \frac{\alpha R_o (T_1 - T_2)}{\lambda} \Delta\lambda \qquad (5.7)$$

Note that the sensitivity for changes in λ is proportional to the temperature difference across the cell. This temperature difference is proportional to I^2R, but since the voltage change is $I\Delta R$, the sensitivity is proportional to I^x where x is very close to 3. There is a limit to how high a temperature the center conductor can be run. For most commercial TC detectors using a hot wire, a practical upper limit is 450°C for continuous operation, and 500° C for short periods of time. This is due to oxidation of wire with trace oxygen in the carrier gas. With large concentrations of sample, a variety of other reactions can occur. Since there is a maximum wire temperature, the maximum achievable sensitivity is lowered by raising the detector temperature, T_2. This means that the maximum current that can be used in a TC detector depends on the detector temperature. The instrument manual usually gives a graph of this dependence.

TYPICAL CONDITIONS. A typical geometry for a TC cell is shown in Figure 5.6. Most general purpose TC detectors use geometries similar to this. A common variation is to support the coaxial filament on posts, both of which are mounted on the same end of the cavity. Coiled filaments rather than straight wire are used in order to get higher resistances. Filament materials include platinum, platinum alloys, tungsten, tungsten alloys, and nickel. Since the resistivity and temperature coefficient of resistance, α, do not differ significantly among the commonly used materials, the choice of filament material depends mostly on chemical inertness and mechanical properties. Table 5.4 lists other normal operating conditions.

OTHER HEAT TRANSFER MECHANISMS. There are four other phenomena that have been considered to remove heat from the filament. These are thermal radiation, conductivity out through the ends of the filament (end losses), heat transfer by mass flow of the gas, and free convection. The radiation losses of the filament are proportional to T^4 ($^\circ$K) according to the Stefan-Boltzman equation. A small fraction of energy is re-radiated back to the filament. Estimating the emissivity of the tungsten oxides which are the usual surface coating of a filament the power loss through radiation can be calculated for the conditions of Figure 5.6 and Table 5.4, as 15 mW. Thus, only at the highest filament temperature does radiation transfer more than a few percent of the power.

Through virtually all of the filament length, the temperature is constant. But at the ends the temperature must drop almost to the cell body temperature. Heat is transferred to the body through the ends. Like thermal conductivity, this loss is proportional to the difference in temperature between filament and body, but unlike thermal conductivity, the heat transfer does not depend on conductivity of the gas. This amounts to 45 mW for the case cited. For the small-diameter wire and high-conductivity carrier gases used today, this heat loss mechanism can usually be neglected.

The mass transfer term for heat loss is a product of the mass flowrate of the gas, the specific heat of the gas, and the difference in temperature between the gas flowing out of the cell and T_2. For the geometry of Figure 5.6, the average temperature of

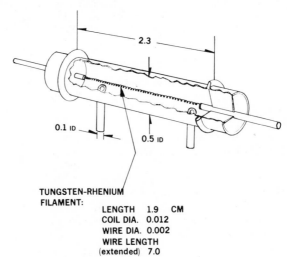

TUNGSTEN-RHENIUM
FILAMENT:
LENGTH   1.9   CM
COIL DIA. 0.012
WIRE DIA. 0.002
WIRE LENGTH
(extended) 7.0

Figure 5.6. Typical geometry of a thermal conductivity detector cell. The feedthroughs supporting the axial filament are insulated from the stainless steel body by alumina ceramic. All dimensions are in centimeters.

the gas exceeds T_2 by $(T_1-T_2) \, d_w/(d_t + d_w)$. Thus in the example, the gas temperature exceeds the block temperature by only 2.4% of the $200^\circ C$ difference across the cell. This gives a heat loss of 7 mW. This calculation is only approximate because it assumes uniform axial flow.

Free convection heat transfer is proportionately more important as the cell gets larger. For some of the prechromatography cells, which were an order of magnitude larger than TC cells today, this term had to be taken into account, for accurate work. Several investigators have examined this experimentally in modern cells and have not found significant effects. It seems to be generally accepted that this effect is quite negligible.

TABLE 5.4   TYPICAL OPERATING CONDITIONS

Temperature	
Body of detector	$150^\circ C$
Filaments	$350^\circ C$
Filaments	
Material	tungsten-rhenium
Temperature coefficient, α	$0.0033/^\circ C$
Resistance, R_o (at $0^\circ C$)	25 ohms
Resistance (at $350^\circ C$)	55 ohms
Electrical	
(for four-element bridge)	
Current	0.3 A
Voltage	16.5 V
Power	4.95 W
(for each filament)	
Current	0.15 A
Voltage	8.25 V
Power	1.24 W
Carrier Gas	helium
Flowrate	$1.0 \ cm^3/sec$
Thermal conductivity, λ	
(at $150^\circ C$)	$4.4 \times 10^{-4} \, cal/(sec\text{-}cm\text{-}^\circ C)$

All of the heat transfer terms considered are summarized in Table 5.5. Except for thermal conductivity of the gas, all the effects contribute only a few percent to the heat transfer. However, if nitrogen were used as the carrier gas, the TC term would be six times lower, and the other terms would be more important

by comparison. As will be shown below, the mass flow term is the
cause of the flow dependence of the TC cell, and so contributes
to the noise and drift of the detector.

TABLE 5.5   SUMMARY OF HEAT TRANSFER (cell of Figure 5.6 and
                        Table 5.4)

Effect	Heat Transfer (mW)
Thermal conductivity (G=3.08)	1130
Radiation	15
End losses	45
Mass flow	7
Free convection	negligible
TOTAL (Measured)	1240

5.4.3 Practical Detectors

Consider once again the typical detector and conditions of Figure 5.6 and Table 5.4. When a rather large peak enters the detector cavity (0.1 mg, half-width 5 sec) changes take place.
They are summarized on Table 5.6. The change of direct interest
is the voltage change on the chart recorder, where a peak height
of 140 mV is registered. Response factors for this detector are
usually calculated as the number of mV for the concentration of
sample in mg/cm^3 at the detector. This factor is sometimes
called the DPS number, after Dimbat, Porter, and Stross (5), who
first proposed it. The response factor of 7000 mV cm^3/mg is
fairly good for a detector which can be operated as low as $150^{\circ}C$.
If greater sensitivity is needed, the filament temperature, T_2,
could be increased somewhat, at the expense of filament lifetime.
Higher-resistance filaments are available, but these are usually
thinner, and thus they burn up faster.

Since large samples give millivolt changes, and small ones
give microvolt signals, the thermal conductivity cell is extremely sensitive to fluctuations of physical variables. Some of
these can be cancelled out if two filament cavities are used
(Figure 5.7), one to detect the sample, the other to serve as a
reference. The resistance R_3 of the sample cell is balanced

against the reference R_4 by means of a bridge circuit. The rest
of the bridge includes two conventional resistors, R_1 and R_2. In
the absence of sample, all of the resistors have the same value,
and no difference in voltage is sensed at the measuring terminals.
If a sample elutes, then R_3 increases, and a voltage is sensed.
If however, a room temperature change affects R_3, then it affects
R_4 by nearly the same amount. This gives no significant change
in voltage at the measuring terminals. Another undesirable
change in the sample cell's resistance is the effect of column
bleed. For this reason, the reference flow into the R_4 cell is
usually set equal to the flow in the analytical column, and
passed through a similar column, so that a matching amount of
bleed is detected by R_4. This is particularly useful with temperature programming.

Several commercial detectors are made according to the circuit
of Figure 5.7. One serious problem that the designers of such
detectors had to solve was the effect of temperature on the conventional resistors, R_1 and R_2. Either resistors with very low
temperature coefficient could be used, or they could be temperature controlled. Another possibility, used in the majority of
detectors meant for industrial use, is to replace R_1 and R_2 with
yet two more filament cells (Figure 5.8). Now two matched cells,
of resistance R_3, each monitor half of the effluent from the
analytical column. Two reference cells, R_4, each monitor half
the effluent from the reference column. Sometimes the matched
cells are two filaments mounted in the same cavity. The benefits
of this approach include a doubling of response factor over the
circuit of Figure 5.7. While the cells of both approaches sense
the same concentration of sample, in the four-cell configuration,
there are two cells that contribute to the change in recorder
signal.

TABLE 5.6   CHANGES TO TYPICAL DETECTOR (SEE TABLE 5.4) DUE TO
SAMPLE

Sample concentration at peak maximum	0.02 mg/cm^3
Thermal conductivity	from 4.4 to 4.14 x 10^{-4} cal/ (sec-cm-$^{\circ}$C)
Filament temperature	from 350 to 362°C
Filament resistance	from 55 to 56 ohms
Voltage across filament	from 8.25 to 8.39 V
Recorder response	140 mV
Detector response factor	7000 mV-cm^3/mg

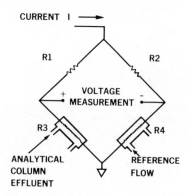

Figure 5.7. Bridge circuit used in a two-cell detect-
or. The reference cell, R_4, compensates for the drift
in the analytical cell, R_3, due to flow and temperature fluctuations.

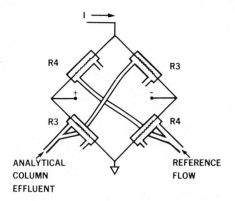

Figure 5.8. Bridge circuit used in a four-cell detector. This approach gives twice the response of that
of Figure 5.7.

Two important parts of the electronics for a TCD are not shown
in Figure 5.8. In practice the four cells of a detector rarely
match. A control is usually provided so that the output voltage
can be nulled before a chromatogram is run. This is usually done
with variable resistors connected in the bridge. Often there are
two controls on the front panel: "Fine" and "Coarse" Balance.
Another necessary control, at least when an integrator is not
used, is an attenuator switch. This reduces the response on the
recorder by various amounts so that the larger peaks can be kept

on scale.

The electrical requirements for the TCD are much simpler than almost any other detector in GC. By contrast, the mechanical requirements, particularly thermal control, are unusually demanding.

This is why TC detectors are made of cells mounted closely together, embedded in metal, with the assembly meticulously insulated. In spite of this effort, the cells are not exactly matched in heat transfer to the metal they are embedded in. This means that it is important to control the temperature of the detector body accurately, In most modern instruments, the TCD temperacontrol circuit produces much better thermal stability (though not necessarily accuracy) than even the chromatographic oven control.

It is also important that the oven not influence the detector temperature. This could happen, not only by thermal conduction, but also by heat transfer through the flowing carrier gas. The second effect is particularly bad because it is one way that variations in gas flow can cause noise or drift in the detector.

Not all of the mechanical problems are in thermal control. The filaments must match, not only electrically but also mechanically, because of the effect on the geometry factor. It is common, in detector manufacture, not only to select matched sets of filaments for each detector but also to select cells matched for geometry factor. The cells must be gas-tight, even at high temperature. Diffusion of air into the cell will give drift and noise, and may lead to an early failure of the filament due to oxidation. Electrical leakage from the filaments to the detector body must be avoided. The problem is not so much loss of current. Rather, it is the variation in such a leakage, as a noise source. Recently, high-density ceramics have been replacing glass as insulators for the filament mounts. Figure 5.9 shows a complete thermal conductivity detector.

5.4.4 Other Variations of TC Detectors

There are three variations sometimes used with TC detectors. These are: thermistors instead of filaments, miniaturized cell geometries, and other carrier gases besides helium.

Thermistors, which are metal oxide beads used as temperaturesensitive resistors, have been used in thermal conductivity detectors since the mid-1950s. They offer several advantages. Their temperature coefficient, α, is larger by an order of magnitude, though it is negative. The resistance, R_O, is also large, usually measured in thousands of ohms. This means that the

response factor is much larger than a corresponding hot-wire detector. However the minimum detectible level is limited by the
ability to control temperature and pressure fluctuations, just
as in a conventional detector. Another advantage is the small
size--thermistors can be made as small as 0.25 mm in diameter,
allowing cells to be made with volumes as small as $5 \times 10^{-5} cm^3$.
This is useful with capillary columns, where the flows are much
lower than with packed columns. In some cases it is an advantage
that the thermistor is a metal oxide glass. The device is almost
inert to oxidizing conditions, such as the analysis of trace components in air. By contrast, thermistors are quite sensitive to
reducing conditions. Sensors with an inert coating have been
developed to ameliorate this problem. But the major problem with
this type of detector is the fact that the sensitivity worsens
rapidly as the detector temperature is raised above $50^{\circ}C$. At
typical operating temperatures for liquid samples, a hot-wire detector is actually more sensitive. Another disadvantage is somewhat greater difficulty in setting the operating conditions for
the detector. Two categories account for most of the applications of thermistor detectors today. Capillary columns require
fast response and, therefore, small detector volumes. Gaseous
samples can usually be run at low temperature, where the thermistor has its best performance, thereby taking advantage of
greater resistance to oxidation.

  The need for small cells with capillary columns is, of course,

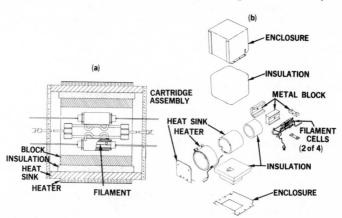

Figure 5.9. Diagram of a commercial TCD. (a) Four
thermal conductivity cells are embedded in a metal
block, which is insulated from the metal heat sink
surrounding it. A heater distributes heat uniformly
over the outside of the heat sink. (b) This assembly
is further insulated on all sides and mounted in a
metal enclosure.

because the response time of the detector must be much less than
the peak widths. There is another term in the response time, because of the thermal response time of the heat sensor itself, but
this term is usually a small fraction of a second. Virtually all
of the TCD applications in North America use helium as the carrier gas. Elsewhere, hydrogen is common. Hydrogen resembles
helium in having a thermal conductivity much higher than virtually everything else. It gives roughly the same response factors
and linearity as helium. Problems with hydrogen carrier include
fire and explosion hazard, destruction of thermistors, and catalytic hydrogenation of some types of samples on hot filaments.

All other carrier gases share several more serious problems:
(a) Some compounds give positive peaks, some give negative peaks.
(b) Response factors are dependent on temperature in unexpected
ways and linearity is often poor. (c) Minimum detectable levels
are usually an order of magnitude worse than with helium or
hydrogen. (d) In some cases, W-shaped peaks result. This is because low concentrations of sample decrease the thermal conductivity until a minimum is reached. Further increases of concentration increase the thermal conductivity.

The carrier should be chosen with a TC either larger than most
of the samples to be run, nitrogen with organic samples, for instance, or smaller, such as argon with some fixed gas analyses.
Temperature dependence and linearity of response should be investigated carefully for all compounds to be analyzed. It should be
remembered that a much lower current is necessary to reach the
same filament temperature in low-conductivity gases. Mixtures of
carrier gas, particularly where one of the constituents is also a
sample compound, lead to further complexities.

5.4.5 Response Factors

One of the best features of thermal conductivity detectors with
helium carrier gas is the ease of quantitative analysis. It has
been shown experimentally that relative response factors, where
sample weight is used, are independent of (a) type of detector
(filament or thermistor), (b) cell and sensor temperature, (c)
concentration of sample, (d) helium flowrate, and (e) detector
current. In addition, relative response factors change only
slightly within a series of homologous compounds. The first systematic study of TCD responses in helium was done by Rosie and
Grob and are summarized in reference (6).

A very useful empirical rule is that all compounds have a
weight response close to that of benzene. There are three exceptions to this. Compounds with heavy metal atoms in them have

unusually low response factors. Tetraethyl lead has one-third
the response predicted by this simple rule. Halogenated compounds also tend to have low response factors. Very light compounds, with molecular weight below 35, tend to have high response factors.

Of the 171 compounds in reference (6) (excluding the three
types of exceptions) for which there are relative response factors, 164, or 96%, are within 30% of the response factor for benzene. In addition, 88% of the compounds examined are within 20%
of the response of benzene.

Response factors have also been tabulated for hydrogen (7) and
nitrogen (8) carrier gas.

5.4.6 Noise and Minimum Detectable Level

It would be very convenient if the TCD had noise levels down in
the nanovolt region. Amplifiers can be built without too much
difficulty which contribute no more than 25 nV of their own noise.
Unfortunately, the TC detector is subjected to many extraneous
influences. It is the fluctuations in these which dominate the
noise and drift seen on the recorder.

Experimental values of the recorder change due to shifts in
some external parameters are given in Table 5.7. The sensitivity
to bridge current changes is measured with a new, well-balanced
detector. The value can be up to an order of magnitude larger,
if poorly matched cells are used, either resistively or geometri-
cally. It should also be remembered that oxidative degradation
of the sample arms of the bridge will tend to increase the sensitivity to bridge current fluctuations.

The flow sensitivity indicates that while a detector may be
accurately balanced in terms of resistance and voltage, the reference flow only reduces the flow sensitivity of the analytical
column flow by a factor of three to four. Much of the drift due
to flow changes comes from the flow controllers. One type commonly used today has been found to have a mass flowrate proportional to absolute temperature. Flow induced noise, however, can
come from column temperature fluctuations. Even if the front of
the column is fed from a perfect flow source, a temperature
change in the column will lead to a viscosity change in the carrier gas. Since the gas is compressible, a transient flow change
occurs in the detector. Needless to say, a fluctuation in column
temperature also leads to a fluctuation in the bleed level, which
affects the recorder baseline by a much more direct process.
Even the temperature coefficient of the pressure regulator on the
gas supply can affect the flow, for high-precision measurement.

Sensitivity to mechanical impulses becomes important, for instance, if the oven fan motor has worn bearings, causing the chromatograph to vibrate.

The large value of the filament temperature sensitivity shows how important temperature control of the detector is. Problems in this area appear most frequently with temperature controllers which oscillate about their setpoint, with a period of a few minutes. If the cells of the detector respond to this change with different delay times, then the oscillation will appear in the baseline.

In spite of all of the effects, a well-designed and operated TC detector is capable of noise levels as low as 2 or 3 μV. This means that a 1-mV recorder on the most sensitive attenuation will barely show noise on the baseline.

At the same time, the sensitivity for the conditions of Table 5.4 allow a response factor of 7000 mV-cm^3/mg to be achieved. This corresponds to 1 ng/cm^3, giving a peak twice the noise level. Better minimum detectable levels can be achieved if even higher currents are used.

5.4.7 Practical Operating Hints

The sections marked with an asterisk (*) also apply to other detectors.

OXIDATION.

1. Don't run out of carrier gas with the filament current on. As the air diffuses into the detector, the filaments will literally burn up. Some instruments have sensors that turn off the filaments if the carrier flow in interrupted, but it is safer not to rely on this feature.

*2. Leaks in the gas lines can be a problem. Air can diffuse in, even while the carrier gas is flowing out. The most common location for leaks is the column fittings. One way to check an entire gas chromatograph for leaks is to seal the detector exits. Check the operating manual for possible pressure limitations. Once the entire flow system is up to the pressure set on the second stage regulator, at the gas cylinder, turn the second stage regulator off. The pressure gauge on the cylinder indicates the pressure at the exit of the regulator. If there are no significant leaks, there should be no noticeable change in this pressure over a 30-min period. Release the seals on the detector exits slowly. A sudden change in pressure can blow the column packing into the detector.

    3. Turn off the current when changing columns, septums, or helium cylinders.

SAMPLE DAMAGE.

    1. Oxidizing, halogenated, or strongly reducing compounds, particularly in large amounts, can change the resistance of the filaments on the sample side. If this is happening, the sample and reference sides of the detector should be periodically reversed to equalize the changes. This problem is indicated by permanent or slowly changing shifts in baseline following a large peak.

    *2. Large numbers of high-concentration samples can give serious condensation in the detector exit tubes. It can even lead to choking off of the column flow. Thermally insulating the exit lines, periodically heating them with a hot air gun, or solvent cleaning all may help.

    *3. Do not condition a column while it is connected to the detector.

    *4. When turning on a gas chromatograph, bring the detector up to operating temperature, then heat up the oven. The reason is that the oven heats very rapidly, the detector more slowly. Otherwise condensation may occur in the detector.

OVERNIGHT.

    *1. Leave the detector at operating temperature overnight, otherwise it may take hours to get a stable baseline the next morning.

    2. For long filament life, turn off the detector current overnight; for the highest stability the next morning, leave it on (along with the oven temperature).

OPERATING CONDITIONS.

    *1. Place the gas chromatograph away from drafts from air conditioning or heat vents, and, away from poorly insulated outside walls and direct sunlight, as all of these things can contribute to temperature fluctuations in the detector (and in the oven).

    *2. If the detector is mounted inside the instrument enclosure, leaving the instrument covers on will give greater thermal stability.

3. Set the detector temperature just slightly above the column temperature (or maximum temperature, for temperature-programming). Higher setpoints sacrifice sensitivity.

4. Guided by the operating manual, set the current higher for better sensitivity, lower for long filament life.

5. Set flows as low as is consistent with good column perform-ance and speed of analysis. Response factors are inversely pro-portional to column flow.

CARRIER GAS.

1. The carrier gas purity required depends somewhat on the application. Generally helium better than welding grade is needed, but rarely as pure as 99.999%.

*2. Oxygen scrubbers, "moisture" traps, and the like in the carrier gas lines are almost always worth the effort, but see below.

*3. An exhausted chemical trap is a source of temperature-induced drift. When the trap is in equilibrium with the carrier gas, if the ambient temperature drops, then the trap is no longer saturated, and it begins to absorb impurities again. This causes the baseline to drift downwards. If the temperature is raised, then the trap adds impurities to the carrier gas. If chemical traps are not going to be maintained, they should be removed from the flow system.

TEMPERATURE PROGRAMMING.

1. Instead of setting the reference flow equal to the flow in the analytical column, it can be optimized for minimum drift dur-ing temperature-programming: (a) Set the flows equal. (b) Make a temperature program run, observing the drift on a recorder. Hold at the maximum temperature. (c) Adjust the reference flow until the recorder pen returns to its value at the beginning of the run. (d) For a more accurate adjustment, repeat steps (b) and (c).

OTHER PLUMBING CONFIGURATIONS.

1. For nondemanding analyses, it is not necessary to use a matched column with the reference cells of the detector. The carrier flow can be passed through the reference cells, then through the injection port, analytical column, and finally through the analytical side of the detector. This removes the

need for a second flow control, injection port, and column.

2. Different types of columns can be connected to the two sides of the detector. To use the other column, it is only necessary to reverse the signal polarity to the integrator or recorder.

3. In some applications involving valving, it is often convenient to use the second side of the detector to analyze a different part of the sample. One example is <u>backflushing</u>. After the more volatile compounds have eluted through one side of the detector in the conventional way, the column flow is reversed, and the less volatile part of the sample is directed from the front of the column through the other side of the detector. Another application is heart cutting, wherein a portion of a column's effluent is valved to a second column for further separation. In both examples given, the flow system is simplified by using the second side of the detector to measure part of the sample.

5.5   FLAME IONIZATION DETECTOR

The flame ionization detector (FID) has become the most commonly used detector in GC. This is because it possesses several outstanding features:

1. It responds to virtually all organic compounds, with roughly the same high sensitivity.
2. It does not respond to common carrier gas impurities such as water or CO_2.
3. It has minimal effects from flow, pressure, or temperature changes.
4. In the absence of sample, it has virtually no response. This gives a stable baseline.
5. Linearity is good for a range as high as 10^8.
6. There are few adjustments to make.

The detector consists of a small hydrogen-air diffusion flame burning at the end of a jet--the tip of a length of capillary tubing (Figure 5.10). When organic compounds are introduced into the flame from the column, electrically charged species are formed. These are collected by applying a voltage across the flame. The resulting current is amplified by an electrometer.

5.5.1 History

Up to late 1957 none of the high-sensitivity detectors had been
developed yet. Higher sensitivity was a problem on everybody's
mind, and many different approaches were being investigated. In

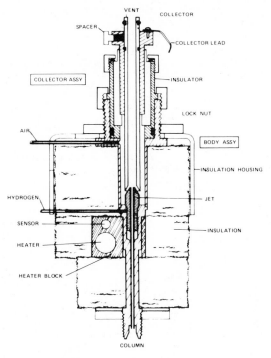

Figure 5.10. Diagram of a flame ionization detector.
Hydrogen enters the detector and flows down the outside of the jet to the column fitting, where it mixes
with the column effluent. The mixture of gases flows
up the jet and burns at its tip. Air flows down the
outside of the long cylindrical collector and enters
the flame at the tip of the jet. The combustion
products vent from the top of the collector assembly.
The base of the detector is temperature controlled by
a sensor and heater embedded in the heater block. In
this detector, the jet is electrically grounded and
the collector operates at high voltage. The collector
lead is connected to the electrometer, for current
measurement, through an electrically floating high
voltage supply.

fact, there were two other flame detectors invented before the
FID. There was the flame emission detector, by which the total
visible light emitted is monitored and the flame conductivity
detector which uses the temperature of the luminescent cone above
a flame to indicate the presence of eluting compounds. It was in
the early 1960s that the reaction that produces ionization in a
flame detector was confirmed. Most of the direct evidence came
from mass spectroscopic examination of the interior of flames,
but other techniques were also used. Most of this work was done
by combustion researchers because of the difficulty of the problem. Recent summaries of flame ionization processes have been
published by Miller and also by Bocek and Janak (9,10).

5.5.2 Flame Chemistry

In spite of its importance, ionization is a very small part of
the mechanism of a flame. Only one out of every 10^5 burning carbon atoms gives up an electron. Since the FID uses a diffusion
flame, the mechanism begins at the tip of the flame jet (Figure
5.11). The hydrogen-containing carrier gas flows out of the jet
at a much higher velocity than the air flow, so it expands outward. At the same time, the flow of air rises around the outside
of the jet, veers outward, and flows alongside the gas from the
jet. As each gas flow approaches the reaction zone (dotted line),
it is met by a back diffusion of heat energy and of radicals produced there. These preheat the gases and cause some breakdown
of the oxygen and hydrogen. If organic compounds are present in
the effluent, they begin to be cracked and stripped of protons
and terminal groups. When the two gas flows mix together at the
reaction zone, they have already been heated enough for reactions
to take place very rapidly. It is here that ion production takes
place. The heat energy and radicals released in the reaction
zone flow upward into the afterglow region, where slow reactions
take place. Here the temperature remains high, since the slow
reactions are also exothermic.
    It is in this afterglow region where most of the light is
emitted. In a pure hydrogen-air flame, virtually the only light
emitted is from the OH radical, which is in the ultraviolet region of the spectrum, and therefore invisible. If hydrocarbons
are present, then the blue, green, and yellow emissions by the
radicals CH and C_2 are visible, and possibly the incandescence of
hot soot particles.
    There is a good balance among all of the flows and diffusions.
If the flame is too small, heat loss mechanisms are relatively
more important, and the diffusion-limited combustion cannot occur

fast enough to keep the flame above the ignition temperature. If
the jet diameter is too large, the flow must convolute itself into turbulence to get the gases together. If the flows are too
low, all of the available gas is burned up, and the flame burns
out before any more gas comes out of the jet. If the flow is too
fast, tongues of flame are lifted off the jet, one after another,
in a turbulent flame. There must also be a balance between the
velocity of air and fuel flows. If this balance does not exist,
all of the flame processes, including ionization of hydrocarbons,
are less efficient.

In the case of a pure hydrogen-air flame, radicals such as H,
O, OH, O_2H, and excited versions of the same are formed. But
there is no detectable ionization. But when hydrocarbons are introduced into this flame, ionization takes place, in strict proportion to the amount of compound. This is due, we know now, to
the radical reaction:

$$CH + O \rightarrow CHO^+ + e^- \qquad\qquad (5.8)$$

There is one such pair of charged species formed for approximately every 100,000 carbon atoms introduced. This proportion
holds true from the impurities in the purest tank of carrier gas,
all the way up to a candle flame, where there is no hydrogen sup-

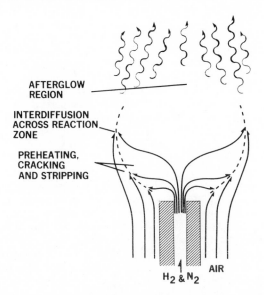

AFTERGLOW
REGION

INTERDIFFUSION
ACROSS REACTION
ZONE

PREHEATING,
CRACKING
AND STRIPPING

H_2 & N_2 AIR

Figure 5.11.   Schematic diagram of the basic flame
processes in the FID.

plied except by protons stripped off the vaporized wax.   (An exception is carbon monoxide, which gives virtually no ionization. Evidently the carbon in this compound burns to CO_2 without going through a CH intermediate.   The same thing is true in organic compounds with an OH or OOH functional group on it.   The adjacent carbons in the molecule only contribute a small fraction of the ionization of the other carbon atoms.)

This reaction is an almost quantitative counter of carbon atoms being burned.   For hydrocarbons, this means that flame response is proportional to the number of carbon atoms, rather than weight or moles.   Compounds with heterogroups on some of the carbon atoms lose some of their ability to form CH.   This is usually corrected for by considering the effective carbon number, that number which would account for the compound's response.

When careful measurements are made in a flame, very little HCO^+ or e^- is found.   This is because these species further react very quickly.   The first loses H^+ (proton abstraction) to any of a variety of species on collision.   The most stable product is H_3O^+ and its polymerized form $(H_2O)_nH^+$.   It is such polymerization about a charged species which makes raindrops.   Such watery charge is what is collected.   The free electron can be captured by radicals, most importantly, OH, but also by H_2O and O_2.   Thus the negative ionic species weighs about the same as the final positive ion.

5.5.3 FID Design

Soon after the flame ionization detector was announced in 1958, many different research groups began systematic studies to optimize the detector.   In only a few years, the device was mature. By the mid-1960s all of the commercial instruments began to resemble each other in the important respects.

JET.   The diameters of most jets used today are roughly 0.5 mm. This is because of the type of flame used.   In a diffusion flame, as opposed to a flame in which the oxidizer and fuel are premixed, the rate of diffusion of the two gases controls the rate of burning.   The velocity of the gas flow out the jet must be set to match the rate of diffusion.   If the jet diameter is a little different from 0.5 mm, the detector can be optimized at a proportionately different flow.   Much beyond a factor of two differences in diameter leads to some other effects, such as thermal transfer or flow stability problems.

Other properties of the jet are important.   Since the diffus-

ion of the gases at the tip of the jet is so central, the detailed shape affects the detector operation. The tip should be
smooth. Differences are seen between jet tips that are squared
off, or beveled. In heavy use situations, samples or bleed may
pyrolyze before passing out of the jet, and condense there, re-
stricting the flow. Solid support particles from the column can
also block the jet.

AIR FLOW. It is tempting to calculate the amount of air flow required in an FID from the amount necessary to convert all of the
hydrogen to water. But in a diffusion flame, not all of the air
enters the reactive part of the flame. Experimentally five or
ten times more air is needed than predicted by the reaction stoi-
chiometry. Doubtless some of the air flow isolates the flame
gases from the walls of the detector. Less air flow is needed in
those designs in which the flow is directed along the jet through
a narrow collector. It has been found important to have uniform
and laminar flow along all sides of the jet. For this reeason,
many instruments introduce the air through a porous diffuser located well below the tip of the jet.

CARRIER AND HYDROGEN FLOW. Like all detectors, the FID requires
that the carrier flow from the column to the detector be temperature controlled, with no cold spots. The flow geometry should
have minimum dead volume. There is the additional requirement
that the hydrogen be thoroughly mixed with the carrier gas before
passing out of the jet. In practice, a few centimeters of tubing
between the mixing point and the jet are sufficient. An explosion hazard exists if the hydrogen flows into the oven. Thus, the
hydrogen flow should be turned off when a column is removed from
the oven. This is particularly important with a dual column
chromatograph having a single hydrogen control. When a single
column is used, the unused detector fitting must be capped
tightly. More than one chromatographer has been injured by an
exploding oven.

EXHAUST FLOW. The exhaust flow system has perhaps the least
stringent requirements of all the parts of the FID. Yet even
here a variety of possible problems have been uncovered. It is
necessary not only to get the gases out of the detector, but also
the heat generated, and in the case of large samples or heavy
column loadings, soot and silica. The soot is a problem if particles of it occasionally fall back into the flame,giving spikes.
The bleed from a silicon-coated column burns to silica which can

condense on the interior of the detector. This gives an insulating coating on which an electric charge can build up, changing the electric fields in the detector. If the exit path from the detector is too cool, water will condense. It is obvious that drops of water running back into the detector will cause a major upset. Air currents in the vicinity of the gas chromatograph ought not to disturb the flows within the detector. This is why most instruments have a mildly tortuous flow path from the flame to the outside. Another problem to be avoided is back diffusion of ambient air into the flame chamber. The author remembers the first flame detector he built, which had difficulties in this area. The detector could sense a cigarette smoker at a distance of 25 feet!

THERMAL CONTROL. There are really two separate kinds of thermal requirements in a flame detector. The front panel control for detector temperature is used mainly to set the temperature of the carrier gas transfer line. After the flow has passed through the flame, there is no need to keep temperatures above column temperature, as long as water does not condense. It is useful, however to keep the body of the detector stable in temperature, as there is a slight effect on the detection mechanism. It is much more important to keep things relatively cool. Any solid surface that is sufficiently heated by the flame may emit electrons by thermal processes. To see this, leave a hot-wire ignitor turned on while monitoring the detector output on a recorder. The insulators, whether on the jet or on the collector, must be kept cool for another reason. Electrical leakage along their surfaces can contribute a very unstable offset to the output current of the detector. Some of the thermal problems in the flame detector can be subtle. Platinum collectors mounted too close to the flame have been found to produce catalytic ionization, giving anomalously increased sensitivity to certain classes of compounds. In other cases, parts of the detector were close to the threshold for significant thermionic ionization. Large samples which normally give negligible response, such as CO or CS_2 increased the temperature, giving a positive detector response. Under these conditions, a compound that reduces the flame temperature, possibly water, should then give a negative peak. The same responses can occur if the easily ionized alkali salts are present as impurities in the construction materials of the detector.

ION COLLECTION. For very small samples, almost any size or shape collecting electrode, with only a few volts, will collect all of the ions of one polarity. Nanograms of compound can be effici-

ently collected with as little as 10 V on the collector. When
very large samples are burned, several things happen. The flame
increases in size to as much as a few centimeters high--it is now
more of a hydrocarbon flame than a hydrogen flame. The location
of ion generation is now much higher than it was before, and
possibly higher than the collector, and far away from it. There
are now high concentrations of both positive and negative charged
species in the gas. These shield the center of the flame from
the electric field due to the collector. The effective mobility
of the ions is also lower in this case. Recombination of positive ions with negative ions is proportional to the square of
sample concentration, so more ions recombine before they can be
collected. All of these effects together produce saturation,
the inability of the detector to give a signal bigger than a certain level. To run larger samples before saturation effects set
in, the following should be done: Use a collector of very large
surface area, extending higher than the largest expected flame.
Use as high an electric field intensity as is practical. This is
generally around 300 V. Higher voltages induce ion multiplica-
tion as in a Geiger-Muller Tube.

In all flame detectors, either or both the collector and the
jet must be electrically insulated. The quality of the insulation must be extremely good. Even 10^{12} ohms of leakage resistance at 100 V polarization will give 10 nA offset of the baseline. In some cases detectors are used at sensitivities as low
as one pA full scale. In this example very slight temperature
fluctuations can change the leakage current many times full
scale. There have been two approaches to satisfying this requirement. One is to arrange for the insulator to be far removed from the hot part of the detector, so that a fluorocarbon
plastic can be used for the insulator. These materials, when
they are clean, allow insulation resistances of 10^{15} to 10^{18} ohms
to be achieved. The other approach is to use a high purity ceramic in the hot part of the detector. Insulation resistances of
10^{14} to 10^{16} ohms are possible at $300^\circ C$ with some aluminas. In
both cases, contamination of the insulation surface can ruin the
performance.

The different parts of a flame detector must be rigid. If
bumping the gas chromatograph moves the collector momentarily by
0.1 mm, the capacitance between the collector and the rest of the
detector may decrease (or increase) by perhaps 0.1 pF. If the
collector is polarized at 300 V, this change in capacitance injects a spike of 30 pC into the electrometer. At the typical
flame sensitivity, this spike is equivalent to a very narrow peak
of half a pg, a readily detectable amount. The coaxial cable
from the detector to the electrometer is also a capacitor sensitive to vibration. It also suffers from triboelectricity (charge

generated by friction at the surfaces of the insulation within it when moved).

ELECTROMETER. Unlike most parts of the flame detector, the electrometer is rarely constructed by the chromatographer. The state of the art for electrometer design has been revolutionized over and over again in the last two decades, as has all of electronics.
There is only a little technique required with electrometers today. When possible, the attenuation control rather than the range control should be used to adjust the signal to fit the span of the recorder. Most instruments are calibrated better for attenuation changes than they are for range changes. There is usually more filtering, and slower response, at the more sensitive ranges. It is often better for linearity to use a less sensitive range setting. These effects are usually small, and can be ignored in less demanding applications.

IGNITORS. An ignitor is a hot wire or spark gap device that is momentarily activated to light the flame. In one sense, it is a convenience, because a match will also light the flame. The problem with a match is more than burning fingers or getting shocked on a high voltage polarizing supply. A fragment from the match may fall into the detector. The FID detects match fragments as well as other organic material.
Ignitors should be adjusted so that they do not protrude into the flame. Otherwise, it is a flame electrode, and may have many of the problems mentioned above.

POLARITY. When all of the pitfalls mentioned in the last few pages are avoided, it makes absolutely no difference what the polarity is for the flame detector. The jet can be grounded, and the collector can be either positive or negative in voltage. The jet can be positive or negative, with the collector connected directly to the electrometer (which is at ground potential). This convenient symmetry allows two flame detectors, opposite in polarity, to be connected to the same input of a single electrometer. In this way, differential dual column operation can be done with a single electrometer.

5.5.4 Response Factors

The early and basically correct understanding was that flame

response factors are very well behaved. First of all, the area
response for a compound does not change with small changes in
carrier flow. This means that the detector is mass flow sensitive, rather than concentration sensitive like the TCD. The proper units, then, for response factors are A-sec/g of carbon, or
coulombs/g(C). In fact, when a detector is well adjusted, most
compounds have about the same sensitivity, 0.015 coulombs/g(C).
The detector is noted for its linearity (Figure 5.4). Very early
commercial detectors could demonstrate linearity within a few
percent over five orders of magnitude.

EFFECTIVE CARBON NUMBER. Exceptions and extensions of these
simple ideas were soon found. The presence of heteroatoms and
different functional groups was found to modify the response predicted by carbon number. Adjustments were proposed, to use an
effective carbon number, that is, the adjusted value of the carbon number for a compound that gives the correct FID response,
relative to some standard compound, such as benzene. This parameter is calculated by adding up the contributions from all of
the atoms or functional groups in the molecule. One set of parameters used to calculate effective carbon numbers is given in
Table 5.8. Experimental response factors are sometimes reported
in terms of effective carbon number. In general, this approach
is useful for predicting relative response factors with about
20% accuracy. Much better accuracy is achieved when compounds
are restricted to members of a homologous series, or to compounds
with a limited number of heterogroups. Several workers have
tabulated large numbers of experimental relative response factors
(11-13).
    Whatever value there is in predicting response factors, it can
be seriously diminished by improperly set flow parameters. In
addition to phenomena mentioned in the last section, relative response changes have also been reported due to water in the air
supply and a high column bleed.

TABLE 5.8   CONTRIBUTIONS TO EFFECTIVE CARBON NUMBER

Atom	Type	Effective Carbon No. Contribution
C	Aliphatic	1.0
C	Aromatic	1.0
C	Olefinic	0.95
C	Acetylenic	1.30
C	Carbonyl	0.0
C	Nitrile	0.3
O	Ether	−1.0
O	Primary alcohol	−0.6
O	Secondary alcohol	−0.75
O	Tertiary alcohol, esters	−0.25
Cl	Two or more on single aliphatic C	−0.12 each
Cl	On olefinic C	+0.05
N	In amines	Similar to O in corresponding alcohols

PART II

DETECTORS

5.6   ELECTRON CAPTURE DETECTOR

In itself, the name of this detector is indicative of the imagin-
ative inquiry gas chromatography (GC) has generated in many
minds. Although the detector does not "capture electrons," compounds with high electronegativity entering the detector do react
with particles emitted there. The operating mechanism has become
one of the most sensitive means the analytical chemist has for
examining complex mixtures.

The detector has been the most significant factor for studies used by scientists for alerting the world population to the spectre of global damage to the ecosystem. For the compounds most amenable to the detector are the chlorinated hydrocarbons: pesticides and polychlorinated biphenyls; substances strongly linked to pathological states. There is a place in history for the effors and achievements of those who made possible the ability to detect 10^{-14}g/sec of material.

5.6.1 Historical

The electron capture detector is the result of a series of developments which were initiated in 1951 by D. J. Pompeo and J. W. Otvos (14) of the Shell Company's development laboratory in California. The device they invented was a beta-ray ionization cross-section detector (Section 5.8). Deal et al. (15) at the Shell laboratory in California and Boer (16) in Amsterdam modified the detector, used originally to monitor effluents of a large scale plant process, for applications in GC. From the limited success of the detector, Lovelock (17) produced the beta-ray argon detector in 1958 (Section 5.8). This modification substituted argon as the carrier gas and placed a potential of 1000 V across the electrodes. Argon passing between the electrodes absorbed radiation and formed a metastable species with energy (11.6 eV) sufficient to ionize most substances. Proposed mechanisms for this process are:

$$Ar \xrightarrow{\beta} Ar^+ + e^- \tag{5.9}$$

$$Ar \xrightarrow{\beta} Ar* \tag{5.10}$$

$$C + Ar* \xrightarrow{\beta} Ar + M^+ + e^- \tag{5.11}$$

$$Impurities \xrightarrow{\beta} Background \tag{5.12}$$

Equations 5.10 and 5.11 are representative of efficient mechanisms whereby each primary electron from the radioactive source can result in the formation of as many as 10^4 organic ions. Equation 5.12 emphasizes strongly the necessity of controlling all sources of contamination in the system.

Because of anomalies present when halogenated compounds were eluted, Lovelock proposed the theory that electronegative species, functionally present in an organic molecule, could capture an electron to form a negatively charged species:

$$CX + e^- \rightarrow CX^- + Energy \tag{5.13}$$

These entities would then cause a reduction in the standing or background current. This phenomenon is known as electron capture and is observed more readily at lower electrode potentials. Essentially a current of electrons is monitored so that when an electroactive solute passes through this stream there is a decrease in the standing current.

5.6.2 Principles of Operation

CELL DESIGN. Through the course of its utilization, the design of the electron capture detector has changed little from the original. The simple arrangement of a chamber containig two electrodes with a source of radiation to induce ionization would seem to lend itself to a well developed theory. This is unfortunately not the case. The real situation is that complex processes are operable and literature reports frequently pertain only to the experimental system generating the data, and thus, modeling should be done cautiously. An excellent review by Pellizzari details these mechanisms (18).

Early cells were usually converted from argon ionization detectors. Detectors of this type are of coaxial geometry wherein the anode laying along an axis is sheathed by the cathode containing a radioactive source. Another design is referred to as the plane parallel electron capture detector (ECD)(Figure 5.12). Figure 5.12 illustrates the parallel alignment of the anode and the cathode. This structural design directs the flow of carrier gas in a direction opposite to the motion of the negatively charged species. These species have a slow drift velocity which becomes impeded and increases the probability of combination with positively charged ions. For this reason it is considered a more efficient design. A third type, the concentric cylinder ECD has the radioactive foil located in the cathode region and the anode in an isolated region. Another design is illustrated by the pin-cup detector, which can be considered a modified chamber of the coaxial ECD geometry (Figure 5.13). Evaluation of these design types has been conducted, but because of the complex nature of the experimental variables, firm conclusions are not possible. A special cell has also been designed to promote gas phase coulometry (Figure 5.14) (19). Coulometry by electron capture occurs when the ratio of electrons captured in a second (coulombs) to the number of molecules through the detector in a second approaches 1. Sensitivity is at a maximum under these conditions.

RADIATION SOURCE. Many sources of radiation have been used, but

today most users employ ^{3}H or ^{63}Ni. Tritium, which is a weak
beta emitter, necessitates a short cell distance of about 2.0 mm
for efficient currents. Radiation of beta-rays from ^{63}Ni have a
range of 8.0 mm. The requirement of small detector volume for
tritium favors efficient chromatographic systems. Temperature
restricts the maximum operating limits of the detector. For
tritium the upper limit is 225°C and for nickel-63 the limit is
400°C. The lower limit is fixed by the column conditions and
should be high enough to prevent condensation of stationary phase
bleed and high-boiling eluants. Within these restrictions, the
ionization chamber should be well insulated so that temperature
fluctuations are within ±0.3°C. Tritium has the advantage that
the flux of radiation is higher and thus ionization is more efficient. Although tritium is more sensitive, it is easily contaminated by adsorbed components on the foil which effectively
shield the weak beta-rays (18keV). As a result some workers prefer the higher durability and energy of ^{63}Ni (67keV). Recent developments have demonstrated the effectiveness of tritium embedded in scandium. Using this source, the maximum operating
temperature could be extended to 325°C and thereby reduce the
possibility of adsorption. Tritium at high temperatures emanates
with the effluent from the detector at a level in excess of what
the Atomic Energy Commission considers safe. A count beyond
$2 \times 10^{-7} \mu Ci/cm^3$ in air constitutes a health hazard. It is recommended that the effluent be vented into a fume hood. Federal licensing is also required of operators using radioactive material
(20).

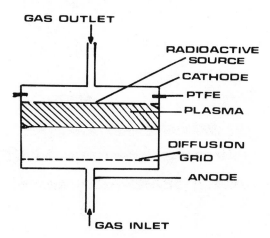

Figure 5.12.   The plane parallel ECD.

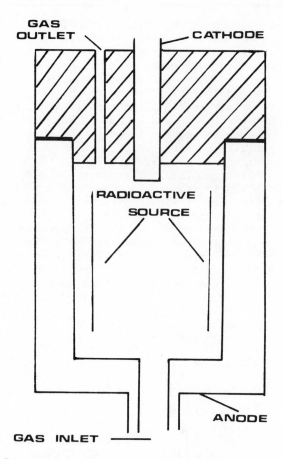

Figure 5.13.   Pin-cup ECD Detector.

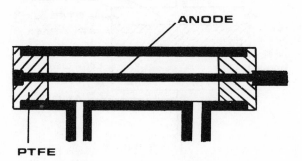

Figure 5.14.   Coulometric ECD detector.

FLOW REQUIREMENTS. The carrier gases used are nitrogen or argon
containing methane at 5 - 10% of the total volume. The methane
reduces the concentration of metastable argon and promotes thermal equilibrium of the electrons. The ECD <u>is/is not</u> a flow-
sensitive detector. Many believe that column bleed and traces of
oxygen in the carrier gas are responsible for flow and temperature dependence. It is prudent to see if the system is dependent.

VOLTAGE REQUIREMENTS. The potential applied to the electrodes
can be accomplished in several ways: (a) at constant voltage,
(b) under pulsed-constant frequency, (c) under pulsed-variable
frequency, and (d) by solute switching and synchronous demodulation.
     Under normal d-c operating voltages, the selection of the
applied potential is determined by the species measured, cell
design, carrier gas composition, and detector contamination.
This point is critical and was emphasized in a study comparing
two detectors. In one report a chlorinated hydrocarbon decreased
the standing current greater at voltages corresponding to 90% of
the standing current in the first case and 20% in the second
case. Good operating practice dictates construction of a voltage
response curve. It is also possible to have competing processes
under d-c conditions (21) so that anomalous behavior results
(Figure 5.15). To overcome these possibilities pulsing the voltage has been successfully employed and is preferred by researchers. This mode has the desired feature of decreasing significantly the buildup of charged zones in the detector. This situation results from differences in the velocity of positively
charged ions compared to the mobility of free electrons. The
positive ions form a charged cloud near the cathode and can contribute to anomalies occurring when a solute enters the chamber
and perturbs the equilibrium. Unlike the d-c mode, the concentration of the electrons is not constant but varies with the
application of the pulse. Another significant difference is that
the driving force of the applied potential field is absent and
the electrons attain thermal equilibrium. The pulse can be visualized as a process for collecting electrons (Figure 5.16). In
general, the pulse is between 30 and 50 V over a time interval of
0.5 - 1.0 μsec and repeated every 100 μsec. Sensitivities are
enhanced by this technique.
     An innovative variation of the pulse mode was to hold the cur-
rent constant and vary the frequency of the pulse. Normally the
electron concentration within the cell is constant in the d-c
mode. When an electroactive solute enters the cell the current
decreases as it does when a pulse is applied. By slowing the

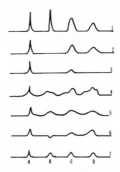

Figure 5.15.   Anomalous responses that can occur with
an electron capture detector operated in the d-c mode
(Ref.182, with permission).   The chromatograms depicted are illustrative of observed phenomena:
(1) Chromatogram that is truly representative of a
mixture; A and D are peaks of electron absorbers, B
is a larger amount of a nonabsorber, and C is an unresolved mixture of absorbing and nonabsorbing compounds.   (2) An ECD operating correctly. (3) ECD
losing peaks as a result of space charge effects.   (4)
ECD with contact potential (the result of material adsorbed on an electrode) enhancing the applied potential.   (5) Contact potential opposing the applied
voltage.   Observe the increased tailing and the false
peak at B. (6) ECD acting as if it were an argon detector or a cross section detector. Note the inversion
of B and the reduction of peak C. (7) ECD operating as
an electron mobility device and an electron capturing
device.   Note the false peak at B.

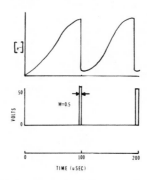

Figure 5.16.   Pulsed mode.

frequency of the pulse when a solute is present, a steady current
can be maintained and the following relationship holds:

$$(Solute) \sim Frequency \qquad\qquad (5.14)$$

The frequency generator is controlled by a feedback loop to maintain the steady state. Amplifiers can magnify the signal and as
a result this method has an improved linear dynamic range and increased sensitivity.

The concept of solute switching and detection by synchronous
demodulation was proposed recently by Lovelock (22). The application of the insight originated from work describing a coulometric electron capture detector (Figure 5.14) which was believed
capable of functioning as a solute switch. A solute switch would
have the capability of removing instantaneously all of the components in the carrier gas without altering the stream. Employing two of these coulometric ECDs in series enabled the first to
be used as the switch and the second as a detector. The switch
then modulates by pulsing the transfer of solute to the detector.
Selection of filters and a phase shifter enabled the signals to
be demodulated, reduced in noise, and amplified. Signal to noise
was enhanced by a factor of 10. The concept of the solute switch
poses prospects for a new generation of detectors and affords experimental design for investigating mechanisms of the process.

5.6.3 Response Factors

In the ECD, electrons react with strong electron-absorbing com-

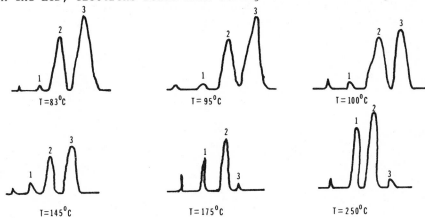

Figure 5.17.   Influence of temperature on response.

pounds. As such, this is a second-order rate mechanism which can be influenced by temperature (Figure 5.17), the electronegativity of the species itself, the presence of other species, and the energy of the electrons. Control of these conditions still does not eliminate variations in comparative detector responses published in the literature. This point has been the focus of some controversy even among competent laboratories. Improvement in the stability of response has been reported and is due to the independence of response factors attributable to carrier gas impurities and column bleed. However, if the forward rate of the reaction is high enough, the mechanism becomes pseudo first order and approaches coulometry. This is accomplished by selection of the right operating parameters, and having a strong electron absorbing molecule. Knowing the kinetics of the reaction would then enable a response factor to be calculated. A more convenient approach was forwarded by Sullivan (23) for pulsed electron capture detection. It was shown that the output frequency (F) and sample concentration (A) are related by

$$F = 1/K(k_1\{A\} + K_d) \tag{5.15}$$

where k_1 is the rate constant for electron capture,
$\quad K_d$ is the pseudo first-order rate for reassociation of electrons with positive ions and is generally small,
$\quad K$ is the constant of proportionality.
The output current (I) can be expressed:

$$I = I_s/K(1-e^{-K}) \tag{5.16}$$

where I_s = the current produced by the radioactive source,
$\quad K = (k_1(A) + K_d)t_p$
$\quad t_p = 1/F.$

Substitution of values for I and I_s in Equation 5.16 and using reiterative techniques yields an approximate value of K, so that

$$k_1 = \frac{KfMx10^{12}Z}{Nr} \tag{5.17}$$

where f is the flow (cm^3/sec),
$\quad N$ is Avogadro's number,
$\quad r$ is the frequency to voltage correction factor for the analog output ($\mu V/Hz$),
$\quad M$ is the sample molecular weight,
$\quad Z$ is the area response factor (μV-sec/pg).
The electron capture detector is a specific detector and as such the response factor will vary by the very nature of the

solute. Some generalizations can be made in this regard. Table
5.9 lists classes of compounds and their relative responses.

TABLE 5.9 RELATIVE RESPONSE VALUES[1]

Chemical Classes	K' [2]	Selected Samples
	0.01	
Alkanes, alkenes, alkynes, aliphatic ethers, esters, and dienes		Hexane Benzene Cholesterol Benzyl alcohol Naphthalene
	0.10	
Aliphatic alcohols, ketones, aldehydes, amines, nitriles, monofluoro, and monochloro compounds		Vinyl chloride Ethyl acetoacetate Chlorobenzene
	1.0	
Enols, oxalate esters, mono-bromo, dichloro, and hexafluoro compounds		cis-Stilbene trans-Stilbene Azobenzene Acetophenone
	10.0	
Trichloro compounds, chlorohydrates, acyl chlorides, anhydrides, barbiturates, thalidomide, and alkyl-leads		Allyl chloride Benzaldehyde Tetraethyl-lead Benzyl chloride Azulene
	300	
Monoiodo, dibromo, and trichloro compounds, mono-nitro compounds, lacramators, cinnamaldehyde, fungicides, and pesticides		Cinnamaldehyde Nitrobenzene Carbon disulfide 1,4-Androstadiene-3,11,17-trien Chloroform
	1000	
1,2-Diketones, fumarate esters, pyruvate esters, quinones, diiodo, tribromo, polychloro, dinitro compounds, and organomercurials		Dinitrobenzene Diiodobenzene Dimethyl fumarate Carbon tetrachloride
	10,000	

[1]From ref.(18) with permission. [2]K' values are based on chloro-
benzene = 1.0.

5.6.4 Linearity of Response

The response of an ECD in the d-c mode is linear with concentration over only two orders of magnitude (10^2). A relationship proposed by Wentworth and his colleagues (24) increased the linear range to about 10^4 by the equation:

$$\frac{I_s - I}{I} = kc \tag{5.18}$$

where I_s is the standing current,
$\quad$ I is the current measured when the electron absorbing
$\qquad$ species is at concentration c,
$\quad$ k is a constant of the cell and the species present.
$\quad$ By employing analog convertors the linear dynamic range can be extended to 1×10^5.

5.6.5 Minimum Detectable Level (MDL)

For a coulometric response, an ideal compound, sulfur hexafluoride, has remarkable sensitivity (10^{-14}g/sec). Again it should be emphasized that this value will be unique for a particular compound and depends on the species affinity for electron capture.

5.6.6 Practical Operating Hints

The following list contains recommendations for the integrity of a gas chromatographic system utilizing electron capture detection.

1. Flow regulators at the column head should not have parts that would contribute to background. A simple needle valve is sufficient.
2. Oxygen, which is an electron absorber, should be scrupulously trapped by molecular sieves.
3. Cure septa in a vacuum oven.
4. Clean all tubing in the system and bake it.
5. Keep column bleed to a minimum.
6. System should be leak tight and not allow diffusion of gases.
7. Exercise the necessary precautions when handling radioactive isotopes.
8. Use another detector to aid in the evaluation of a chroma-

togram.

Adherence to the above list should avoid the detector's severest limitation--susceptibility to contamination.

5.6.7 Applications

The ECD has seen its greatest utilization in the field of pesticide analysis. Chemicals such as dieldrin, aldrin and DDT are amenable to the ECD. In addition, the environmental hazards of polychlorinated biphenyls (PCBs) have been raised as a result of analyses by the ECD. Organometallics are also good electron absorbers. With the aid of halogen derivatization, many classes of organic compounds such as steroids, acids, amines, phenols, and alkenes have been assayed.

5.7   THE HELIUM/ARGON DETECTOR

These two detectors are incorporated under the same section since the design and operating principles discussed apply to both. Both detectors use the same chamber design so that fundamental differences are with respect to the gases themselves. That is, the ionization potential of helium is significantly higher than argon and thus has the capability of ionizing some species which argon cannot. In this sense it is more universal.

As described in Section 5.6.2, argon/helium atoms are excited to a metastable state by beta radiation from a radioactive source. The species formed is then capable of ionizing all compounds with a lower ionization potential. The products formed are then subject to an electric field (500-1100 V) and the change in current measured.

The most significant problem associated with these detectors is the purity of the carrier gases as well as impurities introduced by the chromatographic equipment. This problem is common to all highly sensitive detectors and the guidelines in Section 5.6.6 are applicable in this situation. Another aid in reducing background current associated with column bleed is to employ gas-solid chromatography. This is generally the case since the detectors are most often used in the analysis of the permanent gases.

The part-per-billion sensitivity for the fixed gases makes the helium detector the most suitable for trace analysis of these gases. A minimum detectable level of 4×10^{-14} g/sec and a linear range of over three orders of magnitude is possible.

5.8   CROSS-SECTION DETECTOR

As was mentioned in Section 5.6 this detector was the first of
the beta-ray ionization detectors. Its main advantage is that
the response to a given substance is calculable using published
values of the atomic cross sections of the constituent elements.
In design and construction this apparatus is similar to an argon
detector (Section 5.6.1) and has been evaluated in terms of its
basic principles. In general, ion pairs are formed in the ionization chamber and the applied potential across the detector is
adjusted so that losses by recombination and fragmentation at
high energies are avoided. When a component in the carrier gas
emerges from the column, the current measured (i) is related to;
the mole fraction (X_S) of the sample to the carrier gas; the
ionization cross section of the sample vapor (Q_S) and the carrier
gas (Q_C).

$$i = X_s Q_s + (1-X_s)Q_c \tag{5.19}$$

   The ionization cross section is a constitutive property and is
the sum of the atomic cross sections. Knowing the elemental composition of a gas enables prediction of a response. Hydrogen is
the carrier gas used. The detector is universal and capable of
an MDL of 3×10^{-11} g/sec. It has the advantage of a wide linear
dynamic range of 10^5. Some exception to the linear dynamic range
and predictability of response has been made and certainly rests
in the diversity of experimental conditions.

5.9   FLAME PHOTOMETRIC DETECTOR

The impact of the flame photometric detector (FPD) resides in its
simultaneous sensitivity and specificity for the determination of
sulfur and phosphorus. It is inherently compatible with the FID
and as such affords the analytical chemist a discriminating
ability beneficial to many analyses. In 1966, Brody and Chaney
published data on their design of an FPD (26)(Figure 5.18).

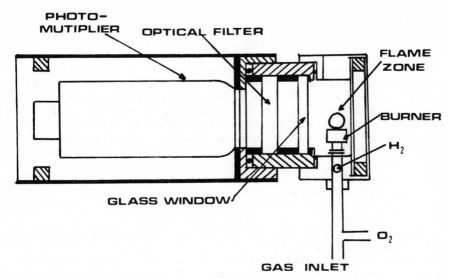

Figure 5.18.   Flame photometric detector.

5.9.1 Principles of Operation

In a hydrogen-rich flame, combustion of samples containing phosphorus and/or sulfur results in the formation of chemiluminescent
species which emit light characteristic of the heteroatom introduced into the flame.   Selection of an interference filter with a
394- or 526-nm bandpass allows selectivities for sulfur and phosphorus respectively.   Recent work by Krost and co-workers (27)
found that a 690-nm filter showed selectivity for some nitrogen-
containing compounds.

     As Figure 5.18 illustrates, the photomultiplier should be pro-
tected from the intensity of the heat generated by the flame.
Design of this detector also shields regions of the flame associated with background noise.   Additionally, early models included
mirrors near the flame in an effort to increase the signal to the
detector.   However, since this also enhanced the noise level, no
real advantage was achieved and mirrors became obsolete.   In
their place another detector was mounted, making it possible to
monitor phosphorus and sulfur simultaneously.   The detector flame
also served as an ionization source and was capable of being used
as an FID.   This was not always completely advantageous in trace
analysis since the collector electrode was not always efficient,
and the applied potential was kept low (22 V) in favor of chemiluminescence.   This prompted one investigator to split the effluent from the column to an FID and an FPD so that full advantage
of both detectors could be attained.

GAS FLOW. In GC using an FID, it is common to mix the carrier gas with hydrogen and burn the mixture in air at the burner tip. With the FPD it has been found advantageous to use nitrogen as the carrier gas and mix it with oxygen at the column exit in a proportion similar to air. Hydrogen is added at the burner to initiate combustion. Carrier flows can vary over wide ranges (20-160 cm^3/min) so long as the total flow of nitrogen and oxygen are maintained near 200 cm^3/min. By analogy to the FID it is apparent that the detector is mass flowrate sensitive and not concentration dependent.

5.9.2 Response Factors

The response of phosphorus in the FPD was determined to be linear while that of sulfur varied such that the square root of the response was proportional to concentration. This prompted researchers to propose a mechanism in which S_2^* was formed and was the chemiluminescent species. The following mechanism has been proposed (28):

$$2RS + (2+x)O_2 \rightarrow xCO_2 + 2SO_2 \qquad (5.20)$$

$$2SO_2 + 4H_2 \rightarrow 4H_2O + S_2 \qquad (5.21)$$

$$S_2 \rightarrow S_2^* \qquad (5.22)$$

The sulfur selectivity to hydrocarbons is 10^4:1. Better selectivities have been reported provided the sulfur was completely separated from other species. Anomalous behavior has been observed, and includes quenching of the sulfur chemiluminescence when sulfur elutes with other compounds, and when sulfur elutes at high concentration self-absorption occurs in the flame, producing deviations from linearity.

Phosphorus selectivity to hydrocarbons is 10^5:1. Effects of quenching and self-absorption have not been detailed in the literature.

Selectivity of phosphorus to sulfur is as low as 4:1 and of course not favorable. Occurrence of the crossover phenomena in the phosphorus mode should always be checked against response in the sulfur mode. Fortunately, sulfur has a selectivity of 10^3-10^4 for phosphorus so a distinction can be made.

In the event that a compound contains both phosphorus and sulfur, the ratio of the response of phosphorus to the square root

of the sulfur response is an excellent diagnostic of their relative amounts (29).

Compound	$R_p/\sqrt{R_s}$
PS	5.2-6.0
PS_2	2.8-3.3
PS_3	1.7-2.3

   These ratios are independent of concentration, column temperature, or retention time.
   As was mentioned, the FID in tandem with either the phosphorus
mode or the sulfur mode is a discriminating technique. A ratio
of the FID signal-to-noise ratio to the FPD signal-to-noise ratio
in either the P or S mode will be greater than 1000 if neither P
or S are present and less than 1 if they are present.
   Response is also a function of the degree of oxidation of the
sulfur such that the response follows the order: $-S - S - > SO_4^{2-}$
$>S=O >SO_3^{2-} >S^2$.

   5.9.3 Minimum Detectable Level

The minimum detectable level for phosphorus is about $2x10^{-12}g/sec$
and that of sulfur is $5x10^{-11}g/sec$.

   5.9.4 Linearity of Response

Phosphorus is linear over a 10^4 range and sulfur, on a log-log
scale, is linear over three orders of magnitude.

   5.9.5 Applications

The flame photometric detector has been found to be versatile in
pesticide analysis, food putrefaction, air pollution, and for
fuel analysis.

5.10   ALKALI FLAME IONIZATION DETECTOR (AFID)/THERMIONIC
        DETECTOR (TID)

The alkali flame ionization detector (AFID) or thermionic detect-
or (TID) is a specific detector used primarily in the trace analysis of pesticides; particularly those containing phosphorus.
In this resepct it fulfills a need comparable to the ECD for
analysis of pesticides containing halogens.

Progenitors of current AFIDs were founded on the observation
that a metal anode when heated in a gas emits positive ions. Use
of a platinum filament showed some sensitivity to halogenated
compounds. An FID was modified such that a probe, with sodium
salt deposited on it, was inserted above the flame. This detect-
or showed a specificity for phosphorus- and halogen-containing
molecules. However, the selectivity was poor, and Karmen (30)
improved this shortcoming by stacking two flames and attained a
phosphorus: hydrocarbon selectivity of 10^5:1. In this experimental design, the first flame burned the eluted materials and
vaporized sodium deposited on a platinum screen located above the
flame. The vapor was then transferred into the second flame
where ionization took place. Many design modifications were proposed to enhance certain characteristics and eliminate variables
(Figures 5.19A and B). Location of the salt around the burner
tip has become a preferential design since the salt is not diminished quickly, equilibrium is established more rapidly, and
sensitivity is better.

5.10.1 Operating Principles

Several theories have been proposed to explain the mechanisms involved in an AFID system (31). In general, thermal energy is required to atomize a particular alkali metal salt. The alkali
metal atoms formed ionize and are subjected to an electric field.
This produces a current proportional to the number of ions. The
presence of halogen, phosphorus, and even nitrogen enhance the
signal. The system is complex and does not lend itself to a
complete theory as intricate surface phenomena are possible. In
addition, there is speculation that photochemical processes
occur and realization that combustion products formed in the
flame can interact to form a multitude of species compound the
difficulty. It has been proven that the process does depend on
thermal energy and not strictly speaking on the products of combustion. For this reason many researchers prefer the term
thermionic ionization.

5.10.2 Flowrates

Utilization of the hydrogen flame necessitates some of the considerations discussed in Section 5.5 for the FID. In addition it
should be emphasized that the thermionic process is temperature-
dependent and as such is a function of the flame temperature and
therefore dependent on the hydrogen flowrate. Considering that
different salts are used, and in the time of their use are depleted, optimization of flowrates is mandatory. In fact, manufacturers set rigid specifications for hydrogen flow regulation.
Figure 5.20 illustrates this fact by plotting the selectivity for
nitrogen or phosphorus as a function of flowrate.

5.10.3 Applied Potential and Polarity of Electrodes

As with the FID, about 300 V is necessary for efficient collec-

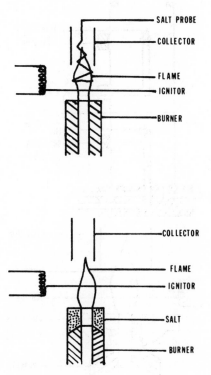

Figure 5.19.   Alkali flame ionization detectors.

tion of ions. Unlike the FID, the collector electrode is maintained as the negative electrode. Position of the electrode
should also be optimized.

    5.10.4 Alkali Metal Salts

Comparison of all the salts used is virtually impossible since
experimental conditions varied so much. It was found that the
reliability of the detector for alkali salts is in the order
K>Rb>Cs, but that sensitivity to nitrogen follows the order
Rb>K>Cs>Na. Sensitivity is found to increase with increasing
atomic number. Purity of the salt used is also critical in terms
of ionization efficiency.

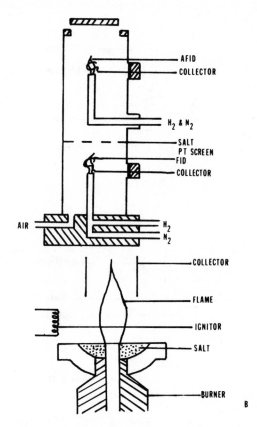

5.10.5 Response Factors

Response factors are related directly to experimental conditions
and vary for different atoms. Although phosphorus has received
the greatest attention, Table 5.10 illustrates relative responses of several species (32). Caution should be exercised in this
type of comparison of heteroatoms since the rest of the molecule,
in that if altered, can change the flame temperature, and influence the response.

TABLE 5.10 TID/FID RESPONSE RATIOS[1]

Compound	Formula	TID/FID
Triphenylphosphine	$(C_6H_5)_3P$	11,000
Triphenyl amine	$(C_6H_5)_3N$	60
Triphenylarsine	$(C_6H_5)_3As$	10
Triphenylstibene	$(C_6H_5)_3Sb$	10
Triphenylbismuthine	$(C_6H_5)_3Bi$	10

[1]From ref. (32) with permission.

Selectivities in the phosphorus/hydrocarbon mode and in the

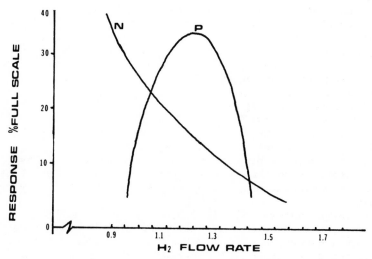

Figure 5.20.   Effect of flowrate on selectivity.

nitrogen/hydrocarbon mode are $10^5/1$ and $10^3/1$, respectively.

5.10.6 Minimum Detectable Level

Limits of detection for nitrogen and phosphorus are 1.2×10^{-13} and 0.5×10^{-13} g/sec, respectively. This is a range of sensitivity comparable to the ECD, and is responsible for use of the detector in pesticide analysis and drug analysis.

5.10.7 Linearity of Response

Again due to experimental design variations, the linear dynamic range varies from 10^2 to 10^4 and also varies with the compound examined.

5.11 FLAME EMISSION DETECTOR

Juvet and Durbin (33) employed a high-energy oxy-hydrogen flame to obtain spectra from the emission of volatile metal chelates. This was accomplished by connecting the gas chromatographic outlet to a Beckman DU flame photometer equipped with a total-consumption burner. Their system exhibited good selectivity and sensitivity, and had the advantages of simplicity and low cost. Selectivity of the detector was 10^3 to 10^4 for metals over organics. Specificity for the metals depended upon the resolution of the monochrometer, but for most elements interference filters were sufficient and increased sensitivity was possible. Minimum detectable levels ranging from 10^{-7} to 10^{-11} moles were observed.

The advantages of the flame emission detector (FED) have been combined with the flame ionization detector. This design features the ability to detect CO, CO_2, N_2O_4, SO_2, N_2F_4, HF and H_2S; gases which respond poorly in an FID. In addition, the system showed qualitative differences in structure attributable to different FED/FID ratios as a function of wavelength for various compounds.

Other investigators used flame emission as a modified Beilstein test for the detection of halogenated hydrocarbons. In such an arrangement, a green flame was produced when halogenated hydrocarbons were burned in the presence of a copper wire. Replacement of the copper with indium improves specificity and

selectivity. The detector differs from the flame photometric
detector in that higher energy is required for excitation of the
light-emitting species which involves electronic transitions for
atoms rather than molecules.

5.12   MICROWAVE PLASMA DETECTOR (MPD)

The microwave plasma detector differs from the flame photometric
detector in that excitation by the microwave plasma involves
higher energies than the cool flame of the flame photometric detector. The energies are such that fragmentation of the molecular species occurs. These fragments are also excited and it is
their emission as well as emission from free atoms or diatomic
molecules that is monitored. The use of a microwave plasma as a
gas chromatographic detector was first reported by McCormack,
Tong, and Cook (34). They reported a sensitivity for hexane of
$2x10^{-16}$g/sec. Refinements were made by replacing argon at atmospheric pressure with helium at reduced pressure. An improved response is obtainable. An apparatus is commercially available
(Figure 5.21). This apparatus is based on investigations of
McLean, Stanton and Penketh (35). They doped the plasma with low
levels of oxygen or nitrogen to prevent carbon from building up
within the discharge tube. The instrument also features multielement detection and improved sensitivity by correcting for

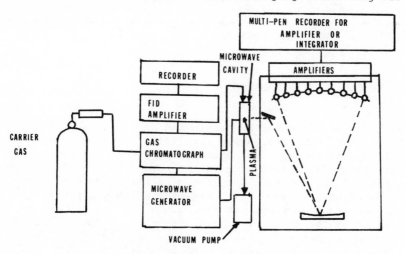

Figure 5.21.   Multielement microwave plasma detector.

background interference. Table 5.11 lists the MDL for several
elements and their selectivities. Linearity is over four orders
of magnitude.

   Requirements for good performance are stringent. Ultra-pure
helium and a clean chromatographic system without leaks are necessary. However, the MPD is capable of very good sensitivity and
has the important feature of aiding structure elucidation. Improvements in design are still possible and may eventually allay
its limited use brought about by high cost.

TABLE 5.11 MPD SENSITIVITY AND SELECTIVITY DATA[1]

Element	Compound	Wavelength	Detection Limit (g/sec)	Selectivity Ratio vs. n-Hexane
Carbon		3883	2×10^{-16}	–
		5165	3×10^{-14}	
Fluorine	C_6F_6	5166	3×10^{-12}	10
		2516	5×10^{-10}	20
Chlorine	$CHCl_3$	2788	8×10^{-10}	20
Bromine	$CHBr_3$	2985	2×10^{-7}	10
Iodine	CH_3I	2062	7×10^{-14}	10^{-4}-10^{-7}
Phosphorus	$(C_2H_5O)_3PO$	2555	1×10^{-11}	100
Sulfur	CS_2	2575	1×10^{-9}	100

[1]From ref. (35) with permission.

5.13   THE GAS DENSITY BALANCE

This detector was invented by Martin and James (1) and is a universal detector. The design was simplified and subsequently made
available by the Gow-Mac Instrument Company. Use of two of these
detectors in a dual column mode is the basis for the direct determination of molecular weight (the mass chromatograph). The
hydrodynamics of the gas density detector as they apply to the
mass chromatograph have been studied (36).

   A diagram of the detector is illustrated in Figure 5.22. A
reference gas enters the detector through an orifice at point A
and is divided so that it sweeps past both detector elements.
The column effluent enters at B. Maintaining the flow of A 15 to

TABLE 5.7 CHANGES TO DETECTOR OUTPUT DUE TO EXTRANEOUS EFFECTS

Effect	Sensitivity
Bridge current	40 μV/mA
Flow	
(one side only)	25 μV/cm^3-min
(both sides)	7 μV/cm^3-min
Cell pressure[1]	
(one side only)	2.3 μV/torr
(both sides)	0.15 μV/torr
Mechanical impulse	10 μV (for 3 g weight dropped 2.5 cm onto detector enclosure)
Filament temperature[2]	
(one side only)	12,400 μV/$^{\circ}$C

[1]For bridge current of 150 mA   [2]From Equation 5.7.

20 cm^3/min faster than B prohibits diffusion of the effluent into the detector filaments (filaments used are those available for thermal conductivity detectors). This feature makes the detector nondestructive and eliminates contamination of the effluent. It is also useful for the analysis of corrosive materials. If the gas flowing from B is of higher density than A, the flow along the downward path is retarded. This raises the temperature, and thus resistance of the lower detector element, and unbalances a Wheatstone bridge producing a signal. The opposite is true if the density of the effluent is less than the carrier gas.

Response for this detector is dependent upon the difference in the molecular weights of the gases used. For the determination of a high molecular weight species, a carrier gas of low molecular weight is best. Conversely, when a low molecular weight compound is determined, a high molecular weight carrier gas is used (ClC_2F_5, Freon-115). Hydrogen and helium are avoided as carrier gases since they diffuse rapidly and can effect the reference stream by back-diffusion of the sample stream. Equation 5.23 reflects the above:

$$A = W K x (M-m/M)$$ (5.23)

where A is the peak area,
      W is the weight of component,
      K is the instrument constant,
      M is the molecular weight of the solute,

m is the molecular weight of the carrier gas.

The mass chromatograph uses two different carrier gases and a standard of known molecular weight. Using a sample splitter at the injection port, the standard and then the sample are chromatographed with both carrier gases. Matched columns and dual detectors are used to keep K constant under both sets of conditions tions. The solution of the following equation for the unknown molecular weight is the basis for the mass chromatograph:

$$\frac{Ax_1 \ (M_s - m_1)}{As_1 \ (M_x - m_1)} = \frac{Ax_2 \ (M_s - m_2)}{As_2 \ (M_x - m_2)} \qquad (5.24)$$

Caution should be used when selecting the carrier gas so that comparison of the standard and sample are done under the best circumstances.

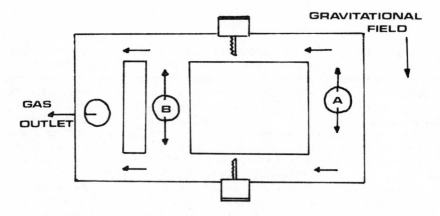

Figure 5.22. The gas density balance.

5.14   ELECTROMETRIC DETECTORS

5.14.1 Coulometric Detectors

Detectors designed according to the principles of coulometry have innate possibilities for selectivity and quantitative capacity. Coulson and Cavanaugh (37) first implemented coulometry to determine chlorine and later sulfur. Martin advanced the state of the art by designing a detector capable of determining trace amounts of nitrogen (38). These techniques were dependent on catalytic combustion of a sample into a coulometrically titratable species (Cl^-, S^{2+}, NH_3). Many modifications were made to the early systems in efforts to improve the efficiency of catalysis, remove inter- ferences, and to improve the sensitivity. Figure 5.23 represents a schematic of the overall process for an automated coulometric titrator as a gas chromatographic detector. Such a detector has four electrodes. One is a reference electrode and one is a sens- ing electrode; the remaining two generate the titrant. The tit- rant, a chemically specific reagent for the effluent, is main- tained constant by the cell. Addition of an active species causes a depletion of the titrant. This is sensed potentiometri- cally and additional titrant is generated until equilibrium is reestablished. The generation of titrant is stoichiometric and the coulombs required are proportional to the moles present in the effluent. Thus the response is absolute. A serious limita- tion to the coulometric detector is its slow response and large dead volume. Sensitivities have been reported at the parts-per- million level and a linear dynamic range extends over two and a half orders of magnitude. Aside from its lower cost it does not offer any advantages over the flame photometric detector for sul- fur determination and generally requires greater experimental control.

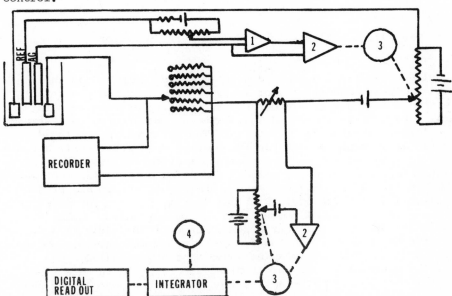

Figure 5.23. Automated coulometric detector. (1) Pre amp, (2) amp, (3) servo motor, (4) sync. motor.

5.14.2 The Reaction Coulometer

The reaction coulometer is among one of the few absolute detectors in GC. In principle it will respond to any compound that
will react with oxygen over a heated platinum catalyst. In practice this is limited to compounds containing C, H, and O; other
elements respond differently and are treated as contaminants. In
the reaction coulometer, an oxygen generator adds oxygen constantly to the gas chromatographic effluent which then passes
over a platinum catalyst. Combustion over the heated catalyst
depletes the oxygen in the stream. The decrease in the oxygen
concentration is sensed by an oxygen-sensitive electrochemical
cell. A feedback loop faradaically generates more oxygen to re-
establish the equilibrium potential set prior to sample reaction.
The main limitation of the detector is the lag present in the operation. Rapidly eluting samples cannot be quantified accurately.
Large samples can saturate the detector and even poison it. The
detector has a sensitivity slightly better than a TCD and is
capable of a detection limit of 10^{-8} g/sec and a linear dynamic
range over four orders of magnitude.

A similar device has been reported using a hydrogen generator
in conjunction with a cell sensitive to hydrogen concentration.

5.14.3 The Electrolytic Conductivity Detector

Coulson (38) first described a detector for GC based on the
electrolytic conductivity of ionic species in water. Material
eluted from a gas chromatograph was oxidized or reduced catalytically to form an ionic species which was transferred to a stream
of deionized water for detection. The water was in a closed cir-
culating system. By constantly removing ions after their detection by means of ion exchange resins, a differential signal was
obtained. The oxidative mode has been found to be ineffective
because of CO_2 which dissolves in the liquid stream and produces
background interference. Hall has developed an improved design
with a sensitivity of better than 0.1 ng for sulfur, chlorine,
and nitrogen, and a linear dynamic range of 10^5 (39). The detector has been used in the fields of pesticide analysis and herbicide analysis. Nitrosamines have also been studied.

The detector should be optimized according to furnace temperature, reactant gas flowrate, solvent flowrate, composition and
surface area of the nickel catalyst, and cell voltage.

5.15   THE ULTRASONIC DETECTOR

Because this detector is found to be universal, have a broad dynamic range, good sensitivity, and a wide choice of carrier gases
it is gaining in acceptance. Today there is an instrument commercially available (Figure 5.24)(40). In this detector, sound
waves are propagated at one transducer and received at another,
A phase meter monitors the signal received and is sensitive to
any changes. A disadvantage of the instrument is the high cost
and sophistication of the electronics necessary for good performance. The changes in the phase angle ϕ with the introduction of
a component in the mobile phase are expressed in the following
mathematical relationship:

$$\phi = 180sf \ (M_1/RT\gamma_1)^{\frac{1}{2}} \ n \ x \ \left[M_2/M_1 \ \{1+(C_{p2}/C_{p1})\gamma_1/(\gamma_2-1)\}-1 \right] (5.25)$$

where ϕ = degrees phase change,
       s   = sound pathlength (cm),
       f   = frequency (cycles/sec),
       M_1 = the molecular weight of the carrier gas,
       M_2 = the molecular weight of the sample gas,
       R   = the gas constant,
       T   = the absolute temperature of the gas,
       γ_1 = specific heat ratio of the carrier gas,
       γ_2 = specific heat ratio of the sample gas,
       n   = mole fraction of the sample,
       C_{p1} = the gram specific heat ratio of the carrier gas at
              constant pressure,
       C_{p2} = the gram specific heat ratio of the sample at
              constant pressure.
    From the above relationship it can be seen that the detector
is sensitive to pressure. In practice a backpressure is maintained in the cell to support propagation of the waves. For
hydrogen, 65 psig is necessary, and for helium only 10 psig is
required. Good control of temperature is necessary, and at the
lower limits of detection 10^{-3} to 10^{-4}°C is required. Variation
in flowrates do not effect response within the range 30 to 80
cm^3/min. Below 10 cm^3/min the change is exponential. The separation distance of the transducers is critical and must be optimized for the particular carrier gas. The cell volume (0.18 cm^3)
influences the resolution only significantly in the case of capillary GC. The detector has a linear dynamic range of six orders
of magnitude which is one of its strongest features, and a sensitivity to $2x10^{-9}$g of hydrogen as measured in helium. This is the
least favorable circumstance for measurement since the response
is directly proportional to the ratio M_2/M_1. Use of this

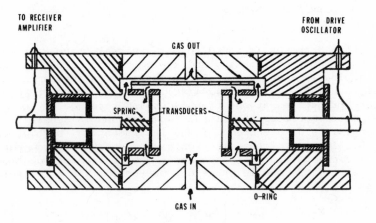

Figure 5.24. The ultrasonic detector.

detector has been restricted to the permanent gases and low-
boiling compounds since the detector does not perform well at
elevated temperatures (200°C).

5.16   THE PIEZOELECTRIC SORPTION DETECTOR

Adsorption of a solute is a technique that has been used as a
detector in GC. King (41) employed a piezoelectric quartz crystal as a detector by measuring the change in the oscillating frequency when a substance was adsorbed on the surface or partitioned in a liquid coated on the surface. This feature enables utilization of a particular phase such that selective or nonselect-
ive mechanisms can be chosen for measurement. This suggests a
great variety of detectors, and with the added characteristics of
low dead volume and rapid response, a viable approach to detection seems in the offering. The crystal is sensitive to temperature effects as expected. It appears possible that this type of
detector, although two to three orders of magnitude away, may
someday rival the FID and the TCD.

5.17   RADIOACTIVITY DETECTORS

Measuring the radioactivity of compounds eluting from a gas

chromatograph has been done by: (a) trapping fractions for subsequent radioassay, (b) measuring radioactivity directly in the
gas phase, (c) condensing and dissolving the effluent in a moving
stream of scintillation fluid which passes through a counting
device.

The use of flow-designed Geiger counters to determine the
radioactivity of products from a catalytic reaction and also flow
through scintillation counters has seen application in the analysis of some radioactive isotopes. Unfortunately, neither system
is capable of measuring the commonly used isotopes ^{14}C and ^{3}H.
To overcome this limitation, Wolfgang and Rowland (42) added
methane to the helium stream at the column outlet and passed the
mixture into a proportional counter. This system could detect
10^{-9}Ci.

Another approach employs the catalytic combustion of organics
over heated copper oxide. The CO_2 formed in the process was
suitable for ^{14}C counting. Passing the water formed in the combustion over heated iron produced hydrogen suitable for tritium
analysis. By this technique, Winkelman and Karmen (43) measured
202 nCi of carbon-14 and 227 nCi of tritium.

Recently Schutte and Koenders (44) described a method engi-
neered to overcome problems associated with trapping methods and
combustion methods. In their design, the column effluent merges
with scintillation fluid and is transported to the counter. As a
result of this design: (a) Sensitivity of 0.2 nCi was attainable; (b) resolution was not lost; (c) ^{14}C and ^{3}H could be dis-
tinguished.

5.18   MISCELLANEOUS DETECTORS

5.18.1 Semiconductive Thin-Film Detector

Changes in the electrical conductivity of a thin-semiconductive
film when an eluate is adsorbed on the surface was used by
Seiyama and co-workers (45). The response on a ZnO film (20-
1000 A) depended upon the nature of the interaction. For electron acceptors, such as O_2, a decrease in the conductivity was observed, while for electron donors such as ethyl alcohol and CO_2,
an increase in the conductivity was measured. Temperatures of
200 $^{\circ}$C or greater were necessary to avoid slow desorption rates
and concomittant loss in resolution. Sensitivities were poor,
but the phenomena are worth further scrutiny in hopes of obtaining materials exhibiting stable characteristics as well as

selectivity.

5.18.2 Dielectric Constant Detector

Johansson (46) designed a microwave circuit that accurately measured the dielectric constant of pure gases as well as detected
components in a mixture. The detector lacked sensitivity but had
the advantage of good thermal stability.

5.18.3 Brunel Mass Detector

Direction of the gas chromatographic effluent into a vessel containing activated carbon attached to an automatic recording
electromicrobalance is the basis for the device known as the
Brunel mass detector (47). This is an absolute analytical method
and requires no calibration, and in fact, can be used to calibrate other detectors which have unpredictable responses. The
sensitivity of the detector is in the same range as the thermal
conductivity detector.

5.18.4 Hydrogen Flame Temperature Detector

This detector, invented by Scott, consists of a thermocouple
placed directly above a flame (48). When a compound is eluted,
an increase in temperature produces a signal. The device is
simple and inexpensive and has the capability of detecting microgram quantities. The signal from the detector is proportional to
the molar heat of combustion of the eluted compound. To improve
the stability of the detector nitrogen is used as the carrier
gas.

5.18.5 The Titration Detector

James and Martin (49) reported on visual and automatic titration
methods capable of detecting microgram quantities of acids and
bases. This type of detection has the distinction of being the
first type of detector used in gas-liquid chromatography. Electronically modified designs for improving the automation of the
detector are possible.

5.18.6 Gas Volume/Pressure Detectors

By using carbon dioxide as carrier gas, Janak (50) demonstrated
that passage of the gas through a solution of potassium hydroxide would remove the carrier gas from the system. Providing the
sample compounds were not water soluble or hydrolyzable, a simple
Dumas nitrogen apparatus or a more sophisticated apparatus could
measure the volume of gas not absorbed by the solution as a function of retention time. Instead of measuring volume, van de
Craats (51) measured pressure changes at constant volume. Novak
and Janak constructed an apparatus that is the pressure analog of
the Wheatstone bridge (52). This instrument does not require an
external source of power and has sensitivity in the range of the
thermal conductivity detector.

5.18.7 Flow Impedance Detector

The flow impedance detector measures the pressure drop of a gas
across a capillary tube located at the column exit. Changes in
the carrier gas composition are due to gas viscosity when using
a true bore capillary and due to density when an orifice restricts the flow. An interesting application of the principle
was used for the direct determination of magnesium, zinc, and
cadmium present in aluminum alloys at temperatures as high as
$1400^{\circ}C$. In the absence of oxygen and using tungsten or graphite
as construction materials, the operation of the detector could be
extended to 3000 $^{\circ}C$.

5.19 THE MASS SPECTROMETER AS DETECTOR (GC/MS)

A gas chromatograph can have a mass spectrometer as a detector,
or a gas chromatograph can be used as in injection system for a
mass spectrometer. In this book the former case is true. In any
event, GC/MS is one of the most powerful techniques available and
concomittantly one of the most technically sophisticated. The
high specificity and sensitivity for thermally stable volatiles
enable unambiguous qualitative and quantitative information. It
is expensive in terms of hardware and the expertise necessary for
reliable operation. It is recommended that interested parties
consider the system in its entirety. McFadden's (53) book is
well suited for such a purpose.

5.20   ULTRAVIOLET/FLUORESCENCE DETECTORS

Ultraviolet spectrophotometers have been used as gas chromatographic detection systems mainly after condensation of the chromatographic effluent.   Systems are capable of detecting naphtha-
lene at 10^{-8}g by scanning every 20 sec from 165 to 220 nm.   Use
of a monochrometer permits selectivity.   Reactions producing
chemiluminescence are known.

Anthracene and pyrene were determined by fluorescence at
$1 \times 10^{-11}$and 5×10^{-12}g/sec, respectively.   The technique and its
principles have been described (54).

5.21   INFRARED DETECTORS

Attachment of an infrared spectrophotometer to the outlet of a
gas chromatograph has been successfully employed as a specific
detector by many analysts.   A text by Welti (54) emphasizes the
advantages of such a system as opposed to trapping techniques.
Welti has also provided evidence asserting the usefulness of vap-
or phase spectra as characteristic of the sorbate.   Several systems capable of sensitivity in the 0 - 100 µg range have been
described and stress the performance and cost reduction as compared to the gas chromatograph/mass spectrograph.   Penzias (55)
described a system capable of producing a quality spectra in 6
seconds from 2.5 to 15 µm.   Recent design improvements permit
MDLs near 50 ng.

Of course, infrared detectors which measure solutes "on the
fly" require an oven for the detector cell.   Initial attempts to
avoid that problem catalytically converted organic compounds to
CO_2.   Sensitivity increased with molecular weight as the number
of carbon atoms per molecule increased.

Fourier transform infrared spectroscopy (FT-IR) has been developing into a viable analytical technique (56).   The use of an
interferometer requires a computer which increases the cost of
the system.   The ability of IR to differentiate geometrical isomers is still an advantage of the system, and computer techniques
such as signal averaging and background subtraction, improve
capabilities for certain analyses.

REFERENCES

1. A. J. P. Martin, in Nobel Lectures, Including Presentation

Speeches and Laureates' Biographies, in Chemistry, 1942-1962, Elsevier, New York, 1964, p. 359.

2. H. A. McLeod, A. G. Butterfield, D. Lewis, W. E. J. Phillips, and D. E. Coffin, Anal. Chem., _47_, 674 (1975).

3. H. Oster and F. Oppermann, Chromatographia, _2_, 251 (1969).

4. L. S. Ettre, J. Chromatog., _112_, 1 (1975).

5. M. Dimbat, P. E. Porter, and F. H. Stross, Anal. Chem., _28_, 290 (1956).

6. D. M. Rosie and E. F. Barry, J. Chromatog. Sci., _11_, 237 (1973).

7. D. Jentzsch and E. Otte, _Detektoren in der Gas-Chromatographie_, Akademische Verlag., Frankfurt, 1970.

8. G. R. Jamieson, J. Chromatog., _3_, 464 (1960) _3_, 494 (1960); _4_, 420 (1960); _8_, 544 (1962); _15_, 260 (1964).

9. W. J. Miller, in _Fourteenth Symposium (International) on Combustion_, The Combustion Institute, 1973, p. 307.

10. P. Boček and J. Janák, Chromatogr. Rev., _15_, 111 (1971).

11. R. Kaiser, _Gas Phase Chromatography_, Vol. III, (trans. P. H. Scott) Butterworths, Washington, 1963.

12. W. A. Dietz, J. Gas Chromatog., _5_, 68 (1967).

13. L. S. Ettre, In _Gas Chromatography_, N. Brenner, J. E. Callen, and M. D. Weiss (Eds.), Academic Press, New York, 1962, p. 307.

14. D. J. Pompeo and J. W. Otvos, U. S. Patent 2,641,710 (1953).

15. C. H. Deal, J. W. Otvos, V. N. Smith, and P. S. Zucco, Anal. Chem., _28_, 1958 (1956).

16. H. Boer, _Vapour Phase Chromatography_, D. H. Desty (Ed.), Butterworth, London, 1957, p. 169.

17. J. E. Lovelock, J. Chromatog. _1_, 35 (1958).

18. E. D. Pellizzari, J. Chromatog., _98_, 323 (1974).

19. W. A. Aue and S. Kapila, J. Chrom. Sci., _6_, 255 (1973).

20. U. S. Atomic Energy Commission, _Conditions and Limitations on the General License Provisions of 10 CFR 150-20, Rules and Regulations_, May 1, 1964, Division of Materials Licensing, U. S. AEC, Wash., D.C.

21. J. E. Lovelock, Anal. Chem., _33_, 162 (1961).

22. J. E. Lovelock, J. Chromatog. _112_, _29_ (1975).

23. J. J. Sullivan, J. Chromatog. _87_, _9_ (1973).

24. W. E. Wentworth, E. Chen, and J. E. Lovelock, J. Phys. Chem., _70_, 445 (1966).

25. M. Borman, and M. Beroza, Anal. Chem., _40_, 1448 (1968).

26. S. S. Brody and J. E. Chaney, J. Gas Chromatog., _4_, 42 (1966).

27. K. J. Krost, J. A. Hodgeson, and R. K. Stevens, Anal. Chem., _45_, 1800 (1973).

28. E. R. Adlard, CRC Crit. Rev. Anal. Chem. _6_, 19 (1975).

29. H. W. Grice, M. L. Yates, and D. J. David, J. Chromatog. Sci., _8_, 90 (1970).

30. A. Karmen, Anal. Chem., $\underline{36}$, 1416 (1964).
31. J. Sevcik, Chromatographia, $\underline{6}$, 139 (1973).
32. L. Giuffrida, N. F. Ives, and D. C. Bostwick, J. Assoc. Offic. Agr. Chemists, $\underline{49}$, 8 (1966).
33. R. S. Juvet and R. P. Durbin, Anal. Chem., $\underline{38}$, 565 (1966).
34. A. J. McCormack, S. C. Tong, and W. D. Cook, Anal. Chem., $\underline{37}$, 1470 (1965).
35. C. A. Bache and D. J. Lisk, Anal. Chem., $\underline{37}$, 1477 (1965).
36. E. Kiran and J. K. Gillham, Anal. Chem., $\underline{47}$, 983 (1975).
37. D. M. Coulson, L. A. Cavanaugh, J. E. DeVries, and B. Walter, Agr. Food Chem., $\underline{8}$, 399 (1960).
38. D. M. Coulson, J. Gas Chromatog., $\underline{3}$, 134 (1965).
39. R. C. Hall, J. Chromatog. Sci., $\underline{12}$, 152 (1974).
40. T. Todd and D. DeBord, Amer. Lab., December (1970).
41. W. H. King, Anal. Chem., $\underline{36}$, 1735 (1964).
42. R. Wolfgang and F. S. Rowland, Anal. Chem., $\underline{30}$, 903 (1958).
43. J. Winkelman and A. Karmen, Anal. Chem., $\underline{34}$, 1067 (1962).
44. L. Schutte and E. B. Koenders, J. Chromatog., $\underline{76}$, 13 (1973).
45. T. Seiyama, A. Kato, K. Fujushi, and M. Nagatani, C. A., $\underline{60}$, 1083d (1964).
46. G. Johansson, Anal. Chem., $\underline{34}$, 914 (1962).
47. S. C. Bevan, T. A. Gough, and S. Thorburn, J. Chromatog., $\underline{43}$, 192 (1969).
48. R. P. W. Scott, Vapour Phase Chromatography, D. H. Desty (Ed.), Academic Press, New York, 1957.
49. A. T. James and A. J. P. Martin, Analyst, $\underline{77}$, 917 (1952).
50. J. Janak, Microchem. Acta, 1038 (1956).
51. F. van de Craats, Anal. Chim. Acta, $\underline{14}$, 136 (1956).
52. J. Novak and J. Janak, Gas Chromatographie, H. P. Angele, and H. G. Struppe, (Eds.), Akademie-Verlag, Berlin, 1963.
53. W. H. McFadden, Techniques of Combined Gas Chromatography/ Mass Spectrometry, Wiley-Interscience, New York, 1973.
54. D. Welti, Infrared Vapour Spectra, Heyden & Son, Ltd., London, 45 1970.
55. G. J. Penzias and M. F. Boyle, Amer. Lab., $\underline{5}$ (10), 53 (1973).
56. K. L. Kizer, Amer. Lab., $\underline{5}$ (6), 40 (1973).

CHAPTER 6

Instrumentation

RÜDER SCHILL

Hewlett-Packard Company

6.1	INTRODUCTION.	290
6.2	CARRIER GAS SYSTEM.	291
	6.2.1 Regulations/Controls.	292
	Pressure Controllers.	294
	Flow Controllers.	297
	6.2.2 Flow Measurement.	300
6.3	SAMPLE INLET SYSTEMS	304
	6.3.1 Gas Sampling	304
	6.3.2 Liquid Sample Inlets.	305
	Septumless Injectors.	309
	Septum-cooled Injectors	310
	Septum Bypass Injectors	310
	Vented Septum Injectors	310
	Inlet Systems for Capillary Columns	311
	6.3.3 Solid Sampling	315
	6.3.4 Automatic Samplers.	317
	Automatic Liquid Samplers	317
	Automatic Solid Samplers	320
	Automatic Gas Samplers	321
6.4	COLUMN OVENS.	321
	6.4.1 Column Temperature Control.	322
	6.4.2 Temperature Programming	326
	6.4.3 Column Oven Size and Configuration.	332
	Column Configurations	333

6.5   DETECTORS. 335
      6.5.1 Detector Gases 336
      6.5.2 Detector Temperature 338
      6.5.3 Electrometers 339
      6.5.4 Dual Detectors 342
            Dual Column Compensation 342
            Two Detectors in Series. 345
            Two Detectors in Parallel. 346
            Two Independent Column/Detector Systems. 348
6.6   RECORDERS 348
6.7   DATA REDUCTION AND HANDLING. 352
      6.7.1 Mechanical Integrators 354
      6.7.2 Electronic Integrators 354
      6.7.3 Computing/Reporting Integrators. 356
      6.7.4 Multichannel Computer Systems. 357
      6.7.5 Digital Reporting Chromatograph. 359
REFERENCES. 361

6.1   INTRODUCTION

Gas-liquid chromatographic apparatus was first described when
James and Martin published a paper (1) describing a system employing a recording burette for the separation and detection of
mixtures of organic bases and acids. The real potential of
applying this technique to the analysis of a wide range of compounds was only realized with Ray's publication (2) of a series
of gas-liquid chromatograms obtained by employing a thermal conductivity detector.

Estimates of gas chromatographs in use today are as high as
150,000. Today there are now well over 30 manufacturers of over
130 different models.

The 1960s witnessed significant progress in the performance
and accuracy of the apparatus. The versatility of the equipment
increased as new detectors and auxiliary equipment were developed.

In the late 1960s and early 1970s the handling of vast data
generated by gas chromatographs received the emphasis. This led
to automation of gas chromatographs and in some cases complication of the instrumentation. In the late 1960s electronic integrators were introduced. These instruments printed out the retention times and areas of the peaks generated by the chromatographs.

The last five years have brought two new trends:

1. Development of moderate cost instruments that provide a high level of performance.
2. Processor or computer control of all or at least some of the chromatograph's parameters.

Today gas chromatographs are available in a wide price range ($1500-$15,000) and tailored to many particular needs. The most appropriate equipment for one's use depends entirely on the specific problems which are to be solved. Guidelines will be presented here to assist the reader in making the proper selection for solving his own specific problems.

For best results in modern gas chromatography (GC) all parts of the equipment must be produced to careful specifications. A poor, low-efficiency column placed in a well designed chromatograph will produce disappointing results and vice versa.

All gas chromatographic apparatus must contain certain basic components as shown in Figure 6.1. In the remainder of this chapter, the components in gas chromatographic equipment will be discussed in the order shown in Figure 6.1.

6.2   CARRIER GAS SYSTEM

The purpose of the carrier is to transport the sample through the column to the detector. The selection of the proper carrier gas is very important because it affects both column and detector performance. Unfortunately, the carrier gas that gives the optimum column performance is not always ideal for the particular detector. The detector that is employed usually dictates the carrier to be used. For instance, an electron capture detector operating in the pulsed mode requires an argon-methane mixture; a thermal conductivity detector works best with hydrogen or helium. The most common carrier gases are listed in Table 6.1.

From a column performance point of view a gas having a small diffusion coefficient is desirable (high molecular weight, e.g., N_2, CO_2, Ar) for low carrier velocities while large diffusion coefficients (low molecular weight, e.g., H_2, He) are best at high carrier velocities (3).

The viscosity dictates the driving pressure. For high-speed analysis, the ratio of viscosity to diffusion coefficient should be as small as possible. Hydrogen would be the best choice, followed by helium.

The purity of the carrier should be at least 99.995% for best results. Impurities such as air or water can cause sample decomposition and column and detector deterioration. In temperature-programmed runs, impurities in the carrier gas such as water can be retained at low temperatures but are then eluted at higher temperatures impairing the baseline (4). Many instrument problems have been traced to contaminated carrier gases. The carrier must also be inert to the components of the sample and the column.

6.2.1 Regulations/Controls

Carrier gases are normally obtained in bottled form at about 2500 psi (150-160 atm). A two stage regulator is recommended at the cylinder. The second stage pressure regulator is usually set at 40-100 psi. After the gas leaves the cylinder it should pass through a molecular sieve trap (grade 13X). The trap will remove

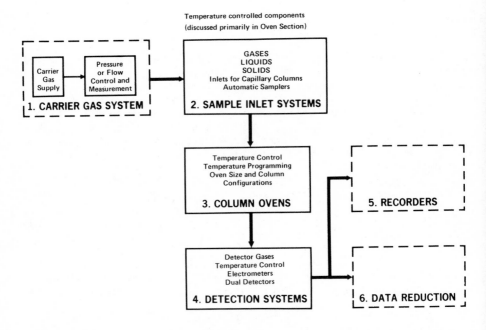

Figure 6.1. Basic components of gas chromatographic apparatus.

TABLE 6.1   PROPERTIES OF COMMON CARRIER GASES

	Molecular Weight	Thermal Conductivity $\lambda \times 10^5$ at $100^\circ C$ (g-cal/sec-cm-$^\circ$C)	Viscosity $\eta \times 10^{-6}$ $100^\circ C$ (μP)
Argon	39.95	5.087	270.2[1]
Carbon Dioxide	44.01	5.06	197.2
Helium	4.00	39.85	234.1
Hydrogen	2.016	49.94	104.6[2]
Nitrogen	28.01	7.18	212.0
Oxygen	32.00	7.427	248.5[3]

[1]At $99.6^\circ C$.   [2]At $100.5^\circ C$ [3]At $99.74^\circ C$

any water vapor or oils that may have been introduced in the filling process since a number of gases are water pumped. The contaminants removed by the trap could otherwise interact with the column packing material to produce spurious peaks (this would be particularly true when temperature programming is employed). In addition the contaminants can cause increased detector noise and drift. The traps should be reconditioned (about twice a year) by heating to $300^\circ C$ for 4-8 hr with a stream of gas passing through it or in a vacuum oven.

Several gas chromatographic detectors are sensitive to changes in the flowrate of the carrier gas. Any changes in flow rate cause the baseline to be displaced. These displacements make quantification quite difficult especially since the response of certain detectors such as thermal conductivity also changes with changes in flowrate. When an accuracy of ±1% in quantitative analysis is required, the flowrate should not fluctuate more than ±0.2 percent (see Chapter 4).

The column represents a resistance to the carrier gas flow and this resistance depends primarily on the column temperature since both viscosity and density of gas change with temperature. Figure 6.2 shows how the viscosity of four common carrier gases changes as a function of temperature.

Column performance is affected by the carrier gas flowrate and there is always an optimum flowrate for every column. Retention times also are affected by the carrier gas flowrate. A 1% change in carrier gas flowrate will cause a 1% change in retention time. For all these reasons it is important to keep the flow of the carrier gas constant. There are basically two ways to assure this:

1. Control of carrier gas inlet pressure.
2. Control of carrier gas flowrate.

   In isothermal operation the means of regulation is immaterial because both means provide constant inlet pressure as well as constant flowrate. In temperature-programmed runs, however, the situation is quite different. If one maintains the inlet pressure constant the flowrate will change (Table 6.2). Therefore, with temperature programming of the column, the flowrate must be controlled.

PRESSURE CONTROLLERS. The carrier gas inlet pressure can be controlled by:

1. The second stage regulator on the cylinder.
2. A pressure regulator mounted in the chromatograph or just before the chromatograph.
3. A needle valve (variable restrictor) mounted in the chromatograph.
4. A fixed restrictor mounted in the chromatograph.

TABLE 6.2   VARIATION OF FLOWRATE THROUGH A COLUMN (6 FEET x 1/4 INCH) PACKED WITH CHROMOSORB W AS A FUNCTION OF COLUMN TEMPERATURE AT CONSTANT INLET PRESSURE.[1]

Column Temperature (^{0}C)	Measured Flowrate Relative to Flow at 50^{0}C= 40 cm^{3}/min (cm^{3}/min - helium)
50	40.0
100	34.4
150	29.6
200	27.8
250	25.6
300	20.4

[1]The pressure was controlled at 13.9 psig at the head of the column by means of a pressure regulator.

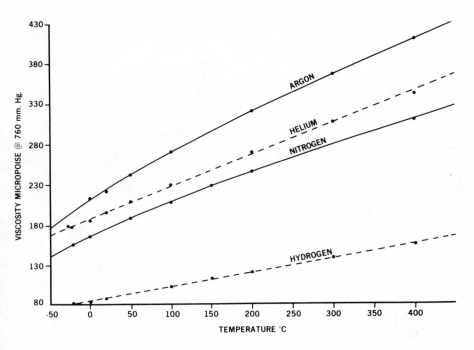

Figure 6.2.   Effect of temperature on the viscosity of four common carrier gases.

Few chromatographs depend on the cylinder regulator alone.
The problem in using that regulator as the only source of regulation for the chromatograph occurs when more than one instrument
requiring different head pressures are connected to the same
source of carrier. The least expensive regulator is the fixed
restrictor. This type is normally used for controlling the
hydrogen and particularly air for the flame ionization detector
(FID). It is advantageous that these chromatographs have a pressure regulator in addition to the fixed restrictor. Figure 6.3
shows a curve for flow of air as a function of pressure for a
fixed restrictor (a 0.004 ± 0.0001 inch diameter orifice). Manufacturers usually supply a curve with the chromatograph since the
curve is seldom completely linear. These fixed restrictors are
convenient for control of flows that are not changed often. If
the flow is to be changed often, as is the case with the carrier
gas, then needle valves are more convenient. Fine metering
valves (Figure 6.4) having a taper of no more than 1-2° are easiest to set, but also are more easily damaged.

It must be pointed out that the flow will be held constant
only if the downstream pressure drop is constant as in isothermal
operation. In addition, the particular restrictor must also be
held at constant temperature.

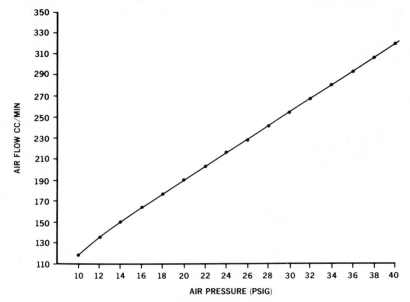

Figure 6.3. Flow of air through a fixed restrictor
(0.004±.0001-inch diameter orifice) as a function of
pressure.

When working with open tubular (small bore) columns the regulation of the inlet pressure rather than the flowrate is recommended. The reasons for this are as follows:

1. The low flowrates used by these columns are difficult to control with the flow controllers available today.
2. Open tubular columns are normally operated with a flame detector. The absolute change of flow is small compared to packed columns (although percentagewise it is similar) and influence on the FID response is less dominant.
3. Splitters requiring pressure regulation are commonly used with open tubular columns.

FLOW CONTROLLERS. A variety of techniques are available for controlling the flowrates in GC. The simplest method for maintain-

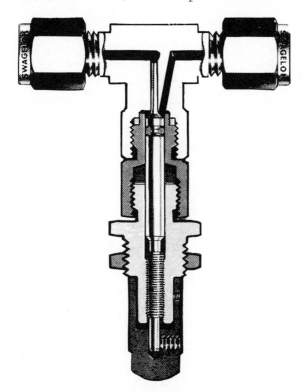

Figure 6.4. Fine metering valve has a 1-2° taper. (Courtesy Nupro Company).

ing a nearly constant flowrate of carrier gas even with increasing column inlet pressure is the use of a high-resistance flow device in series with and ahead of the column inlet. A long fine-bore capillary tube can be used for this control as described by the following equation:

$$F_c = (P_{in} - P_c) \cdot k_{cap} + (P_c - P_o) \cdot k_c \qquad (6.1)$$

The flowrate is virtually constant if $(P_{in} - P_c) \cdot k_{cap}$ is much greater than $(P_c - P_o) \cdot k_c$ and if P_{in} is much greater than P_c where F_c = flow through the column, P_{in} = inlet pressure ahead of capillary, P_c = pressure at head of column, P_o = outlet pressure, k_{cap} = capillary constant, and k_c = column constant.

Flowrate was controlled to better than 2% in temperature programming with a 2 meter by ¼-inch column from 30 to 200°C (5). An inlet pressure about 10 times greater than the maximum pressure drop across the column was required. By going to inlet pressures of 100 atm, the flow should be controllable to ± 0.2%. The main problem with this technique is that if the column or any part of the system should plug, the pressure of the system prior to that point rises until it reaches the primary pressure. This can result in dangerously excessive pressures.

Knox (6) described a constant flow device that ensured a flowrate independent of the pressure at the column inlet while Guild,

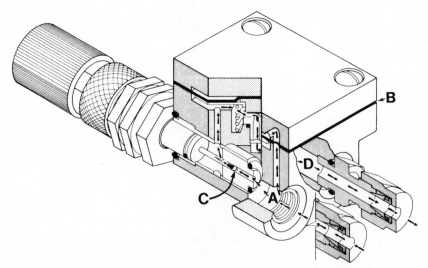

Figure 6.5. Flow controller with cutaway (Courtesy Brooks Instrument Div. - Emerson Electric Co.).

Bingham, and Aul (4) described a mechanical constant flow controller similar to the Knox device. Flow controllers of this
type are now commercially available from several sources, Figure
6.5 shows a cut-away of the Brooks Model 8744 flow controller. A
constant high pressure is applied at the inlet (A), pushing the
spring loaded diaphragm (B) downward. This closes a variable
orifice (not shown). An appropriate opening of the needle valve
(C) allows gas to slowly leak through, increasing the pressure in
(D) (The outlet side of the diaphragm). This pressure increases
until the diaphragm rises, opening the variable orifice and
thereby reducing the pressure in (D). For a given valve setting,
the pressure differential between (D) and (A) is determined by
the tension in the spring and by any pressure exerted by the column on the variable orific. Any changes in downstream pressure
unbalance the diaphragm which controls the opening of the variable orifice to maintain a constant pressure differential across
the manual needle valve (C). This maintains a constant volumetric flow of carrier gas through the flow controller. Table
6.3 illustrates the extent to which such a controller can achieve
constant flow with varying downstream pressure as when the column
temperature is programmed.

TABLE 6.3 VARIATION OF FLOWRATE THROUGH A COLUMN (6 feet x ¼
inch) PACKED WITH CHROMOSORB W AS A FUNCTION OF COLUMN TEMPERA-
TURE AT CONSTANT INLET FLOWRATE.[1]

Column Temperature ($^{\circ}$C)	Measured Flowrate Relative to Flow at 50°C = 40 cm^3/min (cm^3/min - helium)	Column Head Pressure (psig)
50	40.00	16.2
100	40.02	18.0
150	40.04	20.2
200	40.02	22.7
250	39.98	24.6
300	40.01	26.5

[1]The inlet flowrate was controlled by means of a Brooks Model
8744 Flow Controller. Relative flows were measured by means of
the thermal flow sensor shown on Figure 6.7 on a Hewlett-Packard
5840 Gas Chromatograph.

Most flow controller needle valves are manufactured to close tolerances. The needle valve should never be turned off too vigorously, as this can damage the O-ring shut-off seal and the fragile needle. The flow controller needle valve should therefore not be used as an on-off valve.

Spurious peaks could occur due to volatile compounds emitting from the flow controller diaphragm. This problem can be solved by use of metal diaphragms. Some manufacturers have used short molecular sieve traps between the flow controller and the inlet system, but, if a poor inlet system is used, the sample can flash-back upon injection and condense on the trap, later causing spurious peaks or noise and drift as the sample slowly comes back off the trap.

Automatic flow controllers have recently become available which operate in the $1-100$ cm^3/min flow range. These are voltage settable such that thumb wheels switches or potentiometers can be used to set the flowrate. These are specified at a repeatability of setpoint of about $\pm0.2\%$ and an accuracy of $\pm1-2\%$ of full scale. These flow controllers can easily be adapted to give a flow readout. The price of these units is still quite high at about (\$500 - \$600) per channel but all indications are that they will be more widely used in the future. Typical specifications for three such devices are given in Table 6.4.

6.2.2 Flow Measurement

The soap-film flowmeter has long been a popular method for measuring gas flowrates in GC primarily because of its simplicity, low cost, and independence to the type of gas used. This flowmeter is generally assumed to have a 1% or better accuracy. However, it has been shown that the accuracy is in the order of 2-6% for low flowrates of Ar and CO_2 (7). A typical design is shown in Figure 6.6a. The gas to be measured enters at the base of the measuring tube where a soap (Leaktek) film is introduced by squeezing the bulb containing the solution. The time required for the film to move between two calibrated marks is noted and the flowrate calculated. The three segment tube shown in Figure 6.6b conveniently permits the measurement of flowrates in the 1, 10, and 100 cm^3/min and greater ranges.

For most accurate flow measurement (required when retention volumes are to be calculated) a correction for water vapor pressure must be applied. The flow should also be corrected to the column temperature. Dal Nogare (8) gave the equation for correcting the desired flowrate to column temperature T and outlet pressure, p_O, as:

TABLE 6.4 AUTOMATIC FLOW CONTROLLERS.

| Controller[1] | Gases | Flowrates (cm³/min) | Accuracy (% FS) | Repeatability (% FS) | Operating Range | | | Approximate Cost (2 Channels) |
| | | | | | Temp. (°F) | Pressure | | |
						Max (psig)	ΔP (psi)	
1. Brooks 5830-1 (Complete with 0-100 psi gauge)	H_2, He, N_2, Ar	0-5 to 0-5000	±1	±0.15	0-150	150	5-50	$1176.
2. Chromatrol FC with APS-502 Power Supply & Display	H_2, He, N_2, Ar	2-100 to 2-10,000	±2	±0.20	130 (max)	100	20-100	$1100. to $1200.
3. Tylan	H_2, He N_2, Ar	2-10 to 100-5000	±1	±0.20	40-110	150	5-40	$1088.
4. Porter (Nonautomatic) (Inlet Pressure regulator is included)	He	1-100	±3	±0.50	70-250	250	10-200	$ 395.[2]

301

[1] (1) Brooks Instrument Division, Emerson Electric Company, Hatfield, PA 19440. (2) Applied Materials Inc., 3050 Bowers Avenue, Santa Clara, CA 95051. (3) Tylan Corporation, 19220 South Normandie, Torrance, CA 90502. (4) Porter Instrument Co., Inc., Township Line Road, Hatfield, PA 19440.
[2] Does not have a Sensor-Feedback Control Loop. Digital Setpoint Only.

$$F_c = F \; (\frac{T}{T_a}) \; (\frac{p-p_w}{p}) \qquad\qquad (6.2)$$

where F, T_a, and p are measured at the flowmeter. p_w is the vapor pressure of water at T_a. Temperatures are $^\circ K$. For most available flowmeters the pressure drop is insignificant and p reduces to p_o or the outlet pressure.

A less desirable device for measuring flowrates is the rotameter. A float is placed in a tube which is slightly widened at the top to form a cone. The gas flow produces a dynamic pressure keeping the float, which is free to move in the gas stream, at a

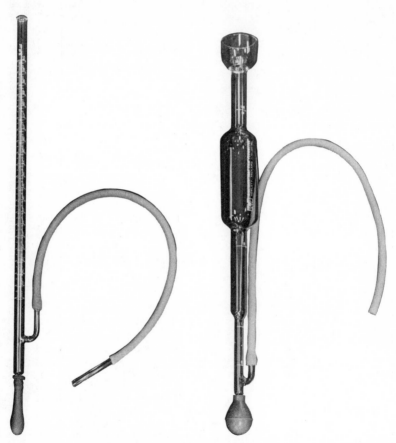

Figure 6.6. Soap film flowmeter. (a) 10-cm^3 size;
(b) three-segment type (1, 10, 100 cm^3).

constant height within the conical tube. The height of suspension is a measure of the dynamic pressure which can then be converted to flowrate by use of a calibration curve. The advantage of the rotameter is that it gives a continuous indication of the flowrate. Unfortunately, this metering device is dependent on the type of gas used, has a high sensitivity to contamination which can bind the float, and is more expensive than the soap-film flowmeter.

Recently, thermal mass flowmeters have been described which give an electrical output. At least two manufacturers of gas chromatographic equipment have made these available, one device being shown in Figure 6.7. The sensing element is a tube wound with a wire having a high temperature coefficient of resistivity. The two ends of the copper tube are kept at a constant temperature. When a potential is placed across the winding the tube heats up but the ends remain constant. As flow enters the copper

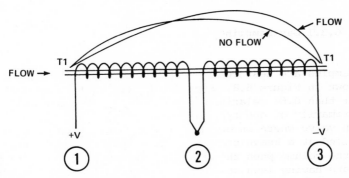

Figure 6.7. Thermal mass flowmeter and sensor element (Hewlett-Packard Co.; available only on 5830/40 series chromatographs).

tube at 1, the side from 1 to 2 cools slightly and the side from
2 to 3 heats slightly. The ends are still held at constant temperature, but the temperature distribution is shifted according
to the flow. Since the average resistance of the winding from 1
to 2 decreases and the average resistance of the winding from 2
to 3 increases, the windings act as two elements of a Wheatstone
bridge. The signal thus produced, proportional to flow, is amplified and linearized. The flowrate range for this meter is
from 0 to 100 cm^3/min with a resolution of 1 cm^3/min or better.

6.3   SAMPLE INLET SYSTEMS

Unfortunately, no universal sample inlet system has yet been designed; the best inlet is the one that works best for the particular application. Columns may be capillaries or packed and
samples may be gases, liquids, or solids containing vaporizable
solids (either neat or in solution). Nonvaporizable samples require some type of pyrolysis inlet which permits pyrolysis of the
sample with subsequent chromatography. In all cases the function
of a well-designed sample inlet system is to receive the sample,
vaporize it instantaneously if not already a gas, and to deliver
the vaporized material to the head of the analytical column in as
narrow a plug as possible. To accomplish this, the inlet must
have the capability of being heated for adequate vaporization of
the sample, the inlet volume must be small, and it must not contain any unswept areas.

6.3.1 Gas Sampling

Gas samples are best injected by means of a gas sampling valve,
as shown in Figure 6.8. Reproducibility with these valves is
better than 0.5% relative for a given loop. Gas-sampling valves
are primarily of one type now--a rotating stainless steel and
Teflon valve where in one position a sample loop of known volume
is filled at a known pressure. This loop of sample is then swept
into the column when the valve is rotated to the second position.
Polymers having less cold-flow properties than Teflon are now in
common use. Some valves use no polymer at all but rather two
metal surfaces. These tend to be more expensive to manufacture
and are also less convenient for routine maintenance such as
tightening. Other valves of this type consist of two linear
sliding surfaces and these are used extensively in process GC.

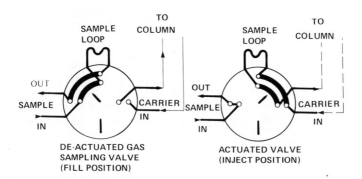

Figure 6.8.   Injection of gas samples by means of a
gas sampling valve.

The other type of valve consists of a cylinder fitted with a
plunger on which small O-rings are fitted. These isolate the
valve into four to eight sections. The plunger has two positions.
As in the first type, the first position is for flushing and
filling the loop and the second position for sweeping the sample
onto the column. This type has been plagued with O-ring memory
effects, dead volume, and excessive maintenance and has practically been abandoned.

For the most sensitive operation with gas-sampling valves,
flows and pressures must be very well balanced in the system.
Even the sample loop should be controlled to the same pressure as
the column side to minimize pressure upsets when switching the
valve. Frazer et al. (9) found that pressure and flow regulation
on all inlets as well as the appropriate micrometering valves
(variable restrictors) used to dynamically balance the carrier
gas flow and pressure allowed them to switch a 20 foot by 1/8-
inch column in or out of a multivalve, multicolumn system with a
resulting change in the baseline signal of less than 5 µV.

One could not discuss gas sampling without some comments on
syringe sampling. Gas-tight syringes are available for injection
of volumes ranging from 0.001 to 50 cm^3. A reproducibility of
±1% can be obtained by very experienced workers. Further discussion on gas syringe sampling is given in Chapter 4.

6.3.2 Liquid Sample Inlets

The sampling of liquid mixtures in GC is usually done by means of
microliter syringes through a self-sealing silicone septum. The

The sample can be introduced into either a flash vaporizer or directly onto the end of the column. The best technique depends on the application, the sample, the column type, and whether the column is heated isothermally or by temperature-programming. Instantaneous vaporization of the sample on injection is the usual method of ensuring a reproducible retention time and maintaining good efficiency of separation. This approach, however, is unsatisfactory for samples containing heat-sensitive compounds (commonly encountered in biomedical applications). Samples that are very dilute and require a large volume to be injected also cause problems.

Instantaneous vaporization requires that the latent heat be very rapidly supplied to the sample after injection (e.g., water requires 0.5 cal/mg). This heat must be supplied by the carrier gas or the material of the injector. The carrier gas is a very poor source. The heat must, therefore, come from the material of the injector usually having a poor thermal conductivity. Thus, the temperature of the injector must be very high or a large hot surface area must be available. Unfortunately, the consequences of high temperatures to heat-labile compounds are disastrous. For these samples, the best systems are those that work at the lowest temperature and accept the sample in the actual column packing at the end of the column. Leibrand and Dunham (10) found that the on-column inlet design shown in Figure 6.9 was superior to all others in terms of efficiency. This design has no unswept volume and has minimum possibilities for back flash. Dressman (11) found that on-column injection into a port which was completely swept by the carrier gas eliminated "memory" or "ghost" peaks in the analysis of aqueous phenol and organic acid samples. Grant and Clarke (12) studied the effects of injection variables on the precision and accuracy of peak height and area ratios. They found that, in general, on-column injection gave significantly better precision for narrow-boiling-range mixtures, but flash vaporization injection gave better results for peak height ratios with wide-boiling-range mixtures. Their results indicated that the best compromise would be to use on-column injection with additional heat supplied by a vaporizer.

Figure 6.10 shows a typical flash vaporizer widely used. The carrier gas is preheated by passing through a short length of small (0.030 inch) i.d. tubing before it picks up the sample. The preheat section acts as a restrictor to keep the sample from back flashing upon injection. This prevents sample from condensing in the cool carrier line where it could later vaporize causing memory peaks. The vaporizer in Figure 6.10 is sometimes referred to as a liner or concentric tube configuration. The inner liner must be designed with a small internal volume--usually 2-4 inches long and 0.020 to 0.030 inches i.d. The advantages of the liner configuration are the following:

1. The convenience and ease of changing the inlet to handle different diameter columns--particularly 1/8 and 1/4 inch.

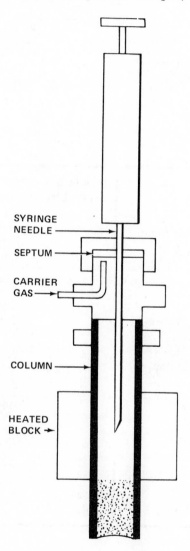

Figure 6.9.   On-column inlet design (Hewlett-Packard
Co.-Biomedical chromatographs).

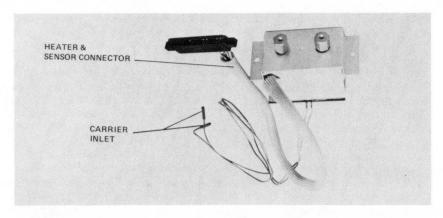

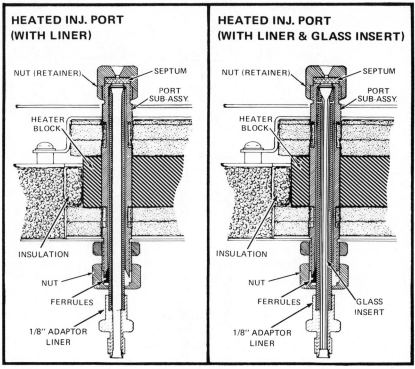

Figure 6.10.   Concentric tube heated injection port
with adaptor for 1/8 and 1/4-inch columns. Glass
liners are used for all-glass systems.

2. The availability of glass liners that fit inside the metal liner thus eliminating practically all contact of sample with metal when glass columns are used (Figure 6.10).
3. The ease of adapting the port to an on-column inlet having an additional heat supply.

Some of the present liner designs can be used for on-column injection by placing the head of the column about 1/8 inch from the septum. The syringe needle can thereby extend at least 2 inches into the column. The injection block temperature should be set at or below the temperature limit of the liquid phase in the column. Greater injector temperatures should be avoided because the liquid phase will be stripped off or decomposed at the front of the column and result in baseline drift or large, skewed peaks.

One slight disadvantage to this type injector for on-column use is that in temperature-programmed operation the port can generally not be programmed with the column. The port must, therefore, be maintained at a temperature slightly higher than that the column will see in the temperature cycle.

One of the biggest problems with present-day septum inlet systems is that of septum bleed. Kolloff (13) was the first to note the bleed of monomers and short-chain polymers (used in the production of the synthetic elastomers from which septa are derived) from gas chromatographic septa. Another problem is sorption of solvents and sample components on the septa. A thorough study of this phenomena was made by Adler (14) related to the use of self-sealing elastomer septa for quantitative operations with volatile laboratory solvents. It was found that a silicone septum could absorb over twice its weight of carbon tetrachloride and chloroform and more than its weight of benzene at $25^\circ C$.

A third problem is leakage of gases in or out through the septum. Improvements or solutions to the problems associated with septa are the following:

1. Septumless injectors.
2. Septum-cooled injectors.
3. Injectors that allow the septum to be bypassed from the carrier gas except during the injection.
4. Vented septum injectors.

SEPTUMLESS INJECTORS.   Van Swaay and Bacon (15) described a septumless injector, but its use is complicated and its reliability is questionable under varying conditions. Evrard (16) described a precolumn arrangement which permitted open column sampling for high temperature analysis.

An injector system was developed which overcame some of the

objections to septum injection. It consisted of a ball valve
through which a capillary glass pipette on a metal holder is inserted. A pressure drop is established across the capillary
which sweeps the sample into the column. The injection of 0.1-
to 0.2 µl samples with good precision is possible with such a
system but large sample volumes cannot be injected. This system
also suffers from an extremely large dead volume, possible losses to the outside, and awkward use.

These approaches could be used for limited manual applications
but better systems are available which can be easily automated.
These are described in Section 6.3.3.

SEPTUM-COOLED INJECTORS. Increasing the tension on the septum
markedly raises the quantity of bleed. Temperature has a greater
effect on septum bleed. This bleed can be decreased by heat pre-
conditioning the septa or by cooling them. Callery (17) found
that the bleed rate was less by a factor of about two at $200^{\circ}C$
than at $350^{\circ}C$. Obviously the septum temperature should be as low
as possible--just high enough to prevent condensation of the injected substance.

Cooling the septum prolongs its life and can reduce the bleed
problem (but never eliminate it). If a septum is always going to
bleed, better solutions than cooling are required.

SEPTUM BYPASS INJECTORS. Zenz and co-workers (18) replaced the
septum with an asbestos septum during the cooldown periods of the
column oven in temperature-programmed runs, only using the silicone rubber septum while the chromatogram developed. This almost
eliminated the ghost peaks seen when the rubber septum was not
replaced. This technique, of course, is laborious for normal operation and cannot be used with automated syringe injectors. An
improvement of this technique is the septum swinger, a device
which removes the septum from the gas chromatographic system except at the moment of injection and replaces it with a smooth
metal surface. This device sells for about $200 and has the
drawback that it must be manually rotated every time one injects.
It cannot therefore, be used with automated syringe injectors.

VENTED SEPTUM INJECTORS. A more radical injector modification is
available. It prevents gas which has had contact with the septum
from entering into the column. Figure 6.11 illustrates such a
system in which the gas contacting the septum is vented by a by-
pass pipe. The carrier gas stream is split, the main portion going to the column while a small portion moves past the septum and

and exits. The danger exists that parts of the sample can be
lost due to the expansion where part of the sample can be flushed
out. To eliminate this problem, the septum purge line can be
closed just prior to injection and opened after an appropriate
period of time. This can easily be automated by use of a solenoid valve. This technique is appropriate even with capillary
columns. Packed columns using 6 to 10 times the flow should operate even better by this technique. Figure 6.12 shows the
effect of the septum purge on the baseline compared to the non-
purged system.

INLET SYSTEMS FOR CAPILLARY COLUMNS. Capillary column inlet systems must also be considered. Interest in glass capillaries as
gas chromatographic columns has been receiving much attention in
the last few years as evidenced by a symposium that was dedicated

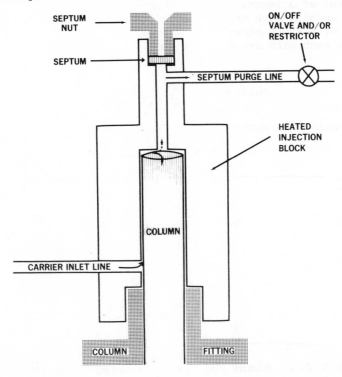

Figure 6.11. Vented septum injector. Any gas coming
in contact with the septum is vented out the septum
purge line.

to glass capillary columns (19). There are basically two inlet
techniques for glass and metal capillary columns as shown in
Figure 6.13:

1. Split injection.
2. Splitless or direct injection.

   The great resolving power of capillary columns is usually at
the cost of sample size. Because of the difficulties of introducing small samples (0.001 to 0.5 µl) into the capillary columns, splitting systems have been used almost exclusively in the
past for sampling. The sample is injected where it must be completely vaporized. The vaporized sample must be thoroughly mixed
and as it flows along it comes to a split in its path. Some of
it goes to the column; the rest is vented. The vent side is usually less of a restriction than the column. It consists of a
larger opening with the end restricted by a hypodermic needle, a
valve, or a short piece of capillary. The flows of pure carrier
through each path are measured and set to give a certain split
ratio:

$$\text{Split ratio} = \frac{\text{Flow (vent)}}{\text{Flow (column)}} \qquad (6.3)$$

Split ratios can vary from about 10 to 1000.

A. RUN WITH SEPTUM PURGE.

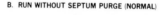

B. RUN WITHOUT SEPTUM PURGE (NORMAL)

Figure 6.12. Blank run using a well-conditioned silicone column and a white septum. A new septum was inserted 30 min before each programmed run was made.
Programmed from 30 to 200°C at 8°C/min. Detector temperature 350°C. Injection port @ 400°C. FID range 10.
Attenuation 16.

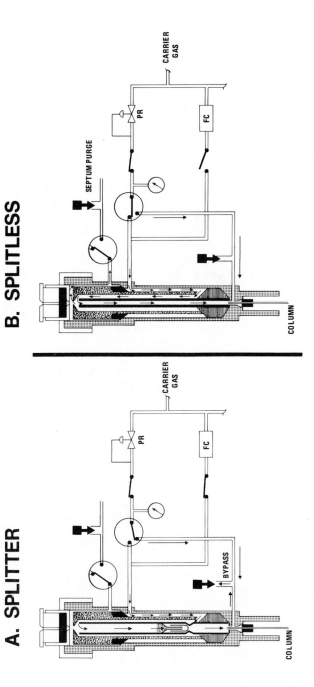

Figure 6.13. Inlet techniques for capillary columns (Hewlett-Packard Co.-Multipurpose glass inlet system). (A) Splitter type. (b) Splitless type with septum purge.

All stream splitters can discriminate toward substances having large differences in molecular weights. Discrimination occurs when all components of a sample mixture are not divided in the same ratio. When capillary inlet splitters were first used, discrimination was considered a real problem. Many times much of the problem could be traced to deficiencies in the rest of the chromatographic system such as response times of electrometers and recorders and errors due to integration. Early peaks in capillary columns are extremely sharp and make stringent demands on the entire gas chromatographic systems. Nonetheless, because of these problems many chromatographers shied away from splitters, except for analysis of complex mixtures of compounds where the molecular weights and boiling temperatures did not vary widely. Most of the problems associated with discrimination have been solved in splitters of recent design installed in modern gas chromatographic systems. Nevertheless, with complex mixtures that vary widely in molecular weights and boiling points, attention must be given to sample splitting. By following the recommendations of the manufacturer as to range of split ratios and temperatures optimum results will be obtained. When following these recommendations, it is possible to obtain results showing less than ±2% discrimination. One also should bear in mind that once the splitter has been calibrated for a particular sample to correct for this possible problem and further samples of the same type (which do not vary widely in composition or concentration from that calibrated) are to be analyzed, the reproducibility of the splitter is the only concern. Acceptable reproducibility of properly designed and used splitters has been proven repeatedly (20).

In the past, capillary inlet splitters were mostly all-metal construction. With the activity in glass capillaries, all-glass splitters have received attention. It is of little use to have an inert column without preventing the sample from decomposing in the injection port. An all-glass splitter system has been designed for the analysis of tobacco smoke in which a charcoal section was used on the vent side to absorb the sample prior to exiting. Recently, Jennings (21) described the performance of an all-glass splitter which is easily disassembled for cleaning and inactivation. An interesting inlet splitter that allows injections of 2- to 6-µl samples uses a short packed column to eliminate aerosol splitting and at the same time to protect the capillary column from solvent shock. The split ratio is adjusted by a flow-pressure relationship established at the head of the column and the by-pass very similarly to the design shown in Figure 6.13. Satisfactory quantitative work was achieved with a column flow of 1-2 cm /min and at split ratios between 5:1 and 10:1 over the 2- to 6-µl sample volumes (22).

Several manufacturers of gas chromatographs now offer all
glass splitters for use with glass capillary columns. Figure
6.13 is one design available. Splitters are generally used for
solventless samples and also in petroleum and petro-chemistry a
and for the analysis of essential oils.

Splitless or direct sampling procedures have been extensively
studied. These methods involved trapping of the high-boiling
materials in a small cooled zone followed by rapid desorption and
introduction to the column. The column itself has been used at a
low temperature to allow the solvent to pass while the high-boiling material was trapped on the front end. Once the solvent has
passed the column is quickly heated to the desired temperature.

There was some concern that the solvent peak might shorten
column life. Double column chromatography with trapping over-
comes this problem and further opens up the possibility of selectively looking at portions of the sample on a capillary column by
means of heart-cutting. These techniques show promise for the
future although present systems appear somewhat complex for the
average user.

The splitless technique is generally useful for trace analysis
or for biomedical applications where the sample has been diluted
in some solvent. Because the entire sample goes to the column,
there is no sample discrimination problems and the sensitivity is
higher because of the larger sample introduced. Temperature programming is essential to elute the sample since the technique depends on the fact that the solvent passes through the column concentrating the sample behind it. The resolution and efficiency
of separation is quite dependent on the initial temperature of
the column. Some of the problems of splitless operation are the
following:

1. Low flowrates in the inlet system require that there be no unswept areas.
2. Septum bleed is worse than in split mode because all the flow
   goes through the column. Most of the splitless inlet systems
   that are presently available use a vented septum technique to
   avoid this problem.
3. High-boiling compounds such as tars and residues can destroy
   a column faster than in split mode.

6.3.3 Solid Sampling

There are several techniques for handling solid samples. The
"best" technique again depends on the particular application one
has. If there is a suitable solvent, one can dissolve the solid

and treat it as a liquid sample. There are at least three requirements for the solvent.

1. The solvent must not react with any of the components of interest in the solid or with the column packing.
2. The solvent must completely dissolve the solid sample unless
   an extraction technique is used. In the latter case, the solvent must completely extract the components of interest out of
   the solid sample. These must be completely miscible with the
   solvent.
3. The solvent must not elute where components of interest in the
   sample elute.

Sometimes more than one solvent has been used. The time spent
dissolving the solid can usually be decreased by weighing it into
a bottle, adding a known amount of solvent and placing the sealed
bottle into an ultrasonic bath. If an internal standard is used,
it can be premixed with the solvent. Heat may be applied if it
does not degrade the sample. This technique has been successfully applied to samples such as detergents, pills, and other
soluble powders.
    Solid samples can be introduced by means of a piece of notched
hypodermic tubing fitted with a handle. The tubing is filled
with sample and then introduced into the heated port. A similar
device is now marketed by the Hamilton Company. Solid samples
can be introduced on small portions of stainless steel or platinum gauze. The sample is quantitatively transferred onto the
gauze by placing the gauze in an indentation on a Teflon plate.
A solution of the sample (0.05 to 1.00 cm^3) is then placed in the
same indentation. On evaporation of the solvent, the solution
concentrates to the point where it is distributed by capillary
action on the gauze. Complete evaporation of the sample leaves
the solid sample on the gauze. The gauze is then introduced into a hot zone while carrier gas is flowing past it, forcing the
sample into the column. A different technique for the analysis
of barbiturates in urine uses a stainless steel holder for a
small piece of filter paper (23). The sample, a purified ethyl
ether extract of acidified urine, containing the barbiturates on
the filter paper. The process can be repeated prior to introduction into the flash heater, thereby concentrating the sample.
The advantages of this technique are:

1. No interference of solvent peaks on the chromatogram.
2. No contamination of column due to nonvolatile compounds in the
   urine extract. These compounds remain on the filter paper.
3. Quite dilute samples can be concentrated on the filter paper,
   thereby increasing the sensitivity of the gas chromatographic

analysis.
Metal, glass wool, quartz, and other materials have been used instead of the filter paper.

The material used as the carrier effects the time required to
evaporate polar compounds whereas nonpolar compounds do not appear to have this problem. A quartz needle is the basis of an all
glass solid-sampling device described for use with open tubular
columns (24).

The above techniques are not satisfactory for solid sample
containing light fractions that will be lost if exposed to ambient conditions of pressure and temperature. Samples like these
can best be handled by cooling the sample and encapsulating them
in small sealed metal--fusible alloy--containers (capsules) or
glass capillaries. The metal must have a low melting point.
Woods metal (m.p. 60.5°C) is one such material. The container
with the sample is then inserted into the chromatograph and
flushed with carrier gas to re-establish equilibrium. The metal
capsule is then melted or the glass capillary crushed and the
vaporized sample goes to the column.

6.3.4 Automatic Samplers

In many chromatographic analysis, the length of the elution time
greatly restricts the number of runs that can be performed in any
working day on a single instrument. The number of runs in a
single day can be increased by:

1. Duplication of instruments.
2. Working a single instrument more than 8hr/day.

The second method can be accomplished by working shifts or by
automating the gas chromatograph so that it can run unattended
overnight or over a weekend. Automatic samplers allow this and
in addition can consistently give more precise results than that
obtained by manual techniques. Some of the different automatic
samplers available are listed in Table 6.5 Generally, single
syringe samplers are more satisfactory for liquid samples because they completely automate the measurement of sample volume
in addition to handling the injection of the sample into the gas
chromatograph. Capsule samplers are generally more satisfactory
for solid samples or samples containing large amounts of nonvolatile material. Gases are generally best handled by gas
sampling valves and stream selector valves or headspace samplers.
These different systems will now be discussed in more detail.

AUTOMATIC LIQUID SAMPLERS. The first successful commercially available automatic liquid sampler was introduced in 1968 (Figure 6.14). This system is based on a single syringe which is filled and flushed a number of times to eliminate any of the previous sample. The syringe is then loaded with sample by automatically pumping the syringe to eliminate air bubbles and then the needle is forced through the septum of the injection port and the plunger of the syringe mechanically pushed down to force the sample into the injection port. The syringe needle is withdrawn and the cycle is repeated after a preset time has elapsed. The system incorporates the timing electronics which also allows setting up replicate runs from the same bottle. A sample tray holds up to 35 bottles which are individually capped and the system can output the tray position (bottle number). The automatic syringe technique has become quite popular for liquid samples and several variations are available from several manufacturers (Table 6.5). Other syringe systems available are based on flushing the sample through the syringe and then into the chromatograph.

Automated systems using sampling valve techniques have been described. Samples are loaded into tubes on a sample rack from where they are individually aspirated through a valve with a small sample loop. After a set time, the valve is activated and the sample forced into the carrier gas stream just as was described for gas-sampling valves. The sample can be used to flush the valve and loop, but carryover effects are large. A small slug of solvent to flush the sample into the chromatograph eliminates most of this problem. The advantage of the valving systems is that no solid support, glass capsules, gauzes, or

Figure 6.14. Automatic liquid samplers (Hewlett-Packard Co. - 7670 series samplers). (A) Horizontal injection. (B)Vertical injection, (C) Control unit.

TABLE 6.5 AUTOMATIC SAMPLERS (U.S.A.)

Manufacturer	Model No.	Sampler Type	Sample Capacity	Inj/Sample[1]	Self-Dosing	Approximate (Cost($U.S.))
Hewlett-Packard	7670/7671	Syringe	35–36	1–3, R	yes	3150–3700
Precision Sampling	4200/4220	Flow-through syringe	42	1 , R	yes	4220
Varian	8000	Flow-through syringe	60 (4 sets of 15)	1–3, R	yes	4000–4510
Perkin Elmer	AS41	Capsule	100 (10 sets of 10)	1	no	8230–8950
Perkin Elmer	F42	Headspace	30	1–3	yes	15,450[2]

[1]R = remote control.
[2]Includes gas chromatograph—generally not sold separately.

319

capillary tubes are required. The system also automatically
takes the aliquot just as the earlier described syringe systems.
The carryover problems, however, have prevented these valve systems from becoming commercially popular except for process applications.

When considering an automatic liquid sampler one should consider the following points:

1. How much work does the sampler take over (i.e., does it take
   its own aliquot or must one still do that manually?)
2. How much sample is required to flush the system? This is especially important when one has very small amount of precious
   sample.
3. How much cross contamination is there between samples?
4. What is the cost of the sampler itself and also the cost of
   consumable parts (i.e., syringes, bottles, caps, etc.?)
5. Is it reliable and easy to use?
6. Can the sampler be mounted on your chromatograph?

AUTOMATIC SOLID SAMPLERS. These techniques are limited mostly to
samples with negligible volatilities at room temperature although
a Barber Colman system was adapted to handle liquid samples. A
system based on aluminum or gold capsules is available from
Perkin Elmer. The capsule is filled with sample and then sealed.
A device is available for loading of liquids and for doing the
sealing. The capsules are then inserted in coded metal maga-
zines. The capsule is then taken by a rotating cartridge-piston
through a three-stage gas interlock. The capsule is then pierced
by a hollow "thorn" in the heated part of the inlet port and the
sample flushed out by the carrier gas.

Such nonsyringe techniques do not eliminate a prior preparative stage and most of these automatic solid samplers do not take
care of measuring the sample volume or quantity that they introduce. These samplers generally are constructed of metal such
that the sample could interact with metal surfaces. Samples
could evaporate when left in the sample racks for long periods.
In spite of these severe drawbacks, there are advantages to such
systems:

1. Solids, suspensions, or highly viscous samples can be automatically introduced into the chromatograph.
2. A well-designed system can eliminate all the problems caused
   by septa.
3. A well-designed system can be completely inert to unstable
   samples.
4. Solvent-free injections can be made.

AUTOMATIC GAS SAMPLERS.   The use of automatic gas-sampling valves
for on-stream monitoring of gas supplies has become an extremely
popular and useful technique.   Pressure reducing units may be necessary prior to the valve when sampling from high (5000 psi)
pressure lines.   Automatic gas chromatographs are commercially
available for analysis of refinery gases, natural gas, ambient
air, and several other streams.   One can even use a rotary stream
selection valve to monitor up to 16 or so streams.   As long as
the sample lines have positive pressure or there is enough sample
so that one can pump the sample through the valve these techniques work well.   When one has only small amounts of many different samples most commercial automatic systems are inadequate
and one has to resort to manifolds or a syringe system where the
small samples are loaded and then sequentially introduced for
analysis.
      Automatic headspace samplers are available from manufacturers
of gas chromatographs.   These devices are based on the technique
of sampling an amount of vapor above the sample itself.   Samples
are sealed, neat or in a suitable solvent, in containers, and
held at a preset temperature in a thermostatted liquid bath.   The
headspace vapor results as a partition equilibrium is established
between the liquid or solid and the gaseous phase of the volatiles.   As each sample is presented to the analyzer, the vessel
is punctured and a portion of the headspace gas is withdrawn by a
pneumatic injection technique and forced into the column.   The
main application for these samplers is in the routine analysis of
low-boiling fractions in samples containing nonvolatile solids or
high-boiling components.   Some of the more popular applications
today are:

1. Analysis of ethanol and for other volatiles in blood (blood
   alcohols).
2. Analysis of residual monomers in polymers (vinyl chloride in
   PVC).
3. Investigations devoted to the determination of compounds responsible for the aromas of food products.

These and many other applications are reviewed by Vitenberg et
al. (25).

6.4   COLUMN OVENS

The primary purpose of the column is to provide a constant temperature bath for the column where the fundamental separation pro-

cess of GC takes place. The column oven should meet certain minimum requirements for optimum performance:

1. The temperature operating range must be large enough to serve
   individual applications. To make use of the whole range of
   possible applications, it is necessary to have temperatures of
   -50^0 (or -75^0) to 400^0C. Since sub-ambient operation is not
   required for most analysis, the sub-ambient capability should
   be optional. However, operation from 5 to 10^0 above ambient
   to 400^0C should be adequate for the majority of applications.
2. The oven compartment temperature should not be unduly affected
   by other heaters, such as those in the inlet or the detector.
   The temperature must be uniform over the whole column area;
   the temperature difference between the highest and lowest mean
   temperatures in the oven (sometimes referred to as gradient)
   when operating at an equilibrated isothermal temperature must
   be minimal.
3. The column oven should be free from the influence of changing
   ambient temperatures and line voltages. The difference between the maximum and minimum temperature observed over a long
   period of time at any one fixed point in the oven (sometimes
   referred to as thermal noise) must be minimal--less than
   $\pm 0.1^0$C.
4. The accessibility to the column and ease of installation of
   accessories is important if columns are to be changed often
   and if various accessories such as valves are required for the
   particular applications the instrument will need to satisfy.

Additional requirements are placed on the oven by temperature-
programming.

        6.4.1 Column Temperature Control

The column temperature control is one of the most important characteristics of good chromatographic performance. Column temperature is easy to change and yet it has such a dramatic effect on
retention values and on-column performance. As a general first
approximation, the retention times double for a 30^0C decrease in
column temperature (26). This rule holds fairly well for most
materials on most columns if the retention time is at least four
times the retention time of a material not retarded by the column
(i.e., air or a low-boiling solvent). The column temperature selected must be high enough to elute the compounds in a reasonable
time without thermally decomposing the sample, without causing
the column bleed to be excessive, and low enough that the desired

separation is obtained. Lower temperatures on a given column
generally will give improved separation due to the higher solubility of the sample components in the stationary phase. In some
cases it is not possible to use the lower temperatures. Carbowax
20M, for example, solidifies below about $50^{\circ}C$ causing it to lose
its separating power. In any case, temperature constancy must be
better than $0.1^{\circ}C$ if truly comparable retention times are to be
obtained.

A vapor jacket was first used to control the column temperature (1). It kept the column at the constant temperature of a
condensing vapor (at the boiling point of an appropriate liquid).
This technique was quickly abandoned because it was cumbersome to
use and was limited to liquids that boiled up to about 150 –
$180^{\circ}C$.

The introduction of forced air circulation in an oven allowed
the control up to 300 – $400^{\circ}C$ and the air bath is the main type
of column oven available today for laboratory gas chromatographs.
Several types of temperature controllers are used to control the
temperature of these ovens. They basically differ in terms of
cost, accuracy, and flexibility. They are usually advertised as
isothermal or temperature-programming controllers but all, in
general, use the following basic types of control:

1. Constant voltage or power.
2. Full power on/off.
3. Proportional power.
4. Software feedback loop (processor control).

The lowest cost means of controlling temperature is to regulate
the power applied to a heater (wire coil) by means of a variable
transformer. Since there is no feedback, there is no means to
compensate for changes in line voltage fluctuations or changes in
ambient temperature, both of which usually are considerable.
Even in a laboratory it is not unusual for ambient temperature to
change $6^{\circ}C$ during a day; the temperature can, therefore, not be
well controlled. In early GC units the column temperature, injector temperature, and sometimes detector temperature were often
crudely controlled by powerstats. Some lower cost instruments
still control temperature by this means.

The earliest feedback type of unit was the full-power on/off
type controller. The selected temperature is set, then the temperature is measured by various types of sensors. The sensor provides the feedback to the power supply, either calling for heat
on or off. The oven can be kept at a given temperature by cycling the heater on and off at full power as called for by the
sensor. The problem with these controllers is that small cycles
in temperature develop, resulting in inconsistent column

operation.

Sensors used are either thermocouples or platinum resistance
bulbs. All early gas chromatographs used thermocouples refer-
enced to ambient. This is unsatisfactory because the oven temperature would change with ambient. The platinum resistance bulb,
not requiring a reference junction, has gained wide acceptance.
A well-referenced thermocouple can be just as satisfactory and
will probably become the standard in the future. The reference
junction must be in a temperature-controlled zone or ambient compensation must be made.

Power proportioning controllers are generally more sophisticated forms of the constant voltage units, employing solid state
circuitry and rapid time-constant sensing circuits. An isolation
transformer isolates the controller from line voltage fluctuations. If adequate insulation of the column from ambient air
changes is provided, the controllers can be quite accurate.
Early controllers of this type used slide-wire resistors for
setting the temperatures. These were soon replaced by digital
controls such as that in Figure 6.15, consisting of stepped resistance values. Digital controls allowed precise resetability
of setpoints and eliminated noise due to slide-wire bounce.
These controllers are commonly used to control the inlet and detector temperature zones.

The latest in temperature controllers is the proportional integral controller where the feedback loop is closed through a
microprocessor, as shown in Figure 6.16. The processor allows
unique new control capabilities that are too complex to be practical in standard control systems. The processor may use three
modes of column oven temperature control: (a) a standard heating mode, (b) a new system designed to mix controlled amounts of
room air into the column oven, and (c) cyrogenic cooling. The
processor may use any combination of these modes resulting in
precise and accurate temperature control.

Operation of the column oven at $50^{\circ}C$ or lower has been a problem in earlier chromatographs because of the difficulty of completely isolating the column oven from other heated components,
such as the detector, injection port, and splitter, and still
having a usable oven. The processor controller described over-
comes this problem by mixing controlled amounts of room air into
the column oven and can control very adequately at temperatures
of about $30^{\circ}C$ without cryogenic cooling. A further advantage of
the processor controller is that the processor normally also can
handle the temperature control of the other heated zones--inlet,
detector, valves, and so on.

The oven temperature controller for temperature-programming
can consist of one of the following:

1. Mechanical drive unit that supplies the required heat by cycling full oven power.
2. Power proportioning unit where heating rates are generally electronically controlled by solid state circuitry.
3. Software feedback loop (microprocessor).

The early temperature-programmed chromatographs used mechanical drive programmers where the rates were selected by various gear ratios. The gears were driven by a constant-speed synchronous motor. A calibration was required to match the temperature read-out on the programmer with the sensing thermocouple or sensor. A feedback loop then allowed the supply of the amount of power required to establish the programming rate. These programmers are plagued with gear alignment, recalibration, and mechanical failures.

The mechanical programmers were replaced by electronic programmers. These depended mainly on different solid state time-logic RC circuits to increase the power to the column oven to supply the desired heating rates. This allowed multilinear pro-

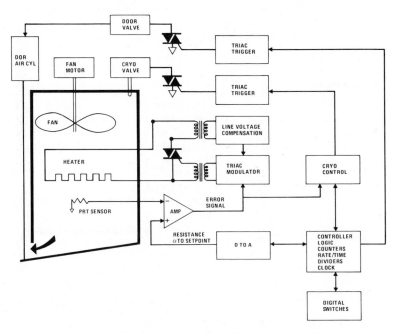

Figure 6.15. Block diagram of digital oven temperature control system (Hewlett-Packard Co.-5700 series chromatographs).

grammers to be introduced. As with the isothermal controllers, temperature programmers then became digital. Now temperature control for temperature-programming is widely done by using microprocessors in the control loop. The processor can generate very accurate and reproducible ramps and control the heater, door, and cryogenic valve to ensure excellent temperature programs.

6.4.2 Temperature Programming

When a chromatogram is run under isothermal operation, two undesirable things can happen:

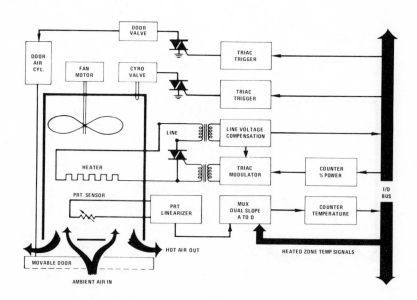

Figure 6.16. Block diagram of processor based oven temperature control system. The processor and most of the digital parts are shared with other temperature zones and with other functions of the gas chromatograph. (Hewlett-Packard Co.-5830/40 series chromatographs).

1. In the case of samples of unknown composition, high-boiling
   compounds may go undetected. If the compounds are eluted,
   they may actually be eluted several runs later. These peaks
   will be very low broad peaks, difficult to accurately
   quantify.
2. Early peaks will be sharp, close together, and poorly resolved
   (while late peaks will be low, broad, and excessively resolved). Only in a relatively small region of the chromatogram is
   the separation optimized.

The above problems can be avoided by a controlled change of
column temperature (temperature-programming) during the course of
a run. Temperature-programming, also called Programmed Temperature Gas Chromatography (PTGC), resulted from the limitations of
isothermal chromatography for the analysis of complex mixtures in
which the sample components have a wide range of boiling points.
The technique has gained wide acceptance as a research and scouting tool. It offers important advantages in these areas, especially in methods development or for the analysis of samples where
the quantity is limited. Temperature-programming is the most
efficient first approach to any sample of unknown composition.
On the first run one can get an indication of the boiling point
range and complexity of the mixture. Quantification and other
qualitative information are obtained. From this scouting run,
the proper isothermal conditions, if desired, can be readily determined. Other reasons for choosing temperature-programming include the following:

1. Analysis time is reduced.
2. Precision of detection limits or peak measurement is improved.
3. Speed of the sample injection need not be as fast as in isothermal operation.
4. Chemical transformation of unstable sample materials can be
   minimized.

Although PTGC was used as early as 1952 for the separation of
alkyl chlorides, the technique did not achieve general acceptance
until after the fundamental work of Dal Nogare and co-workers
(27-29) and the activities of Martin and co-workers (30), who
first made commercial PTGC equipment available and showed how one
could predict retention times based on isothermal data by the use
of graphical methods.
Habgood and Harris (5) derived the following equation showing
the relationship between isothermal retention volumes and retention volumes obtained under linear temperature-programming:

$$r/F = \int_{T_O}^{T_R} dT/(V)_T \qquad (6.4)$$

where r   = linear temperature-programming rate (deg/min),
      F   = flowrate (cm^3/min),
      T_O = column temperature at injection,
      T_R = column temperature at emergence of peak (retention
               temperature),
      $(V)_T$ = the corrected retention volume at temperature T,
      dT   = temperature change.
According to this equation, the retention temperature for any
particular solute and column should depend only upon the ratio
r/F. The retention temperature (T_R) may be calculated from the
isothermal retention volume by graphical integration of the above
equation.

Heating rate, flow velocity, and column length influence resolution and analysis time in PTGC. There exists a significant
temperature, T', which if used as an isothermal operating temperature will give the same degree of peak spreading and resolution
of solute peaks as that obtained from a programmed temperature
run. The significant temperature (T') is related to the retention temperature according to the following equation:

$$T' = T_R - 45 \qquad (6.5)$$

This allows the establishment of isothermal conditions from the
programmed temperature run and vice versa. The resolution is
improved if the column length is increased, up to a certain limit,
but the analysis time is also lengthened. The heating rate
affects the separation: Increasing the heating rate always leads
to an increase in the retention and significant temperatures with
a corresponding loss of resolution. However, the retention temperature increases only slowly as a function of the heating rate
while the analysis time is nearly inversely proportional to the
heating rate. The theory of PTGC is basically similar to the
theory of gradient elution chromatography.

The programmed temperature change in PTGC, in which the temperature of the entire column is uniformly increased with time,
can be of several types (Figure 6.17):

1. Ballistic.
2. Linear.
3. Multilinear.

(Chromathermography, which utilizes a moving temperature gradient in the column, will not be discussed here because it is a technique that has been used mainly in the Soviet Union and has received little attention elsewhere. The interested reader is referred to reviews by Tudge (31) and by Ohline and DeFord (32)).

Ballistic or natural temperature-programming is available on isothermal controllers and is obtained when the isothermal setpoint is suddenly changed to a higher isothermal temperature. The column oven temperature will now rise as rapidly as the controller can supply the heat (Figure 6.17A). This type of programming is usually used for "cooking-out" or conditioning columns after they have been in use for some time.

The more useful and common type of program used is the linear with isothermal holds of selectable durations before and after the temperature program ramp (Figure 6.17B). The duration of one or both of the isothermal holds may be zero.

Multilinear programs (Figure 6.17C) can also be used but there are few applications requiring such sophisticated programs.

With temperature programming the run is started at a relatively low temperature. This initial temperature is selected much the same as for the isothermal analysis of the low-boiling components in the sample of interest. It will normally be lower (by as much as 90°C) than the boiling point of the lowest boiling component in the sample. A lower initial temperature has negligible effect upon the resolution of the higher-boiling components since these solutes are essentially "frozen" at the inlet of the column and are thus unaffected by the early parts of the temperature program. However, the initial temperature should not be set lower than necessary because of the additional time required to reach a column temperature at which the low-boiling solute peaks begin to migrate at a significant rate. If the initial temperature is too high, final peaks elute quickly and resolution is

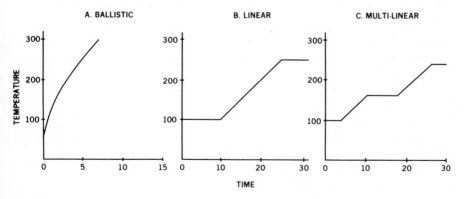

Figure 6.17. Types of programmed oven temperature.

very poor. As the temperature is raised, the remaining compounds
will each automatically select its own ideal temperature in which
to migrate and separate within the column. The effect is a
shorter analysis time than in the isothermal mode, and the peaks
have significantly improved peak shapes (Figure 6.18).

The optimum value for the heating rate is a compromise between
resolution and analysis speed. The heating rate provides the
same function in PTGC as the operating temperature does in isothermal operation. A given change in the natural log of the
heating rate, r, has nearly identical consequences with corresponding changes in the isothermal operating temperature.
Giddings (26) has shown that in order to avoid significant losses
in resolution, the following relationship must be true:

$$r = 12^\circ/t^\circ \hspace{5cm} (6.6)$$

where r   = heating rate,
       t° = time required by an unretained peak to pass through
            the column.
If, for example, the passage time of the air peak (assuming it is
not retained) is 0.5 min, then the heating rate should be kept
below 24°C/min, or serious loss in resolution will be encountered.
Typical heating rates for columns 10 - 20 feet in length and 1/8
or 1/4 inch in diameter are 4 - 7°C/min.

The final temperature selected should be near the boiling
point of the highest-boiling component present in the sample.
Naturally, the practical limitations of the column substrate
volatility must be taken into account.

The flowrate has been mentioned in some of the preceding discussions, but it has a relatively small effect on the analysis
time as compared to the temperature parameters. The flowrate
should be optimized much the same as it would in isothermal operation.

Temperature programming places some demands on the column oven
which are essential for proper operation. These demands, though
not essential for isothermal operation, will in most cases provide better, more stable operation in that mode also. The requirements are as follows:

1. The linear heat up rate is important in temperature-programmed
   units. Temperature rates from about 0.25 to at least 10°C/min
   should be available. The temperature steps at the bottom end
   should be small for capillary columns. Program rates in excess of 10°C/min are rarely useful (33). The penalty of large
   increases in temperature for small decreases in retention
   times does not warrant program rates above about 10°C/min.

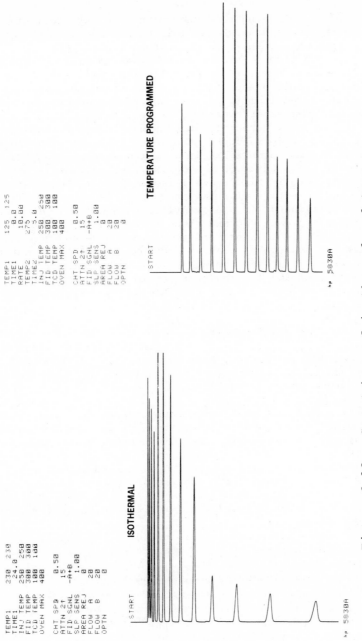

Figure 6.18. Comparison of isothermal and temperature-programmed chromatograms. Sample of C_8 to C_{20}
normal paraffins on 15 foot x 1/8 inch column packed
with 10% OV-1 on Chromosorb W.

However, for "cleaning out" the column, faster "ballistic" heat-up rates are desirable. This puts demands on the oven design (low mass oven) and temperature controller.
2. The cool-down time should be as short as possible as this is unused analysis time. The controller as well as the oven should be capable of automatically resetting to the initial temperature. An indication that the column temperature has re-equilibrated after cool-down is very desirable.
3. The inlet system, detector, and any heated accessories should have separate temperature-controlled ovens. They should not be in the column oven since it is not desirable that the temperature of these parts change during the analysis. These separate ovens should also be well isolated from the column oven since many detectors will have response changes that will result in baseline drift and quantification errors. The connecting lines between oven and these heated parts, especially the detector, should be short and should not have any "cold spots" (areas that are cooler than the maximum temperature that the column will see during the analysis). The inlet system (if not on-column) and the detector should be thermostatted at a temperature slightly higher than the upper temperature seen by the column.

Temperature programming also has demands on other parts of the chromatographic system. These have already been discussed in the areas of carrier gas purity, flow regulation, and injection systems. In addition, columns must be stable at the top temperature of the program ramp and should have a low vapor pressure (low bleed).

Most chromatographs used today have temperature-programming capabilities. One should consider the applications for which an instrument will be used. If temperature programming might be required make sure you obtain an instrument that has this capability or which can easily be updated with this capability. Instruments with temperature-programming include:

1. Separate temperature controllers for the inlet system, column, and detector(s).
2. A column temperature programmer (preferably linear).
3. A column oven capable of rapid heating and rapid, automatic cooling.
4. A flow controller (usually dual channel).

6.4.3 Column Oven Size and Configuration

Column ovens vary considerably in dimensions and geometry from
manufacturer to manufacturer. Many types of columns have to be
accommodated in gas chromatograph column ovens, from short capillary columns to the larger diameter preparative columns. The
"best" oven really depends on the application which dictates the
type of chromatograph. In the past, three configurations have
been common:

1. Small-volume ovens (100-500 cubic inches) for portable units
   and for single column isothermal units. These ovens seldom
   have the capability of rapid heating and cooling.
2. Medium-volume ovens (300-900 cubic inches) for dual column
   units. These ovens usually have the capability of rapid heating and cooling (temperature programming).
3. Large-volume ovens (1000-3500 cubic inches) for U-tube columns,
   especially popular in biomedical GC in the past, and for preparative applications.

An oven size of 30 x 30 x 15 cm (12 x 12 x 6 inches) is adequate for handling two columns, each up to 20 feet in length.
Ovens of this or somewhat smaller size, such as that shown in
Figure 6.19, give the best flexibility and are employed in the
majority of chromatographs of recent manufacture. The smaller
ovens such as that in Figure 6.20 will always have application in
small portable chromatographs or those chromatographs that will
be dedicated to specific applications, do not require temperature
programming, and do not require constant changing of columns.
Large ovens are now considered unsatisfactory.

COLUMN CONFIGURATIONS. The column oven design affects the shape
and configuration of the columns that may be used. Oven designs
that use coiled columns should be considered. The two most
common coiled configurations for packed columns are shown in Figure 6.21. When carrier gas is continually flowing through the
column, the force of gravity pulling downward on the column packing could result in channeling which results in a loss of selectivity and resolution. Thus, the vertically coiled configuration
(Figure 6.21B) would be the better choice since the packing could
only channel at the top and possibly the bottom of the coils.
The horizontally coiled column (Figure 6.21A) can channel
throughout the length of the column. Ideally, the column configuration should be interchangeable such that the column can be
turned end for end (especially glass) or mounted in any of the
available ports. If each coil is separated by an air space to
allow good circulation, gradients along the column are minimized
and provide the best possible uniformity in heating if the oven

heat flow is uniform. This is especialy true when long metal
capillary columns are to be used with temperature programming.
An ideal configuration for metal capillaries is shown in Figure
6.22.

Glass columns are available for most ovens. The small to medium ovens usually take 6-to 9-inch coils, whereas the large ovens
can take the long U-tubes. U-tubes are easier to pack and do
give smaller plate heights when small particles are used.

Figure 6.19. Common oven configuration (Hewlett-Packard Co.-5830/40 series chromatographs).

6.5   DETECTORS

The function of the detector is to sense and respond with an electrical signal when the composition of the gas emerging from the
column changes.   The type of detector used is dependent on the
application.   The most widely used detectors are the thermal conductivity, flame ionization, and electron capture.
    There are certain interactions between the detectors and the
rest of the chromatographic system that must be considered.

1. The type and rate of the carrier gas can affect the operation
   and response of the detector.
2. The type and amount of the sample injected can affect the operation and response of the detector.(A 1-µl injection of carbon tetrachloride into an electron capture detector will yield
   an inoperable detector for several minutes).   Overloading must

Figure 6.20.   Small ovens for portable or isothermal
chromatographs.   These ovens are normally used for
column, detector, and inlet.

be avoided.

3. The temperature and type of control used to thermostat the detector can affect the operation and response of the detector. In general, it is beneficial and advantageous to have separate temperature controls for each detector.

4. The electronics involved in converting the detector signal to usable voltage or digital reading has a large effect on the operation and response of the detector.

6.5.1 Detector Gases

The thermal conductivity detector (TCD) is a concentration dependent detector. Low flowrates are, therefore, beneficial for increased sensitivity. Table 6.1 lists the thermal conductivities of various carrier gases. The larger the difference between the thermal conductivities of the carrier gas and the sample, the greater the detector response. Since most commonly analyzed compounds (large molecules) have lower thermal conductivities, a light carrier gas such as hydrogen or helium provides the best results. Nitrogen is the preferred carrier for the analysis of hydrogen, but the response of most other compounds is greatly reduced. Whenever the carrier gas flow is interrupted in the TCD the filament current should be turned off to prolong the life of the filaments. Switches are available which sense the carrier pressure and automatically turn off the current to the filaments when the pressure drops below a preset value.

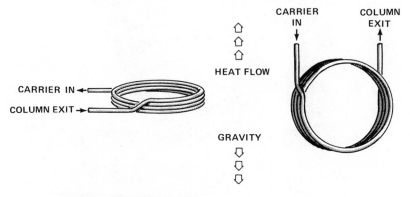

A. HORIZONTAL COIL   B. VERTICAL COIL (IDEAL)

Figure 6.21. Common coil configurations for packed columns.

Different carrier gases affect the response of the FID. The
relative response decreases for the gases argon, nitrogen,
helium, and hydrogen, respectively. Two other gases, hydrogen
and air, enter the flame detector, and these are normally controlled by fixed or variable restrictors. Above a certain minimum value, the flowrate of air has little or no effect upon the
response. The actual value depends on the specific manufacturer
but will usually range from 150 to 500 cm^3/min. Once set, there
is no need to adjust this flow. For a given carrier flow there
is an optimum hydrogen flow. This means that the carrier and
hydrogen flowrates are interrelated and must be considered together. Figure 6.23 shows the relationship for a series of carrier flowrates. At low flowrates, such as those encountered when
using capillary columns, the flame detector response is sensitive
to changes in carrier flowrates and the sensitivity is lower than
optimum. These two problems can be overcome by adding additional
(make-up) gas between the column and the detector or at the detector. It is important that the instrument have the capability
of adding make-up gas if its application calls for low carrier
flowrates (i.e., for capillary column operation). At the higher

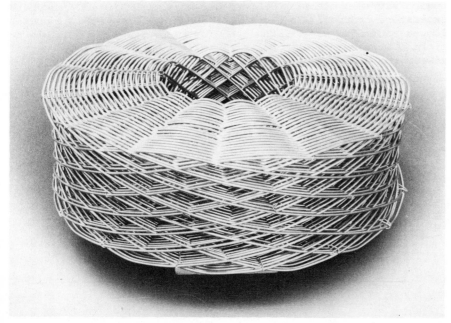

Figure 6.22. Ideal configuration for metal capillary
columns (500-foot coil).

flowrates one should be operating at the plateau where the response is high and where the sensitivity to carrier flowrate
changes is small. However, at high total flowrates (greater than
100 cm^3/min) through the detector, the noise tends to increase
and the linearity may be affected. For this reason it is best to
optimize the flows oneself or to follow the manufacturer's
recommendations.

The electron capture detector (ECD) is also a concentration-
dependent detector, and like the TCD will give a higher response
for a given compound at lower carrier flowrates. Carrier flowrate must be carefully controlled. Usually a 95% argon - 5%
methane mixture is used for carrier gas. Presence of oxygen or
water in the carrier gas results in loss of sensitivity and a
compression of the linear range.

6.5.2 Detector Temperature

All detectors must be thermostatted at temperatures that are sufficient to keep sample and column bleed from condensing. Thermal
conductivity and electron capture detectors require very accurate
temperature control. If the TCD or ECD block temperatures are
allowed to drift, so will the detector baseline. The sensitivity
of the TCD increases with an increase in temperature differential
between the block and the filaments. For this reason it is ad-

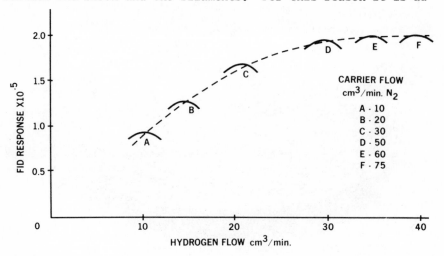

Figure 6.23. The effect of hydrogen and carrier (nitrogen) flows on flame ionization detector response.

vantageous to keep the temperature of the TCD cell as low as practical but never less than $15\text{-}50^{\circ}C$ above the column temperature. The FID is not very sensitive to temperature changes and is generally operated at $300\text{-}400^{\circ}C$. The FID temperature should always be set above $100^{\circ}C$ to keep water, formed in the combustion process, from condensing inside the detector.

The capture process of the electron capture detector can be very temperature-sensitive. The sensitivity may either increase or decrease with an increase in temperature, depending on the compound involved, as illustrated in Figure 6.24 for three benzene derivatives. Since detector temperature may affect sensitivity it is sometimes possible to improve the analysis by operating at a different temperature. The radioactive source determines the maximum temperature limit for the detector which is listed in Table 6.6. Exact values vary with manufacturer.

TABLE 6.6   TEMPERATURE LIMITS OF ELECTRON CAPTURE DETECTOR
            SOURCES

Source	Maximum Temperature ($^{\circ}C$)
Nickel (^{63}Ni)	370
Tritium (^{3}H)	
Titanium Substrate	225
Scandium Substrate	325

6.5.3 Electrometers

The electronics used to convert the detector signal to a usable voltage or digital reading also play an important role in the results achieved. The output signal of the FID and ECD, or any other ionization detector, consists of an extremely small electrical current (10^{-9} to 10^{-12} amperes for FID). To use this small current for a suitable readout device such as a recorder or integrator, an electrometer is required. Early electrometers operated as linear current-to-voltage converters using very high-impedance, low-noise operational amplifiers in conjunction with several high-value (10^{6} to 10^{11} ohms) feedback resistors. By

feeding part of the output back to the input through these resistors the amplifier is stabilized. By selecting a different resistor, the gain (usually referred to as <u>range</u> on the electrometer) or overall amplification is selected. The range switch
usually covers the span in decade steps. Since all detectors
have a background signal (i.e., signal due to flame and column
when no sample is present) and since most electrometers have an
offset signal, an additional resistor and a variable voltage
supply (usually referred to as <u>suppression</u> on the electrometer)
are connected to the input to subtract these offset currents from
the detector signal prior to amplification. Also present in such
electrometers is a <u>balance</u> or <u>zero</u> adjustment which takes care of
any voltage offset in the electrometer amplifier so that the output is zero when an input signal is absent.

    The high-value feedback resistors used in these electrometers
tend to have poor stability with time and temperature as well as
a poor tolerance (a fingerprint, speck of dust, or humidity on
the outside will conduct better than the resistor itself). This
limits the repeatability of measurements and causes large range
to range tracking errors (34). Because of these problems, the
range switch and resistors are usually kept in a sealed compartment containing a bag or can of dessicant. Another problem
associated with this type of electrometer is the large transients
caused by range-switching resulting in baseline upsets (Figure
6.25).

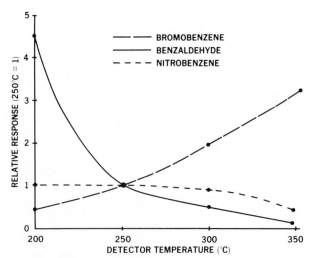

Figure 6.24. Effect of detector temperature on the
response of three compounds on the electron capture
detector.

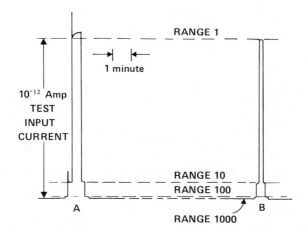

Figure 6.25. Electrometer range-switching transients.
(A) Early conventional resistor feedback electrometer:
large transients. (B) Transistor feedback electrometer: no measureable transients.

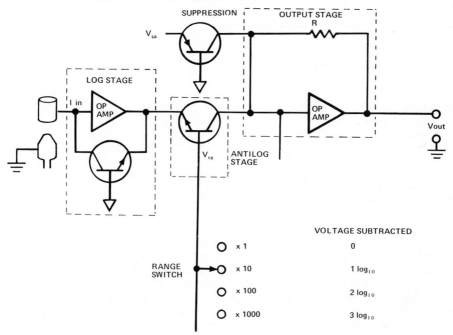

Figure 6.26. Transistor feedback electrometer.

To overcome many of the problems associated with this type of electrometer, the transistor feedback electrometer was recently developed (34). It consists of two amplifiers operating in series as shown in Figure 6.26. The first amplifier has a logarithmic output to span the wide dynamic range of the input current. The second amplifier, an exponential converter (anti-logarithmic output), restores the linearity of the overall response. Ranging is done by adjusting the reference voltage (V_R) in discreet voltage steps. Range-to-range tracking errors are greatly reduced, eliminating the need for sample calibration from range to range. Since the ranging is a set of low-impedence voltage steps, ranging transients are completely eliminated (Figure 6.25B) and remote ranging can easily be provided.

The success of the transistor feedback analog electrometers has led to the development of direct digital electrometers for processor based gas chromatographs (Figure 6.27). These can be ranged automatically without transients and require no front panel controls. The instrument processor can easily remember the offset signal and subtract it from subsequent data points eliminating the need for the suppression control. Autoranging eliminates the range controls. Some of these electrometers continuously integrate the input and peak areas are simply and accurately determined by summing data points.

6.5.4 Dual Detectors

Normally an operating GC can use one, sometimes two detectors, to full advantage. There are several useful applications requiring two detectors. The most common application is for compensation of column bleed in dual column programmed temperature operation such as illustrated in Figure 6.28A. Other applications include the use of two detectors in series (Figure 6.28B); the use of two detectors in parallel (Figure 6.28C); and the use of two independent detector/column systems (Figure 6.28D).

DUAL COLUMN COMPENSATION. Dual column compensation is available on many temperature-programmed chromatographs to alleviate the problem of column bleed resulting in a steeply rising baseline as the column temperature increases. Two nearly matched columns are used and connected to a dual detector or two detectors. As the oven temperature is programmed up, the bleed signals from the two columns should be nearly equal (Figure 6.29). By subtracting the signal of the compensating column from the signal of the analytical column, the upper temperature limit of a stationary phase

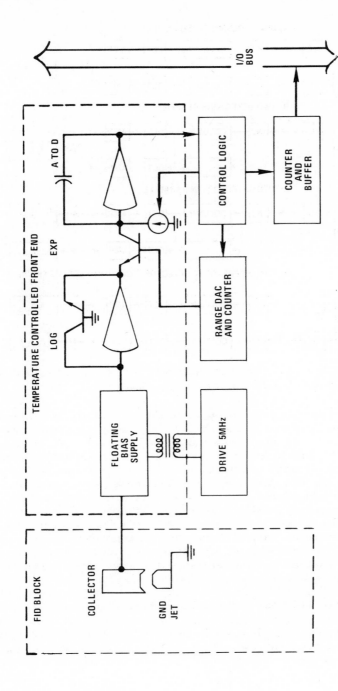

Figure 6.27. Digital electrometer for processor-based
gas chromatograph. Digital feedback loop provides
autoranging, eliminating the need for balance controls.

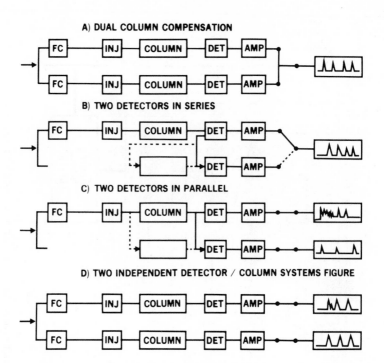

Figure 6.28.   Simplified schematics of practical dual
detector and/or column configurations.

acceptable from the standpoint of baseline control can be extended about 30-100°C (33).   The ability to compensate is a function
of the magnitude of the bleed and the sensitivity of the detector used.   Columns used with FID should have approximately the
same amount of liquid bleed, those used with TCD can tolerate as
much as a threefold difference in the amounts of liquid phase.
     Thermal conductivity detectors usually are built as dual de-
tectors because originally a reference filament formed one leg of
the Wheatstone bridge.   Today most TCD units manufactured still
require a reference gas.   In dual column compensation mode the
compensating column becomes the reference side.
     The compensation can be carried out by two alternate means for
ionization detectors:

1. One electrometer takes the differential signal and amplifies
   it.
2. Two electrometers are used to amplify the two signals independently and then the two amplified and properly ranged signals
   are subtracted.

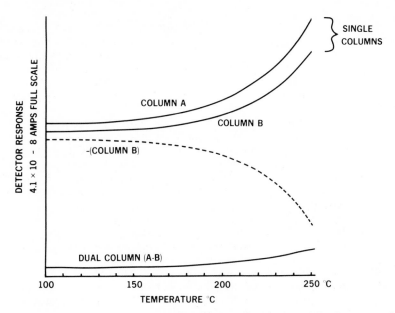

Figure 6.29. Dual column compensation. 15-foot columns packed with 10% OV-1 on Chromosorb with temperature program as shown (10°/min) using flame ionization detectors.

The former has been the more common of the two primarily because of its cost advantage. The latter has the advantage that it has more versatility (it can handle the applications of two independent detector channels as well) and does function as a backup electrometer. In the newer digital chromatographs, the signals from two independent detector electrometer systems are easily and accurately subtracted by the processor.

Only one output device is required for this mode of operation.

TWO DETECTORS IN SERIES. This mode of operation has practical applications in at least two areas:

1. Enhancement of Separation. For some applications, a single column cannot adequately perform the complete separation of all the desired components. By connecting a second column and detector system in series with the normal one, components not resolved on the first column can be selectively switched into the second column for further separation and then detected on the second detector. If the columns are of the same liquid phase, shorter

analysis times can be obtained because the late eluting compounds, if well resolved, do not have to pass through the entire length of column required to separate the early eluted compounds. In addition, compounds which are detrimental to the second column (i.e., carbon dioxide on a molecular sieve column) may be switched to bypass the second column, thereby prolonging its life.

2. Detection of All Components in the Sample. For the analysis of samples containing trace organic compounds as well as inorganic compounds, the column can be connected in series with a TCD and an FID to carry out a complete analysis with a single injection.

An example of this second area of application is the analysis of "dry" natural gas. Since the sample contains nitrogen and carbon dioxide in addition to hydrocarbons, the flame detector alone could not provide a total analysis. By using the signal from the flame detector to detect the low concentrations of the C_3 - C_6 portion and then switching to the TCD for carbon dioxide, ethane, oxygen (if present), nitrogen, and methane, a complete analysis is shown in Figure 6.30. The signal was automatically switched to the thermal conductivity detector at 8 min.

Normally, a single output device is used for this mode of operation. The elution times are controlled so that peaks do not elute from both detectors at the same time; however, two output devices can be used if desired.

TWO DETECTORS IN PARALLEL. This mode of operation has sometimes been referred to as dual channel operation. It has practical applications where additional qualitative information about a peak or sample is desired. There are several ways this additional information can be obtained. The most common ones are:

1. A Single Sample Injected into One Column. Two dissimilar detectors simultaneously monitor the column effluent. The ratio of response of one compound on the two detectors can allow the chromatographer to distinguish between one compound or another even though the two may elute at the same time from the particular column.

2. A Single Sample Simultaneously Injected into Two Dissimilar
   Columns. This is done by means of a simple inlet splitter which allows half the sample to enter one column and the other half of the sample to enter the second column. Two similar

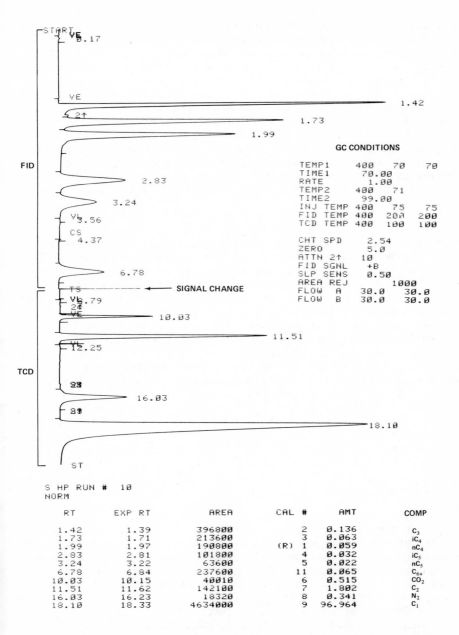

S HP RUN # 10
NORM

RT	EXP RT	AREA	CAL #	AMT	COMP
1.42	1.39	396800	2	0.136	C_3
1.73	1.71	213600	3	0.063	iC_4
1.99	1.97	190800	(R) 1	0.059	nC_4
2.83	2.81	101800	4	0.032	iC_5
3.24	3.22	63600	5	0.022	nC_5
6.78	6.84	237600	11	0.065	C_{6+}
10.03	10.15	40010	6	0.515	CO_2
11.51	11.62	142100	7	1.802	C_2
16.03	16.23	18320	8	0.341	N_2
18.10	18.33	4634000	9	96.964	C_1

Figure 6.30. Use of two detectors in series. Application to natural gas analysis (Hewlett-Packard Co.
5840 chromatograph).

detectors monitor the effluents from the two columns. The ratio
of the retention time of a compound on one column (i.e. nonpolar)
to the retention time on the second column (i.e., polar) allows
the chromatographer to gain additional qualitative information
concerning that compound.

Each detector must have its own electrometer-amplifier, if one is
needed for the particular detector, thereby making it possible to
adjust the sensitivity of each detector as needed or desired.
The amplified signal from each detector is fed to two single pen
recorders or to a dual pen recorder where the simultaneous responses are recorded. Highly selective detectors are commonly
used as one of the detectors in combination with a flame detect-
or.

TWO INDEPENDENT COLUMN/DETECTOR SYSTEMS. This mode of operation
actually provides two independent gas chromatographic systems
operating simultaneously. Each detector has its own amplifier-
electrometer and recorder. The recorders may consist of a dual
pen recorder or two single pen recorders. Since the two chromatograms are normally not related to each other and the starts of
the analyses do not necessarily have to be simultaneous, the use
of two single pen recorders is usually preferred.
    Both column systems share the same oven. Herein lies the maj-
or limitation of this mode of operation. The two columns must be
at the same isothermal temperature or must be temperature-programmed in an identical manner. This mode of operation should
only be used where a common oven is not a problem. Otherwise, it
is preferable to use two separate instruments at a slightly
higher cost.

6.6 RECORDERS

The main function of the recorder is to graphically reproduce the
output of the detector as accurately as possible to obtain a permanent record of the results (chromatogram). The signal to the
recorder can come directly from the TCD or, in the case of ionization detectors, from the electrometer-amplifier. The signal
usually goes through an attenuator (voltage divider) which changes the amplification to the recorder. The attenuator covers the
span of X1 to X1024 or greater in binary steps.
    For many years conventional potentiometric (strip chart) re-
corders have been the standard recording device for gas chroma-

tography.   A block diagram of a typical continuous null balance
potentiometric recorder is shown in Figure 6.31.   An a-c motor,
synchronous with line frequency, is geared through an over-runn-
ing clutch to a take-up clutch and the paper drive sprockets.
This mechanism is independent of the main measuring circuit.   Its
only purpose is to provide a continuous movement of the chart
paper at a known constant rate (cm/min).   Most recorder models
give the user a choice of several chart speeds.   The most common
speeds are 0.5 - 2.5 cm/min.

The differential amplifier compares the attenuated input signal from the detector-amplifier network with the reference voltage, V_0, as illustrated in Figure 6.31.   The resulting error signal, $V_1 - V_0$, is amplified and applied to the servomotor, causing
it to move in one direction if the error voltage is positive.
The slidewire wiper which is mechanically linked to the motor and
a pen moves in the direction that will develop a reference voltage equal to the input voltage.   This continuous balancing action
attempts to cancel the error voltage ($V_1 - V_0 = 0$).   The relative
motion of the pen and the chart paper results in a plotted graph
of response versus time (the chromatogram).

The recorder characteristics which affect the ability of the
recorder to accurately and faithfully plot the input signal are
the following:

1. Range.
2. Dynamic accuracy - step response time.
3. Dead band.
4. Linearity.

The span of the recorder is defined as the algebraic difference between the end scale values.   The range is then simply the
region covered by the span, expressed by stating the two end-
scale values.   This is determined by the d-c voltage across the
slidewire.   An example of these two parameters is:

$$\text{Span:}\quad 1.5 \text{ MV}$$
$$\text{Range:} \quad -0.25 \text{ to } 1.25 \text{ MV}$$

The most common recorder range is 0 - 1 MV, although 0 - 0.1 MV
recorders are frequently used for high-sensitivity operation with
TCDs.

The recorder is a dynamic device; its pen must move as a function of a signal that is changing with time.   The dynamic properties of the pen, then, can become critical when the signal
changes very rapidly as it does, for example in capillary column
chromatography.   Recorder response affects the fidelity of the
plot.   Since the recorder cannot respond any faster than its de-

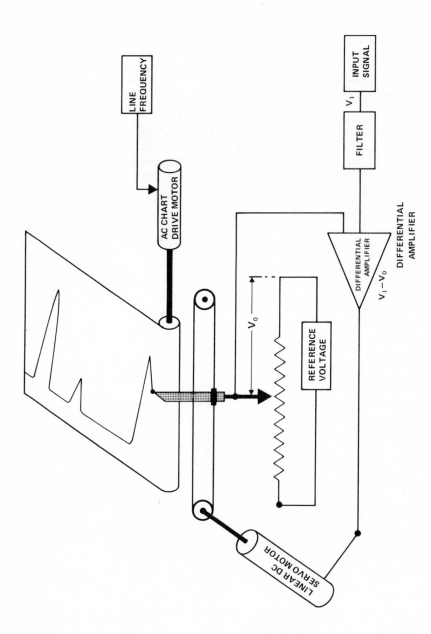

Figure 6.31. Block diagram of conventional continuous
null balance potentiometric recorder.

signed pen velocity, the step response time can give one a rough
determination of whether the instrument is capable of following
a given signal. The step response time is the time required for
the recorder to come to rest in its new position after an abrupt
change to a new constant value has occurred in the input signal.
If the recorder is balanced at zero and a step signal of 1 mV
(full scale) is imposed on the recorder input, the time between
the application of the step and the time the pen comes to rest at
1 mV is the step or pen response time. A step response time of
1 sec is common for most chromatographic recorders. For fast
capillary work 0.25 - 0.50 sec is more desirable. Since the response time is the sum of the initial acceleration time, the
full-speed slewing time, and the final deceleration time, the response to a series of smaller step amplitudes gives a more complete and concise evaluation of the recorder pen response.

Dead band is defined as the range through which the measured
quantity (the input signal) can be varied without initiating response. This is usually expressed in percent of full-scale deflection. Ideally the dead band should be around 0.1% of full
scale. If it exceeds 0.5% of full scale, very small peaks in
capillary GC operation will be missed. Dead band can sometimes
be observed on the chromatogram as a stepping movement on the
tailing edge of a peak and as a rounded or flat peak. The
following contribute directly to dead band:

1. Servo motor torque.
2. System inertia and friction.
3. Resolution of slidewire.
4. Amplifier gain.
5. Source impedence (the resistance the recorder sees at the attenuator output).
6. Interference rejection (a measure of the recorders ability to
   operate without change of calibration in the presence of extraneous a-c signals.

The slidewire determines the linearity of the recorder. The
linearity is limited by the slidewire resolution, its mechanical
construction, and resistance loading effects. In the conventional wire-wound slidewire, the resolution is dictated by the number
of turns. The linearity of the recorder then is equal to the
ratio of a full-scale peak to that of the smallest discernible
peak--one turn of the slidewire. Generally a linearity of 0.1%
is available. By attenuating the input signal the linear range
is effectively extended.

Until recently the recorder played an important role in quantitation. This role is still important for manual means of peak
integration for mechanical integrators also depend heavily on the

performance of the recorder. The trend, however, is towards using the recorder for its main function, a permanent graphic record of the chromatographic analysis. The recorder will never be
eliminated because it gives so much qualitative information of
overall chromatographic performance, particularly as to how well
the column is functioning.

   With the great strides made in semiconductor technology, resulting in small low cost microprocessors and digital chromatographs, a new approach to fulfilling the recorder needs has
evolved. This new approach consists of a microprocessor controlled printer/plotter as illustrated in Figure 6.32. This device has the capability of both inkless, high-resolution plotting
of the chromatographic output signal and alphanumeric printing of
the chromatographic conditions and final analysis reports as
shown in Figure 6.33. While the print head is moved across the
thermal paper, selected dots are energized to form a standard
5 x 7 dot matrix print font. For plotting of the chromatogram,
only one dot is continuously energized. Velocity information in
the processor modulates the heat to produce a high-resolution
inkless trace. Although this printer/plotter is presently not
available as a stand-alone recorder, future availability of simi-
lar-type devices seems inevitable.

6.7 DATA REDUCTION AND HANDLING

The ability to generate a large mass of data in a relatively
short period of time has been one of the problems inherent in gas
chromatographic analysis. It is not surprising then, in light of
today's technology, that the handling of this mass of data has
seen some revolutionary changes, particularly in the last few
years. In the early days of GC the data generated was presented
in analog form on a strip chart recorder. The normal practice
was to then measure, from the recording, peak heights or peak
areas and convert them to concentrations by manual calculations.
Most of the manual procedures which have evolved share the attributes of being tedious, time-consuming, and limited in accuracy
and precision. Today's chromatograph user, who has a small
sample volume and who has no need for maximum accuracy, may find
the manual methods to be the most economical solution to his data
handling problem. He has at his disposal, however, a number of
alternate means for handling his data. The proper choice, of
course, depends upon his needs and resources.

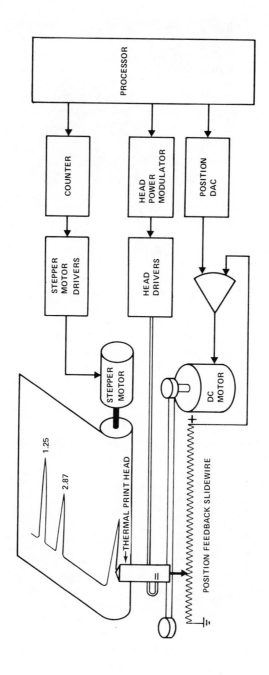

Figure 6.32. Block diagram of processor-controlled printer-plotter (Hewlett-Packard Co.-5830/40 chromatographs and 3380S integrators).

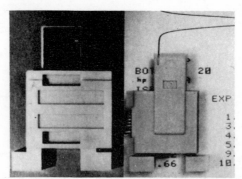

Figure 6.33. High-resolution inkless plotting and
alphanumeric printing is provided by processor-controlled printer/plotter (Hewlett-Packard Co.-5830
chromatograph).

6.7.1 Mechanical Integrators

The first level of automation is the analog ball and disc integrator, which until recently was perhaps the most widely employed
means of integration of gas chromatographic peaks. It is a low-
cost electromechanical device (see Figure 6.34) which automatically integrates peak areas by following the tracking of the recorder pen. It can provide a "trace" readout which is recorded
on the side of the chart as illustrated in Figure 6.34 or a dis-
played readout on a manually controlled counting register. Since
the integrator depends on the recorder for its operation, it requires a good, properly operating recorder. Only peaks which are
on scale on the recorder can be properly integrated. Drifting
baselines can be partially compensated. Retention times have to
be calculated manually. Further calculations to convert the
areas to concentrations are done manually or with a calculator.

In 1971, Feldman et al. (35) compared various methods of integration including the ball and disc and electronic integrators.
They concluded that for the laboratory with a small work load and
limited budget, excellent results can be obtained with the simple
yet effective disc integrator. The precision of this type integrator is in the order of 1 - 5%, depending on the chromatogram.

6.7.2 Electronic Integrators

The next level of automation is the electronic or digital integrator, which automatically gives a printout of retention times

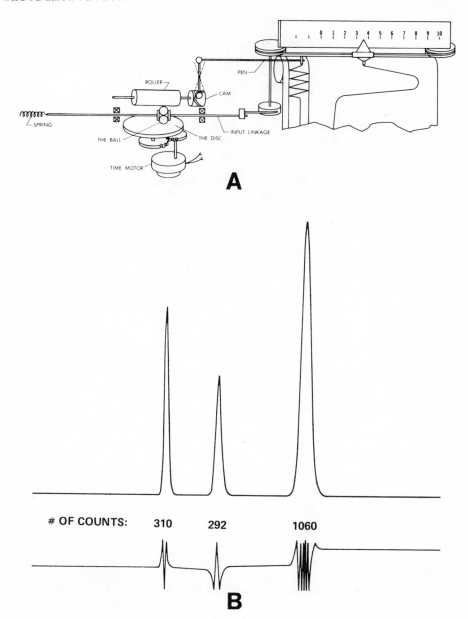

Figure 6.34.   Ball and disc integrator.   (A) Block
diagram. (B) Tracing (no. of counts are manually calculated).

and areas of the chromatographic peaks by directly integrating
the amplified detector signal. The electronic integrator actually uses analog (not digital--as is often implied) logic and
circuitry for peak detection, baseline correction, and retention
time measurements but reports times and areas digitally. Electronic tube-type integrators were first introduced by Infotronics
in 1961 at a cost of $6000 to $7000. With the development of the
transistor, improved solid state versions became available from
several manufacturers at a greatly reduced size and cost. Today
these are available in the $2000 to $3000 range. One of these
is shown in Figure 6.35. All incorporate automatic baseline correction and many have adjustable peak detection limits. Some
have provision for paper tape output to enable the use of offline
automatic data reduction facilities. In general, however, further calculations are done manually with a calculator or with a
computer. These integrators are a labor-saving device compared
to the earlier described methods of integration. With manual
procedures, the data reduction itself was the limitation on precision and accuracy. This is not the case with electronic integrators. The limitation shifts to the chromatographic technique
of the user and to the chromatograph itself. With good technique
and with chromatographs of the electronic era, accuracy and precision in the order of 1% is attainable.

Most manufacturers of electronic integrators will eventually
replace these integrators with microprocessor-based integrators.

6.7.3 Computing/Reporting Integrators

The availability of relatively inexpensive microprocessors, just
within the last few years, made possible the next level of automation: the computing or reporting integrator. The first of
these was introduced in 1971. It can handle one to four gas
chromatographic inputs.

Figure 6.35. Electronic integrator (Hewlett-Packard
Co.-3370 series).

The computing integrator is a microprocessor-based device that combines the functions of peak integration and final calculation of concentrations. The analog signal from the detector or electrometer-amplifier is converted directly to a digital form which may be handled by the processor. Programs and algorithms for integration, peak detection, baseline correction, and final calculations are permanently stored in read only memory (ROM). Such programs are sometimes referred to as firmware. A complete list of all operations and data required to produce the final answers desired can be entered or adjusted by the user through a keyboard or operator control panel and stored in the processors read/write memory. Such a list is called a method. Included in the method are peak identification data, response factors, and so on. Also included in some are instructions, if needed or desired to change on a time basis valve settings, and various integration parameters.

Integration by these devices is clearly more accurate and precise than by manual measurements and precision of results is in the order of 0.5%. Peak identification algorithms approach the reliability of a trained operator. Concentrations are normally calculated by one of four standard calculation procedures.

In 1973 single channel auto calculation versions costing around $5000 were simultaneously introduced by two manufacturers. One of these (Figure 6.36) incorporates a built-in printer/plotter which eliminates the need for a separate recorder. The printer/plotter provides a record of each setting change, draws and labels chromatograms, prints the analytical reports, and on command provides a written list of all integrator settings. Other manufacturers have since introduced single channel computing integrators.

Recently limited versions of the reporting integrators have been introduced. These generally only calculate area percent automatically, but cost only $3000 to $4000. Only some of these can be upgraded to include the autocalculation capabilities found in the higher priced versions. An example of one of these is shown in Figure 6.37.

6.7.4 Multichannel Computer Systems

For the chromatographer who has several (more than four) chromatographs with analog output, multichannel minicomputer-based systems are available from several manufacturers. For the most part, these systems use a minicomputer that is dedicated to the task of data acquisition, data reduction, and reporting of results for a number of chromatographs. Typically, the programs

Figure 6.36.   Auto calculating-reporting integrator
incorporates built-in printer/plotter (Hewlett-Packard
Co.-3380 reporting integrators).

that enable them to do the job are turnkey (completely provided
by the manufacturer) and invisible to the user.   Interaction between the user and the software of the system is normally in dialog through a terminal such as a teletype.   This dialog can be
entirely in terms which are familiar to the chromatographer, such
that no knowledge of programming is required.   These larger systems use an analog-to-digital converter (A/D) which digitizes the
detector signals.   The A/D converter can either be at each chromatograph or there can be a single multiplexed A/D at the computer.   In the latter case, analog signals are transmitted from the
chromatograph to the multiplexer which switches the different
analog signals through the A/D.   In the former case, analog signals are digitized as close to the chromatograph as possible and
the digital signal is transmitted.   The i:tegration and complete
data analysis job is done in the computer.   For example, the HP
3352B laboratory data system provides a background basic language

programming capability. This allows the user to write and exe-
cute his own special programs without deteriorating the real time
system operation. Extended memory-disc or drum capabilities are
also available on some systems. The capabilities vary greatly
from manufacturer to manufacturer and these should be thoroughly
investigated prior to making a choice. Price per channel for
these systems ranges from $3000 to $7000 depending on the capabilities and to a large extent on the number of channels that
must share the initial system cost.

Figure 6.37. Area percent only reporting integrator
can be upgraded to include auto calculation capabilities (Hewlett-Packard Co.-3380S integrator).

6.7.5 Digital Reporting Chromatograph

For the chromatographer who needs to obtain a modern chromatograph and also wants to have data reduction and handling capabilities, a processor based chromatograph of the type shown in
Figure 6.38 should be considered. The key to the operation of
this type of chromatograph is a built-in-digital processor. The
processor is sufficiently fast and powerful to allow digital
multiplexing for simultaneously servicing multiple signals in
addition to carrying out, by means of firmware, the functions required by the gas chromatograph, that is, temperature control.
The processor can be justified on this basis alone. As a bonus
there is sufficient processor time left to perform the function
of converting raw data into usable answers and to perform the
function of instrument controller during the run. Ten different

functions can be programmed to change as a function of elapsed
time by establishing a timed event table through the control con-
sole; a function-oriented keyboard which replaces the analog adjustment knobs of the conventional gas chromatograph. The timed
event table specifies the time at which the event is to take
place and, where applicable, the new value to be established. An
example of such a table as applied to the analysis of natural gas
(Figure 6.30) is shown in Figure 6.39. The processor checks the
table every 0.01 min to see whether the next event should be exe-
cuted. The processor, by performing this function, frees the
user from the constant attention required when this function is
done manually; and, because it is never distracted from the task
at hand, ensures that every analysis is performed exactly as specified.

The reporting gas chromatograph consolidates all the functions
of the gas chromatograph, strip chart recorder, computer automation, and computing and reporting integrator into one unit providing a powerful and cost effective solution in both routine and
research applications.

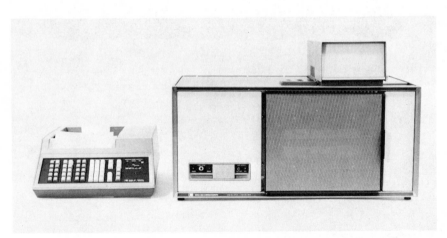

Figure 6.38. Digital reporting chromatograph (HP 5840)
is a unified data generation/data reduction system.
Data output is done on the printer/plotter alongside
the GC unit. Control inputs are entered via the print-
er/plotter keyboard or via the magnetic card in the
main module where the data is generated (Hewlett-Packard Co.-5840 chromatograph).

TIME	EVENT/OR FUNCTION	NEW VALUE
0.10	VLV/EXT	3
0.10	VLV/EXT	1
1.25	VLV/EXT	2
1.60	ATTN 2↑	9
3.50	VLV/EXT-	1
3.80	CHT SPD	0.64
8.00	TCD SGNL	A
8.70	VLV/EXT-	3
8.70	VLV/EXT-	2
9.20	ZERO	5.0
9.30	ATTN 2↑	7
9.80	VLV/EXT	3
12.00	VLV/EXT-	3
15.00	SLP SENS	0.10
15.00	ZERO	5.0
15.00	ATTN 2↑	5
17.00	SLP SENS	1.00
17.00	ATTN 2↑	10
21.00	STOP	

Figure 6.39.   Timed event table used for the analyses
of natural gas (see Figure 6.30) (Hewlett-Packard Co.-
5840 chromatograph).

REFERENCES

1. A. T. James and A. J. P. Martin, Biochem. J., 50, 679-690
   (1952).
2. N. H. Ray, J. Appl. Chem., 4, 21-25, 82-85 (1954).
3. J. C. Giddings, "Advances in Analytical Chemistry and Instrumentation, Vol. 3, C. N. Reilly (Ed.), John Wiley and Sons,
   New York, 1964, p. 339.
4. L. Guild, S. Bingham, and F. Aul, Gas Chromatography, D. H.
   Desty (Ed.), Academic Press, New York, 1958, pp. 226-247.
5. H. W. Habgood and W. E. Harris, Anal. Chem., 32, 451 (1960).
6. J. H. Knox, Chem. Ind. (London), 1085 (August 29, 1959).
7. J. J. Czubryt and H. D. Gesse,, J. Gas Chromatog., 6, 528-530
   (1968).
8. S. Dal Nogare and R. S. Juvet, Jr., Gas-Liquid Chromatography,
   Interscience Publishers, New York, 1962, p. 162.

9. J. W. Frazer, V. DuVal, and R. E. Anderson, An Automated High Sensitivity Three Column Gas Chromatograph, UCRL - 50849, University of California, Livermore, Lawrence Radiation Laboratory, April 21, 1970, p. 2.

10. R. J. Liebrand and L. L. Dunham, Res. & Develop., 24 (9), 32-38 (1973).

11. R. C. Dressman, J. Chromatog. Sci., 8, 265-266 (1970).

12. D. W. Grant and A. Clarke, J. Chromatog., 97, 115-129 (1974).

13. R. H. Kolloff, Anal. Chem., 34 (13), 1840-1841 (1962).

14. N. Adler, Anal. Chem., 36 (12), 2291-2295 (1964).

15. M. van Swaay and J. R. Bacon, J. Chromatog., 19, 604-606 (1965).

16. E. Evrard, J. Chromatog., 68, 47-54 (1972).

17. I. M. Callery, J. Chromatog. Sci., 8, 408-411 (1970).

18. H. Zenz, H. Klaushofer, and H. Szep, Chromatographia, 5 (12), 343-345 (1972).

19. R. E. Kaiser (Ed.), Proceedings of the First International Symposium on Glass Capillary Chromatography including Glass Micro-Packed Columns, Institute Für Chromatographie, Bad Durkheim, West Germany, May 1975.

20. R. D. Condon and L. S. Ettre, Instrumentation in Gas Chromatography, J. Krugers (Ed.), Centrex Publishing Co., Eindhoven, 1968, p. 100.

21. W. G. Jennings, J. Chromatog. Sci., 13, 185-187 (1975).

22. A. L. German and E. C. Horning, Anal. Letters, 5 (9), 619-628 (1972).

23. K. E. Rasmussen, S. Rasmussen, and A. B. Svendsen, J. Chromatog., 66, 136-137 (1972).

24. P. M. J. van den Berg and T. P. H. Cox, Chromatographia, 5, 301-305 (1972).

25. A. G. Vitenberg, B. V. Ioffe, and V. N. Borisov, Chromatographia, 7 (10), 610-619 (1974).

26. J. C. Giddings, J. Chem. Ed., 39 (11), 569-573 (1962).

27. S. Dal Nogare and C. E. Bennett, Anal. Chem., 30, 1157-1158 (1958).

28. S. Dal Nogare and J. C. Harden, Anal. Chem., 31, 1829-1832 (1959).

29. S. Dal Nogare and W. E. Langolis, Anal. Chem., 32, 767-770 (1960).

30. A. J. Martin, C. E. Bennett, and F. W. Martinez, Jr., Gas Chromatography (1959), Noebels, Wall, and Brenner (Eds.), Academic Press, New York, 1961, pp. 363-374.

31. A. P. Tudge, Can. J. Phys., 40, 557 (1962).

32. R. W. Ohline and D. D. DeFord, Anal. Chem., 35, 227 (1963).

33. W. E. Harris and H. W. Habgood, Programmed Temperature Gas Chromatography, John Wiley and Sons, 1966.

34. D. S. Smith, <u>Advances in Chromatography</u>, Vol. 12, J. C. Giddings, E. Grushka, R. A. Keller, and J. Cazes (Eds.), Marcel Dekker, New York, 1975, pp. 177-222.

35. G. L. Feldman, M. Maude, and A. Windeler, Internat. Lab., 47-53 (March/April 1971).

CHAPTER 7

Trace Analysis
by Gas Chromatography

GERALD R. UMBREIT

Greenwood Laboratories

7.1 INTRODUCTION 366
7.2 STAGES OF TRACE ANALYSIS 371
7.3 SAMPLING AND SAMPLING DEVICES 372
   7.3.1 Bulk Products, Tissue (animal and vegetable),
   Crops, Other Solids and Body Fluids. 372
   7.3.2 Water Sampling 372
   7.3.3 Air or Gas Sampling. 374
7.4 ISOLATION AND CONCENTRATION TECHNIQUES 380
7.5 DERIVATIZATION 385
7.6 CHROMATOGRAPHIC TRACE ANALYSIS OF PROCESSED SAMPLES. . 386
   7.6.1 Qualitative Analysis 386
   7.6.2 Quantitative Analysis. 390
7.7 PREPARATION OF STANDARDS FOR TRACE ANALYSIS. 395
   7.7.1 Preparation of Liquid or Solution Standards. . . 395
7.8 MODEL METHODS FOR CONCENTRATION OR TRACE ANALYSIS. . . 397
   7.8.1 Determination of Nonvolatile Solutes in Water. . 398
   7.8.2 Analytical Procedure 399
      Preparation of Solvents and Resin. 399
      Column Preparation 402
      Sample Processing. 402
      Concentration of Eluate. 403

       Separation and Quantification. 404
   7.8.3 Determination of Volatile Organic Solutes in
       Water. 404
7.9   GENERAL SCHEME FOR TRACE ANALYSIS 409
   7.9.1 Achieving Maximum Instrumental Sensitivity . . . 413
   7.9.2 Removing Atmospheric Pressure Aliquots from
       a Closed Gas Sampling Container. 414
7.10 ANALYTICAL ETHICS AND STATISTICAL VARIATION. 416
REFERENCES 418

7.1   INTRODUCTION

A modest perusal of the literature that may be pertinent to the
topic of trace analysis can easily convince the reader that this
is a field reserved to witchcraft and black magic. It would
appear also to be reserved to those who have inordinate amounts
of time on their hands. In part this is most certainly due to
the nature of trace analysis since it clearly involves seeking
the proverbial "needle in the haystack." However, in this case
not only are all the sought needles different, each of the hay-
stacks is also different. Thus, it tends to imply that for each
specific problem of trace analysis brought to the analytical
laboratory, a specific procedure must be developed or employed.
This particular circumstance is not precisely stated by the individual contributor to the literature. It requires an overview
of many of these contributions to derive that conclusion.

However, if science is to serve in one of its primary capacities, that of unifying the observations of nature, then this circumstance too should be subject to a considerable number of generalizations which tend to be broadly applicable. The conclusion
previously suggested, based on literature review, could and
should likewise be derivable solely on the basis of fundamental
scientific reasoning and the application of good common sense.
Common sense should tell us that on a microscopic level those
things, other than the components of particular concern as
traces, which are part of a bulk material, must loom large by
comparison. Therefore, even though two samples of meat, for example, are to be analyzed for some particular trace component, it
should be expected that there will be other, larger variations in
the composition of these two samples. One sample may have more
or different types of fats or protein, more or different types of
inorganic components, or may indeed contain one or more organic
components not present in the other sample. As a rule, it should
be expected, therefore, that the variety of minor components that

may be present in otherwise similar samples, will involve variations in quantity of some of these components which are far
greater than the quantity of the specific chemical or chemicals
sought by the analyst. This circumstance has to be responsible
in some fair measure for the difficulties encountered in reproducing trace analytical procedures in different laboratories and
occasionally within one laboratory.

This, however, is clearly not the only reason for such a problem nor is it the only reason for the inordinately tedious and
involved procedures which are often reported and recommended. It
is also true that if trace components can be of concern in a
sample, then trace components can also be of concern in all the
items involved in processing that sample. Thus "other traces"
are not necessarily introduced into the system in the same quantity as the sought components. In many cases, they can be expected to exceed that quantity many-fold. As a consequence the
sample, as finally prepared for measurement, can almost never be
considered as a single, pure component suspended in a single
pure matrix. It must still be a highly complex mixture. Were
this not so, the elegance of instrumentation and technical training would be a considerable waste since the only need, under
those circumstances, would be a precisely sensitive and accurate
balance.

To proceed further with this discussion, one should recognize
the common factors existing among analytical measurements that
can be described as trace analysis. The determination of single
natural constituents in body fluids or tissue is an extremely
common analytical need. Similarly, in medical and pharmaceutical
research as well as forensic science, measurements of drugs and
metabolites of drugs are sought in these same body fluids or
tissues. However, equally recognizable as trace analysis is the
need to determine additives, catalysts, residues of starting
materials, and other similar analytical needs in industrial products. Also recognizable for their similarity are those measurements that may be considered as pollution analyses. These analyses can often involve the very same sample materials suggested
above. For example, DDT in an animal system is a pollutant just
as much as is DDT in water. Therefore, to apply the reasoning of
chemical science and provide unification, it should be recognized
that each of the measurement necessities described above involves
simply the measurement of one chemical compound (or a specified
group of compounds) in a vast mixture of other chemical compounds. There is no sacred difference between a biologically derived material and an industrial material. They all involve
chemicals solely.

There is, however, one primary exception to this generaliza-
tion and, in hope of reducing fears of analytical complexity, it

should be pointed out that measurements of pollution or contamination of air and water should be the very simplest of trace analysis problems. To overstate the simplicity of this point for analysts, it can be put thus: Anything detected in a water sample which is not water must be considered a pollutant or contaminant. The same statement, of course, is true for air. Other samples mentioned previously cannot be considered so simply since the question "what is blood?" can hardly be answered even at this stage of advanced knowledge of biochemistry. Therefore, one cannot say that in measuring a blood sample, "all that is not blood is pollutant or contaminant." Clearly the problem is not quite this simple, since many things that are in water are natively present in water; and man, in fair measure, does not have direct control over these components. Likewise, some of the materials that man introduces into his environment are clearly of far greater concern than others; therefore, we cannot simply seek in a water sample all things that are not water.

Given the problem of determining the quantity of a component present in trace amounts, there is presently available to the analytical chemist a number of different means of obtaining the final measurement. Certainly a high degree of sensitivity is a prime requisite of any method of measurement, but an equally important requisite is a high degree of selectivity, remembering that the sample, or any processed portion of it, is still a grand mixture.

In a classic paper titled "The Chemical Analysis of Things as They Are," G. E. F. Lundell (1) correctly points out that "many talks and articles on analytical subjects deal with the chemical analysis of things as they are not." As a further part of the discussion in that paper, he points out that many classical analytical methods for the determination of one specific ion or another do not take into account the unquestioned fact that in a sample taken from the "real world" that ion is never found alone, nor in many cases is it the major part of the sample. This leads Lundell to the suggestion that an analyst's Valhalla would be a laboratory that contained 92 analytical reagents, each one absolutely specific for one of the 92 then known chemical elements. However, since he was discussing inorganic analyses, what is clearly omitted from this Valhalla are the several million specific reagents that would be needed to account for the several million different organic chemicals that we know to exist.

For these reasons, chromatography in its many forms has been the most widely used tool in both sample processing and measurement for trace organic analysis. Of the various chromatographic techniques, gas chromatography (GC) has been the method of unquestioned widest use for final measurement. This fact clearly recognizes that gas chromatographic techniques presently provide

the best available combination of selectivity and sensitivity.
The superiority of this technique arises from three fundamental
facts. First, the essence of any chromatographic technique is
its ability to deal with materials that are mixtures and to isolate one component from others, thereby providing a high degree
of selectivity. Second, additional selectivity is further provided by the nature of detectors available for GC in which certain compounds can be selectively detected, even though unresolved or inadequately separated from other components that are not
detected. Third, with this high degree of selectivity, GC combines an unusually high sensitivity when compared to other methods of quantitative chemical measurement.

We have reached the comfortable state of achieving a detection
sensitivity which commonly equals our needs for that sensitivity;
that is, in many cases a single component present in an acceptable matrix can be directly detected at levels in the range of
parts per million to parts per billion. Where this limit of
sensitivity is not directly available, it is generally not so far
removed from the limits desired that a sample cannot be appropriately processed to provide the components sought in adequate concentration for proper measurement. Thus, to close this section
with a further unification, we will define trace analysis as being that portion of analytical chemistry that is concerned with
the measurement of single components present in the matrix of any
given sample, in quantities in the range of 1 ppb to 100 ppm.
This statement is not intended to exclude measurements of components present below 1 ppb, but such extreme traces are not commonly sought.

Given the preceding discussion, the purpose of this chapter is
to attempt to remove some of the fears of entering the apparent
morass of this field, to provide guidelines in some detail to
avoid the pitfalls of the field, and to define many of these pitfalls so that they can be recognized. Throughout this text there
will be a continuing emphasis on the need for the maximum application of common sense. Common sense dictates first that if one
is to make measurements of the quantity of a material, he must
either be precisely correct or, more likely, he must know the extent to which he may be incorrect. He needs to know also the degree of correctness required to provide the basis for decisions
that must be made as a consequence of the measurements. For
example, it is clear that if a concern exists over a potential
explosion from some explosive vapor, the analysis that indicates
this concentration to be less than one tenth the explosive limit
need not be more accurate than a factor of two, and heroic efforts to provide a relative accuracy of ±1% represent a complete
waste of time. On the other hand, the measurement that suggests
the concentration to be 96% of the explosive limit, must clearly

have a relative accuracy to better than ±3%! Therefore, the
needed precision and accuracy of the analysis are defined by the
nature of the problem and not by any arbitrary rule or the
classically taught desire to achieve 0.1% relative accuracy.

Also, while emphasis has been placed on the selectivity of
chromatographic methods, and GC in particular, it should be remembered that this single technique does not represent the hypothetical several million individual totally selective reagents.
Therefore, on many occasions it becomes necessary to establish by
alternate or related means that the compound measured is, in fact,
the compound sought and only that compound. Once again, the nature of the problem as presented to the analyst should provide him
with the basis for determining the degree of certainty he must
achieve in identifying the measured component. If, for example,
the component of concern is critical when present in excess of
1 ppm, it is obviously not critical to identify the component
apparently measured as that compound which shows a concentration
of 0.1 ppm.

While it has been strongly implied that significant correlations exist among the many varieties of trace analysis problems
that may be brought to the analytical chemist, he must recognize
the variations in required depth of investigation which each presents. It is for this reason that the leading statement of this
chapter is made, and cautions are thus introduced as to adopting
without question analytical procedures for trace analysis which
are available in the literature. The chemist should be aware
(again, using common sense) that the more complex the method, the
more likely it is to introduce difficulties in analysis and particularly losses of the very component sought. Therefore, an
additional general rule of trace analysis is that the best analytical method is the simplest and most rapid that can be devised or applied. There are many reasons for this general rule.
Included among these are: (a) The average person lowers his
attention when engaged in tedious, obscure details, (b) the
greater the number of manipulations, the greater is the likeli-
hood of introducing materials not initially a part of the sample,
(c) the more a material is manipulated and the longer it is held,
the more likely are both chemical and biological reactions to
proceed to an extent sufficient to cause significant error. This
last fact is particularly true when the component sought is initially present at trace levels.

Any presentation such as this, which is intended to guide the
analyst through trace analysis, must begin with describing those
factors that can introduce errors so that the application of a
general approach can be properly exploited by the analyst with
his particular sample problem. However, before he can evaluate
which of these factors must be of concern to him, he must have

adequate answers to a related series of questions. He must know:

1. What specific chemical compound(s) are of concern?
2. What concentration range of these compounds is of concern?
3. What is the nature of the sample matrix; that is, what are the known major parts of its composition?
4. How much sample can be made available?
5. Where and how is the sample taken?
6. What is the nature of the bulk material intended to be represented by the sample?
7. What decisions are to be based on the measurement; that is, why is the measurement required?

The analyst must have reasonable answers to these questions in order to participate intelligently in decisions involving the number of samples required, the accuracy necessary, and the variability to be expected so he can provide analytical measurements adequately tailored to the importance of the decision required?

7.2   STAGES OF TRACE ANALYSIS

All analytical problems that can be defined as trace analysis have certain recognizable similar stages. Following an assurance as to the total nature of the problem indicated earlier, a decision must first be made as to the nature of the technique to be used for the final measurement. This decision will affect all other decisions related to sample processing. In this chapter we must assume that GC is the chosen method of final measurement. Given this situation, and a knowledge of the sensitivity limits to be achieved with the detector system intended, then the size of samples and the degree of concentration required is defined. The number of samples and the method of obtaining the samples is a decision that must also be made. Following this, adaptation of the sample to suit the measured portion to the method of measurement is involved. This adaptation may take the form of concentrating the sought component or isolating it from a variety of other constituents of the sample. In many cases, both alterations are necessary, that is, partial isolation and concentration. Further, there may be a need or desire to convert chemically the sought components either to improve their detectability or their chromatographic behavior. This stage also can conceivably be combined with the steps of concentration and/or isolation. Alternatively, in favorable circumstances, no further modification of the sample is required.

Subsequently, appropriate standards must be prepared in order
to establish the true minimum sensitivity of detection and to
calibrate the response of the system. The final step in all
cases is measurement of the prepared sample portion. In GC this
implies choice of chromatographic column, operating temperature,
type of detector, and alternatives to these choices if the prepared sample portion presents chromatographic or detection difficulties.

7.3   SAMPLING AND SAMPLING DEVICES

7.3.1 Bulk Products, Tissue (animal and vegetable), Crops, Other Solids and Body Fluids

With samples of this nature, the sampling devices are either dictated by the nature of the sample source such as body fluids, or
can otherwise be the simplest types of containers. The problem
that exists is that of being assured that the sample is either
sufficiently homogeneous to remove a representative small portion comfortably, or that a sufficiently larger sample is taken
randomly and this larger sample reduced in particle size by one
of the number of means available (grinding, blending, etc.); and
finally a representative portion of blended material taken for
the analytical procedure. The primary concern in these circumstances is that the container and other utensils used to obtain
the sample be scrupulously clean and that the sample, once taken,
is stored until processing so that no deterioration or chemical
change occurs. Freeze-drying is widely used as a preliminary
treatment for tissue and crop samples where this process is
available and applicable.

7.3.2 Water Sampling

For pressurized or public water systems, the only real concern in
obtaining the sample is to avoid the portion that has been standing quiescent in the system prior to sampling. In other words,
water that may have been standing for an unknown period of time
in contact with valve packing, check valves, pump fittings, and
other lubricants may have dissolved materials from this exposure
in proportions much greater than the bulk flowing water when in
use. Therefore, the system should be run for some short period
of time before the sample is taken. The size of the sample may

be dictated by the known sensitivity limit of the analytical
method and the anticipated level of contamination, or the specified allowable upper limit of contamination. Again, the sampling
device is generally not critical other than the clear necessity
for cleanliness. For this type of analysis, glass containers are
superior to any other alternative. The cap, or stopper, or other
closure should be metal-foil lined or Teflon (Du Pont trademark
for polytetrafluoroethylene film) lined to avoid any concern over
leaching of organic material from the closure. As before, the
time of holding before processing and the conditions of storage
are important. If processing is not to be carried out immediately, the sample should be refrigerated in the interim. Preservatives such as mineral acids, toluene, or boric acid, which
are sometimes used for clinical analyses, should be avoided since
they generally complicate the subsequent analysis. Because of
the ever present potential for bacterial action to change the
sample composition, analysis should follow obtaining of the
sample in a relatively short time period, ideally 1 day or less.

Sampling of waste or natural streams is generally a more complex problem. In the case of waste streams from manufacturing
plants, the composition of components of concern can change dra-
matically over short time periods. In addition, in many systems
different waste lines enter the same stream at different points
and in varying volume and velocity. This requires that the analyst have some knowledge of these entries to the stream and their
locations. Recognition must be taken of the possibility of incomplete mixing of these various streams at the point chosen or
available for sampling. For the materials sought or suspected,
the relative solubilities and densities should be noted. If the
suspected or possible concentration exceeds this limit, there
will obviously be a tendency for the excess either to float to
the surface or sink to the bottom; and a sample taken solely of
flowing solution below the surface will not necessarily represent
the true level of contamination. Similarly, if insufficient mixing has occurred, there may be laminar portions of the flowing
stream which vary significantly from other portions.

If single or "grab" samples are taken, the sampling containers
again are relatively simple, though in this case some effort must
be exerted in an attempt to obtain proportionate volumes from
different parts of the cross-section of flow if this appears necessary or desirable. This type of sampling can also be accomplished automatically with commercially available instrumentation.
In general, the automatic systems will not provide for as large a
sample as may sometimes be desired. Usually, an automatic system
will sample from one single point in the stream and it will be
necessary to assure that this sampling point is representative.
The samples thus obtained can be analyzed singly and the vari-

ation of concentration with time subsequently determined. This type of measurement can be extremely useful even in cases where the compounds detected are not fully identified. For example, if each day at a certain hour there is a marked increase of one or several components, this may be correlated to the operation of a leaky plant valve and the economic loss which this represents corrected before it becomes serious. Conversely, if each single sample is too small for adequate analysis, the automatically collected group of samples can later be combined in one container (a composite sample) and all, or a portion of this composite sample can be subsequently analyzed. This, of course, eliminates the ability to recognize variations with time. Some automatic systems are designed to provide a composite sample directly, that is, each portion that the system removes from the stream is transferred to a single container that accumulates all samples taken over some fixed period of time.

In the case of natural waters, a flowing stream can usually be treated in much the same fashion as a plant stream, again taking cognizance of the entry of subsidiary streams. The most diffi-cult water sampling problem involves quiescent ponds, lakes, marshes, and so on. In this case the exchange of composition from one portion to another can be extremely slow so that a sample taken at one point will yield an analysis markedly differ-ent from a sample taken at another location in the same body of water. This is even more true in circumstances where the contam-inant sought is incompletely soluble. If it floats on the sur-face, then wind action will tend to concentrate the material against the downwind shore. If the contaminant tends to sink, then the concentration of this component nearest its point of en-try to the body of water will generally be higher than at any other point. These last circumstances present the greatest diff-iculty in assuring that a sample is properly representative. This also is a circumstance in which subjective judgment enters before any analytical processing can occur. In such a circum-stance, multiple sampling at a variety of locations is highly desirable, if not imperative.

7.3.3 Air or Gas Sampling

For this type of sample, the problems, as well as sampling de-vices, are generally more complex. The primary reason is the fact that the sample itself is invisible and is quite likely in-homogeneous. Two circumstances that are fundamentally different are encountered. One involves sampling of plant stack output, which is usually done from some point within the stack or immedi-

ately over the top. Under these circumstances, the sample itself
is essentially driven past the sampling point and some opportunity exists to assure mixing ahead of sampling or to design mathematically a variation in locating the sampling point.

The second circumstance is in quiescent or slowly moving air
such as encountered in testing pollution in the region of a high-
way or on plant grounds. Here, as in the case of the pond, a
single sample taken at one point in time is unlikely to represent
a correct picture of the total circumstance. The sampling point
should be varied and samples should also be taken over a significant period of time before the analytical results can be expected
to have significant meaning. It is obvious, for example, that
the pollution level generated by motor traffic will be highly
variant throughout the normal day.

For both types of air or gas sampling, a wide variety of sampling devices is available, or can be devised from available
items. The simplest must be the system used for "grab" sampling.
This may be nothing more than a sealed, evacuated glass container
(Figure 7.1) which can be opened at the site chosen for sampling
and reclosed after the sample has entered the container. A manual or electric pump can also be used to draw sample into the
container after opening on site. This type of device can be obtained with or without stopcocks at both ends. While the stopcocks are convenient, they are not certain insurance against
leaks or alteration of the sample by diffusion. Rubber septum
seals tend to be better for this purpose, though less convenient.
Teflon stopcocks are strongly recommended for this type of closure. The lubricant required for glass stopcocks must always be
suspected of preferentially absorbing or dissolving some one or
another of the sample components that may be of critical concern.

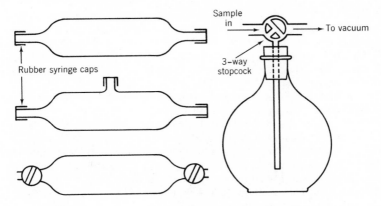

Figure 7.1. Types of gas-sampling containers.

This same possibility exists with rubber closures, though not to the same degree. If the compounds sought in the sample are clearly not oil soluble, then lubricated stopcocks can be used. Olefins and aromatic compounds are generally more readily soluble and more easily lost under these circumstances than are saturated hydrocarbons or oxygenated compounds.

Plastic bags are also used to obtain gas sample (2 - 5). The most reliable of these are Teflon, though Mylar, and Tedlar (Du Pont trademarks for polyester film and polyvinyl fluoride film) are also used. The simplest means of obtaining the sample is to enclose the empty bag in a cardboard or other container that is reasonably well sealed. The entrance to the bag is brought through the box at some point and sampling tube attached if a tube is used. The bag is filled by creating a partial vacuum within the box. This forces the bag to expand either to the interior volume of the box or to its capacity limit. The operation is illustrated in Figure 7.2. Additional information may be found in Chapter 4.

It should be noted in all cases that the driving mechanism causing the air or gas to enter the sample container is the creation of a vacuum or partial vacuum. This contrasts to pressuring the sample and driving it into a container. In other words, the driving mechanism for transferring the sample from its source to a container should be downstream from the sample container in the flow pattern. The purpose is to avoid inadvertent contamination of the sample with lubricants or other chemicals associated with the pump or other devices. A device that is designed contrary to this general rule (6) is a battery-powered, programmable device for taking multiple air samples in plastic bags at predetermined times. The pumps used are of special construction in which all gas-contacting surfaces are of Teflon.

A large syringe in many cases can be an adequate device for obtaining and holding a sample. It can also serve as the means for drawing a sample into a container as long as the container volume is not significantly larger than the volume of the syringe. At least three container volumes should be displaced to obtain a proper sample.

Each of the sampling methods described to this point is primarily aimed at gas samples in which the components sought are directly detectable without concentration. However, there will be circumstances where the components sought may be present in a concentration that cannot be directly detected or measured. Thus sampling must also include a concentration process. For this purpose, a variety of approaches have been used, some of which are illustrated in Figures 7.3 - 7.6. Alternately, trapping systems in series which cause the sample to bubble through a liquid are also used (Figure 7.7). The liquid solution that results

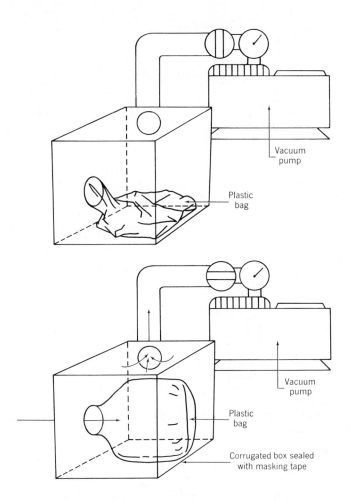

Figure 7.2. Obtaining a gas sample by expansion of a plastic bag.

when sampling is completed is then analyzed. The cautions that should be observed with these combined sampling-trapping systems are discussed in conjunction with concentration techniques.

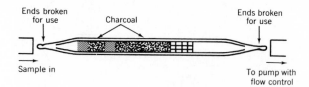

Figure 7.3. Approved absorber tube for vinyl chloride monomer sampling.

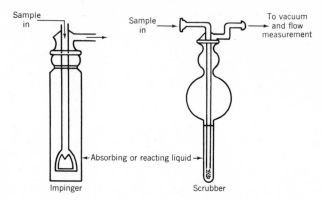

Figure 7.6. Devices for concentrating gas samples in a liquid absorber.

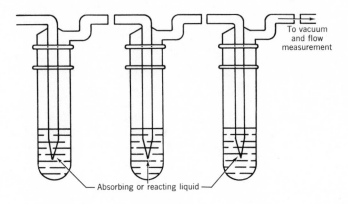

Figure 7.7. Traps in series for concentrating gas.

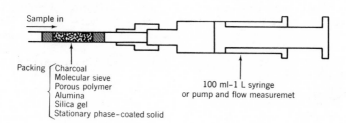

Sample in

Packing ⎧ Charcoal
⎪ Molecular sieve
⎨ Porous polymer
⎪ Alumina
⎪ Silica gel
⎩ Stationary phase-coated solid

100 ml–1 L syringe
or pump and flow measuremet

Figure 7.4.   Absorber concentrator for atmospheric
sampling.

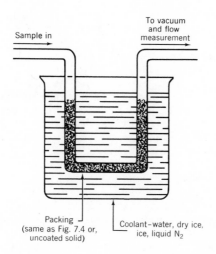

To vacuum
and flow
measurement

Sample in

Packing
(same as Fig. 7.4 or,
uncoated solid)

Coolant–water, dry ice,
ice, liquid N_2

Figure 7.5.   Cooled "U" tube trap for concentrating
gas samples.

7.4   ISOLATION AND CONCENTRATION TECHNIQUES

Concentration involves the transfer of components sought in a
sample from the volume of the original sample matrix to a smaller
volume, usually of a different matrix. The techniques to be applied to a given sample will be determined by the relative volatility of the component sought and the matrix, or major components of the matrix, and also the chemical nature of the component
sought relative to the chemical nature of the matrix or major
components of the matrix.

For liquids and solid samples, the obvious concentration techniques familiar to all chemists are extraction, evaporation of
the matrix, distillation, and precipitation. For trace analysis,
extraction and/or evaporation of the matrix are generally the
only two of these techniques likely to avoid severe losses of the
component sought.

All processing steps are also aimed at the goal of isolating
the components sought from other portions of the sample. In the
case of extraction, for example, a large portion of the sample
components remains behind. This step can also involve the initial use of an extracting volume significantly smaller than the
original sample volume and/or be followed by evaporation of all
or a large part of the extracting solvent. Given a circumstance
in which significant reduction in volume of a solvent is involved
in sample processing, then clearly, analytical checks should be
made, not on the solvent as received, but on a portion of the
solvent similarly reduced in volume.

In common use for the combined steps of isolation and concentration are a variety of chromatographic techniques and trapping
systems. The simplest of these would be applicable only to air
or gaseous samples and consists of a trap immersed in a cold bath
with the intent of condensing the components of interest while
the remainder of the sample flows freely through the trap. While
this has been used, it is rarely satisfactory and even more rarely, quantitative. A trap of this sort will be clearly more effective if it is packed with particulate matter, thereby providing
a larger surface on which condensation can occur, as illustrated
in Figures 7.3 - 7.5. However, in most cases not only is the
trap packed to provide additional surface contact with the flowing sample, but an active adsorbent or absorbent is used. Among
the materials that have been used for this purpose are alumina,
silica gel, charcoal, porous polymers, and liquid-coated solid
support materials. In other words, many materials commonly used
in GC as column packings are usable for trap packing. With these
types of packings it is not always necessary that the trap be
cooled below ambient temperature. The retentiveness of the pack-

ing and the nature of the components sought will determine whether cooling is necessary.

In any case, following the concentration of sought components (along with others not necessarily sought or desired), it is necessary to remove the trapped materials from the trap and transfer them to the gas chromatograph for final measurement without serious loss. This is commonly accomplished by plumbing the trap in the carrier gas flow line of the chromatograph, followed by heating the trap. Normally, this heating cycle involves a rapid rise to temperatures approaching $200^{\circ}C$, and commercial instrumentation is available for this type of trapping and processing (7, 8). At the present time the materials in most common use for this purpose are the porous polymers such as Chromosorb 101, 102, and so on (Johns-Manville trademark); XAD 1, 2, and so on (Rohm and Haas trademark); and the Porapak materials (Waters Associates trademark).

For water-based samples in which organic compounds are sought as trace components, both charcoal and porous polymer systems have been used. Again, currently the porous polymer systems are being more widely used. The customary mechanism is to pour the sample through a column of absorbing material in the fashion of liquid column chromatography, the intent being that the components of concern will be retained by the column packing while the bulk of the sample and other components will pass through. The succeeding step must then involve a proper choice of second solvent which will quantitatively remove the retained components in a volume significantly smaller than the original sample volume.

For liquid or solid samples more complex than water, a combination of techniques is commonly required. Certainly a first step involves a need to obtain the components of interest in a solution phase. This may either involve leaching of a solid or extraction of a liquid sample with or without concurrent concentration. If the components of interest are then obtained in a water system, the techniques applicable to water analyses are immediately available. Conversely, if the extraction is into a nonmiscible organic solvent and the components sought can be reextracted into water by appropriate choice of pH, then again the techniques of water sample processing can be used.

However, a larger portion of materials which may be of concern as trace components will not be extractable or generally soluble in aqueous systems. This presents the necessity to concentrate components contained in organic solvents. When the organic solution is too complex, or the components sought are too volatile to allow concentration simply by evaporation, then more complex means must be pursued. Under these circumstances, liquid column processes are again commonly used. Alumina, silica gel, Florisil, other adsorbents, and mixtures of adsorbents have been

used in a variety of specific cases. Usually the attempt is made
not only to concentrate the portions of concern by this means,
but also to further reduce the complexity of the prepared portion
of the sample. Two alternative mechanisms can be involved in
this process.

The adsorbent can be chosen to retain a variety of components
that represent unwanted materials and potential interferences in
the desired measurement. In this case, the components of interest pass through the column and are included in the collected
effluent. This does not generally result in concentration,
though it may then provide a circumstance where evaporation of
solvent now leaves a concentrated sample portion that is free of
interfering additional components. In the alternative mechanism,
the column may retain the component sought along with others and
pass additional components as well as the bulk of the solvent.
Under these circumstances, two subsequent possibilities exist.

One possibility suggests that an appropriate small volume of
correctly chosen solvent will elute all of the component of interest and all or some of other components that do not subsequently interfere in the analysis. A more complex, but common
possibility is that no single solvent will elute the components
of interest without other components that clearly interfere in
the analysis. This circumstance requires a true practice of
liquid chromatography and the cutting of fractions of the resulting eluate. The purpose would be to discard those fractions that
contain interfering components while retaining fractions that
contain the components of concern. Here is where the individual
differences among samples and sample types will require the most
careful efforts of the analyst.

The wide variation shown in the literature for this type of
process confirms the need for caution and sound technical knowledge. To illustrate this point, Table 7.1 is reproduced (9).
In this case, the compound sought and the nature of the sample
matrix was the same for every investigator or group. Nevertheless, for reasons best known to them, wide variations in the type
of adsorbent or mixture chosen were involved; but more than that,
wide variations in recovery and, therefore, correctness of the
analysis are shown. This is true for systems in which different
adsorbing materials were used and for systems in which apparently
the same adsorbing system was used. Therefore, a chemist with
responsibility for trace analysis must view this circumstance
with great suspicion and he is cautioned against using literature
methods verbatim without careful testing of each step.

The primary reason for the variations illustrated by this tabulation is the fact that a given adsorbent, by its name alone, is
not defined as to its physical activity. This adsorbent activity
is highly variant and dependent upon the nature of pretreatment,

TABLE 7.1 PERCENT PESTICIDE RECOVERY WITH VARIOUS CLEANUP METHODS[1] (9)

Cleanup Method	Parathion Methyl A	L	Malathion A	L	Malaoxon A	L	Parathion A	L	Paraoxon A	L	Ethion A	L
Beckman and Garber (12) Absorbent: Florisil Eluents: Benzene, ether-benzene, acetone, methanol	100	97	96	97	100	80	99	94	93	87	100	85
EEC Method, Versino et al. (11) Absorbent mixture: Na$_2$SO$_4$, Florisil, Celite 545, Attaclay Nuchar C Eluent: Chloroform	89	84	99	99	81	84	100	100	98	94	95	88
McLeod et al.(14) Absorbents: Nuchar C-190 N + Solka Floc BW-40 Eluents: Acetonitrile-hexane, chloroform, benzene	100	93	91	85	65	88	100	88	100	93	94	80
Nelson (15) Adsorbent: Florisil Eluents: Ether, dioxane petroleum ether	20	90	0	53	–	–	66	92	–	–	–	–
Samuel (16) Adsorbent Mixture: Celite 545, Nuchar C-190N, Attaclay Al$_2$O$_3$	100	95	100	96	20	0	96	89	93	91	100	95

383

Method												
Eluent: Dichloromethane Watts et al. (17) Adsorbent Mixture: Celite 545, Nuchar C-190N, Sea Sorb 43	99	77	45	39	0	tr	101	92	80	51	100	90
Eluent: Ethylacetate- benzene Kadoum (18) Adsorbent: Silica gel	97	94	83	100	0	0	100	103	0	40	98	103
Eluents: Benzene-hexane, ethylacetate-benzene Wessel (13) Adsorbent: Florisil Eluents: 6,15,50% ether in petroleum ether	100	96	50	40	0	0	97	89	0	0	100	87

[1]Average recoveries from (A) apples and (L) lettuce.

storage, and use, as well as less obvious factors.   Therefore,
even with the best attempts by the manufacturer to limit batch-
to-batch variations, when they are used for these purposes the
variations are nearly overwhelming.   Thus, it could not be
assumed that the system reported in the literature, which adsorbs
a given compound from a large volume sample, will release that
same compound with the passage of the precise volume of eluting
solvent indicated by the author.   That particular volume works
for that author with his batch of adsorbent treated in his precise manner.   For another analyst or laboratory, the true eluate
volume in which the compound of concern will be found, must be
determined independently with the materials in use in the second
laboratory.

It is for these reasons that a recommendation is made to avoid,
wherever possible, concentration and cleanup steps that involve
"column chromatography" of this nature, that is, chromatography
where an in-line detector is not involved.   This distinction is
made to eliminate any implied criticism of instrumented high
pressure liquid chromatographic systems which have not to date
been used extensively for this particular purpose.

An analagous process for isolating and concentrating portions
of a sample is the use of thin layer chromatography.   Here, however, it is common practice, in the treatment of an unknown
sample, to place a known standard on the same plate and to run
the two systems parallel so that differences in performance of an
adsorbing system from one laboratory to another are accounted for
by basing isolation of the desired components from the plate on
the determined position of the known standard, run on the same
plate.   The normally limited capacity of thin layer chromatographic systems tends to exclude their wide use for the combined
isolation-concentration steps of trace analysis, but circumstances in which this technique may be useful should not be overlooked.

7.5   DERIVATIZATION

There are a number of reasons for converting a compound sought at
trace concentration levels to another compound.   Any attempt at
chemical conversion of the sample must be justified on the basis
of providing a faster, more convenient, or more accurate final
analysis.   These goals may be served by forming a new compound
that is more readily extractable, more readily chromatographed,
more accurately or sensitively measured, or more easily separated
from interfering components, either in the steps of sample

processing or in the final chromatography of the prepared sample. As a general rule, extractability, as well as achievement of most nearly ideal chromatography, is accomplished by conversion of more polar compounds to less polar compounds. One of the most dramatic examples would be the conversion of carboxylic acids to the corresponding ester derivatives. The acids tend to have a moderate water solubility and present generally severe problems in achieving good quantitative accuracy with GC. This is particularly true at trace levels. Other polar compounds that present similar problems are alcohols, amines, and aldehydes. Fortunately, these compounds are also the ones most readily derivatized to some less polar materials. Silylether derivatives can generally be prepared from organic compounds which contain the functions —OH, —SH, or —NH. The —OH and —NH compounds can be converted to acetates or other acyl derivatives with some ease. Carboxylic acids are commonly converted to esters, most commonly, methyl esters, by a wide variety of methods, each of which may have merit in a specific circumstance (10). Carbonyl compounds can be converted to oximes by reaction with hydroxylamine or derivatives. They can also be reacted with primary amines to form Schiff bases, or with alcohols in certain cases to form acetals or ketals. A number of references that review a variety of derivatization methods are available (11-14).

For trace analysis it is common to include in a derivatization step the formation of a derivative that not only aids the desired chromatographic or separation factors, but also provides sensitivity to one or another of the specific detectors available. It is common to use a halogen-substituted derivatizing reagent to provide simultaneously decreased polarity or reactivity, better chromatographic separation, and more selective, sensitive detection using electron capture. Thus, for example, trichloroacetic anhydride is used to form acetate derivatives instead of using acetic anhydride.

7.6   CHROMATOGRAPHIC TRACE ANALYSIS OF PROCESSED SAMPLES

7.6.1 Qualitative Analysis

In most problems of gas chromatographic analysis, there is a quantity of sample available which is sufficient for isolation of important compounds by a variety of techniques to provide for unequivocal identification. However, in the case of trace analysis, great effort has already been spent to provide a sample

normally sufficient only for detection or measurement with the
extreme sensitivity of gas chromatographic detectors. Therefore,
such activities as fractional distillation, crystallization, or
identification techniques such as infrared or nuclear magnetic
resonance spectrometry are completely ruled out.

It is reasonable to consider chromatography of the sample and
a standard of the suspect compound on two or more clearly different chromatographic columns as a means of tentative identification. The details and cautions of this approach are well-defined
in Chapter 4.

An additional mechanism of identification, again dependent on
standards and on reasonable prior knowledge of likely identities,
has been reported by Beroza and Bowman (15-18). The technique
involves a measurement of the distribution of the unknown component between two nonmiscible liquids. This work was primarily directed toward pesticide residue analysis. Dealing with pesticide
residue analyses, it is accepted that sample processing will
ultimately provide a sample for measurement in hexane or isooctane solution. Occasionally this may be in another solvent.
In any case, the prepared sample extract is chromatographed using
appropriate instrumental sensitivity settings to generate a properly measurable chromatogram. The remainder, or a portion of
the prepared extract is then equilibrated with the same volume of
a nonmiscible solvent. For example, acetonitrile, or an acetone-
water mixture has been used. Again, a portion of the hexane or
isooctane phase after equilibration is chromatographed under precisely the same conditions. The peak height or area of the component of concern is determined for both chromatograms (Figure
7.8). The ratio of signal following equilibration divided by the
signal before equilibration has been defined as the "p-value":

$$\text{p-value} = \frac{\text{Peak height (or area) after partition}}{\text{Peak height (or area) before partition}}$$

The p-values for standards which have the same retention time
under the chosen chromatographic conditions are then determined.
The p-value for the unknown component and one of the standards
should be the same if the unknown and the standard are the same
compound. If two or more standards have closely similar p-
values, then the experiment should be repeated with a different
solvent pair.

For convenience, the p-value was selected to designate the
component distribution in solvent systems of equal volumes. If
different volumes of the two solvent phases are used, then appropriate corrections must be made (18). Ideally, the solvent systems should be so chosen that the p-values for components of interest range between 0.25 and 0.75 to provide for greatest pre-

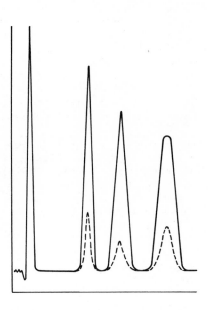

Figure 7.8.   Experimental determination of "p-values."
_____Chromatogram before equilibration. - - - Chroma-
togram after equilibration.

cision and assurance of identity or nonequivalence with the standard.

The p-values for over 100 pesticides and related substances
were established by studying the extraction behavior of those
compounds in a wide range of binary solvent systems. Gas chromatography—based on electron-capture detection—was used to obtain
the experimental data. As a result of these studies, Beroza and
Bowman reported that:

- Each pesticide exhibits a characteristic distribution ratio
  (i.e., p-value).
- Distribution ratios are practically independent of pesticide
  concentration over any range of concern in trace analysis.
- Other components extracted from the original sample do not
  appreciably affect this ratio.
- Compounds other than pesticides can be identified using this
  p-value and GC technique.

The use of distribution coefficients or their simplified equivalents as p-values is not new and is based on sound chemical principles (19). Its particular value, however, is that it can be applied as a confirming means of identification where the component of concern is not available in sufficient quantity for the more common identification techniques such as infrared spectroscopy, elemental analysis, or physical property measurements (20).

A technique that is becoming more common for the identification of compounds present at trace levels is the combination of gas chromatography/mass spectrometry. The general application of this technique is discussed in Chapter 4. For trace analysis, however, there are some cautions. As a rule, in order to have enough of a single component from the gas chromatograph effluent to obtain a complete mass spectrum, the compound of concern must be present in quantity sufficient to be detectable by flame ionization. In other words, components sensitive to electron-capture detection, and comfortably measured by this means, will likely be inadequate in quantity for mass spectral scanning. In some cases, sensitivity and tentative identification can be obtained on these lesser quantities by "single-ion monitoring" with the mass spectrometer. In this case, an assumption as to the identity of the suspect component is made. The single ion produced in greatest quantity from that compound is chosen and the spectrometer set to monitor only that particular mass-to-charge ratio. Because it is possible to create an equivalent fragment with the same mass-to-charge ratio from an unrelated compound, this method for qualitative identification must be regarded as tentative or supportive, but not conclusive.

Often overlooked is the possibility that derivatization also provides qualitative information. For example, assume the suspect compound is a herbicide such as 2,4-D (2,4-dichlorophenoxy acetic acid). As such, this compound cannot be reasonably chromatographed and detected at trace levels. Therefore, it is customarily converted to the corresponding methyl ester and analysis is based upon comparison to standards of the corresponding methyl ester. The qualitative information available from this sequence is:

1. The component measured has the same retention time as the standard.
2. The component measured is responsive to electron-capture detection, as is the standard.
3. The component measured has been converted from its original state to the component which satisfies the first two statements above (this assumes that a chromatogram of the processed sample before methylation lacks a component with this

corresponding retention time).

4. As a final addition to this collection of circumstantial evidence, a parallel portion of the sample can be converted to
   ethyl, propyl, or butyl ester, and a new correspondence of retention time with the appropriate standard can be established.
   With a sequence of circumstantial evidence all positive, such
   as this, the identity of the compound is almost certainly established.

This discussion has been aimed solely at qualitative methods
that may be applied to unknown components present in a sample at
trace levels. Reference should be made to some of the more general principles and limitations of qualitative analysis by GC
described in Chapter 4.

7.6.2 Quantitative Analysis

Here also, the fundamental principles are described in Chapter 4.
The discussion will be limited to those aspects that are unique
to trace analysis, or which present particular difficulties not
encountered except in trace analysis. The cardinal rule will be
that standardization should be accomplished with component levels
corresponding to those sought or found in samples. It is an
unforgivable sin to measure a standard that is 100 times more
concentrated than the components found in the sample and extra-
polate from the standard measurement to the level found in the
sample. Also, because of the extensive processing sometimes required for trace analysis, it is proper to prepare the standard
in the matrix of the sample, that is in water for a water-based
sample, or in clean soil for a soil-based sample, or by standard
additions to the sample itself. The standard thus prepared
should be carried through all the processing steps that are applied to the sample. By this means, losses that are consistent
in sample processing are accounted for by comparable losses of
the standard and computation from the standard chromatogram does
not require any so-called correction factors. It is important to
know what losses do occur. Therefore, comparison to a clean
standard prepared in the final sample solvent is also appropriate.

While the use of internal standards is generally recommended
for gas chromatographic analysis, this technique is not as uniformly applied in trace analysis. In many cases, the chromatogram of a processed sample is too complex to provide for the
additional peak from an internal standard. In cases where selective detectors are used, an internal standard which is comfort-

ably chromatographed within the sample chromatogram and also has
a comparable signal to weight ratio, is difficult to find. Further, since processed samples for trace analysis cannot generally
be predicted to contain the same number and magnitude of extraneous peaks, one cannot be certain that the chosen internal standard is properly applicable to all samples, even though they are
otherwise assumed to be similar.

Since trace analysis also includes air or gas samples, it is
appropriate to point out that proper addition of an internal
standard to this type of sample is difficult. This difficulty
lies, not in the mechanical problem of transfer, but in the difficulty of knowing that the precisely intended volume has properly
been transferred. However, the internal standard technique is
still not widely used here for the same reason it is not generally used in trace analysis. This reason again is because the
analyst normally has no prior knowledge of the variation in composition from sample to sample. The continual risk exists that
any given sample in a series will have a component, not present
in others, which elutes with the internal standard. This occurrence would introduce significant error into the quantitative calculations which result.

For both qualitative and quantitative trace analysis, the problem of adsorption or degradation on column packing material and
other parts of the chromatographic system that contact the sample
is severe by comparison to analysis of more concentrated samples.
It becomes important to determine in advance whether chromatographic problems of this nature are existent in the system chosen
and to correct or avoid these problems if at all possible. A
general rule will be to use glass columns in preference to metal
columns, or until it is shown that the metal column system does
not affect the correctness of the analysis. This is because
glass is clearly a less reactive substrate than metal. The effects of adsorption and degradation can be recognized by chromatography of standard solutions. Adsorption, when serious, is
always recognizable by peak tailing. Adsorption in gas chromatographic systems in which gas-liquid partition is intended is
the result of a fixed, small number of accessible sites. In a
sample with larger-than-trace concentration this effect may not
be recognizable since it involves perhaps 1% of the total quantity of the sought component. However, if the total quantity of
the sought component injected into the system is reduced to part-
per-million levels, then it is possible that all of the component
injected will be affected by adsorption. The result of this is a
severe broadening of the resultant peak, and possibly complete
loss of the ability to detect it. Assuming the peak is still detectable, a concomitant effect will be an apparent increase in
retention time. This effect can be of sufficient magnitude to

lead the analyst to a completely erroneous conclusion as to the identity of the compound, or conversely, to conclude that the sought component is not present. This is illustrated in Figure 7.9. The proper use of standards, that is in corresponding concentration, would prevent these erroneous qualitative conclusions, but would not prevent difficulties in quantitative measure.

Where adsorption effects are not as dramatic as described above they can be recognized by quantitative tests. In this case, it is necessary to measure standards covering a concentration range of five- to tenfold, which includes the range of concentration expected in processed samples. If adsorption problems are absent, then the calibration curve that results will be a straight line which passes through the origin as forecast by both theory and good practice. On the other hand, if the resulting calibration graph is not a straight line, but curves so that it apparently intercepts the quantity axis, implying that a finite quantity of component will give zero signal, then adsorption will present quantitative problems. This comparison of calibration curves is shown in Figure 7.10.

If adsorption is shown to exist as a problem by this test with the chromatographic system chosen, then every reasonable effort should be expended to avoid this problem. Normally, many or all of the following alterations of the system are available to avoid this problem.

1. Replacement of the stationary phase with a more polar phase.
2. Replacement of the column with an equivalent column, but containing more carefully prepared support, more carefully

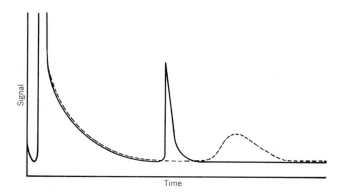

Figure 7.9. Effect of severe chromatographic adsorption at differing concentration levels.

coated. (This should show improved column efficiency, i.e.
sharper peaks.)
3. Replacement of the column with its equivalent, but using a
higher proportion of the same stationary phase (this will not
necessarily produce higher efficiency but may reduce adsorption).
4. Use of a higher operating temperature. This may require a
corresponding decrease in carrier flow to maintain useful retention times.
5. Substitution of a less reactive support material. For example,
the use of Teflon or glass beads in place of diatomaceous
earth.
6. Derivatization of the sample to provide the sought component
as a less polar compound.
7. Treatment of the installed column with silylating reagents.
8. Substitution of a more polar stationary phase in lower concentration to avoid excessive retention time or temperature limitations. A shorter column with the new phase may also be
used.
9. Further attempts to provide a processed sample of higher concentration.

The chosen alteration of the system can again be tested with
the same series of standard solutions to determine readily whether the problem has been resolved. If reasonable attempts to
correct the problem do not yield a solution, then the analyst
must proceed with great care. The most appropriate manner to

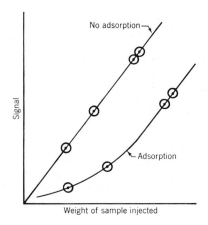

Figure 7.10.   Effect of adsorption on standard calibration curve.

provide proper analyses in the face of an adsorption problem is to bracket each sample chromatogram with a standard, ideally one that produces a peak for the sought component as near exactly the same as the sample as possible. In other words, if a series of similar samples is to be analyzed, the chromatographic sequence should show standard followed by sample 1, standard, sample 2, and so on. Also it is particularly appropriate under these circumstances that duplicate or multiple chromatograms of each sample be generated and that all measurements be included to compute an average.

The older literature, and some practices in use today, suggest a process known as "priming the column." Almost without exception, this process will create more problems than it will solve and is to be avoided at all costs. Priming involves passing a large (relative to a normal sample) quantity of the compound to be analyzed, or a more polar compound through the chromatograph before injecting samples or standards. The intent of this step is to saturate, or occupy all adsorptive sites prior to sample analysis; thus component loss and peak distortion are presumably eliminated. However, the adsorption process is not irreversible, but only more slowly reversible than the partition process. Therefore, the proportion of active, adsorptive sites occupied when the sample is injected will vary with the quantity of material used for "priming," the chemical nature of that material, and the time lapse between "priming" and analysis. As a result, second and third sample analyses after priming will be increasingly in error, and some question will exist as to the accuracy of the

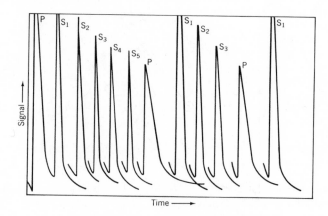

Figure 7.11. Effect of "priming" with severe adsorption. P = priming compound, S = sample at concentration 1/10 of prime.

first analysis.   An excellent example of the nature of this problem, when severe, is illustrated in Figure 7.11.   The peaks designated "P" result from injecting the priming compound.   Those
designated "S" result from equal sequential aliquots of a sample
containing one-tenth the amount of a similar, less polar compound
for analysis.   It should be clear that the accuracy of all analyses of "S" would be in doubt under these circumstances.

7.7   PREPARATION OF STANDARDS FOR TRACE ANALYSIS.

A cardinal rule of trace analysis by GC (and almost certainly by
any other method of measurement) requires that standardization
involve quantities of the test materials equal to, or closely
approximating the quantities to be measured in samples.   In the
case of standard materials that are liquids or solids, a properly
measurable weight or volume of the material can be appropriately
diluted.   To avoid the use of excessive solvent volumes, serial
dilutions can be made, as long as reasonable care is exercised.
    It is with gaseous materials that preparation of extremely di-
lute standards presents real difficulties.   For example: how can
one be certain that a syringe pulled back to 1 cm^3 volume in fact
contains 1 cm^3 of the test gas and not 0.5 cm^3 of that gas, plus
0.5 cm^3 of air; or 1 cm^3 of the gas at reduced pressure?   Furthermore, since the concern is for quantities in the parts-per-
million range, a 1-cm^3 volume of a standard gas is not a reasonable volume to be considered since it would imply the use of a
1000-liter container.   Therefore, a variety of alternative methods has been devised, though the serial dilution of a known gas
volume is also used.   This is defined as a static gas standard,
while methods for the preparation of dynamic flowing concentrations of trace levels of the sample gas in a carrier have also
been well defined.   The variety of techniques for preparation of
gas or vapor standards is fully covered in Chapter 4.

        7.7.1 Preparation of Liquid or Solution Standards

To prepare standard solutions of solid or nonvolatile materials,
normal good laboratory practice is all that is necessary.   In
most cases, even with the availability of a microbalance, it is
convenient to prepare an initial standard solution with a concentration significantly higher than that ultimately intended for
measurement.   Standards for calibration are then prepared by

dilution or multiple dilution where necessary. Where the standard is a liquid itself, then it is clearly more convenient to
prepare the standard solution by volume. The weight is then computed from the known density of the material. It should be noted
that for trace analyses, the microsyringes in common use are
quite adequate for preparation of standards. The addition of
1 µl volume of standard to 1 cm^3 of solvent will give a 1000 ppm
standard solution on a volume basis. The use of microsyringes in
this fashion contrasts somewhat with their common use for injecting samples into the chromatograph. The basic difference lies in
the fact that the chromatograph injection area is normally heated
far above room temperature. As a result, the volume of solution
delivered to the chromatograph includes not only that displaced
by the syringe plunger, but also the volume vaporized from the
needle. In the preparation of standards, the only volume delivered is that displaced by the plunger and this addition can be
accurate to a degree approaching ±1%. This accuracy is far
better than can normally be achieved with trace analysis.

More commonly, the nature of the analysis suggests a strong
need for standards prepared in water. This provides a standard
that can be carried through the processing given the sample and
thereby account for certain repeatable losses. In many cases the
compound of concern is not sufficiently soluble to prepare an
initial higher concentration. Attempting to add the necessary
minute volume of a liquid sample, or the minute quantity of a
weighed solid standard directly to a large volume of water will
often introduce errors of much greater magnitude. This is because the material added will not go into solution immediately
and, most commonly, will float on the surface. The result of
this, in many cases, is a strong tendency to adsorb a significant
portion of the standard compound on the walls of the container.
For these purposes, it is preferable to prepare the standard in
greater concentration in a water-miscible solvent which will not
interfere in the subsequent analysis.

This choice of solvent is not difficult when it is recognized
that the subsequent dilution in water to provide the desired concentration level will reduce this solvent concentration to the
order of 1000 ppm or less. When the analysis intended will use
electron capture detection, then any number of solvents, including the light alcohols and acetone, will cause no difficulties
but will provide for a ready dispersal of the standard compound
in the water system. Where flame detection is to be used, there
are some additional risks on choice of solvent since it will be
detected. In this case, if the standard compound is relatively
nonvolatile, the solvent, though detected, will be a part of the
normally expected "solvent peak" and will not interfere. In
other cases, where the standard compound is significantly more

volatile, it will be desirable to choose the initial solvent to
be highly water soluble. This accounts for the fact that most
analyses of this type will require an extraction step at some
point and the solvent initially used to dissolve or disperse the
standard compound will remain in the water phase or will be
transferred into the extract in significantly decreased concentration.

It is neither necessary nor desirable to prepare large volumes
of the working standard solution since there is always a question
of the long-term stability of such solutions. Likewise, in most
cases, the consumption of standard solution is in microliter
units; therefore, if it is desired to retain a stock solution, it
should be the more concentrated original solution in small volume
carefully kept in a freezer. From this, convenient small portions of working standard solutions can be prepared as needed.

7.8   MODEL METHODS FOR CONCENTRATION OR TRACE ANALYSIS

This portion of the chapter has as its purpose the description in
some detail of methods that have been or are currently applied to
some specific problems of trace analysis. They have been selected by the author as models because of their completeness, simplicity, or potential wide applicability. The selection of these
methods very definitely reflects the author's bias in evaluating
methods. In addition, a detailed guide to the development of a
trace analysis method is included.

It should be noted that the Environmental Protection Agency
(EPA), the National Institute for Occupational Safety and Health
(NIOSH), and the AOAC Methods Book (published every 5 years by
the Association of Official Analytical Chemists, and currently in
its twelfth edition), are sources for methods which are applied
by these organizations to measurements of concern to them (21-
24). Valuable information on sampling, standard preparation, details of sample processing, and very useful bibliographies are
included. Some of the methods indicated by these groups have
been evaluated in a number of laboratories and their reliability
is reasonably established. However, it is this author's opinion
that a number of such methods are cumbersome or more complex than
necessary. In the literature methods cited below, reference to
the original complete text is strongly recommended; however, it
is intended that sufficient detail be included here for a reproduction of the mechanics of the indicated analytical process.

7.8.1 Determination of Nonvolatile Solutes in Water (25)*

One of the most effective and thoroughly examined methods for concentrating trace organic contaminants from water has been described by Fritz et al. in a series of papers (25-30). The essence of this approach involves the use of "AMBERLITE XAD-2" resin. A short column containing this resin carefully prepared is used to remove contaminants from water samples up to 150 liters in essentially a quantitative fashion. The organic components that are retained by the resin are then removed with a very small volume of diethyl ether. The resultant ether solution can be analyzed directly, dried prior to analysis, or further concentrated by evaporation in order to provide the most sensitive means of analysis. This last step runs the risk of losing some part or all of volatile to moderately volatile components. The authors recommend drying; in their work the ether solution was further concentrated by evaporation. In a general presentation such as this, one should not overlook the possibility that the components sought are in sufficient concentration in the ether solution to be directly detected without further concentration. It is also true that some components may still not be sufficiently concentrated for proper detection; thus, evaporation to a minimum volume may be necessary or desirable.

One of the particular values of this work was the investigation of adsorption and recovery of an unusually wide variety and number of organic compounds. These were tested in concentrations realistically approximating the concentration ranges customarily sought in this type of analysis, as shown in Table 7.2 A scale drawing of the extracting apparatus is shown in Figure 7.12.

Based upon summarizing the work reported in this series of papers, a general procedure for the isolation and concentration of organic compounds would involve the use either of the apparatus illustrated or, as an alternative, a similar column fitted with an appropriate hose coupling (Figure 7.13) for direct attachment to a standard pressurized water supply. The latter device has been used for sampling large volumes of water up to 150 liters, while the apparatus depicted is most readily used with samples which have been collected in the field and brought to the laboratory.

*This section concerning the work of Fritz cites extensively from ref.(25) as indicated; also from other work of the author. The text has been reviewed for correctness and application by Professor Fritz. Table 7.2 and the quoted portions are reproduced by permission of the author and of the Journal of Chromatography.

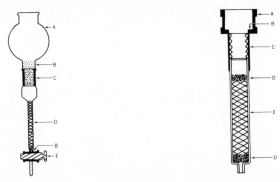

Figure 7.12. Apparatus for extracting organic solutes
from water. (A) 5-liter reservoir, (B) glass wool
plugs, (C) 24/40 ground glass joint with Teflon sleeve,
(D) 8x140 mm glass tube packed with ~ 5 ml 40-60 mesh
XAD-2 resin, (E) Teflon plug stopcock (25).

Figure 7.13. Apparatus for extracting organic solutes
from water. Apparatus for extracting organic solutes
from finished drinking water samples. (A) standard
garden hose coupling, (B) Teflon washer, (C) 1/2 inch
i.d. Teflon tubing, (D) glass wool plugs, (E) 1/2 inch
o.d. x 4 in. long glass tube packed with ~5 ml, 40-60
mesh XAD-2 resin (25).

7.8.2 Analytical Procedure

PREPARATION OF SOLVENTS AND RESIN. Organic-free water for preparation of standards and columns is prepared by passing distilled
water through a column containing clean XAD-2 resin.

All solvents used are either spectrograde or analytical grade.
The analytical grade solvents are further purified by fractional distillation whenever blank determinations suggest the
presence of impurities detectable by flame ionization GC.
    The macroreticular resin, XAD-2 is obtained from Rohm and
Haas (Philadelphia, Pa. U.S.A.). The fines are removed by
slurrying in methanol and decanting. The remaining resin
beads, predominantly 20-60 mesh, are purified by sequential
solvent extractions with methanol, acetonitrile, and diethyl
ether in a Soxhlet extractor for 8 hr per solvent. The purified resins are stored in glass-stoppered bottles under
methanol to maintain their high purity. In some cases the
resins as received from the supplier are ground to smaller

TABLE 7.2   USE OF MACRORETICULAR RESINS IN ANALYSIS OF WATER (29)
OVER-ALL RECOVERY EFFICIENCY OF THE POROUS POLYMER
METHOD OF ANALYSIS FOR ORGANICS IN WATER AT THE 10- TO
100-ppb LEVEL[1]

Compounds Tested	Efficiency (% Recovery)[2]	Compounds Tested	Efficiency (% Recovery)
Alcohols		Acids (acidified)[3]	
Hexyl	93	Octanoic	108
2-Ethylhexanol	99	Decanoic	90
2-Octanol	100	Palmitic	101
Decyl	91	Oleic	100
Dodecyl	93	Benzoic	107
Benzyl	91		
Cinnamyl	85	Phenols[3]	
2-Phenoxyethanol	102		
		Phenol	40
Aldehydes + ketones		o-Cresol	73
		3,5-Xylenol	79
2,6-Dimethyl-4-		o-Chlorophenol	96
heptanone	93	p-Chlorophenol	95
2-Undecanone	88	2,4,6-Trichlorophenol	99
Acetophenone	92	1-Naphthol	91
Benzophenone	93		
Benzil	97	Ethers	
Benzaldehyde	101		
Salicylaldehyde	100	Hexyl	75
		Benzyl	99
Esters		Anisole	87
		2-Methoxynaphthalene	97
Benzyl acetate	100	Phenyl	91
Dimethoxyethyl			
phthalate	94	Halogen Compounds	
Dimethyl phthalate	91		
Diethyl phthalate	92	Benzyl chloride	88
Dibutyl phthalate	99	Chlorobenzene	95
Di-2-ethylhexyl		Iodobenzene	81
phthalate	88	o-Dichlorobenzene	88
Diethyl fumarate	92	m-Dichlorobenzene	93
Di-2-ethylhexyl		1,2,4,5-Tetrachloro-	
fumarate	84	benzene	74
Diethyl malonate	103	α-o-Dichlorotoluene	96
Methyl benzoate	101	m-Chlorotoluene	80
Methyl decanoate	95	2,4-Dichlorotoluene	71

Methyl octanoate	98	1,2,4-Trichloro- benzene	99
Methyl palmitate	70		
Methyl salicylate	96		
Methyl methacrylate	35	Nitrogen compounds	
Polynuclear aromatics		Hexadecylamine	94
		Nitrobenzene	91
Naphthalene	98	Indole	89
2-Methylnaphthalene	95	o-Nitrotoluene	80
1-Methylnaphthalene	87	N-Methylaniline	84
Biphenyl	101	Benzothiazole	100
Fluorene	84	Quinoline	84
Anthracene	83	Isoquinoline	83
Acenaphthene	92	Benzonitrile	88
Tetrahydronaphtha- lene	62	Benzoxazole	92
		Pesticides + Herbicides[4]	
Alkyl Benzenes		Atrazine	83
		Lindane	95
Ethylbenzene	81	Aldrin	47
Cumene	93	Dieldrin	93
p-Cymene	92	DDT	96
		DDE	81

[1] The ppb designation corresponds to parts of organic solute by weight. Thus 10 ppb corresponds to 10 µg/l of water.

[2] The average reproducibility of these values is ±12%. This uncertainty is high because it includes results accumulated by a number of different analysts and results where rigid procedural control was not yet established. With dedicated control of all variables discussed in the text the reproducibility limit may be decreased to ±∿5%.

[3] The water was acidified by adding 5 cm^3 of concentrated HCl prior to adding the organic solute. XAD-4 resin was used.

[4] All pesticides and herbicides except atrazine were tested at a concentration of 20 parts per trillion in water.

particles, sieved and then purified by Soxhlet extraction as described above.

COLUMN PREPARATION. The apparatus used for removing trace organics from water is shown in Figure 7.12. With the upper 1- to 5-liter reservoir detached, insert a clean glass wool plug near the stopcock of the glass column. Add the purified XAD resin as a methanol slurry until a resin bed approximately 6 cm high is obtained (1.5-2.0 g dry resin), then insert a second, clean glass wool plug above the resin. Drain the methanol through the stopcock until the level just reaches the top of the resin bed, then wash the resin with three 20-cm^3 portions of pure water. For each portion stop the flow when the liquid level reaches the top of the resin bed.

SAMPLE PROCESSING.

1. Grab Samples. "Grab" samples are taken for all surface waters using clean amber 4-liter solvent bottles. After settling overnight, the clear water is decanted into the 5-liter reservoir shown at the top of Figure 7.12. The stopcock (E) is adjusted to deliver a flowrate of 25-50 cm^3/min. When the water level reaches the upper glass wool plug the sediment from the bottle is transferred to the reservoir using several rinses with organic-free water. After all the water has passed through the resin, the stopcock is closed, the reservoir removed and 15 cm^3 of diethyl ether is added to the resin. About 5 cm^3 is allowed to flow through the resin and collect in a 60-cm^3 separatory funnel. The stopcock is then closed for 15-30 min after which the remaining 10 cm^3 of ether is collected in the separatory funnel. This elution procedure is then repeated with a second 15 cm^3 portion of ether which is combined with the first. The water layer is drained from the separatory funnel and the final traces of water are removed from the eluate by adding 10-15 cm^3 of petroleum ether and 2-3 g anhydrous sodium sulfate. The mixture is shaken for ∿30 sec and the liquid extract transferred quantitatively to a concentration flask.

2. Composite Samples. The apparatus shown in Figure 7.13 is used for composite sampling of a ∿24-hour period. The standard garden hose coupling is attached to a suitable faucet and the water flow adjusted to deliver ∿150 cm^3/min. After 24 hr the XAD-2 column is removed from the coupling

and Teflon sleeves are used to connect a 25-cm^3 reservoir
and a Teflon stopcock to appropriate ends of the column.
The column is then eluted with diethyl ether and the eluate is treated as described above for the "grab" samples.
Either sampling procedure may be employed dependent on
whether one is interested in instantaneous (grab) or average (composite) concentrations over an extended time
period.

CONCENTRATION OF ELUATE. Add a small boiling chip to the concentration vessel and attach a three-cavity Snyder column as
shown in Figure 7.14. Add about 2 cm^3 of diethyl ether to the
top of the Snyder column and tap gently to distribute the
ether into the three cavities. Apply heat from a hot plate or
steam bath so that the boiling action is vigorous enough to
agitate the balls of the Snyder column continuously. A solvent evaporation rate of 0.5-2.0 cm^3/min should be attained.
When the volume of solution in the calibrated appendage of the
concentration vessel is about 0.5 cm^3, remove the apparatus
from the heat and immediately spray acetone over the outside
walls of the concentration vessel. The condensation of the
ether vapor causes an automatic sequential washing of the in-

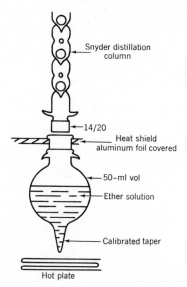

Figure 7.14. Apparatus for concentrating eluate for
determination of organic solutes in water.

side walls of the vessel with the ether held in the three cavities of the Snyder column. The volume of liquid in the calibrated section of the concentration vessel should now be
≤ 1.0 cm^3. Remove the Snyder column and add ether if necessary
so that the solution volume is exactly 1.0 cm^3. Cap the
vessel with a 14/20 stopper and swirl to mix the solution.
Other appropriately chosen techniques for concentration of
this eluate can be used. Proceed with the GC analysis of the
concentrate as soon as possible.

SEPARATION AND QUANTIFICATION. Separation and quantification
of components in mixtures extracted from real samples is
accomplished by using standard quantitative GC procedures.
The amounts of separated components are determined from peak
height or area measurements and the volume of water used in
step 3.

7.8.3 Determination of Volatile Organic Solutes in Water*

The procedure of Fritz described in Section 7.1 of this chapter
has as its primary limitation the fact that volatile components
would be lost to a greater or lesser extent, particularly in the
final steps of evaporating the eluting solvent. A method for
analysis of these volatile materials which is elegant in its simplicity has been proposed by McAuliffe (31). The essence of
McAuliffe's method involves equilibration of a portion of the
aqueous sample with an equal volume of an inert gas. This may be
helium, nitrogen, air, or other water insoluble gas. Those components in the sample which have limited water solubility and
significant vapor pressure will be transferred in a reproducible
proportion to the gas phase. A portion taken for analysis then
totally eliminates the normal chromatographic problem of a broad
solvent peak and associated solvent impurities. This contrasts
with the common practice of extracting organic components from
water samples with a nonmiscible organic solvent followed by analysis of a portion of the extracting solvent.
   To illustrate the advantage of this technique, compared either
to direct chromatographic analysis of the water system or of an
organic solvent extract, we may consider the circumstance of
hexane as a hypothetical contaminant. This compound is distributed to the extent of 96% into the gas phase. The implications

*This discussion of the work of McAuliffe has been reviewed by
Dr. McAuliffe for correctness and completeness.

of this distribution for trace analysis are as follows:

If it is assumed that a 10-cm^3 volume of sample is equilibrated with a 10-cm^3 volume of gas and a 1-cm^3 aliquot of the gas removed for analysis, then nearly 10% of the original total sample content is made available for the actual analysis. A 1-cm^3 gas sample is very commonly and comfortably handled with a laboratory gas chromatograph. On the other hand, if an attempt is made to analyze the water directly and an equivalent liquid sample taken using 1 μl of liquid sample, that is, one which is common in use and comfortably handled by the same instrumentation, then only one ten-thousandth of the total sample is made available directly for the analysis. In other words, for this comparison, the method of McAuliffe provides the equivalent of a thousandfold concentration with no loss. For compounds with a smaller percentage distribution to the gas phase, this concentration factor is not as favorable, but is still highly significant. For a further comparison, one can customarily extract many organic compounds from 10 volumes of an aqueous sample into one volume of organic solvent. Using the previous reasoning, a 1-μl aliquot removed from this organic extract then contains one-thousandth of the originally available quantity of component (assuming 10-cm^3 sample volume and 1-cm^3 extract volume). This leaves McAuliffe's approach still with the advantage of an equivalent one hundred-fold concentration, again without loss, where the extraction process will not necessarily achieve 100% extraction efficiency under these circumstances.

There are several further advantages of McAuliffe's approach to the analysis of volatile organic components in water. Among these are the fact that water-soluble or nonextractable components may be analyzed by this method. Further, the quantity of the component transferred to the gas phase can be varied and, in this case, increased by conducting the equilibration at elevated temperature and by increasing the ionic content of the water. By these means compounds that may be highly water-soluble, such as acetone or alcohol, can be transferred in significant proportion to the gas phase for analysis.

Finally, and as important as all of the features of this method, the sample can be re-equilibrated with successive fresh volumes of gas and each successive phase analyzed. The result of these series of analyses will be to define the distribution of the particular compound or compounds involved, thereby providing a measure of another physical property of the compound involved. This amounts to a qualitative mechanism of identification of trace components as a part of the primary analysis. It also provides for an extrapolation of this accumulated data to a zero equilibration stage which would be indicative of the original concentration, even though the process of taking and transporting

the sample may have resulted in the loss of the equivalent of the
first equilibration. Even without a knowledge of the potential
identity of the compounds detected, the nature of the distribution coefficient indicated by successive equilibrations provides
clues to the nature of the compound. For example, alkanes are
essentially removed with two equilibrations. Cycloalkanes are
distributed in lower proportion to the gas phase and can be
carried through five or six equilibrations. Aromatics have a
still lower distribution and other compounds of potential concern
will show similar variances usually with lower distribution into
the gas phase. These statements are illustrated in Figures 7.15-
7.17.

The chlorinated hydrocarbon compounds which have recently become of significant interest in chlorinated water supplies can be
determined by this method very elegantly and accurately without
interference by other potential contaminants using electron-
capture detection. With the combination of selectivity and extreme sensitivity of this approach, the parts-per-billion concentrations that are of concern can be directly measured, thus eliminating all concern for loss or introduction of confusing

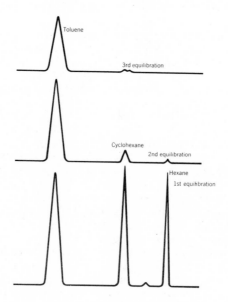

Figure 7.15. Chromatograms of successive equilibrations illustrating the behavior of different volatile
solutes in water.

impurities as a result of the more common types of sample processing.

For hydrocarbons, using a 5-cm^3 gas sample injected into the
chromatograph, McAuliffe indicates

> the method to be capable of detecting alkane and cycloalkane
> hydrocarbons in water if they are present in amounts of 1-3
> parts in 10^{12} parts of water by weight. Aromatic hydrocar-
> bons, because of their lower partitioning into the gas phase,
> can be detected if present in concentrations of 4-12 ppt.
> Reasonable accuracy can be obtained if the aqueous concentra-
> tions are 20 to 30 times these values.

McAuliffe's presentation does not precisely define the equipment used for processing and analyzing a sample, though it is
apparent this is neither critical nor complex. This process has
been used with very satisfactory results when a 30 - 100-cm^3 standard glass syringe was employed as the container for equilibration. The syringe is loaded with or without a needle from the
bulk sample to one-half or less of its calibrated volume. The
needle hub is then sealed with a septum cap and the desired equal
volume of inert gas added through this septum with a second

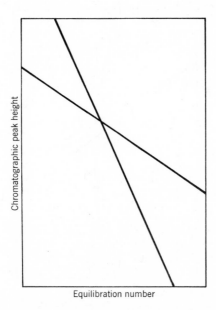

Figure 7.16. Graphic representation of the difference
between two solutes with successive equilibrations.

syringe. Clearly, the simplest, cleanest approach is to load the
inert gas directly from the chromatograph in use simply by pierc-
ing the inlet septum with the second syringe and allowing it to
fill to the desired volume with carrier gas. The sample, with
liquid and gas phase, can then be vigorously shaken, rolled, or
otherwise mixed to achieve proper equilibration. Following this,
the syringe plunger can be used to transfer the necessary gas
volume to a gas-sampling valve which is a part of the chromatograph; or alternately, the desired volume can be removed with an
appropriate gas-tight syringe and transferred to the chromatograph for analysis. A syringe that has an integral closing valve
is extremely convenient for this purpose. Since it is a simple
matter to have sufficient equilibrated gas volume available, replicate analyses can be made without concern for loss of gas
pressure in the sample. Following this, should it be desirable
to examine successive equilibrations, the remainder of the gas
phase for the first equilibration can be displaced by removal of
the septum seal, or by inserting an open needle through the septum, and subsequently a fresh volume of clean gas re-introduced.
It is not even critical if a small portion of liquid sample is
lost in this step. It is only necessary that an equal volume
portion of equilibrating gas be introduced for the second and

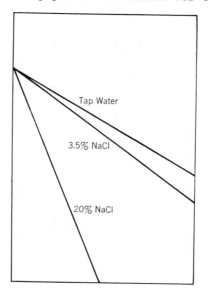

Figure 7.17. Effect of salt content of water on vapor
distribution of toluene. Successive equilibration of
toluene (equal starting concentrations) between helium
and water of differing salt content.

successive equilibrations. In other words, if the first equilibration should have involved a 20-cm^3 volume of sample with a
20-cm^3 volume of gas, and in displacing this first equilibration,
5 cm^3 of the liquid sample should be lost, the second equilibration can effectively be carried out using 15 cm^3 of equilibrating
gas.

Quantitative analysis in this fashion is dependent on the
physical parameters of the sample being properly reproduced in
prepared standards. With water samples that are essentially
clean, that is, drinking water, an appropriate standard can readily be prepared in distilled water. However, if the sample of
concern is questionable in salt content and likely deviates significantly from distilled water, it would then be appropriate to
make an overwhelming addition of salt to both sample and standard
prior to calibration and analysis.

7.9   GENERAL SCHEME FOR TRACE ANALYSIS

This section is intended to provide a detailed outline to a consistent series of steps which will provide an analytical method
for trace analysis. If properly applied, using much common
sense, a proper understanding of the need for, or effect of each
of the processing steps chosen is provided. It is primarily
aimed at the determination of compounds that are obtained in solution or can be processed to provide a solution. Thus, the application to air or gas analysis is somewhat less precisely defined. Though exact mention of processing steps is not stated,
it is intended that any of such processes that may be applicable
should not be excluded from consideration. For example, in a
specific analytical problem, the analyst may properly feel that
derivatization is an appropriate step to be included in the total
sampling processing. It is intended by the outline that the necessary testing of this stage of processing will be indicated.

(1) Determine appropriate chromatographic conditions for the
components sought. This may be done by reference to literature,
or on the basis of experience in choosing an appropriate column,
operating temperature, flow, and detector system. This choice is
then appropriately tested with standards either of the compounds
sought or of compounds which are reasonable substitutes, both
chemically and physically. The operating conditions can then be
optimized to provide the best chromatography of these standards
prepared as solutions in concentrations wherein instrument sensitivity does not limit the ability to examine the effects of

chromatographic variations. At this point, other known components of the sample which can be expected to appear in the final processed extract should be tested to determine whether any problems of serious overlap or interference exist. If such problems are evident, then consideration must be given either to changing the chromatographic system or modifying anticipated sample processing to eliminate the interference.

Finally, when the full set of chromatographic parameters are determined to be most suited for the proposed analysis, then the standards can be serially diluted to provide for a final estimation of the lower limit of detection with the system used.

(2) Choose from literature or devise an appropriate workup of the sample, recognizing the concentrations that must be achieved to provide detectability. At this point, solvents, reagents, and equipment which are anticipated for use in sample processing should be checked appropriately for any contribution of interfering materials. For example, glassware may be rinsed with the solvent used for final processing and the resulting solution checked. Also, if it is intended that an organic extract be concentrated by evaporation, then a portion of that solvent should be concentrated to an equivalent degree before examining the resultant solvent concentrate. It cannot be assumed that a solvent that is apparently free of interfering components as removed from the reagent bottle will likewise be free of interfering components when concentrated by a factor of 10 or 50 to 1. If problems arise here, then alternatives must be sought or additional processing given the equipment, reagents, or solvents such as recrystallization or redistillation.

(3) Choose an appropriate substitute for the sample matrix. Water would be the choice for essentially aqueous systems. Hydrocarbons; pure or mixed distillates, or triglycerides; pure or natural oils such as cottonseed or coconut oil would be appropriate substitutes for petroleum- or fat-based systems, respectively. Air or inert gases are appropriate substitutes when air pollution measurements are intended.

Following this choice, the intended sample processing should be carried out completely with the matrix substitute to determine that processing alone does not provide or create interfering materials. Two possible results follow at this point. If no interfering peaks are shown under the specified chromatographic conditions, using the highest sensitivity intended, then the succeeding step can be immediately undertaken. Alternately, there may now be peaks shown that would interfere with the measurement of the desired components. Under these circumstances, one must check containers; for example, avoid plastic vials, vial caps,

stopcock grease, and so on, or check these materials by exposing
them independently to the various processing steps and examine
the resulting solutions. Similarly, check each solvent or reagent involved. If evaporated to dryness or concentrated in the
procedure, duplicate the process for this check. Again, alternatives must be sought or corrective action taken. In some
cases, it will be impossible, or nearly impossible, to eliminate
all potential interferences. It must then be established that
the interference is essentially repeatable, and that it does not
correspond to an overwhelming proportion of the component finally
measured, that is, if the upper limit of concentration intended
or allowed is 1 ppm, then under no circumstances should the interference correspond to more than an equivalent of 1 ppm.
Ideally, this should be much less than the upper limit intended
and more nearly equivalent to the lower limit of detectability.

(4) Spike the matrix substitute with a large amount of the
test component and carry this mixture through the intended procedure as developed to this point. This large amount should
correspond to 10 - 100 times the quantity ultimately sought. The
purpose of such a large spike is that many processing steps may
result in loss of component and if a significant quantity is initially present, the amount remaining in the final extract will
still be enough to detect adequately and to evaluate clearly the
magnitude of losses. With this level of initial concentration,
where significant losses are shown to occur, there will be a sufficiently high concentration at various intermediate steps of
processing to provide for examination of portions of the sample
available at these intermediate steps. This will then allow for
the identification of those processing steps in which significant
losses occur. Modifications of those steps should be considered
to eliminate or minimize the loss. Under the most favorable circumstances, where significant losses occur and no convenient
alternative exists, then it must be established that this overall
processing loss is highly repeatable. Once problems in this step
have been satisfactorily solved or accepted, then step 5 logi-
cally follows.

(5) Spike the matrix substitute with a realistic, expected
amount of test compound and repeat the process of step 4. If
there are significant losses evident in step 4 and these have not
been eliminated, then a procedure modification is already indicated, since a higher overall concentration factor will be required to reach the predetermined limit of sensitivity or lower
limit of acceptable concentration of the test compound. At any
rate, it will be necessary to establish that the realistic quantity of test compound is handled by the workup procedure in ess-

entially the same manner as the large quantity previously tested.

(6) Spike the genuine sample with the correspondingly large amount of test compound and analyze the resultant mixture. The purpose of this step is to re-establish with the sample itself that there are no unanticipated losses due to the presence of components in the sample which are not present in the matrix substitute chosen. These components may result in additional losses by altering the extractability of the test compound, by saturating adsorbent columns, slurries, or thin-layer plates, or by tenaciously holding the test components because of the presence of a strong adsorbent in the sample itself. Chemical reaction may also occur with components of the sample under the conditions of sample processing which do not occur naturally and do not occur with the matrix substitute. Each of these items must be considered as a possible factor if the resultant measurement does not account for all or nearly all of the spike plus the quantity of the sought component which may natively be present in the sample. Obviously, if any of these problems results in significant error or loss, further modification of the process is indicated before proceeding.

(7) Spike the genuine sample with a realistic, expected amount of test compound and analyze. Again, this step should result in an apparent overall recovery of the spike of 100% or more since the final measurement will include any of the component natively present in the sample. This computation of 100% must, of course, account for any established, reproducible losses.

(8) Finally, process the genuine sample without additions and carry through the analysis. In many cases not all the steps suggested previously will be necessary, the most obvious being problems of air pollution analysis in which the matrix can customarily be considered inert. However, in some plant atmospheres or other unusual circumstances, there may be single or multiple components present which interfere with the desired measurement and cannot be anticipated until a realistic sample is examined. It should not be assumed that the mere trapping of a sample in any of the variety of containers which may be available will necessarily provide for a proportionate removal of sought component or other components from the container. Therefore, the deliberate addition of the sought component to a contained sample is highly desirable in order to evaluate whether such a problem exists and to what extent. In the case of air pollution samples, the period of time over which a contained sample can be reliably analyzed should be established.

(9) Use the power of GC. Sample the system at each of the
various processing steps to determine whether it is necessary to
process further. Since the chromatograph is capable of measuring
one component in a mixture, it is clearly not necessary to process a sample to the point where it contains only solvent and the
component of interest. Wherever simplification of sample processing can be achieved, it is likely to improve not only the
convenience of the method but the overall reliability.

7.9.1 Achieving Maximum Instrumental Sensitivity

First, attention should be called to the discussions in Chapters
2 and 3 that are concerned with column efficiency. In trace analysis the narrowest peak (i.e., highest column efficiency) that
can be achieved for the component of concern will correspondingly
give the highest peak for a unit quantity of sample. The net result of approaching this desired goal will the improvement of
sensitivity, since it is the displacement above baseline, (i.e.,
the detectable height) whether the measurement is made manually
or automatically. For this same reason, achieving the lowest
possible detector noise level is also desirable. With standard
flame detectors, the use of higher than normal air flow tends to
decrease noise. A detector design that may customarily use 180
cm^3/min of air as support for combustion and to sweep the detect-
or volume will generally show a decrease in noise level when the
air flow is increased to as high as 300 cm^3/min. Detector geometry that customarily uses 300 cm^3/min of air will normally show
decreased noise level when this flow is increased to 350 or 400
cm^3/min. This effect is dependent on detector design and should
be tested for applicability with the instrument at hand. Correspondingly, detector noise and sensitivity are decreased by a decrease in hydrogen flow. Some compromise in hydrogen flow adjustment to achieve the best signal-to-noise ratio is proper to
examine with the chromatographic system which is available.
    With pulsed electron capture detectors, some instruments can
use the necessary detector-gas mixture (usually argon with 5 -
10% methane) either as a carrier gas or as a purge gas entering
the detector at the column exit and adding to the normal carrier
flow. Maximum sensitivity will be achieved with this type of detector system when the detector gas is used as carrier gas with
no purge. Likewise, maximum sensitivity is achieved when the
pulse interval applied to the detector is increased; that is,
the time between pulses is lengthened. The result of this change
is a distinct increase in sensitivity with a corresponding loss
in the range of linear response. However, since maximum sensi-

tivity is the goal, the loss of linearity at higher levels is not generally a problem, though a sample which has an unexpectedly high amount of the component sought may be underestimated because of this.

With all electron capture detectors which use tritium as the source of beta radiation, alcohols and water should be rigorously avoided as solvents for the sample and oxygen or air as a matrix for a gas sample should similarly be avoided if possible. These precautions are mentioned because each of these materials can exchange or react with the tritium in the cell resulting in decreased sensitivity or decreased detector life.

Where flame detection is used, carbon disulfide is a solvent of choice if it is applicable to the analytical problem since the flame detector is essentially insensitive to this compound. Of course, the solvent available should be checked for potential interfering impurities. The second choice solvent is carbon tetrachloride, because of its relatively low sensitivity to flame detection. An additional assistance in trace analysis is the use of a solvent and chromatographic system in which the solvent is eluted after the component to be measured. This avoids the primary problem of distortion of the peak of interest or of the potential need for measurement on the sharply sloping side of a solvent peak. Obviously this option is available only for a limited number of problems. With any detector system, the narrowest achievable peak for the sought component will give the highest sensitivity. Therefore, in addition to optimizing column efficiency, the shortest usable retention time for the sought component should be used. This will be most commonly achieved with increased column temperature, but lower stationary phase loading or a shorter column should also be considered.

7.9.2 Removing Atmospheric Pressure Aliquots from a Closed Gas Sampling Container

A large proportion of atmospheric samples will be obtained in containers similar to those shown in Figure 7.1. It should be recognized that these samples may have been taken under circumstances in which the temperature of the sample was much higher than the laboratory in which analyses are to be made. An immediate result of this would be that the sample contained, as available to the analyst, is present at an unknown pressure less than atmospheric. In addition, some samples are taken by using evacuated containers, then opening them at the site to transfer sample to the container. If the container was not opened for a sufficient period of time, the contents may still be below atmospheric

pressure. Finally, even if the sample container has sample (or
standard) at atmospheric pressure, the removal of successive
aliquots for analysis (or calibration) will result in a continuing decrease in sample pressure within the container. As a general rule, the container should be large enough that the removal
of several analytical aliquots does not seriously change the internal pressure and therefore does not seriously affect measurement of sequential aliquots.

Nevertheless, it is most convenient and proper to deal with a
measured sample known to be at atmospheric pressure. It is
nearly always desirable to know whether the sample, as it is
available in the laboratory, is at a positive or negative pressure. This is most readily determined by use of a standard glass
syringe of 1-2 cm^3 capacity in which the piston moves easily in
the barrel. The syringe needle is inserted into the sample container, with care taken to hold the plunger. As hand pressure is
released slightly from the plunger, a sample under positive
pressure will displace the plunger measurably. On the other
hand, if releasing hand pressure on the plunger does not result in
in plunger motion, then the sample is either at or below atmospheric pressure. Pulling the plunger out part way and noting the
effort required to do so will indicate whether a slight vacuum or
significant vacuum exists. Releasing the plunger after this
operation will also indicate the degree of vacuum by the extent
to which the plunger returns to its undisplaced position. In
either case, where the sample is at atmospheric pressure or below, the removal of any part by means of a syringe will leave the
operator with a measured volume in the syringe having an unknown
pressure. Therefore, the analyst will not know how much sample
he has in fact withdrawn. Further, as soon as the needle is
withdrawn from the container, laboratory air will enter the syringe to compensate for the pressure difference between atmospheric pressure and the syringe content, thereby diluting the sample.
To avoid these problems two convenient methods are available.

The simplest of these is to insert a 1 - 10-cm^3 syringe and
then heat the sample container until the heat expansion results
in displacement of the syringe plunger. This requires that the
sample be near atmospheric pressure so that excessive heat is not
required. Displacement of the plunger indicates a small positive
pressure, but with the bulk sample content at an elevated temperature which is not conveniently measurable. However, the portion
of the sample that is displaced into the syringe which is not
heated will be at atmospheric pressure and at room temperature.
This temperature differential should not be great enough to cause
condensation of liquid in the syringe. As soon as the desired
aliquot volume has been displaced by this means, the syringe
should be removed and the sample container removed from the heat-

ing source. The most convenient heating source is an infrared
lamp, and if necessary the opposite side of the sample container
can be sheathed with aluminum foil to concentrate the heat within
the sample.

If the analyst is concerned about the integrity of a standard
glass syringe, the plunger can be pressed back to the bottom of
its travel to maintain all the sample within the container, then
a gas-tight syringe inserted to remove the sample portion for
analysis. Following removal of the aliquot by this means, the
heat source is removed from the sample. In this case the sample
aliquot actually removed is at a pressure slightly above atmospheric and when the syringe needle is removed from the sample
container, the displacement to create atmospheric pressure within
the syringe results in travel of sample out of the syringe rather
than travel of laboratory air into the syringe.

The second alternative uses a gas-tight syringe with a valve
closure, is equally convenient, and does not require the creation
of a positive pressure in the sample container. In this case, it
is desirable once again to use a standard glass syringe but with
a valve at the end. The syringe chosen should be larger than the
desired sample aliquot volume. For example, a 2-cm^3 syringe may
be used for a 1-cm^3 aliquot. In this case, the plunger is pulled
back to the 2-cm^3 mark, the valve is closed, and the plunger is
allowed to move to a rest point. If the sample pressure is co-
incidentally ½ atm, then the plunger will move from the 2-cm^3
mark to the 1-cm^3 mark and the 1 cm^3 which is now contained in
the syringe will be at 1 atm (this is illustrated in Figure
7.18). If the sample pressure is less than ½ atm, obviously it
will require a bigger syringe to attain a 1-cm^3 portion of sample
at atmospheric pressure. It is also true that a reasonably accurate estimation of the true sample pressure can be obtained by
this means.

Once again, the use of gas-tight syringes is preferred. The
reason that they are not used to test sample pressure initially
is that the plungers do not move freely. In any case, if it has
been determined that a 2-cm^3 volume must be taken from the sample
at its contained pressure to obtain a 1-cm^3 volume at atmospheric
pressure, then this same reasoning and process can be applied using a gas-tight syringe. It should be remembered that the gas-
tight syringe must be supplied with a shut-off valve. Such a
device is available through most GC suppliers. The volume of
needle plus hub plus valve will be important, particularly if
this volume is significant in proportion to the sample aliquot
intended for use.

7.10   ANALYTICAL ETHICS AND STATISTICAL VARIATION

In this society we have reached the point of enacting laws providing rules limiting at trace levels the discharge of a wide
variety of specific compounds into our environment and have provided penalties for exceeding such limits. As a consequence,
even though there is reasonable uncertainty in the identification
of specific compounds at trace levels, regulatory action is taken
on the basis of analytical measurements, in many cases without
regard to an evaluation of the accuracy of these measurements.

When one considers the distribution of trace components in
dynamic systems, it must also be accepted that the distribution
will not necessarily be uniform either in space or time. Nevertheless, most people are attuned to a sense that analytical measurements should be highly repeatable. In consequence, there is
far too often a strong tendency to discount or eliminate from
consideration, measurements in a sample series which deviate

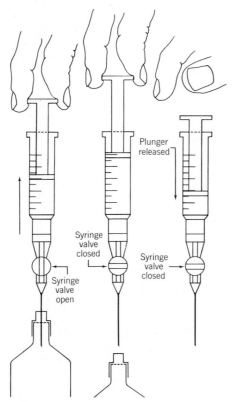

Figure 7.18.   Removal of an atmospheric pressure aliquot from a reduced pressure sample.

significantly from the average of a group of related samples.
The discounting or eliminating from consideration of this apparently abnormal measurement is one of the most dangerous practices
exs tent in the general field of trace analysis and particularly
in pollution analysis.

If the analyst cannot cite a definite physical reason for
failure to include a specific sample measurement, such as evident
instrument malfunction, sample container leakage, or sample loss
on transfer or injection, then there is no justification for
deleting a measurement from subsequent evaluation of a total
sample series. If, on injection of a sample into the chromatograph, the operator <u>believes</u> that septum leak is evident, the
resulting chromatogram <u>should</u> be marked immediately and never
measured or included in the subsequent results. Conversely, if
he cannot cite such a reason, the resulting chromatogram <u>must</u> be
measured and included in the overall evaluation of the study.

It is this latter case which is truly most important, for the
variant analysis should be a warning to the investigator that
either there is an operating variable in the system that is not
being controlled, and therefore will have an overall and longer-
term effect on all analyses of the system, or that a short-term
physical variation in composition has, in fact, occurred. In
either case, including this analysis in the statistical evaluation of the total study provides a proper basis for decisions resulting in regulatory action. This provides for a statistical
evaluation which allows a clear-cut decision on new samples as to
whether they truly differ from an earlier series or must be considered equivalent. On this basis, a variation in a component
otherwise unidentified can still be evaluated as significant or
insignificant. This inclusion of all relevant data removes this
type of decision from the subjectivity of either operators or
interpreters. Normal simple statistics are easily applied to
nearly all measurements which can be categorized as trace or pollutant analyses. No justification exists for failing to make
this type of evaluation.

REFERENCES

1. G. E. F. Lundell, Indust. Eng. Chem. Anal. Ed., $\underline{5}$, 221-225
   (1933).
2. A. P. Altshuller, A. F. Wartburg, I. R. Cohen, and S. F.
   Sleva, Int. J. Air Water Pollut. $\underline{6}$, 75 (1962).
3. C. A. Clemons and A. P. Altshuller, J. Air Pollut. Control
   Assoc., $\underline{14}$, 407 (1964).
4. H. Drasche, L. Funk, and R. Herbolsheimer, Staub-Reinhalt
   Luft, $\underline{32}$, 20 (1972).

5. Carborundum Plastics, Inc., Protective Plastics Div., 117
    State Road, Avondale, PA 19311

6. S. O. Farwell, H. H. Westberg, and R. A. Rasmussen, Anal.
    Chem., 47, 1490-1492 (1975).

7. Chromalytics Corp., Div. of Spex Industries Inc., Unionville,
    PA 19375.

8. SKC Inc., Pittsburgh, PA 15220 (P.O.Box 8538).

9. H. P. Burchfield and E. E. Storrs, J. of Chromatog. Sci., 13,
    205 (1975).

10. A. J. Sheppard and J. L. Iverson, J. of Chromatog. Sci., 13,
    448-452 (1975).

11. M. Beroza and R. A. Goad, "Reaction Gas Chromatography," in
    L. S. Ettre and A. Zlatkis, (Eds.), The Practice of Gas
    Chromatography, Interscience, New York-London-Sydney, 1967,
    pp. 488-492.

12. J. C. Cavagnol and W. R. Betker, "Sampling," in L. S. Ettre
    and A. Zlatkis (Eds.), The Practice of Gas Chromatography,
    Interscience, New York-London-Sydney, 1967, pp. 72-106.

13. K. Hammarstrand and E. J. Bonelli, Derivative Formation in
    Gas Chromatography, Varian Aerograph 6/68: A-1006, Varian
    Aerograph, Walnut Creek, CA 94598, 1968.

14. Pierce Chemical Company, Handbook of Silylation, Handbook
    GPA-3 (1970), Rockford, IL 61105.

15. M. Beroza and M. C. Bowman, Anal. Chem., 37, 291 (1965).

16. M. Beroza and M. C. Bowman, J. Assoc. Off. Agr. Chemists, 48,
    358 (1965).

17. M. C. Bowman and M. Beroza, J. Assoc. Off. Agr. Chemists, 48,
    943 (1965).

18. M. Beroza and M. C. Bowman, Anal. Chem., 38, 837-841 (1966).

19. B. B. Brodie, S. Udenfriend, J. E. Baer, J. Biol. Chem., 168,
    299 (1947).

20. G. R. Umbreit, "Qualitative and Quantitative Analysis by Gas
    Chromatography," in Theory and Application of Gas Chroma-
    tography in Industry and Medicine, Grune and Stratton, New
    York-London, 1968, pp. 54-67.

21. The Industrial Environment—Its Evaluation and Control, U.S.
    Dept. of HEW, Public Health Service, Center for Disease
    Control, Nat'l. Inst. for Occ. Safety and Health, U. S. Gov't.
    Printing Office, Washington, D.C. 20402, 1973.

22. W. Horwitz (Ed.), Official Methods of Analysis of the Associ-
    ation of Official Analytical Chemists, 12th edit., AOAC,
    Washington, D. C. 20044, 1975.

23. Handbook for Analytical Quality Control in Water and Waste-
    water Laboratories, EPA, Tech. Transfer, Analytical Quality
    Control Laboratory, Nat'l. Environmental Research Center,
    Cincinnati, Ohio, 1972, Chap. 8.

24. NIOSH Manual of Analytical Methods, HEW Publication No.
    (NIOSH) 75-121, Superintendent of Documents, U. S. Gov't.
    Printing Office, Washington, D. C. 20402.
25. G. A. Junk, J. J. Richard, M. D. Grieser, D. Witiak, J. L.
    Witiak, M. D. Arguello, R. Vick, H. J. Svec, J. S. Fritz,
    and G. V. Calder, J. of Chromatog., 99, 745-762 (1974).
26. A. K. Burnham, G. V. Calder, J. S. Fritz, G. A. Junk, H. J.
    Svec, and R. Willis, Anal. Chem., 44, 139-142 (1972).
27. A. K. Burnham, G. V. Calder, J. S. Fritz, G. A. Junk, H. J.
    Svec, and R. Vick, Journal AWWA, Nov., 722-725 (1973).
28. C. D. Chriswell, R. C. Chang, and J. S. Fritz, Anal. Chem.,
    47, 1325-1329 (1975).
29. J. J. Richard and J. S. Fritz, Talanta, 21, 91-93 (1974).
30. J. J. Richard, G. A. Junk, M. J. Avery, N. L. Nehring, J. S.
    Fritz, and J. J. Svec, J. Pest. Monitoring, 9, 117-123
    (1975).
31. C. McAuliffe, CHEMTECH, Jan., 46-51 (1971).

CHAPTER 8

Selection of Analytical Data from a Gas Chromatographic Laboratory

HARVEY L. PIERSON

DANIEL J. STEIBLE, Jr.

ICI United States, Inc.

8.1   WHY USE ELECTRONIC INTEGRATION? • • • • • • • • • • • • •   422
       8.1.1 Chromatographic Analyses without an Integrator •   422
       8.1.2 Some Advantages of Automatic Integration • • • •   422
       8.1.3 Some Disadvantages or Potential Problems • • • •   423
       8.1.4 The Importance of Planning. • • • • • • • • • •   423
8.2   INTEGRATOR - MICROPROCESSOR - COMPUTER. • • • • • • • •   424
       8.2.1 An Expanding Field in Analytical Instrumentation.   424
       8.2.2 Sorting Integrators by Size. • • • • • • • • • •   424
             Simple Integrators (noncomputer-based) • • • • •   424
             Microprocessor Integrators. • • • • • • • • • •   424
             Minicomputers. • • • • • • • • • • • • • • • •   426
             Midicomputers and Bigger. • • • • • • • • • • •   426
8.3   CHOOSING FOR EFFICIENCY AND ECONOMY • • • • • • • • •   426
       8.3.1 Planning before Purchase. • • • • • • • • • • •   426
       8.3.2 A Planning Checklist. • • • • • • • • • • • • •   427
             General. •   427
             The Existing Situation. • • • • • • • • • • • •   428

        The Projected Situation in 2-5 Years.   430
        Technical Considerations.   430
        Economics   431
        Hardware Starting Point   433
  8.3.3 Testing and Comparing Vendor/Hardware/Software
        Alternatives   435
        General   435
        Evaluating Simple Integrators   435
        Evaluating Microprocessor or Computing
        Integrators   436
        Evaluating Minicomputer Systems   437
        Composing Alternatives for Comparison   439
  8.3.4 Selling Management.   440
  8.3.5 Planning Philosophy and One Recommendation. . . .   441
8.4   ELEMENTS OF TROUBLESHOOTING   442
  8.4.1 The Mental Background (Theory).   442
  8.4.2 The Practical Side.   444
  8.4.3 Summary of Troubleshooting.   448
REFERENCES.   448

8.1   WHY USE ELECTRONIC INTEGRATION?

    8.1.1 Chromatographic Analysis without an Integrator

Ordinarily the task of the laboratory chromatographer is first to
obtain a separation of various components in a mixture and then
to use that separation to carry out quantitative analysis of a
number of samples. That task may be routine or quite challenging
depending on the number of components to be separated, their
relative concentrations, and the accuracy demanded of the analysis. Adequate accuracy may not be available from simple peak
height data; yet more accurate measurements of peak areas from a
series of chromatograms may be overly time consuming.

    8.1.2 Some Advantages of Automatic Integration

Automatic integration can yield the following:

• Precision--within the limits of the ability of the integrator's hardware (and software) to process the signal. No integrator can guarantee that the chromatograph is giving the

right peak (or any peak at all)!
- Standardization in integration of chromatograms.
- Relief from tedious manual measurement integration and calculations.
- A convenient means of collecting, collating, and reducing results.
- Report capability--a typewritten copy of the results of any calculation based on the chromatogram, as required by the analyst (1).

All these factors together mean enhanced accuracy and speed.

8.1.3 Some Disadvantages or Potential Problems

If automatic integration can relieve the chromatographer of some tedious chores, it also places on him a number of responsibilities:

- The care and maintenance of an additional piece of electronic equipment.
- The requirement of learning to apply this instrumentation to his problems.
- If the acquisition is not well planned, the possibility of owning a piece of equipment ill-suited to the needs of his laboratory. Some integrators are inadequate for the complex chromatograms often encountered in some laboratories; but others are more powerful (and expensive) than ever required for the clean separations often encountered in other laboratories.
- A false sense of security. Armed with an automatic integrator, an analyst may be sorely tempted to use separations that are not as sharp as they could be, or to use an instrument not as well-maintained as it might be.

8.1.4 The Importance of Planning

The acquisition of an integrator and then planning its application to chromatographic problems is the best route to let the advantages outweight the disadvantages. Detailed planning becomes more crucial whether you consider cost-effectiveness, corporate growth capability, or even the avoidance of daily headaches in the laboratory. Sections 8.2 and 8.3 present some ideas pertinent to forming these plans for a chromatographic analysis system.

8.2   INTEGRATOR-MICROPROCESSOR-COMPUTER

   8.2.1 An Expanding Field in Analytical Instrumentation

The chromatographer can choose from an enormous range of commercially available instruments (2). Some rather simple devices can
be purchased for a few hundred dollars, whereas a turnkey minicomputer system designed to collect and process data from a half
dozen or so chromatographs operating simultaneously will cost
tens of thousands of dollars. The field is changing too rapidly
to discuss manufacturers, models, and features in any detail.
Any attempt would too quickly be obsolete.

   8.2.2 Sorting Integrators by Size

The wide range of commercially available devices could be sorted
in several overlapping ways: by kinds of electronic function, or
by cost (either total cost or cost per chromatograph), to name a
few. We prefer to sort them by the number of chromatrographs
they can serve. This approach tends to emphasize the economic
aspects of selecting an automatic integrator. Unfortunately,
this way of sorting deemphasizes the fact that some devices which
can service only one chromatograph are of similar electronic
function (and perhaps computer logic) to other devices which can
handle data from perhaps four chromatographs simultaneously.

SIMPLE INTEGRATORS (NONCOMPUTOR BASED). This category includes
electromechanical devices, and instruments with largely conventional analog electronics. Examples of this class include the
Disc integrator, available with a number of strip chart recorders; and the analog Model 78 Laboratory Integrator (K & L Research Co.). In general these service only one chromatograph and
are adequate for relatively simple chromatograms (i.e., well-
resolved peaks, little tailing, no peaks riding on a solvent,
only small baseline shifts, drifting, etc.). These simple integrators are much less expensive than the more sophisticated and
versatile integrators to be discussed in the following sections.

MICRO-PROCESSOR INTEGRATORS. The microprocessor based integrator
may also operate one-on-one with a chromatograph; but some may be
extended to integrate as many as four chromatograms simultan-

eously (the results are then stored and printed one at a time).
The microprocessor is capable of integrating fairly complex chromatograms. The so-called computing integrator will also provide
concentration data based on analyst-supplied response factors or
on a chromatogram of a standard solution. There are entries in
this field from most manufacturers of chromatographs and a number
of firms more oriented toward electronics. Examples of this
class of integrators includes the Minigrator and Autolab series
from Spectra-Physics and the Supergrator series from Columbia
Scientific Instruments. Versatility may seem slightly limited by
the nature of a microprocessor--a dedicated small computer--or by
the different algorithms chosen to treat the complex parts of a
complex chromatogram. For example, different manufactures may
choose distinct techniques to automatically obtain tangent skimming of a peak riding on the tail of a larger (or solvent) peak.
There may also be different limitations on the way in which an
analyst may vary the parameters that control operation of the integrator. For example, there may be differences in the formalism
used to control peak sensing--a measure of slope (rate of change
of input signal) or a measure of peak width. In general, these
differences are trivial and could be thought of as a matter of
personal preference or convenience. This triviality would surely
disappear if several different integrators were in use in the
laboratory. Any chromatograms for which a reasonable formalism
or algorithm fails are probably the sort that should send the
chromatographer back to column selection and basic laboratory
work on sample preparation to get a chromatogram suitable for
quantitative work.

Some integrators may have a facility that can override the
automatic logic of the microprocessor to fit such special circumstances, such as unavoidable baseline shifts, special treatment
of certain fused peaks or of "negative peaks."

Some units are designed to be connected between the chromatograph and its conventional strip chart recorder or to contain
their own strip chart recorder. Thus, the integrator might mark
the chromatogram at the points it has taken as starts and stops
of the peaks it has detected. Other units indicate their activity by a series of panel lights or by codes printed with the
integrator's numeric output. Such aids may be valuable in applying the integrator to a chromatogram or in troubleshooting.

The form of the printed output varies from a simple paper tape
with retention times and areas (perhaps followed by calculated
concentrations) up through a fairly complete report which adds a
sample designation, peak compound names, and so on. The results
shown on a 2-inch paper tape are adequate for laboratory use, but
may require transcription to a final report and typing by a sec-
retary. These steps introduce the possibility of clerical

errors, transposed digits, and so on. When there is a large volume of output, a complete page report format which can be photo-
copied and attached to a report or laboratory notebook may be
especially beneficial.

MINICOMPUTERS. The next step up is a minicomputer based integrator, which may service up to perhaps two to three dozen chromatographs simultaneously. Examples of this class are the 3352-B
system of Hewlett-Packard and the PEP-2 system of Perkin-Elmer.
The minicomputer may have multiple input/output devices to service two or more locations independently, and may make the computer available (through a language like BASIC) to do further
calculations on chromatographic results or to do general laboratory calculations.

MIDICOMPUTERS and BIGGER. The final steps fall outside the scope
of this chapter. These include the midicomputer (to service more
chromatographs than minicomputer), process control instrumentation, "full" laboratory automation, or massive installations utilizing time sharing and/or batch processing through a large disc-
oriented computer. An example of this class of computer is
Varian's 200L Data System.
   Midicomputers for chromatograph applications are just big
minicomputers. They carry all the versatility of minicomputers
plus improved calculating capability, increased data storage, and
an attractive off-line storage capability on media such as magnetic tape or discs. On the other side of the coin, a computer
system is likely to be shared with other applications, thus diluting its impact on chromatography even to the point of marginal
system performance. Furthermore, the "eggs-in-one basket" syn-
drome may arise if maintenance were to become a problem.
   The large computer system will not be discussed further. The
volume of relevant technical information necessary to evaluate
such systems is totally outside the scope of the chromatography
laboratory. The professional analyst and his management are best
referred to computing professionals for assistance.

8.3   CHOOSING FOR EFFICIENCY AND ECONOMY

    8.3.1 Planning before Purchase

Whether the purchase is a simple integrator for one instrument or a medium-sized computer system to integrate the signals from dozens of chromatographs, it's important to plan the system. Planning can help to insure that the electronic integration equipment purchased is:

- Computationally adequate for present needs,
- Reliable,
- Capable of expanding or adding modules to satisfy projected needs,
- Economical,
- Cost justified.

Planning effort should be proportional to the expenditure involved. If a chromatographer is contemplating the purchase of a single simple integrator for one instrument which is processing simple separations, his plan might be quite simple: to buy a standard model integrator from a major manufacturer. If he is considering the automation of 60 or more chromatographs in a half-dozen laboratories spread over 10 acres and a parking lot, his plan might be quite complex: to design his own computer system complete with hardware, software, and professional computer staff.

This latter case, in fact, is so complex as to be beyond the scope of this section. This section will instead be concentrated on that middle ground between a simple integrator and a medium-to-large size computer system; that is, a system which will service a few (two or more) laboratories with six or more chromatographs between them. This is a realistic situation faced by many chromatographers and one that can be met in a practical fashion by the abundance of hardware available. The alternatives will be discussed in sufficient depth to begin their evaluation in the situation where such hardware is appropriate, whether measured by economic or technical requirements.

8.3.2 A Planning Checklist

GENERAL. The following checklist is suggested as an overall planning guide for an electronic integration system for one or more chromatography laboratories. Factors such as physical space limitations, and portability, which may be particular to individual laboratory situations can be added as appropriate. The various factors on the checklist will be discussed to highlight and develop the key considerations.

THE EXISTING SITUATION. Establish the existing situation on detail. Determine the:

1. Number of Chromatographs. It is important to know their manufacturer and age. Vendors may require different hardware to connect the chromatograph analog output signal to the electronic integration hardware. In the worst case, there may be as many different wiring arrangements (interfaces) as there are different instruments (gas chromatographs or liquid chromatographs, perhaps a few with several optional detectors).
2. Number of Locations to Be Serviced. In the simplest case, there is only one location and that is all there is ever going to be. But just as often, there are two or more laboratories that might use a system profitably or have a need to coordinate with one another. A quality control laboratory, for example, might need to communicate with or use the same techniques and hardware as the methods development laboratory. The benefits of a system that services the needs of both laboratories and allows them to communicate are obvious.

In conjunction with the number of laboratories, the distance between them is an important consideration when a system other than a collection of single unit integrators is being considered. The distance determines whether analog or digital signals can (should) be transmitted and dictates the hardware configuration. Technically it is easier to transmit a digital signal over a long distance without distortion than it is to transmit its analog counterpart (3,4).

As long as the laboratories are right next to one another; that is, in the same building and less than 100 feet apart, distance need not pose a problem. Signal distortion and noise pickup from adjacent electrical equipment may pose a problem instead. When more than 100 feet separates the laboratories or they are in separate buildings, then the actual cable distances (through conduits) is the distance to be considered, not the physical distance between them. In practice 500 feet of cable might be required to link two laboratories which are only 100 feet apart, especially if there is a major thoroughfare or a busy area within a manufacturing plant between them. Also if the laboratories are quite far apart, some form of telecommunications, data transfer by telephone lines, may be considered.

Prepare a simple schematic sketch of your multiple instrument situation, as in Figure 8.1, a sample which would be adequate for preliminary planning. The sketches will be helpful in applying the options presented in this chapter or in discussing options made available by various vendors.

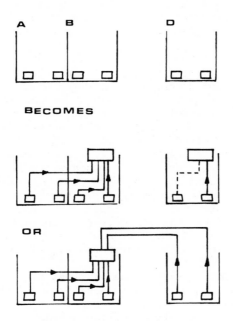

Figure 8.1. A sample diagram showing proposed chromatography system layouts.

3. Number of Tests Per Instrument/Location/Day/Month. Attempt an
estimate of the average workload on each instrument.

What are the peak load periods? Are the fluctuations season-
al? When the plant is in trouble? When a push is on a research
project? Should the electronic integration system be designed
for the very worst case--or for the routine case, an average or
projected workload?

An estimate of the number of tests also is necessary for estimating computer core memory requirements if the data reduction
workload is so heavy that computer data storage seems attractive.

4. Number and Kind of Autosamplers. Autosamplers, injecting
samples with mechanical precision and timing, in many cases make
necessary the use of some automated form of digital integration
simply because of the volume of data they can generate. The
question arises as to whether one simply wishes to collect and
analyze the data that has been collected from an instrument with
an autosampler or whether one wishes to utilize computer control
of the sequence of events of the autosampler injection cycle--
syringe washing injection of sample or standard, advance to another sample, etc. The hardware and core memory requirements

are different in each case. Feedback and control of an auto-
sampler(s) is more complex and expensive since it requires more
computer power--core memory, hardware, and software--than mere
data collection.

5. Type of Service. The type of service--quality assurance,
methods development, or routine testing--for each instrument or
laboratory is a general consideration, as is the question of
whether several types of service will be required of the system.
A quality assurance laboratory associated with a chemicals production facility has far simpler needs in terms of analytical
capability than a methods development laboratory or one that normally analyzes biological samples for pharmacologically active
compounds at sub-ppm or even ppb concentrations. On the other
hand, the data storage and reduction needs of a quality assurance
laboratory are usually much more pressing than those of a methods
development laboratory.

6. The Quality of the Samples to be Analyzed and the Accuracy
   and Precision Demanded of the Analysis. It is poor planning
to employ expensive instrumentation where less expensive equipment will do the job (5).

A chromatographic analysis that suffers from variation due to
extraction losses, incomplete derivatization, column losses, extraneous peaks, and so on, will probably not be greatly aided by
automatic integration. The precision enhancement comes largely
from elimination of elements of manual methods that tend to reduce precision.

In some kinds of work, such as identification or semiquanti-
tative screening, no great quantitative precision is either
available or expected. Good planning would tend away from expensive integrators; unless the volume of samples and paperwork
made automation attractive.

THE PROJECTED SITUATION IN 2-5 YEARS. The very same factors just
considered in the existing situation should be considered to the
extent possible. Some provision for growth should be allowed,
even if this only be to assure that the hardware you purchase can
be modified--at a reasonable cost and downtime--to accommodate
the growth needs. Buying a system that is stretched to the very
limit of its capacity is questionable practice. But individual
circumstances vary, particularly if cost is an overriding consideration. Perhaps the system diagram could indicate provision
for growth.

TECHNICAL CONSIDERATIONS. At the preliminary overall planning
phase technical details pertaining to how an integrator system

integrates peaks is a matter of only secondary importance. The state of the art of electronic integration has progressed to the point where, although purists and manufacturer's representatives would quibble, there really isn't that much difference between them. To illustrate, it's possible for anyone to call up several vendors and have them bring to your site an off-the-shelf integrator, computing integrator, or minicomputer which will acquire, integrate and process your chromatogram. The chromatogram for which this cannot be done or for which special tricks and manipulations are necessary for the integrator to recognize and properly integrate a peak is an exceptional one. If many of your chromatograms are thus exceptional, and you are convinced that they represent useful quantitative chromatography, and your major goal for digital integration is to process those troublesome chromatograms, then the entire planning process may be reduced to a search for a system that can be made to handle those chromatograms. Other sections contain some more thoughts on troublesome chromatograms.

ECONOMICS. Usually, the prime economic reason for buying an electronic integration system for a chromatography laboratory is to reduce operating costs. These reductions may be realized by both manpower or instrument savings immediately or over a long term.

Manpower savings may result particularly in the areas of instrument supervision (monitoring and perhaps adjusting the chromatograph), calculation and report preparation.

Instrument saving occurs by reducing the number of gas chromatographs required for a given workload. Faster sample throughput is provided by the electronic integration system and it is possible to reduce the number of gas chromatographs applied to a particular project. Of course, these savings will vary from laboratory to laboratory, and the laboratory supervisor must make the evaluation.

Data have been generated to show guidelines for expected savings as a function of the number of instruments, number of operators, or number of samples per day (see Tables 8.1 and 8.2). These are estimative guidelines. Appropriate figures for a particular laboratory can be generated only by someone familiar with individual laboratory operation, the personnel, and the existing hardware.

By developing the figures one can determine what type of system is justified. Table 8.3 gives order of magnitude estimates of the systems costs associated with the various hardware alternatives. When the economics and situation indicate that a minicomputer is justified, generally it is less expensive in total cost--manpower, hardware, software and systems development--to purchase a "turnkey" system from a vendor than it is to develop a system. Turnkey systems are those in which one vendor quotes an all-inclusive price and takes total system responsibility, including hardware, software, installation planning, interfacing, system installation, documentation, training, service, and ongoing development. The vendor develops the system and turns the key over to the purchaser. With a turnkey computer system, it is much easier for laboratory supervision and management to weigh the costs against the projected savings and make a decision. Home-made systems tend to be open-ended in cost for their development and maintenance.

TABLE 8.1 COST SAVINGS BASED ON MANPOWER (2,5). Personnel Requirements for 16-Hour/Day Operation.

No. of GC's	Manual	Computer	Net Savings/Year
3 - 5	3	2	$ 8,000 - 10,000
5 - 10	7	3	32,000 - 40,000
10- 15	10	4	48,000 - 60,000
15- 20	13	6	56,000 - 80,000
20- 30	17	7	80,000 - 100,000

TABLE 8.2 COST SAVINGS BASED ON SAMPLE THROUGHPUT (2,5)

	Operator Costs[1]		
	Manual	Integrator	Computer
Instrument supervision	$ .50	$ .10	$ .05
Area measurement	3.00	.30	.30
Composition calculation	1.67	1.67	.05
Report writing	1.00	1.00	.05
Total/sample	$ 6.17	$ 3.17	$ .45

[1] Based on 10 peaks; 10-min chromatograms, $10/operator-hr.
Savings/year = (Samples/day) x (savings/day) x (workdays/year).

TABLE 8.3   ORDER OF MAGNITUDE COSTS FOR HARDWARE ESTIMATES (6)

	Applications	Costs
Pulse-type integrators	Well-defined simple separations	$ 500 - 1,000
Simple integrators	Well-defined simple separations or complex methods development work	1,000 - 2,500
Computing integrators	Complex separations	2,500 -12,000
Minicomputers	Complex separations, multiple instruments and/or locations, information storage and data reduction	12,000 -50,000[1]
Midicomputers	32 or more GCs, mass data storage, extensive postprocessing and data reduction	50,000+

[1]Packaged turnkey system costs.   Detailed costs will vary in the ranges indicated depending upon specific configurations and features.

   In addition to the economic factors associated with manpower and instrument savings, there are additional benefits of electronic integration.   In some cases it is possible to make an economic evaluation of these as well.   The added benefits include:

- Improved accuracy and precision (see Figure 8.2).
- Minimization of human error.
- Less personnel supervision.
- Release of research personnel from some routine tasks.
- Faster turnaround time--greater throughput.
- Improved performance from automated equipment such as auto-samplers.
- Standardized procedures for the integration of complex chromatograms.

HARDWARE STARTING POINT.   Table 8.4 presents a suggested first choice for investigation from among the various hardware alternatives one might consider, based on the type of service and physical consideration but independent of cost.   It must be noted

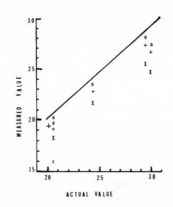

Figure 8.2.   Comparison between actual and measured
values for analyses results obtained by computer and
manually.   (Solutions prepared by serial dilution;
fixed amount of internal standard added. (7)) o, From
minicomputer system. x, Triangulation procedure (manual). +, Peak height procedure (manual). Size of symbol
indicates standard deviation.

TABLE 8.4   FIRST CHOICE FOR INVESTIGATION[1]

	Quality Assurance	Routine Well-Developed Methods	Methods Development Sophisticated Chromatography
1 - 4 Chromatographs 1 lab or 2	Computing integrator	Computing integrator	Simple integrators[2]
4 - 16 Chromatographs 1 - 4 labs	Minicomputer	Minicomputer	Minicomputer
1 lab	Minicomputer	Minicomputer	Minicomputer
16 Units			
Many labs	Midi-system or 2 or more minicomputers	Midi-system	Simple or computing integrators

[1]Based on type of service and physical consideration but indep-
ent of purchase price.
[2]Until the validation step. (Collaborative assay testing with the
end-user laboratory.)

that these are merely guidelines. Individual circumstances
(auto-samplers, available resources, available justification,
time constraints, multishift operations, etc.) may suggest that
another choice is a better system for first consideration. Many
laboratories may not even fit very well into the three categories
we have chosen. For example, if one has two chromatographs
equipped with auto-samplers running 5-min chromatograms 24 hr/day,
a minicomputer with some form of auxilliary data storage may be a
better first choice for evaluation than a computing integrator,
because of probable data storage and reduction requirements. A
computing integrator may be preferable to a simple integrator due
to the speed with which one obtains his final printed results,
especially in a "go"/"no-go" test situation, where the results
are required to ship product immediately; or for "in process"
testing, where results are used to determine the next step in a
plant process.

8.3.3 Testing and Comparing Vendor/Hardware/Software
Alternatives

GENERAL. Armed with the knowledge of what he is trying to accomplish, economic justification, and an overall strategy, the
designer can proceed to the matter of evaluating various vendor/
hardware/software alternatives. The complexity and depth of
testing and comparison should be proportional to the expected
purchase price of the final system. As a general guideline, no
more than 10 % of the expected systems cost should be spent in
evaluating alternatives. If it is expected that the system will
cost $50,000, it's worth another $5,000 in effort, travel and
testing to assure that· it will do the job for which it is being
purchased.

EVALUATING SIMPLE INTEGRATORS. If after an analysis of needs, a
simple integrator appears to be sufficient for the necessary
tasks, the following steps are suggested for testing and comparing simple integrators.

1. Review the literature and select a few integrators that have
   the features you want. Stay with major manufacturer's
   standard models whenever possible. Unknowns and special
   models pose long-term maintenance hazards.
2. Arrange for a demonstration of the equipment at your site,
   connected to your instruments, if possible. In this way, you
   can have first hand observation of the problems to be

encountered in connecting the device to your equipment. Actual operating practice on your chromatograph in your laboratory may differ from the showcase examples in a manufacturer's literature.

3. If practical, arrange for use of the equipment at your site for a test period. Manufacturer's representatives can often leave a unit with you for a week or two for you to evaluate it. It is suggested that you avail yourself of this test period whenever possible to run as many samples as are representative of your workload in order to become familiar with the integrator and to become familiar with the problems you may encounter in using it. Some areas that you may wish to give special attention are indirectly suggested in Sections 8.2 and 8.4. Depending upon the schedule and workload in your laboratory, you may wish to choose a particular separation (sample and column) to test integrators. This would be part of the formal comparison of different models exemplified by Table 8.6.

4. Before your purchase, take the time to call some other people who already have this model integrator. Ask a few pertinent questions pertaining to hardware reliability, maintenance, and service. You may wish to reconsider your choice if you find that the printer mechanism is a constant maintenance problem or that you have to ship the unit cross country, paying insured postage rates both ways; or you may find out that local service is available but unreliable.

EVALUATING MICROPROCESSOR OR COMPUTING INTEGRATORS. The more complex and expensive the digital integration equipment the more features there are to check. In addition to those items suggested to evaluate simple integrators, other checks pertinent to microprocessor systems are suggested here:

1. First, how many chromatographs will the device support? If it will only support one chromatograph at a time and you have two GCs in your laboratory at present and expect to have four eventually, then you may have to purchase two units initially and two more later. Or more probably, you can buy two of this model now and two of some other model and manufacturer later.
   In general, if the chromatographs are in close proximity to one another and the integrator is capable of integrating two or more different chromatograms simultaneously, then a multiple unit device is best, particularly if the device is not at the limit of its capacity. If the chromatographs are remote from one another, however, then single unit devices are preferable. It is preferable to remain with one manufacturer

whenever possible. This provides back-up in kind in the event
that one unit requires repair. Thus personnel are spared the
burden of learning the operating procedures and syntax of a
second system, a system that they will need to use infrequently or under emergency circumstances.

2. Are there any special limitations to the integrators? Are
   there limits to how many peaks the device can store and report? Are there limits on the analog signal that any of your
   detectors cannot meet?

3. Ease-of-use is another consideration. How close is the instrument to being self-explanatory? Can you come back to the
   computing integrator after 6 months away from it and engage in
   a self-documented dialog to define your integration requirements? Or do you have to refer to a more complicated (although possible more powerful or versatile) procedure where
   you have to look up which response or actions you want from
   among dozens of responses and enter that as a complex code or
   arrangement of switches which tells the microprocessor what to
   do? You may find trade-offs between ease of learning and ease
   of use. If you're working with the microprocessor system all
   the time then ease-of-use may be less of a consideration than
   if use is infrequent.

4. Computing integrators that support multiple chromatographs of
   necessity have some limited amount of raw data storage. Consider in addition to this storage the unit's ability to store
   data on some peripheral device such as a magnetic tape cass-
   ette or a magnetic disc. If you are doing developmental work,
   you may wish to calculate concentrations from a chromatogram
   in different ways. Additional storage, although not critical,
   is beneficial in such instances.

EVALUATING MINICOMPUTER SYSTEMS. As the number of instruments,
the technical requirements, the workload of data storage and reduction increase, then minicomputers come under serious consideration. All of the factors pertaining to evaluating simple integrators and computing integrators apply to evaluating minicomputer systems, as well as a few more.

One of the first choices is whether to buy a turnkey minicomputer system for GLC, a "kit" that provides 90% of the software needed for acquiring and processing data from chromatograms,
or whether to start from scratch in assembling your own hardware/
software system (3). This is a key decision and must be made
very early in the evaluation.

Table 8.5 presents cost and time estimates derived from an
actual investigation.

Although the cost figures for hardware show the purchase price

of a turnkey system to be somewhat more expensive than either of
the other alternatives, the development costs and delay time
demonstrate some of the numerous advantages one receives with a
turnkey system:

• Chromatography software is developed by chromatography
  "experts."
• On-going vendor software development.
• Information exchange through vendor user's groups.
• On-call chromatography expertise.
• Professional documentation, which is kept current.
• Vendor training courses.

The purchase of a turnkey system also promotes an increase in
productivity in that the time that would be devoted to developing
a "system" can be devoted to performing the chromatography function or otherwise earning money.

In addition, there are some very good reasons for not assembling or developing your own system.

• It diverts time and attention from analyzing and interpreting
  results to the mechanics of obtaining results.
• In-house computer expertise may be lacking, and outside help
  may be required. External consultants are more expensive
  than in-house personnel.
• The laboratory that develops the software (or assembles it)
  assumes the burden for maintaining it.

It seems that self-made software, for a chromatograph or for anything else, whether developed from scratch or assembled, tends to
be poorly documented. If the software is documented, the documentation is not maintained. Either situation, when coupled with
personnel turnover, can and does lead to costly long-term software maintenance.

There may be a compelling reason for assembling or developing
one's own minicomputer system:

• To take advantage of existing computer capacity (along with
  chromatography and computer systems expertise).
• To link chromatographic results with other experimental data
  to each other and a data bank through a hierarchical arrangement of computers.
• To input data to a larger computer system which provides data
  storage, reduction and statistical analyses.

Unless the chromatographer has extensive computing experience, it
is best to enlist the aid of computer professionals either from

TABLE 8.5   DELIVERED SYSTEMS COSTS FOR VARIOUS MINICOMPUTER
            OPTIONS (6) (Basis: 2-laboratory system, 4 chroma-
            tographs per laboratory located in two separate
            buildings, 500 feet apart with comparable hardware
            configuration)

	Hardware	Software	Development	Total	Timing[1]
Vendor Package 1	25,000	2,000	2,000	29,000	3 months
Vendor Package 2	37,000	–	2,000	39,000	3 months
Vendor Kit 1	22,500	500	18,000	41,000	4.5 months
Vendor Kit 2	25,000	2,000	18,000	45,000	6 months
Start from scratch with existing hardware	2,000	500	18,000	20,500 minimum	8 months minimum
Start from scratch with new hardware	25,000	2,000	18,000	45,000 minimum	6 months minimum

[1]Estimated timing from start of project until a "delivered"
working system is on the job.

within his own organization or consultants. He should take advantage of software that may be available from computer manufacturers, a manufacturer's users group, independent consulting
firms, and so on; and exhaust this search before attempting to
write software himself, particularly if he must share the computer with someone else.

   Intellectual curiosity is commendable and self-rewarding, but
it is not cost justified from many organizational viewpoints.

   The authors consider the turnkey system to be the preferred
alternative unless there are compelling reasons for doing otherwise such as the availability of computer hardware, and chromatography and computer systems expertise. Several packages are
available and a chromatographer can evaluate them in much the
same manner as he would evaluate microprocessor integrators.

COMPOSING ALTERNATIVES FOR COMPARISON. Table 8.6 presents a
sample guide table for preparing one's own comparison of digital
integration alternatives. This is a quasi-objective comparison
method which can help in evaluation of alternative systems. Obviously different factors are important under different circumstances. Some factors, such as the requirement for the employ-
ment of external consultants, might be considered as negatives in

the selection of a computer system. It's all a matter of choosing what's important or unimportant to the individual.

TABLE 8.6   A SAMPLE ANALYSIS SHEET FOR COMPARING DIGITAL
            INTEGRATION ALTERNATIVES

	Importance (1-5)	Rating (1-10)		
		Unit 1	Unit 2	Unit 3
Integrators/computer				
Cost				
Performance				
Reliability				
Maintenance				
Computing integrators				
Multiple units				
Ease-of-use				
Data storage				
Features/timed events				
Recorder feedback				
Report formats				
Minicomputers				
Make or buy decision				
Add on cost per instrument				
Shared system				
Outside consultants				
TOTAL				

8.3.4 Selling Management

The elements of planning discussed previously have been applied
with success to the purchase of hundreds of thousands of dollars
worth of laboratory equipment in a medium-sized company with many
laboratories and various requirements. Following this somewhat
rigorous approach and documenting it in the form of a report and
recommendation has led to early acceptance of proposals and

speedy purchase, installation, and implementation of digital integration system for chromatography applications.

If a thorough analysis is presented to supervision or management, there is a greater chance of proposal acceptance than there is if a less rigorous approach is used. If nothing else, management is confident that the project has been well researched and the proposal is serious and reasonable, substantiated by facts, figures, and logical arguments. If the payback, return on capital investment, meets current guidelines for the company or institution, it presents an excellent professional argument for digital integration.

8.3.5 Planning Philosopy and One Recommendation

Planning does not guarantee a successful digital integration system, that is, one which is precise, reliable, cost effective, and flexible and modular enough to meet short-term needs and longer-term expansion needs. But it does provide some protection against failure (a system that is expensive, hard-to-use, subject to frequent break-downs thereby impeding productivity, difficult to maintain, or incapable of expanding to handle the workload in a few years' time).

Certainly the effort of following a checklist cannot be avoided. A less rigorous alternative is to call several manufacturers, have their representatives stop by, visit their plants or customer locations and witness demonstrations of their equipment, and then decide which seems to best meet one's needs. Both this approach, which seems to be the standard in industry, and the more rigorous approach to overall planning suggested here, will yield a system. But the rigorous approach is most likely to yield a successful system and it is often an approach that is easier to sell to supervision and management.

Based on experience and the state of the art of digital integration as applied to chromatography, the authors recommend the purchase of existing vendor-supplied minicomputer packaged systems wherever economically practical. Experience has shown that they can process more than 90% of the chromatography workload accurately and reproducibly. The remaining portion, for which quantitative work obviously was troublesome to begin with, is still handled manually. In this case the expense and effort of developing a system oneself to automate the integration of those troublesome chromatograms may not be justified. Vendor support and maintenance of the hardware and software of these purchased systems packages has generally ranged from adequate to excellent.

8.4   ELEMENTS OF TROUBLESHOOTING

8.4.1 The Mental Background (Theory)

There are many parts to a mental background that must be acquired
as an automatic integrator is applied to chromatograms. There
must be a general notion of the workings of the integrator and a
firm knowledge of the roles played by the variable parameters
made available in your particular model. Successful application
and/or troubleshooting must begin with the basics of chromatography (8) and integrator electronics and logic.
    What does a digital computer do to a chromatogram? Consider
a single well-resolved peak, as in Figure 8.3. A typical approach would be to convert the analog signal to a digital one (a
number, in area units, microvolt-seconds, etc.) at specified and
sufficiently frequent intervals or grouped intervals. Then these
numbers can be summed by peaks and corrected for baseline points.
The integrator's logic is faced with the task of choosing how
many intervals to add together to call a peak (i.e., where to
start and stop) and how to correct these areas for the more complex situation illustrated in Figure 8.4. Peak detection (starting and stopping) will be based in some way on relative changes
from one interval to the next, or on relative changes through a
series of intervals; for example, continual increase for a period
of time as a signal to start adding areas to form a peak--then
continual no-change for a period of time or (a simple return to
baseline signal) as a signal to stop adding for that peak. How
great the increase must be to start counting or how rigid the no-
change must be to stop counting is controlled by the manufacturer
or user through variable parameters or switch settings. The complication for the chromatogram of Figure 8.4 is that the no-
change period after peak A is practically nonexistent. The integrator's logic with or without much user control must determine
how to split the areas of these poorly resolved peaks on a shifted baseline, if at all. This may take place automatically, based
on relative heights of the two, relative depth of the valley between them, various slope tests, and so on. Alternatively, the
user may be able to impose his own choice of area allocation
through variable parameters, or even take the total area only
thus treating the two peaks as a single component. The form of
the options: drop-line splitting, tangent skimming, and so on,
varies with the algorithms and logic the manufacturer has chosen
to employ. Figures 8.5 and 8.6 illustrate two possibilities for
the extreme case of a small peak not entirely resolved from a
very large (solvent) peak. Rather clearly, the skimming of

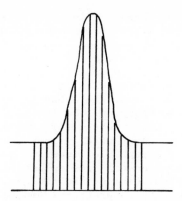

Figure 8.3.   Assigning digital values.

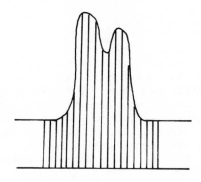

Figure 8.4.   Where to start and stop?

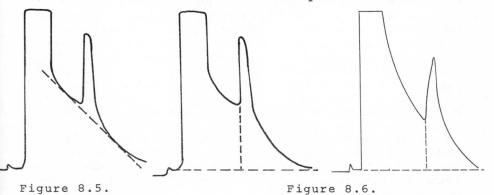

Figure 8.5.

"Tangent skimming."

Figure 8.6.

"Drop-line separation."

Figure 8.5 is to be preferred. For the situation of peaks more
similar in size, as in Figure 8.4, some knowledge of the chromatographic behavior of the separate components may be required to
properly choose a means of allocating areas to each of the two
peaks. If the earlier eluting peak is known to tail, then tang-
ential skimming may be preferred. If both peaks are symmetrical,
then a drop-line splitting of the areas is probably a better
choice. The chromatographer must be familiar with the parameters
he may use to control the integrator's treatment of both simple
and complex situations, so that a correct and consistent treatment is possible. If acceptable linearity or accuracy cannot
be obtained by either means of area splitting, the final option
is to increase the resolution of the two peaks, either by altering the conditions of the separation (temperature, carrier flow,
column length, etc.) or by employing a different column packing
altogether.

8.4.2 The Practical Side

Occasionally a problem will be elevated to the status of trouble
and interfere with the advantageous use of an integrator. These
troublesome situations can be grouped into those that occur
suddenly in the middle of what was thought to be a routine day's
work and those that are encountered while working on a new separation problem. Alternatively, problems could be grouped in a
more operational (even if overlapping) way: accuracy, reproducibility, linearity, and "noise." Both viewpoints will be represented in this discussion.
    The problems that appear suddenly may be the easiest to diag-
nose and cure, principally because they may have nothing to do
with the integrator. Troubleshooting the integrator and chromatograph should start with the chromatograph. This may sound like
a strange piece of advice in a discussion of integrators, but in
the authors' experience, it has been an extremely useful, fruit-
ful approach. Furthermore, the more automatic is the integrator,
the more likely it is to disguise a problem in the chromatograph.
The automatic integrator seems to be rather forgiving and toler-
ant. It may lead an unwary analyst to some strange analytical
results.
    Consider a sudden loss of injection-to-injection reproduci-
bility. With only a little experience, the analyst without an
integrator will look to injection technique, a leak almost any-
where (syringe, septum, column fitting, external gas connection,
etc.) or perhaps a contaminated column. Where one looks first
depends on experience with the instrument and the analysis and

the appearance of the chromatogram. Adding an integrator to the
chromatographic system does not relieve an analyst of the ordinary care of his chromatograph.

Consider the discovery of disagreements between the integrator's area measurement and the "area" shown on the strip chart
recorder. After all, tailing peaks are difficult to estimate by
hand by any technique so some small disagreément may be occasionally expected. But gross differences(e.g., finding the larger
peak on the recorder to be reported to have the smaller area)are
not expected. A number of times this problem has been cured by
repair and maintenance for the recorder: readjusting the recorder's gain, cleaning a dirty, sticky pen carriage, and so on. Any
results from a poorly responding strip-chart recorder will be
poor, regardless of the presence or absence of an integrator.

If the chromatograph and recorder seem to be innocent, then
perhaps the integrator is the culprit afterall. The peak detection system or baseline tracking circuit or some other control
may require only a small adjustment to return the instrument to
normal operation. Perhaps there is something subtle in the operating parameters chosen for this series of samples which permitted good quantification until a few atypical samples introduced an unexpected impurity peak. By simple misfortune of retention time the unexpected impurity might lead an integrator into a significant baseline error, thus spoiling the results.

The X in Figures 8.7 and 8.8 denotes a point in time when an
integrator might be automatically stopped by an interval timer.
The clean peak (the standard solution?) in Figure 8.7 appears to
have nearly the same area as the large one in 8.8 But if the
integrator automatically stops at the same point X, the baseline
it must use is skewed, and the area it will report for that large
peak will be too small.

Another possibility, probably out of the analyst's domain, is
an electronic breakdown in the integrator. All the analyst can
usually do is to check out and eliminate potential problems in
the chromatograph and in the application of the integrator to the
chromatograms at hand, determine whether the integrator seems to
be functioning at all, and then call for the gentleman with the
test meters and replacement circuit boards.

Difficulties in the application of the integrator are sometimes started or stopped when the integrator is purchased. There
are some chromatograms that cannot be adequately measured with
some integrators; there are some chromatograms that cannot be
adequately measured with any integrator. Poor resolution is
still poor resolution; accuracy and precision will be hurt by
poor resolution regardless of the power of the best integrator
system. One must either expect less than the best accuracy and
precision,or return to the ruler, planimeter, or whatever and

expect less than the best accuracy and precision. The experienced eye may do fairly well compared to an instrument when faced
with a messy chromatogram. The solution is to clean up the chromatogram! The analyst should begin with the chromatograph and
some other columns, or perhaps a more efficient prior clean-up,
and change the analysis and change the chromatogram enough to
make it suitable for the integrator. The chromatogram will also
then be better suited for the ruler, planimeter, or whatever. If
the chromatogram cannot be improved reasonably, this demonstrates
that the laboratory has invested in an integrator without profit-
able use. Some planning in the laboratory will give some assurance that the integrator will be inadequate very infrequently.

     Many of the difficulties encountered in application of an in-
tegrator to a particular chromatogram have come from an analyst's
unintentional misapplication of the instrument's operating parameters. As an example, consider a chromatogram with a mildly
noisy baseline. Left unfiltered, an integrator would falsely
trigger and count and report an annoying number of noise peaks.
Furthermore, the integrator may be triggered into splitting the

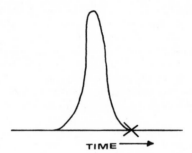

Figure 8.7.   Integrator stop point on clean peak.

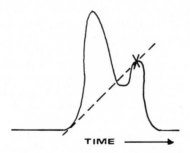

Figure 8.8.   Integrator stop point for peak with unresolved peak.

important peaks sought and reporting them as several peaks, thus obscuring the information the analyst seeks. There are three basic routes to filtering the noise. First, eliminate it at the source, the chromatograph. Clean the detector, recondition or repack the column, or whatever else is required. Second, set up the digital filter, the minimum area or number of counts required to report a peak's area. Thus noise peaks are counted, but not reported because they are small. Third, employ electronic filtering or signal conditioning. This is intended to smooth short-term noise sufficiently that the integrator's peak sensing logic will not trigger on the noise peak. An unwary analyst, relying too heavily on only one technique, could spoil his results. Too great a reliance on digital filtering, the area reject approach, could mean lost peaks--peaks of interest not reported because they were too small. Too great a reliance on electronic filtering may distort a peak's shape. Simply desensitizing the peak sensing logic may bring about late peak recognition and/or early stops and so create poor reproducibility and loss of linearity of response. With care the analyst can find a compromise on filtering that fits the instrument to his chromatogram and can then enjoy the benefits.

Another misapplication could arise from a kind of signal mismatch. Consider a recorder whose electronic zero has been displaced a few inches from its accustomed place. If the attenuated signal from the chromatograph were balanced and zeroed to meet this nonzero, the unattenuated signal (often presented to the integrator) may be out of the input range of the integrators amplifiers. Poor reproducibility and loss of linearity will surely result.

A misapplication of the special features of an integrator could easily result from a failure to follow the fine print portion of the instructions. Of course, sometimes the instructions, manuals, and explanations of the complete functions of various operating parameters are less than lucid. Poorly phrased instructions could result from language problems--"the electronics man who cannot talk to the computer man who cannot talk to the chemist," and where the sales representative fits is anybody's guess. A cursory explanation could be the result of the reluctance of an instrument manufacturer to reveal too much about some feature he chooses to consider proprietary. Any such situation is either unfortunate or inexcusable, depending on viewpoint. The best recourse the analyst has is to ask as many detailed questions as possible, while the instrument is on demonstration loan in the laboratory, being applied to a variety of typical chromatograms encountered in his laboratory.

8.4.3 Summary of Troubleshooting

A thumbnail sketch of steps to take in integrator troubleshooting
should include:

1. Troubleshoot the chromatograph and recorder first. Keep up as
   much regular maintenance as applicable. An analyst being
   helped by an integrator may tend to laxity with inevitable
   disappointing results.
2. Assure yourself that the chromatogram and all the sample
   preparation that went before were suitable for quantitative
   work.
3. Check all the operating parameters for settings truly appropriate for the chromatogram. The best integrator will only do
   what it is told.
4. Only finally turn to the integrator's electronic components or
   thoughts of computer errors. The only exception is the day
   when the instrument stops functioning altogether.

REFERENCES

1. I. G. Young, "A Review of Post-run Calculation Techniques in
   Gas Chromatography, Parts I, II, and III," Am. Lab., 7, 27
   Feb. 1975), 7, 37 (June 1975), 7, 11 (August 1975).
2. "Chromatography," in Laboratory Instrumentation, International Scientific Communications, Green Farms, CT, 1974. This
   is a collection of reprints from American Laboratory. The
   cost data is now obsolete and much of the instrumental
   sophistication has been surpassed, but the papers contain
   many valid considerations for planning a modern chroma-
   tography/integration system.
3. J. Finkle, Computer Aided Experiments: Interfacing to Minicomputers, John Wiley and Sons, New York, 1975.
4. D. G. Larsen and P. R. Rony, "Computer Interfacing: Precision
   of Data Transmission: Analog Versus Digital Data," Am. Lab.,
   6, 67 (June 1974).
5. G. L. Feldman, M. Maude, and A. Windeler, "Comparison of
   Integration Methods for Gas Chromatography," Am. Lab., 2, 61
   (Feb. 1970); reprinted in ref. (2).
6. Various manufacturer's price lists.
7. From internal reports, ICI-United States.
8. Other chapters in this book.

PART THREE

APPLICATIONS

Science is nothing but trained and organized common sense, differing from the latter only as a veteran may differ from a raw recruit: And its methods differ from those of common sense only as far as the guardsman's cut and thrust differ from the manner in which a savage wields his club.

Thomas Henry Huxley (1825–1895)
Collected Essays, iv., The Method of Zadig

The applications of the gas chromatographic technique are many. To write chapters on the various areas of application would result in a very large monograph. The areas of application presented are those which the authors feel represent a happy medium resulting from the years of presenting the short course. Other areas of application have been covered in books and monographs currently available to the student of GC.

Sample types not discussed should be readily analyzed by use of the basic information presented in parts one and two and a variation of conditions presented in this part.

CHAPTER 9

Gas Chromatographic
Analysis of Food

HERBERT L. ROTHBART

Eastern Utilization Research Center, USDA

9.1   LIPIDS. 451
        9.1.1 Fatty Acid Methyl Esters. 453
        9.1.2 Triglycerides 460
        9.1.3 Phospholipids 464
        9.1.4 Trends. 465
9.2   PROTEINS 465
        9.2.1 Volatile Derivatives 467
        9.2.2 Protein Sequence Determination. 473
        9.2.3 Trends 476
9.3   CARBOHYDRATES 476
        9.3.1 Methyl Ethers 478
        9.3.2 Silyl Ethers 479
        9.3.3 Esters 484
        9.3.4 Elimination of Anomers. 484
        9.3.5 Aldonitrile Acetates 485
        9.3.6 Trends 489
REFERENCES. 490

World food production and famine, food composition, and its relationship to nutrition, naturally occurring or added toxicants in food and their effects upon health: almost daily comments on these topics appear in the mass media. United Nations conferences on food production have been described on the front pages of major newspapers during 1974 and 1975; the entire May 9, 1975 edition of Science (1) was devoted to food production and malnutrition. The separation efficiency and sensitivity of gas chromatography (GC) has made it a valuable tool for the determination of the major components of food and (particularly when allied with mass spectroscopy) has made us aware of trace constituents and contaminants in our food supply. Concern about the consequences has led the National Academy of Sciences to consider the presence of toxicants in foods (2). Recurring and sometimes conflicting epidemiological studies relating diet to degenerative diseases such as cancer and atherosclerosis have also increased concern regarding food composition.

Gas chromatographic methods for determination of components of the major classes of compounds in foods (lipids, proteins, and carbohydrates) are considered in this chapter. Methods for trace components (steroids, vitamins and pesticide residues for example), are described in other chapters. One of the major difficulties in a review such as this is that man is nearly omnivorous, an important factor in our survival no doubt, and diversity and complexity of our food supply makes generalizations difficult. Progress in GC has made this part of the analysis possibly the fastest and easiest step of the process. Preliminary isolation and other pre-gas chromatographic steps often require demanding techniques that account for a significant fraction of analysis time. Most of the major food components, other than water, are relatively nonvolatile or decompose upon heating. Detailed analyses often require hydrolysis of large molecular species to produce solutes of lower molecular weight. These materials must often be derivatized to reduce hydrogen bonding and increase volatility.

Use of GC in food analysis will almost certainly grow in the next few years, since composition is a significant determinant of nutritional, functional, and organoleptic properties of foods. In part this will occur to satisfy consumer questions as to what is in the food supply and may lead to extensive labeling declarations. It may even extend to detailed analyses of the composition of new varieties of fruits and vegetables. Chemists and food scientists utilizing sophisticated chromatographic, spectrometric, data acquisition, and processing techniques are well equipped to provide this information.

9.1   LIPIDS

The heterogeneous class of compounds marked by solubility in so-
called lipid solvents (acetone, hydrocarbons, ether, etc.) and
relative insolubility in water, has traditionally been called
lipids (3). This historical classification, based upon isolation
procedures from natural products, is obviously too broad for
simple generalizations since it includes triglycerides, fatty
acids, phospholipids, sterols, sterol esters, bile acids, waxes,
hydrocarbons, fatty ethers and hydrocarbons. For the purposes
of this chapter, we will consider lipids to be fatty acids and
their derivatives.

Triacyl esters of glycerine, usually called triglycerides, are
the most abundant lipids in our food supply and as fats comprise
about 20 - 30% of the U. S. dietary (on a dry weight basis) (4).
The acyl groups (fatty acid moieties) range from 2 to about 24
carbons in chain length. About 30 fatty acid groups are commonly
encountered, leading to 27,000 triglyceride molecular species.
If only the 10 most common fatty acid groups are considered there
are still 1000 triglycerides to be separated in analytical work.
Most of the acid groups are "straight chain" with branched species occurring in lipids of animal and microbiological origin.

The names of triglycerides depend upon their constituent fatty
acids. For example, cis- 9-octadecenoic acid, commonly called
oleic acid, can be referred to as C18:1 (Δ9, cis) and octadeca-
noic acid, stearic acid, can be referred to as C18:0. If all of
the acyl groups are the same, for example,

$$H_2C - OR_1 \qquad\qquad R_1 = CH_3(CH_2)_n \overset{\overset{O}{\|}}{C} -,$$

$$H-C - OR_2 \qquad\qquad CH_3(CH_2)_x CH = CH(CH_2)_y \overset{\overset{O}{\|}}{C} -$$

$$H_2C - OR_3$$

Figure 9.1

C18:0, the triglyceride would be called trioctadecanoin or tristearin (or SSS). Mixed acid nomenclature is somewhat more complex, for example, if R_1 and R_3 in Figure 9.1 are C18:0 and R_2 is
C18:1 (Δ9, cis), the resulting triglyceride is Glycerol 2-cis,
9-octadecenoate-1,3-dioctadecanoate or more simply 2-oleo-1,3-
distearin (SOS) (5).

Triglycerides are among the least volatile materials separated
by gas-liquid chromatography (GLC). For example, molecules of
tristearin contain 57 carbon atoms, 54 of which are in the acyl

groups; the molecular weight is 891, and the vapor pressure at
315°C is only 0.05 torr (6). Requirements for successful chromatography include short columns (less than 2 m), low loading of
stationary phases, pretreatment of columns with silanizing agents
to reduce adsorptive sites, and elevated temperatures (5). A
triglyceride separation is depicted in Figure 9.2 Each peak is
denoted by the number of carbon atoms in the acyl portion of the
triglyceride (carbon number). For example, tristearin would be
eluted under peak 54; unfortunately, so would the triply unsaturated lipid triolein and most other triglycerides with 54 carbon

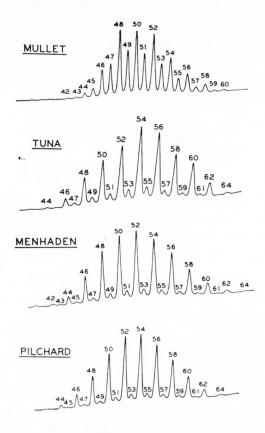

Figure 9.2
Gas chromatograms of fully hydrogenated fish triglycerides on a
1.83-m glass column, 3% JXR on 100/120 mesh Gas Chrom Q, flame
Ionization detector; injector 350°C; detector 320-360 C,
temperature programmed at 4°C/min (210-375°C) (7).

atoms in the acyl chains. In the separation depicted the fat has
been hydrogenated and so any triolein initially present would be
converted to tristearin. In studies of relatively simple mixtures of pure triglycerides some partial separations have been
achieved based upon the degree of unsaturation of triglycerides
with the same carbon number. Similarly triglycerides with
greatly differing acyl chain lengths but the same carbon number
have also been separated. In separations of complex natural
materials, the small degrees of separation of these diverse species results in peak broadening but no useful analytical separation of such closely related species. The state of the art is
that triglycerides with carbon numbers differing by 1-2 can be
separated.

Good analyses of triglycerides require careful attention to
reproducible conditions and the use of standards and correction
factors. Liquid phases that have been used with some degree of
success all have high temperature limits in excess of 300°C and
include some long-chain hydrocarbons as well as silicone polymers,
usually polysiloxanes such as OV-1, SE-30, OV-17, SE-52, and
JXR (8).

9.1.1 Fatty Acid Methyl Esters

The limited applicability of triglyceride separation is compensated by the excellence of fatty acid methyl ester separations by
GLC. Triglycerides are rapidly converted to fatty acid methyl
esters and glycerine by interesterification with sodium or potassium methylate in methanol. Luddy et al. have developed a
simple micromethod in which the methylation reaction takes place
in a capped vial. The esters are usually concentrated and extracted into carbon disulfide, which has a low response when
flame ionization detectors are utilized (9). Separation of a
natural product is depicted in Figure 9.3. Polyester liquid
phases such as the condensation polymer (EGS) formed from ethylene glycol and succinic acid, are the most common materials used
for methyl ester separations. Columns from 1 - 3 m are usually
used, although longer columns, including open tubular columns in
excess of 45 m (11) have been reported. Excessive sorption of
the type noted for triglycerides is not a major problem even at
the 200°C temperature levels utilized in methyl ester separations
tions. The highest degree of accuracy is attained when good
chromatographic procedures including the use of internal standards and response factors are used. Figure 9.4 demonstrates that
the saturated esters are an homologous series whose retention
times on a logarithmic scale, are proportional to the number of

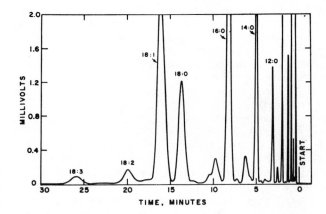

Figure 9.3   Gas chromatogram of methyl esters prepared from
milk fat on a 2.4-m stainless steel column, 25% EGS on 42/60
mesh Chromosorb, thermal conductivity detector, injector 325°C,
detector 225°C, column 200°C(10).

carbons in the species. The monounsaturated esters are a similar
homologous series with longer retention times than the corresponding saturates.

    Excellent separations have been designed on the basis of in-
formation such as that in Figure 9.4. Preliminary identification
of intermediate members of the series has been suggested through
use of this type of figure but structural confirmation, by mass
spectrometry for example, is still required.

    Haken has considered the applicability of "Rohrschneider/
McReynolds constants" for the classification of stationary phases
for the separation of fatty esters (13). He concluded that the
approach was limited since the measurements used to determine the
aforementioned "constants" are made at 100°C and most fatty acid
methyl ester separations are carried out at about 200°C. He had
previously shown significant variation in the, what will now be
called, Rohrschneider/McReynolds coefficients, with temperature
(14). Polar polysiloxanes such as XF-1150 demonstrated greatest
variability in the coefficients and nonpolar types such as
SE-30 demonstrated least variation. Supina pointed out that the
X factor in the McReynolds coefficients should be indicative of
extent of interaction with olefinic substituents (15). Figure
9.5 demonstrates the utility of this approach; the 18:3 and 20:0
methyl esters are used as markers for the consideration of

variation in X. At the lowest X value the 18:3 compound elutes
prior to the 20:0, at intermediate X they overlap and at the
highest X values the 18:3 substituent interacts so strongly with
the stationary phase that it is eluted considerably later than
20:0. Table 9.1 lists commonly used stationary phases for separation of fatty esters and McReynolds coefficients. An interesting new cyanopropyl polysiloxane phase shows promise for the separation of olefinic fatty esters (Figure 9.6) (16). This stationary phase is particularly useful for the study of high unsaturated plant lipids which have been partially hydrogenated. The
figure shows that the cis and trans isomers are well separated
for the most part in the figure and detailed insight into the
composition of, for example, shortenings and margarines may be
obtained. This information of course can be related to properties of these materials and appropriate conditions may be devel-

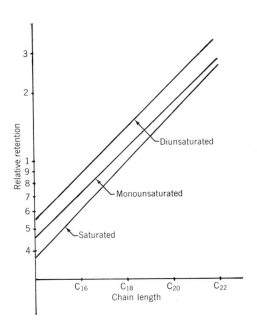

Figure 9.4   Log relative retention of fatty acid methyl esters
vs. chain length. Column: 1.8 m x 0.6 cm o.d., 20% EGS, 42/60
mesh Chromosorb, 223°C (12).

TABLE 9.1   McREYNOLDS COEFFICIENTS OF SOME LIQUID PHASES (13)

		McReynolds Coefficients				
	Composition	X	Y	Z	U	S
Polyesters						
NGA	Neopentyl glycol adipate	234	425	312	402	438
NG Sebacate	Neopentyl glycol sebacate	172	327	225	344	326
NGS	Neopentyl glycol succinate	272	469	366	539	474
BDS	1,4-Butanediol succinate	370	571	448	657	611
EGA	Ethylene glycol adipate	372	576	453	655	617
DEGA	Diethylene glycol adipate	378	603	460	665	658
EGS	Ethylene glycol succinate	451	706	567	904	769
DEGS	Diethylene glycol succinate	496	746	590	837	835
SP-222	Polyester	632	875	733	1000	–

Siloxanes
238-149-99

XF-1150

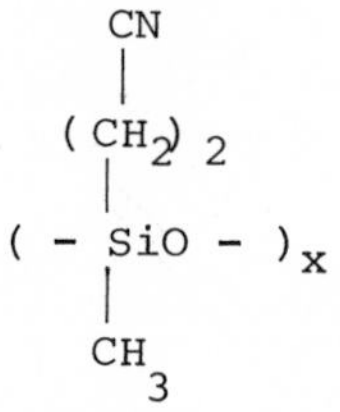

$$\begin{array}{c} CN \\ | \\ (CH_2)_2 \\ | \\ (-SiO-)_x \\ | \\ CH_3 \end{array}$$

292   476   454   705   495

XE-60

$$\begin{array}{cc} & CN \\ & | \\ CH_3 & (CH_2)_2 \\ | & | \\ (-SiO- & SiO-)_x \\ | & | \\ CH_3 & CH_3 \end{array}$$

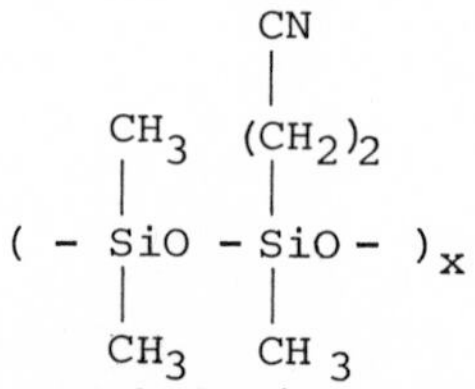

204   381   340   493   367

| EGSS-X | Methyl siloxane modified Low% | 484 | 710 | 585 | 831 | 778 |
| EGSS-Y | Methyl siloxane modified Medium % | 391 | 597 | 493 | 693 | 661 |

SILAR 5CP

SP2300

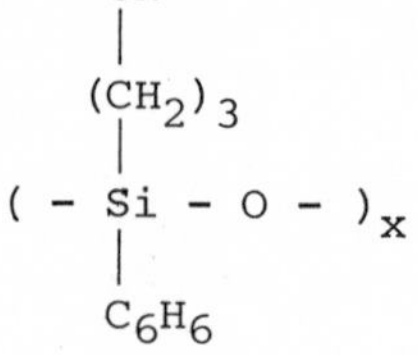

$$\begin{array}{c} CN \\ | \\ (CH_2)_3 \\ | \\ (-Si-O-)_x \\ | \\ C_6H_6 \end{array}$$

316   495   446   637   530

SILAR 7CP SP 2310	70% Groups γ-cyanopropyl 10% Groups phenyl	440	637	605	840	670
SILAR 9CP SP2330	90% Groups γ-cyanopropyl 10% Groups phenyl	489	725	630	913	778

SILAR 10C
SP2340

$$
(- \underset{\underset{\displaystyle CN}{\displaystyle |} }{\overset{\underset{\displaystyle |}{\displaystyle (CH_2)_3}}{\overset{\displaystyle |}{\underset{\displaystyle (CH_2)_3}{Si O}}} } -)_x
$$

		520	757	659	942	800
OV-275		781	1006	885	1177	1089
Carboranes 300 GC	Carborane/methyl siloxane	37	78	113	154	117
400 GC	Carborane/methyl/phenyl siloxane	72	107	118	166	123
410 GC	Carborane/methyl/β- cyanoethyl siloxane	71	286	174	249	171

oped to design fatty materials of desired composition and properties.
   Marine lipids with their diversity of unsaturated and branched
chain acid moieties are a difficult class of materials to analyze.
Ruminants (sheep, goats, cows, etc.) have a bacterial "factory"
in the rumen which is able to produce branched-chain partially-
hydrogenated lipids from ingested plant lipids. These lipids are
incorporated into the milk and meat of the animals and eventually
into animals which feed upon the ruminants. As a rule animal
lipids are highly complex in comparison to plant materials. Although the branched chain materials are usually present in low
concentration when compared to the common fatty acid moieties,
complete description of these fats requires more sophisticated
GC and thus long open tubular columns in tandem with mass spectrometry and computer analysis of the data has become an important approach. Even with a 100-m column, subcutaneous lipids of
barley-fed lambs were so complex that prior fractionation with
urea adducts was necessary (17).
   An excellent interlaboratory analytical study has been used to

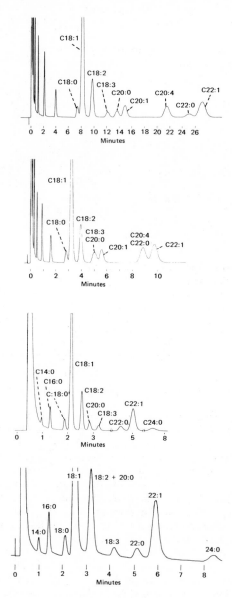

Figure 9.5   Comparison of the X factor in the Rohrschneider/ McReynolds coefficients with retention order of fatty acid methyl esters (15).

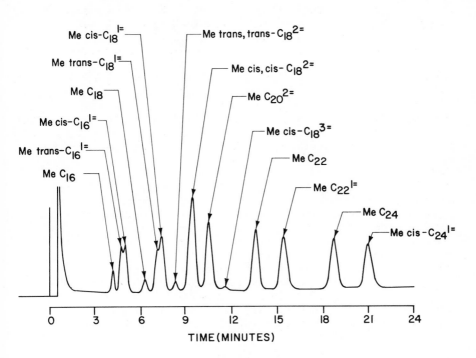

Figure 9.6   Separation of fatty acid methyl esters on a 1.8 m
x 4 mm i.d. glass column packed with 10% Silar 10 C on 100/120
mesh Gas Chrom Q.   Column temperature: programmed from 180° to
200°C at 1°C/min.   Detector: flame ionization at 2×10^{-9}AFS.
Detector temperature: 250°C.   Injection temperature: 250°C.
N_2 Flowrate: 40 cm^3/min (16).   Courtesy of F. Henley and
S. Ramachandran.

demonstrate the agreement between laboratories and the accuracy
and precision achievable by GLC of fatty acid methyl esters (18).
Partly as a result of this study standard mixtures of fatty acid
methyl esters as well as fats and oils have become commercially
available. A researcher who wishes to compare analyses with those
done by the collaborators in the study may: (1) obtain and analyze the standard, (2) evaluate the deviations from the reported
composition for each component in the standard, (3) sum the absolute value of the deviations, and (4) compare the sums with those
found in the collaborative study. If the sum of the deviations
is not greater than 2 times the sum of the deviations reported in

the collaborative study, it is probable that the accuracy of the
experimenter's analyses would be within 2 times the sum of the
deviations found in the study for that particular oil.

The results of analysis of a wide variety of fats and oils has
resulted in the suggestion that international trade in these
commodities be regulated in part by composition determined by GLC
(19). A set of ranges for each fatty acid found in the oils, at
the 0.1% level or greater, has been specified (Table 9.2). In
the proposal the oil in question is analyzed by GLC and the fatty
acid composition in percent is determined. If a particular fatty-
acid value falls outside of the specified range the absolute deviation from the range is added to the deviations of the other
fatty-acid values from their specified ranges. If the arithmetic
sum of the deviations is 2% or less the sample is accepted as
having the claimed identity. If the total deviation is greater
than 2% the composition of the oil is compared with other oils of
proposed composition ranges in an attempt to find the oil(s) it
most nearly resembles. An upper limit of 5% total deviation has
been suggested, above which an oil should not be "identified"
even though its composition may resemble a standard. This approach will probably supplant correlations based on less sophisticated approaches for the identification of oils such as iodine
and saponification numbers (20) and the use of specific reagents
for precipitation of certain types of sterols, followed by microscopic evaluation of crystals (21,22).

9.1.2 Triglycerides

Gas-liquid chromatography has become the standard method for determination of the fatty acid composition of foods; however, the
arrangement of fatty acid moieties in a triglyceride should not
be minimized. Cocoa butter, the fat that gives chocolate its
melting properties, has a fatty acid composition much like mutton
tallow but tallow has a broad melting range extending $20^{\circ}F$ above
human body temperature. In contrast, cocoa butter melts over a
short temperature range and is completely melted at $37^{\circ}C$. The
difference lies in the arrangement and distribution of fatty
acids in the component triglycerides. Cocoa butter consists primarily of triglycerides which contain two saturated and one unsaturated fatty acid per molecule. Mutton tallow has a large
proportion of fully saturated triglycerides, disaturated-mono-
unsaturated glycerides, as well as a significant proportion of
triglycerides with a greater degree of unsaturation. As a result
the "mouth feel" of mutton tallow is vastly different from the
"mouth feel" of cocoa butter.

    Many methods exist to separate triglycerides into fractions on
the basis of their degree of unsaturation. These include fractional crystallization from solvents, and separation by column
and thin layer chromatography (TLC). The classes of triglycerides may then be studied and after methylation the fatty acid
content determined by GLC. Yet even this is not the whole story
since the position of fatty acids on the triglyceride molecule
may uniquely affect the physical and biological properties of the
lipid.

    An ingenious approach for the determination of triglyceride
structure based upon enzymatic specificity has been developed.
Pancreatic lipases, secreted by the human and porcine pancreas,
show specificity for primary ester linkages (23) (Equation 9.1).
The reaction is usually carried out at 37-40°C at pH 8 in the
presence of Ca^{2+}. In some cases bile salts are added to promote
emulsification and removal of fatty acids. The reaction is
quenched prior to the formation of glycerol, by extraction of the
lipids with diethyl ether. Products are chromatographed by preparative TLC using silica gel impregnated with 8% boric acid
which prevents migration of the acyl groups. Each class of compound (di-, or monoglyceride, and fatty acid) may then be converted to the appropriate fatty acid methyl esters and determined
by GLC. If this approach is carried out after preliminary separation of triglyceride classes, virtually the entire triglyceride
composition can be calculated.

$$
\begin{array}{l}
H_2COR_1 \\
HC-OR_2 \\
H_2C-OR_3
\end{array}
\xrightarrow[\text{pancreatic lipase}]{\substack{Ca^{2+} \\ pH\ 8}}
\begin{array}{l}
H_2COR_1 \\
HC-OR_2 \\
H_2C-OH
\end{array}
+
\begin{array}{l}
H_2C-OH \\
HC-OR_2 \\
H_2C-OR_3
\end{array}
\begin{array}{l}
+R_1O- \\
\\
+R_3O-
\end{array}
\longrightarrow \}
$$

$$
\} \longrightarrow
\begin{array}{l}
H_2C-OH \\
HC-OR_2 \\
H_2C-OH
\end{array}
\begin{array}{l}
+R_1O- \\
\\
+R_3O-
\end{array}
\longrightarrow
\begin{array}{l}
H_2C-OH \\
HC-OH \\
H_2C-OH
\end{array}
\begin{array}{l}
+R_1O- \\
+R_2O- \\
+R_3O-
\end{array}
\tag{9.1}
$$

TABLE 9.2 FATTY ACID COMPOSITION OF FATS AND OILS DETERMINED BY GLC[1] (19)

Fatty Acid	Arachis	Cottonseed	Lard and Rendered Pork Fat	Maize	Mustard Seed	Premier Jus and Edible Tallow	Safflower Seed	Sesame Seed	Soybean	Sunflower Seed
C<14	<0.1	<0.1	<0.5	<0.1	<0.5	<0.1	<0.1	<0.1	<0.1	<0.1
C14:0	<0.1	0.5-2.0	0.5-2.5	<0.1	<1.0	1.4-6.3	<1.0	<0.5	<0.5	<0.5
C14:1			<0.2			0.5-1.5				
C15:0			<0.1			0.5-1.0				
C15:ISO			<0.1			<1.5				
C16:0	6.0-15.5	17-29	20-32	8.0-19	0.5-4.5	20-37	2.0-10	7.0-12	7.0-12	3.0-10
C16:1	<1.0	0.5-1.5	1.7-5.0	<0.5	<0.5	0.7-8.8	<0.5	<0.5	<0.5	<1.0
C16:2						<1.0				
C16:ISO			<0.1			<0.5				
C17:0			<0.5			0.5-2.0				
C17:1			<0.5			<1.0				

C17:ISO

C18:0	1.3-6.5	1.0-4.0	5.0-24	0.5-4.0	0.5-2.0	6.0-40	1.0-10	3.5-6.0	2.0-5.5	1.0-10
C18:1	36-72	13-44	35-62	19-50	8.0-23	26-50	7.0-42	35-50	19-30	14-65
C18:2	13-45	33-58	3.0-16	34-62	10-24	0.5-5.0	55-81	35-50	48-58	20-75
C18:3	<1.0	0.1-2.1	<1.5	<2.0	6.0-18	<2.5	<1.0	<1.0	4-10	<0.7
C20:0	1.0-2.5	<0.5	<1.0	<1.0	<1.5	<0.5	<0.5	<1.0	<1.0	<1.0
C20:1	0.5-2.1	<0.5	<1.0	<0.5	5.0-13	<0.5	<0.5	<0.5	<1.0	<0.5
C20:2			<1.0		<1.0					
C20:4			<1.0			<0.5				
C22:0	1.5-4.8	<0.5	<0.1	<0.5	0.2-2.5		<0.5	<0.5	<0.5	<1.0
C22:1	<0.1	<0.5			22-50					<0.5
C22:2					<1.0					
C24:0	1.0-2.5			<0.5	<0.5					<0.5
C24:1					0.5-2.5					<0.5

[1]These ranges, tentatively adopted by the FAO/WHO Codex Alimentarius Committee on Fats and Oils, refer to typical commercial samples of bona fide fats and oils. A range of <0.1% indicates that the fatty acid is not normally present in a quantifiable amount, whereas a blank indicates that the fatty acid is not normally present.

9.1.3 Phospholipids

The phosphoric acid esters of diacyl glycerides, phospholipids,
are important constituents of cellular membranes. Lecithins
(phosphatidyl cholines) from egg white or soybeans are often added to foods as emulsifying agents or to modify flow characteristics and viscosity. Phospholipids have very low vapor pressures and decompose at elevated temperatures. The strategy for
analysis involves preliminary isolation of the class, for example
by TLC, followed by enzymatic hydrolysis, derivatization of the
hydrolysis products, and then GC of the volatile derivatives. A
number of phospholipases are known which are highly specific for
particular positions on phospholipids. Phospholipase A_2, usually
isolated from snake venom, selectively hydrolyzes the 2-acyl
ester linkage. The positions of attack for phospholipases A_1, C,
and D are summarized on Figure 9.7 (24). Appropriate use of
phospholipases followed by GC can thus be used to determine the
composition of phospholipids.

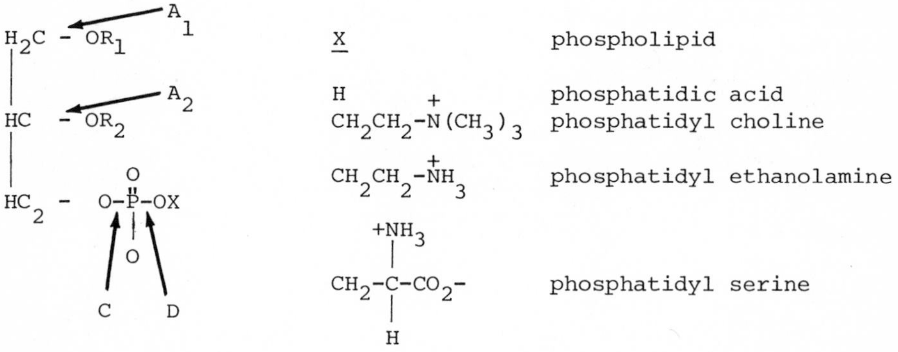

	X	phospholipid
	H	phosphatidic acid
	$CH_2CH_2-\overset{+}{N}(CH_3)_3$	phosphatidyl choline
	$CH_2CH_2-\overset{+}{N}H_3$	phosphatidyl ethanolamine
	$CH_2-\underset{H}{\overset{+NH_3}{C}}-CO_2-$	phosphatidyl serine

Figure 9.7

In some cases separation of lecithins from phosphatidyl ethan-
olamines and phosphatidyl serines is carried out prior to enzymatic hydrolysis. Composition of the phospholipids is virtually
always determined finally by GLC of their fatty acid methyl
esters.

9.1.4 Trends

GLC of fatty acid methyl esters is firmly established as the major method in the analysis of lipids. Two recent issues of the Journal of Chromatographic Science are devoted to the analysis of fatty acids and their esters, in which a variety of techniques are discussed (25). There may be further expansion in industrial specification of fats and oils by compositional criteria rather than statement of source of the lipid. This will likely be of major significance in international trade. In order for triglyceride separations by GLC to be more useful, columns of greater selectivity, efficiency, and thermal stability than are currently available will be required. High-performance liquid chromatography may provide an alternate approach to the separation of triglycerides and phospholipids, although a major problem in this approach is the lack of a detector that is sensitive to these lipids but not their solvents. Separation of phenacyl esters of fatty acids by liquid chromatography utilizing a bound-hydrocarbon stationary phase and acetonitrile as a mobile phase has recently been reported (26). The phenacyl group acts as the ultraviolet -absorbing chromophore and remarkable separations are achieved on a 90-cm column. Tandem systems such as liquid chromatography followed by automatic injection of specific peaks into gas chromatographic columns may also prove useful. Microreactors in the tandem system may also be desirable for enzymatic hydrolysis of the lipids and methylation prior to GLC. At the present time GLC is an important tool for the analysis of lipid composition but not sufficient because of the complexity of lipids in foods.

9.2   PROTEINS

Although there are only about 20 commonly found amino acids, the resulting proteins are vast in scope. For linear peptides with 500 amino acid residues, 20^{500} different proteins are possible. Peptides of significance range from a dimer of aspartic acid and phenylalanine which has remarkable sweetening properties through bradykinin, a nonapeptide which is a smooth-muscle hypotensive agent, insulin (51 residues), the recently synthesized enzyme ribonuclease (124 residues), hemoglobin (574 residues), to large proteins such as myosin (3600 residues), or yeast fatty acid synthetase (20,000 residues with a molecular weight of 2,300,000) (24).

High molecular weight and extensive interaction of hydrogen-bonding and hydrophobic-binding sites of proteins with their environment results in negligible vapor pressure of these compounds. Protein isolates must be degraded to their component amino acids by treatment with 6N HCl at $105^\circ C$ for 24 hr. Most well-dispersed proteins can be hydrolyzed by such treatment. Particularly sensitive amino acid residues such as tryptophan are destroyed and other residues, serine and threonine, may be partially destroyed by these conditions. Improved tryptophan stability has been observed when 3-(3-indolyl) propionic acid was present in the HCl (27). Use of 3N p-toluene sulfonic acid with traces of 3-(2-aminoethyl) indole has been suggested for hydrolysis, and tryptophan stability in this system was noted (28). Glutamine and asparagine are converted to NH_3 plus glutamic and aspartic acids, respectively. There are some examples of incomplete hydrolysis; for example, the carboxyl groups of valine, leucine, and isoleucine are often sterically hindered and may require longer treatment. Some of the errors due to partial destruction may be corrected by the use of empirical factors. Basic hydrolysis with 4N NaOH at $100^\circ C$ for 8 hr is sometimes used for analytical release of tryptophan but this treatment results in decomposition of other amino acids (24,29).

The greatest activity in amino acid analysis of proteins has involved ion exchange resins and liquid, rather than gas, chromatography. Most large medical and research centers have at least one commercial amino acid analyzer at costs that range up to $60,000 for a system with a dedicated minicomputer. Separations are achieved with a series of buffer solutions containing sodium citrate in which the concentration and pH are increased. The eluted amino acids are treated with ninhydrin, in many systems, to provide derivatives which may be quantitated spectrometrically. Analyses typically take a few hours. The expense of instrumentation, inconvenience of preparation and storage of buffer and reagent solutions, and relatively long analysis time are "powerful stimulants" for research on gas chromatographic methods for the determination of amino acid mixtures (30-32). There is considerable effort being expended on this approach, yet no gas chromatographic method has been widely accepted. Each of the methods requires hydrolysis of proteins, sample clean-up, some isolation of the amino acids, and derivatization of the hydrogen-bonding amine, and carboxyl groups (33-36).

Recent approaches to the amino acid analysis of foods require maceration of the sample, hydrolysis of the proteins with HCl, and filtration. The resulting material may still include proteins (and fragmented peptides), carbohydrates, salts, urea, and lipids. The solution is then passed through a small column of cation exchanger (with a nominal cross-linking of at least 8%) in

the H^+ form. The cation exchanger sorbs the amino acids, some
peptides, and the cations in the mixture. The amino acids and
peptides are eluted from the resin by 2M NH_3 solution, although
cations are retained. The solution is evaporated to dryness and
the samples derivatized (37).

9.2.1 Volatile Derivatives

Popular derivatives of the carboxyl groups of the amino acids are
currently n-butyl (38-40) or n-propyl esters although other esters have been suggested. Esterification reagents are prepared by
distillation of HCl from concentrated H_2SO_4 into n-butanol until
it is 3M in HCl, or n-propanol until it is 8M in HCl (Equation
9.2). The dried amino acid mixture is placed in a small screw
cap vial, the esterification reagent added, and dimethoxypropane
is added to scavenge traces of water present or formed during reaction. Heating at $110^\circ C$ for 20 min usually suffices and the
vial is opened and excess reagent is evaporated.

$$(9.2)$$

Acetylation of amino groups is necessary to promote volatility and trifluoroacetyl derivatives (N-TFA) formed from trifluoroacetic anhydride or acyl derivatives formed from acetic anhydride have both been used with success (Table 9.3). Excess reagent must be removed with care at this point since the derivatives are highly volatile. Gehrke and co-workers have studied a
number of stationary phases in detail as summarized in Figure 9.8
(41). They have suggested the use of Apiezon M and a rapid GLC

separation of the N-TFA n-butyl esters of amino acids derived
from a hydrolyzed sample of the enzyme lysozyme as shown in Figure 9.9 There are three regions of significant overlap: the
threonine-alanine-glycine-serine region at 15 min, the leucine-
isoleucine pair at 20 min, and the phenylalanine-lysine-tyrosine
region near 30 min. In addition there are two peaks corresponding to histidine, the first overlaps the aspartic acid ester, and
the second occurs at about 32 min.

TABLE 9.3 COMMON DERIVATIVES OF AMINO ACIDS USED IN GLC

Amine Derivative	Carboxyl Derivative	Reference	Stationary Phase
$CF_3-\overset{\overset{O}{\|}}{C}$	$CH_3CH_2CH_3CH_2O-$	41	10% Apiezon M (See Figs. 9.8, 9.9)
		42	Tabsorb (TA-33)
		43	0.3% EGA
Histidine imidazole as	$C_2H_5O\overset{\overset{O}{\|}}{C}-$	44	1% OV-210
	$CH_3CH_2-\overset{\overset{H}{\|}}{\underset{\underset{CH_3}{\|}}{C}}O-$	45	(Dexsil 400 GC Ucon 75H90,000)
$\overset{\overset{O}{\|}}{CH_3C}$	CH_3O-	31,46,47	1%(XE-60, QF-1, MS-200)
	$CH_3CH_2CH_2O-$	48	(0.3% Carbowax 20M, 0.3% Silar 5CP, 0.06% Lexan)
$CF_3CF_2CF_2-\overset{\overset{O}{\|}}{C}-$	$(CH_3)_2CHCH_2CH_2-O-$	49	3% SE-30
	$(CH_3)_2CHCH_2O-$	50	3% SE-30
	$CH_3CH_2CH_2O-$	30	3% SE-30
$(CH_3)_3Si-$	$(CH_3)_3Si-O$	51	
	(structure)	52	5% SE-30

Reproducibility and quantitation obtained with 0.5 and 1.0 μg
of each amino acid derivative chromatographed is reported in

Table 9.4.   In the regions of overlapping solutes, glycine, and
lysine have low relative weight responses.   In addition, methionine, ornithine, histidine, arginine, tryptophan, and cysteine
have significantly low responses.

TABLE 9.4   RELATIVE WEIGHT RESPONSES (RWR)[1] OF AMINO ACID N-TFA
             n-BUTYL ESTERS ON 10% APIEZON M[2] (41)

Amino Acid	RWR					
	0.5 µg[3]		1.0 µg[3]		Av.	RSD(%)
Alanine	0.944	0.946	0.946	0.945	0.945	0.11
Threonine	0.915	0.916	0.909	0.911	0.913	0.36
Glycine	0.836	0.848	0.878	0.872	0.859	2.31
Serine	0.961	0.939	0.938	0.943	0.945	1.34
Valine	1.018	1.019	1.014	1.014	1.016	0.26
Leucine	0.985	0.983	0.977	0.981	0.982	0.36
Isoleucine	1.017	1.019	1.023	1.019	1.019	0.26
Hydroxyproline	0.948	0.949	0.950	0.952	0.950	0.18
Proline	1.000	1.000	1.000	1.000	1.000	−
Methionine	0.713	0.711	0.723	0.720	0.717	0.78
Ornithine	0.680	0.687	0.718	0.714	0.700	2.71
Histidine	0.329	0.335	0.358	0.348	0.342	3.83
Aspartic acid	1.140	1.122	1.110	1.106	1.120	1.37
Phenylalanine	1.102	1.106	1.101	1.103	1.103	0.20
Lysine	0.798	0.812	0.859	0.854	0.831	3.66
Tyrosine	0.960	0.947	0.950	0.954	0.953	0.59
Glutamic acid	1.065	1.060	1.101	1.102	1.082	2.09
Arginine	0.410	0.431	0.450	0.431	0.431	3.78
Tranexamic acid	1.062	1.056	1.089	1.100	1.077	1.96
Tryptophan	0.625	0.634	0.625	0.619	0.626	0.99
Cystine	0.261	0.243	0.238	0.260	0.251	4.70

[1]RWR = (peak area of amino acid/weight of amino acid taken)/
(peak area of proline/weight of proline taken).
[2]2.5 m x 2 mm glass column, 80-100 mesh Chromosorb W.
[3]Weight of each amino acid taken.

Histidine forms a diacyl derivative that is apparently unstable.   Moodie has suggested the use of ethoxyformic anhydride
to form the stable N^a-trifluoroacetyl-N^{im}-carbethoxy n-butyl
histidinate with no accompanying unstable by-products (44).

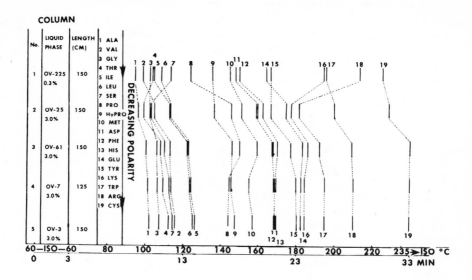

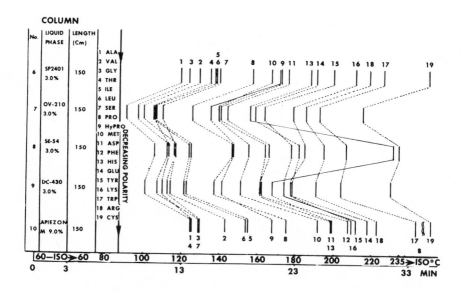

Figure 9.8   Effect of liquid phase polarity on the separation of the N-TFA n-butyl esters of amino acids. Instrumental conditions: initial temperature 60°, delay 3 min, 6°/min, and final temperature 235°C. Nitrogen flowrate: 50 cm³/min (41).

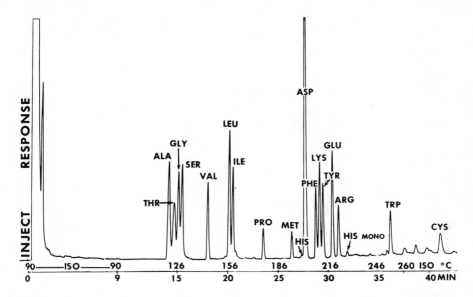

Figure 9.9   Single-column GLC analysis of the N-TFA n-butyl
esters of lysozyme, with cleanup. Column: 10% (w/w) Apiezon M on
80-100 mesh HP Chromosorb W, 2.5 m x 2 mm i.d. glass. Sample:
6.0 mg lysozyme. Hydrolysis: 5 cm^3 6N HCl with 5% TGA, 21 hr,
110°C. Instrumental conditions: initial temperature 60°C, delay
9 min, 6°/min, and final temperature 260°C (41).

$$(9.3)$$

N-trifluoroacetyl amino acids have been reported as stable at
-10°C over a 2-week period (53) and the TFA n-butyl esters are
stable for about 72 hr when stored at 0°C in a nitrogen atmosphere (32). Reasonable stability of derivatives is necessary for
repetitive analyses when many samples are prepared and stored
prior to chromatography. Instability of silyl derivatives to
hydrolysis has been a major problem which has led to decreased
activity in research on these materials. The efficient treatment
of amino acids with N,O-bis(trimethylsilyl)acetamide (51) and
subsequent GLC of the derivatives is now used only occasionally
as a result of instability of the derivatives.

$$
\begin{array}{c}
\underset{\substack{|\\ NH_2}}{\overset{\substack{H\quad\ \overset{O}{\overset{\|}{}}\\}}{RC - C - OH}} + CH_3C\!\!\nearrow^{OSi(CH_3)_3}_{\searrow NSi(CH_3)_3}
\end{array}
$$

(9.4)

$$
R - \underset{\substack{|\\ H_xN(Si(CH_3)_3)_{2-x}}}{\overset{\substack{H\quad\ O\\ |\quad\ \|}}{C - C}} - OSi(CH_3)_3 \qquad ; \ x = 1,0
$$

    Trends towards fortification of foods with synthesized amino
acids, which are either mixed with or bound to proteins, have resulted in techniques for the chromatographic analysis of amino
acid optical isomers. Biological consequences of incorporation
of significant quantities of D-amino acids are unknown although
mixtures of D- and L-amino acids and D-amino acids bound to
L-peptides have properties which differ from pure "L systems."
Searches for amino acids of particular stereochemistry in meteor-
ites (45), samples of Lunar and Martian soils and marine sedi-
ments (54) have been part of the U. S. space program. Presumably,
the presence of optically active amino acids would support the
hypothesis of biogenetic origins of these materials. Techniques
for the separations have taken two directions. Gil-Av and associates have synthesized optically active stationary phases such as
carbonyl bis(N-L-valine isopropyl ester) (55). The materials
have been used for the GLC separation of the TFA esters of the
amino acids in question. The alternate approach utilizes GLC
separation of diasteriomers such as TFA-D-2-butyl esters (45).
Iwase has used TFA-L-prolyl (56) and TFA-L-hydroxyprolyl derivatives (57) for the separation of diasteriomers. Separation of
about 200 N-perfluoroacyl dipeptide esters was studied in detail
(58). It is clear that the aforementioned dipeptide derivatives

and some N-acetyl permethylated tripeptides have sufficient volatility and stability for GLC (59). The utility of such an approach is limited since 20 amino acids could form 8000 tripeptides
and a separation designed for all of these materials would be
prohibitive in time, if achievable. Determination of the primary
structure of a protein fortunately does not require the separation of peptides.

9.2.2 Protein Sequence Determination

GLC is an important adjunct to protein sequence determination.
Automatic "sequenators" based upon the approach developed by
Edman are available and have been described in detail by Niall
(60). The Edman degradation, summarized in Equation 9.5, makes
use of methyl or phenylisothiocyanate which reacts with the
N-terminus of a peptide. Exposure of the isothiocyanate derivative of the protein to acid results in cleavage of the terminal
amino acid as a thiaxolinones and exposure of the next amine
group on the peptide. Thus, the process can be repetitively
carried out, each amino acid removed from the peptide, in a
sequential manner. Thiazolinones rearrange in acid medium to
form thiohydantoin derivatives of amino acids, some of which may
be directly gas chromatographed; others must be derivatized typically as trimethylsilyl derivatives.

Separation of the phenylthiohydantoins has not been as straightforward as is desired. Pisano and Bronzert have grouped the
phenylthiohydantoin derivatives of amino acids on the basis of
their gas chromatographic characteristics (61). The derivatives
of alanine, glycine, valine, leucine, isoleucine, methionine,
proline, and phenylalanine may be gas chromatographed directly.
Phenylthiohydantoins of aspartic acid, s-carboxymethyl cysteine,
cysteic acid, glutamic acid, lysine, serine, and threonine must
be silylated prior to chromatography. Asparagine, glutamine,
tyrosine, tryptophan, and histidine phenylthiohydantoins are relatively   and lead to tailing peaks. Current practice
utilizes a single gas chromatograph with two columns. PTH derivatives formed in each step of the protein sequence determination
are chromatographed on a DC-560 (a chlorophenylmethyl polysiloxane now considered obsolete) column. Silyl derivatives are then
prepared and injected onto the column. Histidine is an exception
which must be removed as an aqueous extract, dried, and derivatized. The sequence determination is further complicated by
multiple silyl derivatives of some of the PTH derivatives which
leads to multiple peaks for some components. If each step in the
Edman degradation went to absolute completion this would be of

N-terminal amino acids   (9.5)

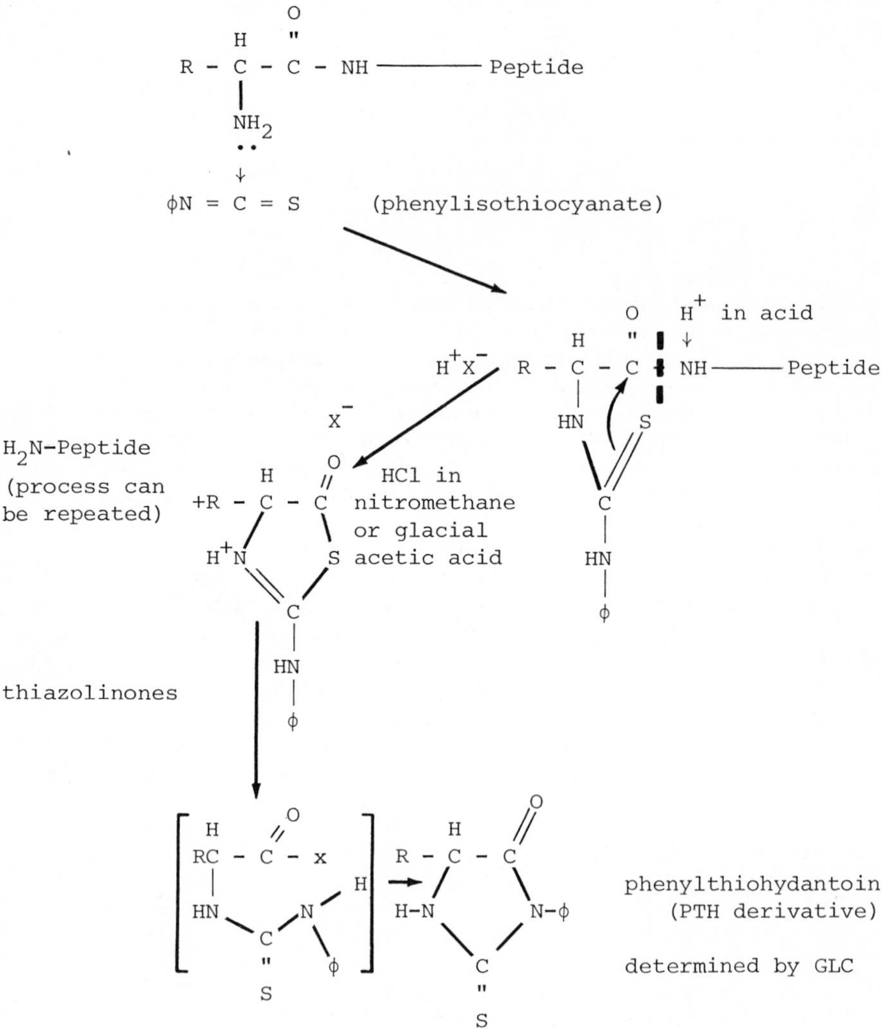

$$R - \underset{\underset{\overset{\cdot\cdot}{NH_2}}{|}}{\overset{\overset{H}{|}}{C}} - \overset{\overset{O}{\overset{\|}{}}}{C} - NH \text{————— Peptide}$$

$$\downarrow$$

$\phi N = C = S$ (phenylisothiocyanate)

H^+X^- $R - \underset{\underset{S}{\overset{\|}{C}}\underset{\overset{|}{HN}}{}}{\overset{\overset{H}{|}}{C}} - \overset{\overset{O}{\|}}{C} \vdots NH$ ———— Peptide $\overset{H^+ \text{ in acid}}{\downarrow}$

X^-

H_2N-Peptide
(process can
be repeated)

$+R - \underset{\underset{\overset{|}{HN}}{}}{\overset{\overset{H}{|}}{C}} - \overset{\overset{O}{\overset{\|}{}}}{C}$ HCl in
nitromethane
or glacial
acetic acid

thiazolinones

phenylthiohydantoin
(PTH derivative)

determined by GLC

Sequence Determination by the Edman Degradation.

relatively minor consequence. Unfortunately, each step results
in a number of amino acids that ought to be, but are not cleaved.
After a few steps amino acid derivatives that should have been
removed in prior steps are found in the reaction products. Then
questions as to which PTH derivative is formed in a particular
step must be resolved by quantitative determination and reference
to all prior steps in the sequence. Multiple peaks for amino
acids, incomplete derivatization, and degradation lead to a situation of "chemical noise" in which the information is confusing
and equivocal.

Eyem and Sjöquist have suggested the use of short (4.5 m)
glass capillary columns and reported the separation of 19 of 20
silylated methyl thiohydantoin (MTH) derivatives of the amino
acids (62). The histidine derivative can be separated on the
same column by starting at a higher temperature. Separation of
some of the silylated PTH derivatives was also demonstrated,
though not in as much detail as the MTH derivatives.

The "chemical noise" problem requires the breakdown of proteins into polypeptides through the use of enzymes of known specificity. For example, trypsin is known to cleave proteins at
lysine or asparagine residues; peptides thus produced may be sequenced and a "puzzle" is thus produced which must be solved by
laying sequences over one another and the use of chemical insight
to fit the pieces together.

A complete sequence determination might proceed in the following fashion. The number of N-terminal groups is determined by
derivatization with 1-fluoro-2,4-dinitrobenzene (Sanger's reagent) followed by protein hydrolysis and determination of the derivatized terminal end groups. Sulfide cross links are destroyed
by reagents such as mercaptoethanol or dithiothreitol. The resultant peptides are isolated and sequences for each are determined as far along the chain as practicable. If required, a
fraction of each peptide may be enzymatically degraded, the resulting smaller peptides sequenced, and repeating units on each
fragment located and correlated with the proteolytic specificity
of the enzymes utilized. For small biologically active peptides
synthesis may be achieved and biological activity evaluated.
With large molecules X-ray crystallographic structure determination may be desirable prior to synthetic activities. Incredibly,
ribonuclease, insulin, and a number of smaller peptides have been
synthesized.

An alternate approach to this sequencing method involves the
use of the proteolytic enzymes, dipeptidylaminopeptidases (63).
Dipeptidylaminopeptidase I (DAPI) cannot hydrolyze peptide bonds
that contain a proline residue or an N-terminal arginine or lysine residue. A related enzyme, dipeptidylaminopeptidase IV
(DAPIV), hydrolyzes peptides from the N-terminus when the next-

to-last residue is proline or terminal lysine or arginine residues. However, it does not hydrolyze the peptide if proline is
the third residue from the N-terminus. A mixture of the two de-
grades a protein into a mixture of dipeptides until the point
when a proline residue third from the end is reached. Hydroly-
sates are converted to methyl esters and then the N,O-perfluoro-
propionyl derivatives. The dipeptide derivatives are separated
by GLC and identified by mass spectroscopy.

9.2.3 Trends

It seems that the limiting factor in acceptance of GLC for amino
acid analysis is the derivatization procedure rather than the
chromatography. The significant advantages in equipment costs,
broad applicability of the chromatographic equipment to separations other than amino acids, high sensitivity of the detector
systems, potential for tandem mass spectrometry, and avoidance of
buffer solutions are among the important aspects of GLC approaches. Lack of a totally suitable derivatization procedure and
the many expensive liquid chromatographic (ion-exchange based)
amino acid analyzers already in use will retard the probability
of purchase of new equipment. As older systems become "senile"
and as microcomputers are introduced with gas chromatographic
systems there may be a shift away from liquid chromatographic
amino acid analysis.
   Oddly, one area in which GLC is used for separation of amino
acid derivatives, appears to have great potential for applications of liquid chromatography. The phenylhydantoins produced in
sequence determination have ultraviolet absorbance and are effi-
ciently separable by high-performance liquid chromatography (64).
Their GLC separations are less than optimum and it is likely that
new developments in this area will include liquid chromatographic
separations.

9.3   CARBOHYDRATES

The presence of carbohydrates in all cells, similarities in
structure of the monomeric sugar units, and the significant functional differences in the polymeric species have been a challenge
to all researchers concerned with these materials. Carbohydrates
are of great abundance and play many roles in biology from protective coats and structural materials to primary energy sources

of cells. Monomeric and dimeric species are usually soluble in
water but higher oligomers and polymers often are not. There are
a variety of enzyme systems in animals which are used to break
down polysaccharides, such as starch, into soluble, biologically
transportable forms. Glucose, the monomeric unit in starch also
forms the basic repeating unit of cellulose which is undigestible
to most animals due to the β-linkage, although symbiotic micro-
organisms assist in some species. Additionally, there is considerable current interest in the health aspects of undigestible
fiber, bulk in human diets. There has been much work to determine the monomeric composition, relate this to structure, biological properties, and function. There are also mono- and oligo-
saccharide materials in the diet, the consequences of which have
not yet been fully evaluated.

This section is concerned with isolation of carbohydrates,
methods used to degrade carbohydrates to their primary structural
units, derivative formation and GC of the volatile derivatives.

Isolation of simple sugars from food is usually straightforward and involves grinding in a blender, dialysis, passage
through anion and cation exchangers followed by freeze-drying
(65). Separation of polymers may be as simple as pressing of
tissues to isolate structural materials or as complex as the acid
sulfite process for pulping paper which is a cellulose isolation
process. It is usually desirable to limit enzymatic destruction
of carbohydrates in the food by denaturation or removal of proteins early in the work-up. This may be accomplished by precipi-
tants such as $(NH_4)_2SO_4$ or solvents used to extract the lipid
materials. The fatfree extracts may be treated with proteases,
solubles dialyzed out, and the carbohydrate isolated for further
study. A recent review has cited four types of information for
further characterization of the polymeric carbohydrates (66).
These are determination of molecular weight, type of chain (linear or branched), nature of the glycosidic bond, structure of the
individual units. The last of these is determined after degradation of the polymer through the action of strong, though dilute
acids (1-2N) to prevent decomposition of the monomers. The molecular species may be determined readily as described in subsequent sections. Under some conditions selectivity of degradation
may be achieved due to differing chemical lability of specific
types of bonds and of some structural information inferred (67).
A particularly attractive approach is summarized in Equation 9.6.
Methylation converts any free hydroxyl groups to methyl ethers
and the methylated carbohydrates are degraded by acid to mono-
and oligomers. The free hydroxyl groups are then chemically
modified, as trimethylsilyl ethers for example, the products determined by GLC, mass spectrometry, NMR, and other chemical or
physical methods (68). In this manner the linkages (1, 3 in the

(9.6)

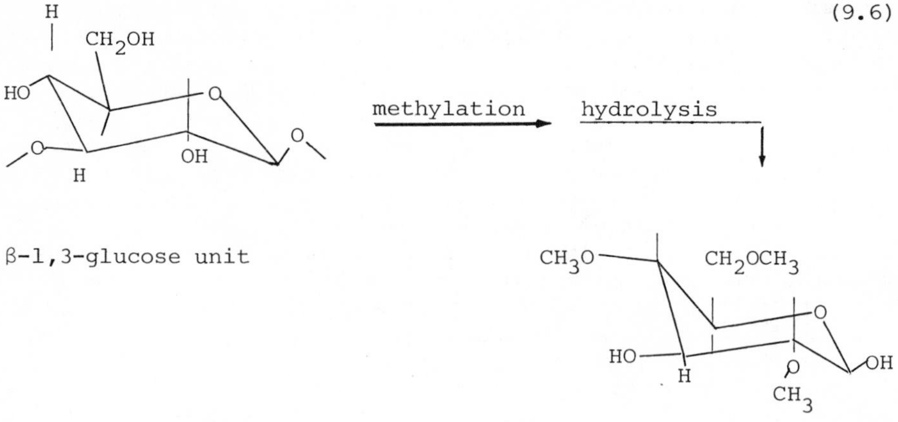

β-1,3-glucose unit

2,4,6-trimethylglucose

figure) may be determined. If the degradation is carried out in
a mixture of acetic acid, acetic anhydride, and sulfuric acids,
the previously unreacted hydroxyl groups are converted to acetyl
esters. These may be separated chromatographically and the position of linkage and numbers of cross links determined from the
relationships of methyl ether groups to acetyl groups. The specificity of enzymes is used to develop information concerning the
nature of linkages in the polymers. For example, cellulases are
specific for the β linkage in their substrate cellulose, and will
not hydrolyze the α linkages in the related glucose polymer
starch. In analogy to enzymatic studies in the foregoing sections, polysaccharides may be degraded and the products of degradation evaluated.

Even isolated mono- and disaccharides are capable of extensive
hydrogen bonding, which results in extremely low vapor pressures.
Attempts at vaporization of these materials results in decomposition. Volatile derivatives must be prepared and the approaches
are similar to those described for amino acids. The most common
derivatives are methyl or trimethylsilyl ethers and acetyl or
trifluoroacetyl esters. A highly detailed review of applications
of GLC to carbohydrates including over 1000 references concerning
derivatization and types of columns used has been written by
Dutton (69,70).

9.3.1 Methyl Ethers

Fully methylated sugars have high volatility and even oligomers, such as trisaccharides, have been separated from one another (71). The Hakomori method is most often reported and utilizes a solution of sodium hydride in dimethylsulfoxide (72). A methylsulfinylmethyl anion is generated in the reagent which abstracts protons from the hydroxyl groups of the sugars (73). The reagent is added to dried samples of carbohydrates in screw-capped vials, and after reaction for 30 min small increments of methyl iodide are added (Equation 9.7). The reaction is quenched by addition of water, and the permethylated compounds extracted into benzene and chromatographed.

$$R(OH)_x + NaH \xrightarrow[\underset{\displaystyle CH_3SCH_2^-}{\overset{\displaystyle O}{\|}}]{\overset{\displaystyle O}{\overset{\displaystyle \|}{CH_3SCH_3}}} R(ONa)_x \xrightarrow{CH_3I} R(OCH_3)_x \qquad (9.7)$$

The procedure is limited to materials that dissolve, or in the case of polymers, which swell in dimethylsulfoxide. Jones has reviewed the GLC of the methyl ethers and pointed out that the free reducing methyl sugars adsorb on columns and are thus of limited utility (74). Methods to avoid this problem will be discussed in section 9.3.3. It has also been pointed out that mixed ether acetyl ester derivatives are useful for the determination of bonding in polymeric carbohydrates. This also will be treated in section 9.3.3.

9.3.2 Silyl Ethers

There are a great number of silyl compounds capable of reacting readily with hydroxyl groups of carbohydrates; Pierce has published an extensive monograph describing silylating agents (75), and Birch has reviewed their applications in GLC of carbohydrates (76). Ease of preparation of derivatives, their thermal stability, and toleration of traces of water in the reaction mixture has made trimethylsilyl ethers popular derivatives for GLC of food components (77-79). A number of the most common reactants are summarized in Equation 9.8. The first two reagents, hexamethyldisilazane (HMDS) and trimethylchlorosilane (TMSCl) were used as a mixture 2:1 (w/w) by Sweeley et al. in one of the early papers in the field. They reported that HMDS alone resulted in almost no reaction and that 50% of the maximum yield resulted from the use of TMSCl alone. Reaction of a mixture of the two

(9.8)

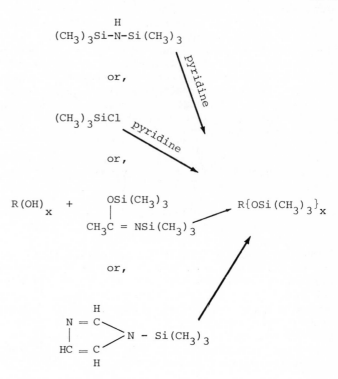

reagents in pyridine with methyl-α-glucopyranoside for 5 min at
room temperature resulted in a 90% yield of methyl (tetra-O-tri-
methylsilyl)-α glucopyranoside (80). Traces of water in these
systems may result in mixtures of partially trimethylsilylated
derivatives and multiple peaks which can be avoided if the reaction is carried out in a mixture of N,N-dimethylformamide and
dimethylsulfoxide (81,82).

N,O-bis(trimethylsilyl) acetamide mixed with HMDS and trimethylsilyl amine have been suggested as a reagent for precolumn
microreaction. An aqueous solution of the carbohydrate is injected into the microreactor followed by the silylating reagent
mixture, and the products are automatically passed onto the column (83). Another silylating agent depicted in Equation 9.8,
N-(trimethylsilyl)imidazole, is particularly useful in the presence of water and reacts virtually quantitatively with glucose
in 50% water solutions (75). Dimethylsilyl ethers generated from
$(CH_3)_2SiHCl$ have also been suggested and provide greater

volatility due to their lower molecular weights (81). The trimethylsilyl ethers, although of higher molecular weight and lower
vapor pressure than the methyl esters, have sufficient volatility
so that di-, and trisaccharides have been separated by GLC at
210°C (Figure 9.10). The relatively nonpolar stationary phase
SE-52 serves well to separate sugar ethers of widely differing
sizes or properties but a more polar stationary phase such as EGS
is needed to separate more closely related ethers (Figure 9.11).

Examination of Figure 9.11 demonstrates an excellent separation of α-glucose from β-glucose which points out the power of GLC
for the separation of these materials as well as a significant
complication in the study of carbohydrates. It is well known
that solutions of some carbohydrates undergo anomerization and
that an initially pure form of a sugar may result in an equilibrium mixture (mutarotation) as in Figure 9.12.

TMS derivatives are relatively stable to anomerization compared to the original sugars and thus derivatization "traps" the
anomers. Although four peaks per hexose may be formed, two are
usually seen and these correspond to the α and β anomers at the
primary carbon atom. Figure 9.13 is a typical gas chromatogram
and the difficulties introduced by multiple peaks become obvious
(82). It is more demanding to develop complete separations in
such cases; quantitation requires matching of appropriate peaks
and addition of their areas sometimes with differing response
factors for each peak. In the early work of Sweeley et al. (80)
it was demonstrated that for some sugars such as glucose and
xylose refluxing in pyridine did not markedly change the composition of the initial anomeric forms (less than 5% change); however,

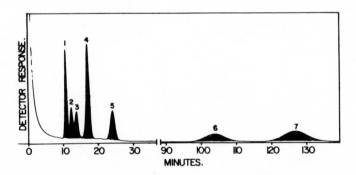

Figure 9.10   GLC of di- and trisaccharides as TMS ethers.
Column: 3% SE-52, 1.8 m x 0.64 cm o.d., 210°C (isothermal).
Compounds: 1 = sucrose, 2 = α-maltose, 3 = β-maltose, 4 = β-
cellobiose, 5 = gentiobiose, 6 = raffinose, 7 = melezitose (80).

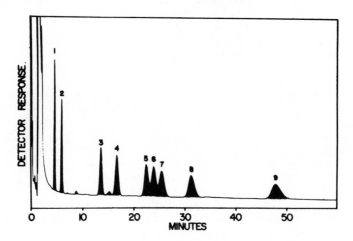

Figure 9.11.   GLC of monosaccharides as TMS ethers. Column:
15% EGS, 2.4 m x 0.64 cm o.d., 150°C(isothermal). Compounds:
1 = α-rhamnose, 2 = α-fucose, 3 = fructose, 4 = sorbose, 5 = β-
allose, 6 = α-glucose, 7 = α-galactose, 8 = β-mannose, 9 = β-
glucose (80).

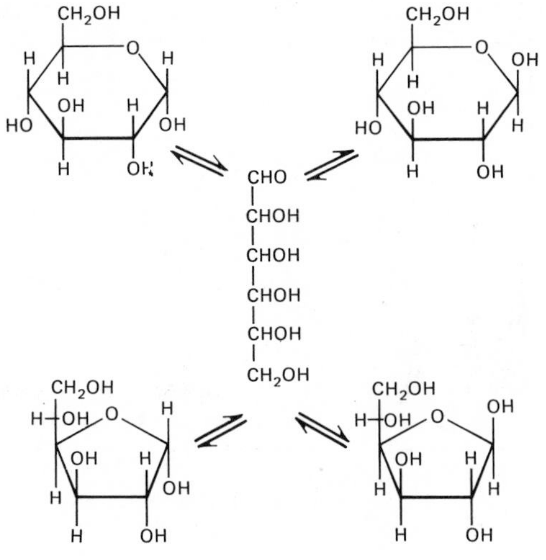

Figure 9.12.   Anomerization (mutarotation) of sugars into
complex mixtures.

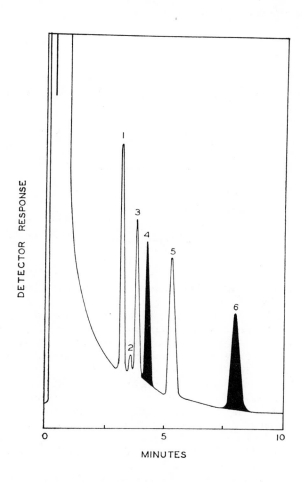

Figure 9.13.   GLC of equilibrated mixture of TMS ethers of
galactose, glucose, and methyl α-D-mannopyranoside (2:2:1,w/w).
Column: 25% Carbowax 20 M on Chromosorb W 80/100 mesh, 1.8 m x
2 mm i.d., injector 220°, column 145°C.   Compounds: 1 - methyl
α-D-mannopyranoside, 2 = furanose form of galactose (unconfirmed),
3 = α-D-galactopyranose, 4 = α-D-glucopyranose, 5 = β-D-
galactopyranose, 6 = β-D-glucopyranose (85).

the anomeric composition of arabinose, xylose, and galactose
changed significantly. It was suggested that derivatization be
carried out at as low a temperature as practicable to avoid the

complications of reequilibration.

9.3.3 Esters

Other derivatives such as acetyl (88) on trifluoroacetyl esters
(TFA) (87) have been used (Equation 9.9)

$$R(OH)_x + (CH_3\overset{O}{\overset{\|}{C}})_2O \quad \xrightarrow[\text{pyridine}]{\text{anhydrous}} \quad R(O\overset{O}{\overset{\|}{C}}CH_3)_x \qquad (9.9)$$

$$R(OH)_x + (F_3C\overset{O}{\overset{\|}{C}})_2O \quad \xrightarrow[\text{pyridine}]{\text{anhydrous}} \quad R(O\overset{O}{\overset{\|}{C}}CF_3)_x$$

even though anhydrous conditions are required for the reactions.
The use of partially acetylated sugar ethers for the determination of polysaccharide bonding has already been mentioned. Table
9.5 lists the relative retention times for a number of partially
methylated and/or acetylated sugars, and methyl 2,3,4,6-tetra-O-
methyl- α-D-glucoside is the reference standard (74). It is
apparent that efficient separation can be effected because of
column selectivity for the isomeric species.

   TFA esters have the particular advantage that an electron
capture detector may be used, and with the large number of fluor-
ine atoms per molecule of sugar high sensitivity may be achieved.
It has been reported that the TFA derivatives are not stable and
may not be stored prior to chromatography (88). These ester derivatives also result in multiple peaks related to anomerization
and generally are more difficult to prepare than the corresponding trimethylsilyl derivatives. For these reasons there is
little reason to choose them as derivatives when other materials
can be used.

9.3.4 Elimination of Anomers

An important approach that reduces the number of peaks in the
chromatogram involves reduction of sugars to corresponding
alditols (Equation 9.10). The alditols are converted to TMS or
TFA derivatives prior to chromatography although acetate deriva-

tives apparently are best resolved. Dutton (70) has rationalized
this in terms of chromatography of the acyclic derivatives which
are formed after the reduction of carbonyl groups. Some information is lost by this since a number of sugars such as D-glucose
or L-sorbose result in identical alditols; conversely the approach does allow a fairly straightforward way to get an overview
of the variety of monomeric sugars present in a mixture, for example from trioses through hexoses and even higher members of the
class. Highly polar phases such as ECNSS-M (89) or XE-60 (90) at
the 3% level are recommended. Another class of derivatives, cyclic butaneboronic acid esters, has been suggested for GLC of
alditols (Equation 9.11). The esters of D-glucitol, D-mannitol,
and D-galactitol were resolvable on a 3% OV-17 column at $200^\circ C$
although related TMS ethers were not (91). Alternatively, completely methylated alditols and related compounds may be separated (92). The distinct advantage of greater volatility of
methyl ethers allows the extension of, the technique to di-, (93)
and even trisaccharides (71) and direct mass spectrometric examination of the methyl ethers after chromatography. Relative re-
tentions of some permethylated alditol saccharides are listed in
Table 9.6.

9.3.5 Aldonitrile Acetates

A newer technique which shows promise but has not yet been fully
tested is based upon the preparation and separation of aldonitrile acetates of reducing sugars (Equation 9.12). The dried
sugar is dissolved in pyridine and treated with hydroxylamine
hydrochloride in a sealed glass ampoule at $90^\circ C$ for 0.5 hr. The
mixture is cooled, acetic anhydride added, and another 30-min reaction is carried out in the resealed ampoule. Actual formation
of the nitrile may occur in the injector of the gas chromatograph.
Nonreducing sugars apparently do not form the nitrile but are
present as the

$$\diagdown C = N\text{-}O\overset{\overset{\displaystyle O}{\|}}{C}\text{-}CH_3$$

derivatives. Separation of standard compounds and sugars found
in gum arabic are shown in Figure 9.14.

TABLE 9.5   RETENTION TIMES OF METHYL O-METHYL-D-GLUCOSIDES

O-Methyl	O-Acetyl	Anomer	Retention Time[1]			
			a	b	c	d
2,3,4,6		α	1.00	1.00	1.00	1.00
		β	0.74	0.75	0.67	
2,3,5,6		α				
		β				
2,3,4		α	1.38		2.19	1.84
		β	1.00		1.78	
	6	α		3.07		
		β		2.19		
2,3,6		α			3.56	2.59
		β			2.48	
	4	α		3.42		
		β		2.42		
2,4,6		α			3.56	2.59
		β			2.34	
	3	α		5.05		
		β		3.33		
3,4,6		α			2.65	2.18
		β			2.62	
	2	α		2.15		
		β		3.11		
2,3		α				6.66
		β				5.30
	4,6	α		8.80		
		β		6.35		
2,4		α	2.48			
		β	1.75			
	3,6	α		7.25		
		β		5.75		
2,6	3,4	α		7.80		
		β		5.26		
3,4	2,6	α		6.10		
		β		9.47		
3,6	2,4	α		6.05		
		β		9.21		
4,6		α				4.92
		β				6.10
	2,3	α		6.01		
		β		6.62		
2	3,4,6	α		20.43		

		β	13.70
3	2,4,6	α	15.80
		β	23.30
4	2,3,6	α	17.98
		β	20.00
6	2,3,4	α	9.65
		β	12.20

[1](a) 15% Butane-1, 4-diol succinate polyester on Chromosorb G, 175°. (b) 15% LAC-4R-886 on 60-80 mesh Chromosorb W, 190°, argon (190° cm^3/min). (c) 4 Ft, 20% butane-1,4-diol succinate polyester on alkali-washed Celite 545, 150°, argon (60 cm^3/min). (d) 25% Diethylene glycol succinate polyester, 220°, helium (100).

TABLE 9.6   RELATIVE RETENTION TIMES OF PERMETHYLATED ALDITOLS (92)

Compound	Liquid Phase		Polarity Increases	
	10% SE-30	1% OV-17	3% QF-1	1.5% XE-60/ 1.5% EGS
Alditols	(140°)	(130°)	(110°)	(130°)
Glycerol	0.11	0.07	0.08	0.11
Erythritol	0.34	0.26	0.26	0.25
Ribitol	0.83	0.70	0.60	0.65
Arabinitol	1.00	0.95	1.05	1.00
{Xylitol	1.00	1.00	1.00	1.00}
Glucitol	2.46	2.82	2.46	2.49
Mannitol	2.47	2.80	2.84	2.62
Galactitol	2.80	3.41	3.88	3.38
Disaccharides[1]	(250°)	(210°)	(200°)	(195°)
Laminaribiitol	1.07	1.10	1.00	0.95
Cellobiitol	1.02	0.96	1.04	0.92
Maltitol	1.00	1.00	1.00	1.00
Lactitol	1.01	1.18	1.10	1.20
Isomaltitol	1.29	1.43	1.38	1.48
Melibiitol	1.36	1.79	1.55	1.82

[1]Ref. (93).

$$
\begin{array}{c}
\text{HC} - \text{OH} \\
| \\
\text{HC} - \text{OH} \\
| \\
\text{HO} - \text{CH} \\
| \qquad\qquad \text{O} \\
\text{HC} - \text{OH} \\
| \\
\text{HC} \\
| \\
\text{H}_2\text{C} - \text{OH}
\end{array}
\xrightarrow{+2\text{H}}
\begin{array}{c}
\text{H}_2\text{C} - \text{OH} \\
| \\
\text{HC} - \text{OH} \\
| \\
\text{HO} - \text{CH} \\
| \\
\text{HC} - \text{OH} \\
| \\
\text{HC} - \text{OH} \\
| \\
\text{H}_2\text{C} - \text{OH}
\end{array}
\xleftarrow{+2\text{H}}
\begin{array}{c}
\text{H}_2\text{C} - \text{OH} \\
| \\
\text{C} - \text{OH} \\
| \\
\text{HO} - \text{CH} \\
\text{O} \qquad | \\
\text{HC} - \text{OH} \\
| \\
\text{CH} \\
| \\
\text{H}_2\text{C} - \text{OH}
\end{array}
\qquad (9.10)
$$

$$
\text{D-Glucose} \qquad\qquad\qquad \text{Sorbitol} \qquad\qquad\qquad \text{L-Sorbose}
$$

$$
\begin{array}{c}
| \\
\text{HC} - \text{OH} \\
| \\
\text{HC} - \text{OH} \\
|
\end{array}
+ \text{C}_4\text{H}_9\text{B(OH)}_2
\xrightarrow{\text{Pyridine}}
\begin{array}{c}
\text{O} \\
\text{HC} \diagdown \\
\qquad \text{BC}_4\text{H}_9 + \text{H}_2\text{O} \\
\text{HC} \diagup \\
\text{O}
\end{array}
\qquad (9.11)
$$

$$
\begin{array}{c}
\text{H} \\
| \\
\text{C} = \text{O} \\
| \\
\text{H} - \text{C} - \text{OH} \\
| \\
\text{X}
\end{array}
+ \text{HCl} \ \text{H}_2\text{NOH}
\xrightarrow{\hspace{2cm}}
\begin{array}{c}
\text{H} \\
| \\
\text{C} = \text{NOH} \\
| \\
\text{H} - \text{C} - \text{OH} \\
| \\
\text{X}
\end{array}
\xrightarrow[\text{pyridine}]{\text{acetic anhydride}}
$$

$$
\begin{array}{c}
\text{H} \\
| \\
\text{X-C-C} \equiv \text{N} \qquad (9.12) \\
| \\
\text{O} \\
| \\
\text{C} = \text{O} \\
| \\
\text{CH}_3
\end{array}
$$

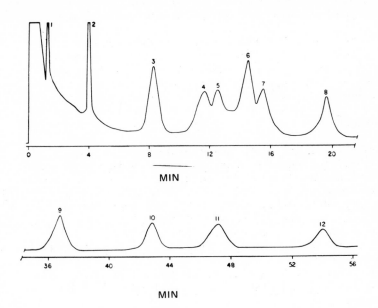

Figure 9.14   GLC of aldononitrile acetates. Column: 10% LAC-
4R-886 (polyester wax), 100-200 mesh Chromosorb W, 1.5 x 0.3 cm
i.d., isothermal 190°C, injector 230°C. Compounds: 4 = fucose,
5 = ribose, 6 = lyxose, 7 = arabinose, 8 = xylose, 9 = mannose,
10 = 2-deoxygalactose, 11 = glucose, 12 = galactose (94).

9.3.6 Trends

Although the use of aldonitrile esters results in reduction of
the number of peaks found in GLC without loss of information
which occurs when derivatized alditols are used, the chromatography is inefficient. It is to be expected that new chromatographic techniques will be worked out to improve this situation
and that they may utilize better column as well as improved derivatization procedures.

GLC is widely used for the study of carbohydrates; however,
there have been far fewer applications to food chemistry (76).
This situation will undoubtedly change as more information regarding the dietary and its impact upon health is requested. For
example, the realization that a large percentage of the world
population is intolerant to lactose has led to the development of

processes for the enzymatic hydrolysis of lactose in foods and
gas chromatographic procedures for the evaluation of the original
disaccharide and the glucose and galactose hydrolysis products
(85). The nature of the interaction of carbohydrates with proteins, for example in the study of loss of lysine availability
after heat processing of food, has required greater consideration
of complex molecular species that can be solved in part by GLC.

REFERENCES

1. Anon., Science, 188, 501 (1975).
2. Anon., Toxicants Occurring Naturally in Foods, 2nd edit.,
   National Academy of Sciences, Washington, D. C. 1973.
3. A. White, P. Handler, and E. L. Smith, Principles of Biochemistry, 4th edit., McGraw-Hill, New York, 1968.
4. J. E. Kinsella, L. Posati, J. Weihrauch, and B. Anderson, CRC,
   Crit. Rev. Food Technol. 5, 299 (1975).
5. C. Litchfield, Analysis of Triglycerides, Academic Press,
   New York, 1972.
6. D. Swern, K. F.Mattil, F. A. Norris, and A. J. Stirton,
   Bailey's Industrial Oil and Fat Products, Wiley, New York,
   1964.
7. C. Litchfield, R. D. Harlow, and R. Reiser, Lipids, 2, 363
   (1967).
8. R. J. Hamilton, J. Chromatog. Sci., 13, 474 (1975).
9. F. E. Luddy, R. A. Barford, S. F. Herb, and P. Magidman, J.
   Am. Oil Chem. Soc., 45, 549 (1968).
10. P. Magidman, S. F. Herb, R. A. Barford, and R. W. Riemen-
    schneider, J. Am. Oil Chem. Soc., 39, 137 (1962).
11. R. G. Ackman, Progr. Chem. Fats and Other Lipids, 12, 165
    (1972).
12. P. Magidman, private communication.
13. J. K. Haken, J. Chromatog. Sci., 13, 430 (1975).
14. J. R. Ashes and J. K. Haken, J. Chromatog., 84, 231 (1973).
15. W. R. Supina, The Packed Column in Gas Chromatography,
    Supelco Inc., Bellefonte, PA, 1974.
16. Anon., Gas-Chrom Newsletter, Applied Sciences Laboratories,
    Inc., 1973.
17. A. Smith and A. K. Lough, J. Chromatog. Sci., 13, 486 (1975).
18. S. F. Herb and V. G. Metzger, J. Am. Oil Chem. Soc., 47, 415
    (1970).
19. G. F. Spencer, S. F. Herb, and P. Gormisky, J. Am. Oil Chem.
    Soc., 53, 94 (1976).
20. J. Falbe and J. Weber, Methodicum Chimicum, Vol. 1, Part B,
    Academic Press, New York, 1974, pp. 970-990.
21. AOAC Methods, 11th edit., 1970.

22. F. L. Hart and H. J. Fisher, Modern Food Analysis, Springer-Verlag, New York - Berlin, 1971.
23. H. Brockerhoff and R. G. Jensen, Lipolytic Enzymes, Academic Press, New York, 1974.
24. H. R. Mahler and E. H. Cordes, Biological Chemistry, Harper and Row, New York, 1966.
25. R. G. Ackman and L. D. Metcalfe, (Eds.), J. Chromatog. Sci., 13, 397, 453, (1975
26. R. F. Borch, Anal. Chem., 47, 2437 (1975).
27. L. C. Gruen and P. W. Nichols, Anal. Biochem., 47, 348 (1972).
28. T. Y. Liu and Y. H. Chang, J. Biol. Chem., 246, 2842 (1971).
29. A. L. Lehninger, Biochemistry, Worth, 1970.
30. M. A. Kirkman, J. Chromatog., 97, 175 (1974).
31. A. J. Cliffe, N. J. Berridge, and D. R. Westgarth, J. Chromatog., 78, 333 (1973).
32. E. J. Conkerton, J. Agr. Food Chem., 22, 1046 (1974).
33. B. Kolb, New Tech. Amino Acid, Peptide, Protein Anal., 129, (1971).
34. B. Kolb, Method. Chim., 1020, (1974).
35. A.Darbre, Biochem. Soc. Trans., 70, (1974).
36. J. R. Coulter and C. S. Hann, New Tech. Amino Acid, Peptide, Protein Anal., 75, (1971).
37. J. P. Ussary, Food Prod. Develop., 84, 86, 88, (1973)
38. M. Feuillat, J. Demaimay, and E. Poitrat, Ann. Technol. Agr., 21, 131 (1972).
39. G. Baerwald and J. Prucha, Brauwissenschaft, 24, 397 (1971).
40. L. Campos, M. Renard, and M. Severin, An. Inst. Super. Agron., Univ. Tec. Lisboa, 31, 151 (1968-1970).
41. G. W. Gehrke and H. Takeda, J. Chromatog., 76, 63 (1973).
42. H. Hediger, R. L. Stevens, H. Brandenberger, and K. Schmid, Biochem., J., 133, 551 (1973).
43. F. Raulin, P. Shapshak, and B. N. Khare, J. Chromatog., 73, 35 (1972).
44. I. M. Moodie, J. Chromatog., 99, 495 (1974).
45. G. E. Pollock, Anal. Chem., 44, 2368 (1972).
46. A. Islam and A. Darbre, J. Chromatog., 71, 223 (1972).
47. M. Butler, A. Darbre, and H. R. V. Arnstein, Biochem. Soc. Trans., 1, 610 (1973).
48. R. F. Adams, J. Chromatog. 95, 189 (1974).
49. J. P. Zanetta and G. Vincendon, J. Chromatog., 76, 91 (1973).
50. S. L. MacKenzie and D. Tenaschuk, J. Chromatog., 97, 19 (1974).
51. P. Fantozzi and G. Montedoro, Am. J. Enol. Vitic., 25, 151 (1974).
52. P. Husek, J. Chromatog. 91, 475 (1974).
53. T. Matsuda, Nara Igaku Zasshi, 23, 396 (1972).

54. K. A. Kvenvolden, E. Peterson, J. Wehmiller, and P. E. Hare, Geochim., Cosmochim. Acta, 37, 2215 (1973).
55. B. Feibush, E. Gil-Av, and T. Tamari, J. Chem. Soc., Perkin Trans., 2, 1197 (1972).
56. H. Iwase and A. Murai, Chem. Pharm. Bull., 22, 8 (1974).
57. H. Iwase, Chem. Pharm. Bull. 22, 1663 (1974).
58. H. Iwase and A. Murai, Chem. Pharm. Bull., 22, 1455 (1974).
59. D. H. Calam, J. Chromatog., 70, 146 (1972).
60. H. D. Niall, Methods in Enzymology, XXVII, Academic Press, New York, 1973, p. 942.
61. J. J. Pisano and T. J. Bronzert, J. Biol. Chem., 244, 5597 (1969).
62. J. Eyem and J. Sjöquist, Anal. Biochem., 52, 255 (1973).
63. R. M. Caprioli and W. E. Seifert, Biochem. Biophys. Res. Commun., 64, 295 (1975).
64. E. W. Mathews, P. G. H. Byfield, and I. MacIntyre, J. Chromatog., 110, 369 (1975).
65. G. A. Reineccins, T. E. Kavanagh, and P. G. Kenney, J. Dairy Sci., 53, 1018 (1970).
66. O. Hockwin, Method. Chim., 1, 1006 (1974).
67. B. Lindberg, J. Lönngren, and S. Svensson, Chem. Biochem., 31, 185 (1975).
68. D. Albersheim, Sci. Am., 232, 80 (1975).
69. G. G. S. Dutton, Advan. Carbohyd. Chem. Biochem., 28, 11 (1973).
70. G. G. S. Dutton, Advan. Carbohyd. Chem. Biochem., 30, 9(1974).
71. J. Kärkkäinen, Carbohyd. Res., 17, 1 (1971).
72. S. Hakomori, J. Biochem. (Tokyo), 55, 205 (1964).
73. H. Björndal, C. G. Hellerqvist, B. Lindberg, and S. Svensson, Angew. Chem. Int. Edit., 9, 610 (1970).
74. H. G. Jones, in Methods Carbohyd. Chem., R. L. Whistler and J. N. Bemittler (Eds.), Vol. 6, Academic Press, New York, 1972, p. 25.
75. A. E. Pierce, Silylation of Organic Compounds, Pierce Chemical Company, Rockford, IL, 1968.
76. G. G. Birch, J. Food Technol., 8, 229 (1973).
77. J. I. Varns and R. Shaw, Potato Res., 16, 183 (1973).
78. R. B. Holtz, J. Agr. Food Chem., 1, 1272 (1971).
79. D. A. Heatherbell, J. Sci. Food Agric., 25, 1095 (1974).
80. C. C. Sweeley, R. Bentley, M. Makita, and W. W. Wells, J. Am. Chem. Soc., 85, 2497 (1964).
81. R. Bentley and N. Botlock, Anal. Biochem., 20, 312 (1967).
82. W. C. Ellis, J. Chromatog., 41, 325 (1969).
83. G. G. Esposito, Anal. Chem., 40, 1902 (1968).
84. W. R. Supina, R. F. Kruppa, and R. S. Henly, J. Am. Oil Chem. Soc., 44, 74 (1967).
85. J. H. Copenhower, Anal. Biochem., 17, 76 (1966).

86. J. S. Sawardeker, J. H. Sloneker, and A. R. Jeanes, Anal. Chem., $\underline{37}$, 1602 (1965).

87. D. Anderle and P. Kovac, J. Chromatog., $\underline{49}$, 418 (1970).

88. R. Varma, W. S. Allen, and A. H. Wardi, J. Neurochem., $\underline{34}$, 471 (1968).

89. H. Björndal, B. Lindberg, and S. Svensson, Acta. Chem. Scand., $\underline{21}$, 1801 (1967).

90. G. O. Aspinall and I. W. Cottrell, Can. J. Chem., $\underline{49}$, 1019 (1971).

91. Frank Eisenberg, Jr., Carbohyd. Res., $\underline{19}$, 135 (1971).

92. J. N. C. Whyte, J. Chromatog., $\underline{87}$, 163 (1973).

93. J. N. C. Whute, Can. J. Chem., $\underline{51}$, 3197 (1973).

94. R. Varma, R. S. Varma, and A. H. Wardi, J. Chromatog., $\underline{77}$, 222 (1973).

CHAPTER 10

Clinical Applications
of Gas Chromatography

JOSEPH C. TOUCHSTONE

MURRELL F. DOBBINS

Hospital of the University of Pennsylvania

10.1   INTRODUCTION. 497
10.2   CLINICALLY IMPORTANT ORGANIC COMPOUNDS. 499
        10.2.1 Estriol. .499
               Amniotic Fluid 503
        10.2.2 Urinary Pregnanediol 506
               Hydrolysis and Extraction 508
               Gas Chromatography 509
               Calculation of Results 509
               Recovery and Reproducibility 509
        10.2.3 Urinary Pregnanetriol. 510
               Hydrolysis and Extraction. 510
               Preparation of Trimethylsilyl Ether. 511
               Derivatives of Sample and Internal Standard
               Gas Chromatographic Analysis 511
               Calculation of Results 511
               Recovery and Reproducibility 513
        10.2.4 Urinary Cholesterol. 514
               Extraction 515
               Acetylation. 515

         Standards. 516
         Gas Chromatographic Analysis 516
         Reproducibility and Recovery 516
10.2.5 Catecholamines and Metabolites in Urine 518
         Urine Extraction 518
         Preparation of the Methyl Ester-
         Trimethylsilyl Ether 519
         Gas Chromatographic Analysis 519
         Quantitative Aspects 520
         Discussion 520
10.2.6 Determination of Blood Alcohol 520
         Preparation of Sample 521
         Gas Chromatographic Parameters 524
         Calculations 524
         Discussion 524
10.2.7 Anesthetics in Blood 526
         Preparation of Samples 527
         The Chromatographic Parameters 527
         Calculations 528
         Discussion 529
10.2.8 Urinary Aromatic Acids 529
         Extraction of Urine 530
         Preparation of Derivatives 531
         Gas Chromatographic Analysis 531
         Quantitative Analysis 531
         Discussion 531
10.2.9 Amino Acids in Blood 532
         Sample Extraction 534
         Preparation of Cation-Exchange Resin 535
         Isolation of Amino Acids from Deproteinized. . .535
         Samples
         Preparation of Derivatization Reagents 535
         Derivatization of the Ion-Exchange Cleaned . . .535
         Physiological Fluid
         Gas Chromatographic Analysis 536
         Procedure Considerations 537
10.2.10 Drugs of Abuse 537
         Amphetamines 539
         Extraction 540
         Preparation of Standards 540
         Gas Chromatography 541
         Preparation of Ketone Derivative 541
         Quantification 541
         Recovery and Reproducibility 542
         Barbiturates 543
         Reagent Preparation. 545
         Sample Extraction 546

Preparation of Standards 546
Gas Chromatography 546
Quantification 547
Recovery and Reproducibility 547
Discussion 547
10.3 CONCLUSIONS 549
REFERENCES 550

10.1 INTRODUCTION

The utility of gas chromatography (GC) as a separation method was
apparent from its beginnings when the first papers appeared in
1949. However, it was not until about 10 years later that other
developments changed the scope and nature of this separation tool
for use in the clinical laboratory. In 1960 it was shown (1)that
steroids could be separated by GC using highly stable polysiloxane liquid phases and high-sensitivity detectors. During the
period following, many other groups of naturally occurring compounds were quantified by gas chromatographic methods. Argon
ionization and hydrogen flame ionization detectors were perfected
during the decade of the 1960s. More recently specific ion detectors, that is, N, P, and electron capture detectors, have been
developed.

The application of GC for use in the clinical laboratory is
relatively new, due possibly to the shortage of qualified personnel to perfect and standardize GC as a routine clinical laboratory procedure. This occurred, even though highly sensitive and
specific methodology of GC was available for many determinations.
The methods presented in this chapter are taken from current literature and from the authors' laboratory. Although the methods
presented are not the only ones used in clinical chemistry, in
the authors' opinion they are the simplest and most reliable.

At present few hospital laboratories use gas chromatographic
methods because colorimetric, fluorometric, and other spectrophotometric methods can give data more rapidly, are less expensive, and are often of sufficient accuracy and precision for
clinical evaluation of the patient. Radioimmunoassay procedures
are available, but these are limited because of availability of
the necessary antibodies and a sufficient demand to justify the
expense of setting up the laboratory. For versatility and adaptability, many different analyses can be done on a single gas
chromatograph. The only requirement is the proper column and
choice of operating conditions. Furthermore, the techniques, in
many cases, are now automated. The utility of GC coupled with

mass spectrometry (GC-MS) may soon be realized by those working in clinical laboratories and become a standard technique of analysis.

The preceding chapters present the principles of GC. Therefore, it is not necessary to repeat this material, except where necessary for clarity in any particular method. However, the manual provided with the instrument should be carefully examined for the use intended. Each method will include an appropriate bibliography which describes alternative methods as well as discussion of the clinical aspects of the results obtained.

It should be strongly stressed here that the chemist obtain a set of "normal" values, for the so-called "normal" may vary somewhat in different regions due to ambient conditions as well as differences in instrumentation and technique. Proper technique, and use of internal standards, should result in agreement among laboratories. The methods in this chapter represent the current state of the art in clinical GC. Nevertheless, results are not guaranteed, and the validity of the results is the responsibility of the individual laboratory employing the procedure. Each chemist should determine the applicability of the methods and the verification of materials and standards used in his own laboratory.

For success in most biomedical applications there are a number of general prerequisites which must be considered. These are discussed below.

1. The injection system and columns should be glass. It has been well established, particularly with steroids, that the active surfaces of metal columns may retain or destroy certain organic compounds, particularly at the temperatures usually employed in GC. Some exceptions to these requirements will be seen in the following discussions of various methods.
2. Before packing, glass columns should be cleaned well followed by silanization to remove any areas (active sites) which may adsorb or react with sample molecules. See Chapter 3 for further information.
3. The stationary phases used should be coated on a fine mesh (80/100; 100/120), high-quality support which has been washed with acid and base, and silanized. This treatment will provide a nearly inert support for the liquid phase. It is sometimes convenient to purchase supports already treated, or to purchase the entire packing made with deactivated support. Details for packing columns are given in Chapter 3.

10.2   CLINICALLY IMPORTANT ORGANIC COMPOUNDS

   10.2.1 Estriol

The influence of pregnancy on the level of estrogenic activity
was noted as early as 1927, when Smith (2) reported that estrogenic activity increased during pregnancy and decreased after
delivery.   In 1953 Spielman et al. (3) found blood estrogens
markedly decreased on fetal death.   Many workers, using chemical
methods, showed that the amount of estrogens excreted every 24 hr
was related to the well-being of the patients.
   The use of GC for the determination of estriol was suggested
in the early years of GC.   But it was not until later that gas
chromatographic methods of urinary extriol were established.
Among the first was that reported by Touchstone et al. (4).
Methods continue to appear for the determination of estriol using
GC.
   Estriol (estra-1,3,5(10)-trien-3,16-α,17-β -triol) is the
major estrogen metabolite found in the urine.   It is excreted in
the form of its conjugate with glucuronic acid.   The determination of this steroid as an index of placental function has become one of the most widely used endocrine determinations.   As
pregnancy progresses, the excretion increases and reaches very
high levels near term.   In abnormal fetoplacental function, the
levels of estriol will fall in some cases.   The fall is usually
progressive, and, because of this, serial determinations of urinary estriol must be carried out.   The drop in estriol can be
taken as evidence of placental insufficiency, and close watch by
the physician is indicated, as a continued drop may necessitate
Cesarean section to save the life of the infant.
   Any discussion of sample preparation for determination of
metabolites of steroids in urine must include the effect of the
medication of the patient on extraction of the metabolites.   The
following are known examples in the case of estriol.
   Glucose, mandelamine, and hydrochlorothorazide decrease the
amount of estriol found when acid hydrolysis is used to release
estriol from its conjugate with glucuronic acid in urine.   The
destruction of estriol in diabetic urine can be monitored by the
inclusion of internal standards of labeled estriol glucuronide
and reduced by prior dilution of the urine.   When glucuronidase
is used to release the estriol, the results are not affected when
glucose of diabetic urine is present.   When this is done the internal standard is not necessary.
   Mandelamine is broken down by the acid hydrolysis to give
formaldehyde which markedly reduces the amount of estriol found.

When glucuronidase hydrolysis is used, the interference with
estriol is overcome.

The effectiveness of estriol determinations is reduced by
hydrochlorothiazide when acid hydrolysis was used. When glucuronidase hydrolysis was used along with internal standards, the
effect of hydrochlorothiazide was overcome. There is no difference between enzyme hydrolysis and acid hydrolysis over the
clinically important range of 0-10 mg/24 hr.

Certain basic principles in clinical evaluation must be considered.

1. A single estrogen determination is meaningless because of
    widespread difference in individuals as to the "normal" value.
2. While a trend in values is significant, relatively constant
    high or low values are not meaningful. A progressive decline
    from a high value might be a source of concern for the attend-
    ing physician, as would a decline from a low one.
3. The same assay procedure should be used for specimens from an
    individual. Some background in the use of urinary estriol as
    an index of placental function may be found in a number of
    reports on the use of this valuable clinical chemistry tool in
    obstetrics. More recently, Wolff et al. (5) reviewed the var-
    ious pathologies and abnormalities of pregnancy versus values
    of estriol levels in urine.

The method for the determination of estriol in the urine of
pregnant women was described by Touchstone (4,6), and has been
used successfully in commercial and clinical laboratories. Without some baseline, a high value for one might be low for another.
The important consideration is not the value itself, but the behavior of the values upon serial daily determinations. Figure
10.1 shows that the range itself is wide, but shows increasing
concentration as gestation progresses. Values as high as 50 mg/
24 hr are not uncommon.

Recently values and procedures for the determination of estriol in amniotic fluid were described (7, 8). In some instances,
it may be easier to obtain a sample of amniotic fluid than a 24-
hr urine, depending on experience. There is a correlation between urinary and amniotic fluid estriol levels. After 20 weeks
gestation, amniotic levels show a progressive increase with
slopes similar to the urinary estriol which increases throughout
gestation, and for this reason, the methodology is included here.

1. Urine Collection. Urine collections are on a 24-hr basis for
proper quantitative evaluation, as well as to overcome variation
during the day, and should be collected in bottles containing a

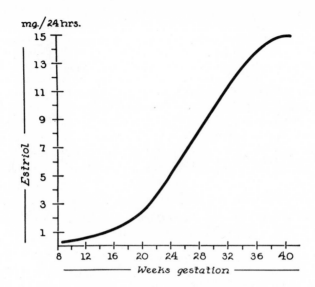

Figure 10.1   Estriol excretion in normal pregnancy.   The curve
represents the average of 20 normal pregnancies.   Individual
values differed from the average as much as 50 percent.

few drops of chloroform, then stored in a refrigerator until time
for analysis.   If the urine is not analyzed the same day, it
should be frozen.

2. Extraction Procedure.   Urines are extracted after hydrolysis
of the estrogen conjugates as outlined below:

   Dilute 5-cm^3 aliquot of urine to 20 cm^3 with water.
   Add 3 cm^3 of concentrated HCl, 15 cm^3/100 cm .
   Reflux 15 min, timing from start of boiling.
   Cool in an ice bath.
   Extract three times with equal volumes of diethyl ether.
   Wash the combined extracts once with 5 cm^3 of sodium bicarbon-
   ate, 5 g/100 cm^3.
   Extract two times with 10-cm^3 portions of sodium hydroxide,
   1 mol/liter, and discard the ether.
   Adjust the pH to 2- 4 with concentrated HCl.
   Extract two times with equal volumes of diethyl ether.
   Wash the ether once with 5 cm^3 of sodium bicarbonate,
   5 g/100 cm^3, and once with 10 cm^3 of water.

Evaporate.
Dissolve the residue in methanol and quantitatively transfer
it to a 1.5-cm^3 centrifuge tube; then evaporate the methanol.
Add 50 μl of t-butanol and inject 2- to 4- μl aliquots into
the gas chromatograph with a 10-μl Hamilton syringe.

3. Determination of Recovery. For evaluation of the recovery
during the total extraction procedure, labeled ^{3}H-estriol is
added to the urine before the hydrolysis step. After the extract
has been prepared in preparation for injection into the gas chromatograph, an aliquot is taken for counting and the recovery calculated.

4. Preparation of the Standard Curve (Calibration). A standard
solution of estriol dissolved in t-butanol (1 μg/1 μl) (1 mg/
1 cm^3) is kept on hand at all times for external standardization
and evaluation of the gas chromatograph performance from day to
day. External standardization, as used in these experiments,
represents alternate standard-sample injection into the gas
chromatograph. In this way the sample curve can be interpolated
against the standard which has a known concentration. The detector response can be determined from day to day and correction
for deviation from the standard curve can be made. Serial
amounts, 0.5-, 1-, 2-, 3-, and 4-μl aliquots, are injected into
the gas chromatograph in order to obtain a standard curve from
which samples are interpolated. Estrone can be added to the extract before injection if an internal standard is desired.

5. Instrumental Parameters. The instrument used in the determination of estriol can be a gas chromatograph with a flame ionization or a β-ionization detector. The column generally recommended for this procedure contains a mixed phase of OV-1 (2.5%) and
OV-210 (5%). The one used in the authors' laboratory is a glass
column packed with Supelcoport (80-100 mesh) coated to the above
concentration. For this particular use, the column was conditioned at 270°C for at least 24 hr, although operating temperature
is 240°C. Carrier gas (helium or argon) flow during conditioning
is 40 cm^3/min. The injection port temperature can be 260°C.
Generally, the detector temperature should be slightly higher
than that of the column, for example, 260°C.
   The carrier gas flow is increased to 50-60 cm^3/min (30 psi at
the cylinder). Now the injection of the standard sample (1-2μl)
can be performed. The column has to be primed with the substance
under consideration.

6. Standard Curve. After priming of the column (same size peak after repeated injections of the same concentration standard), inject 0.5, 1, 2, 3, 4 µl of the standard to prepare a standard curve. The quantitative parameters are based on the area under the curve produced by each injection. The area under each peak (usually height x width at half-height) can be plotted versus the weight of standard under consideration. The standard curve should be reproducible from day to day provided the conditions of the instrument are kept comparable.

7. Analytical Aspects. Analysis of the sample is done by injecting an aliquot from the solution of extract. Values are determined by interpolation on the standard curve after calculating the area under the peak. Correction for recovery is determined by correcting for the aliquot taken for this purpose. Correction for the dilution is based on the fact that 5 cm^3 of urine was taken for analysis from the 24 hr collection.

The amount of estriol in the 24-hr sample can then be calculated using the following formula:

$$\text{Estriol in 24 hr} = \frac{(C_e)\ (V_b)\ 100\ (V_s)}{(V)\ (R_e)\ (5)} \tag{10.1}$$

where C_e = amount of estriol in sample as calculated from interpolation on curve (µg),

V_b = volume of butanol used to dissolve the sample (50 µl in this case),

V_s = volume of the 24-hr sample (cm^3),

V = volume of aliquot injected (µl),

R_e = recovery of added labeled estriol,

5 = volume of urine analyzed,

Answer will be in µg.

Figure 10.2 shows a typical gas chromatogram of estriol isolated from urine and the standard estriol on the same column under the conditions just described.

AMNIOTIC FLUID. Determination of estriol in amniotic fluid using the preceding procedure is somewhat different due to the fact that amniotic fluid contains protein, and it is more difficult to remove the steroids because of protein binding.

The modified extraction procedure is as follows:

Dilute the amniotic fluid, 5 cm^3, with an equal volume of water. To this add 10 cm^3 of saturated $(NH_4)_2SO_4$ solution, and adjust the pH to 1.0 with concentrated HCl. Extract this mixture

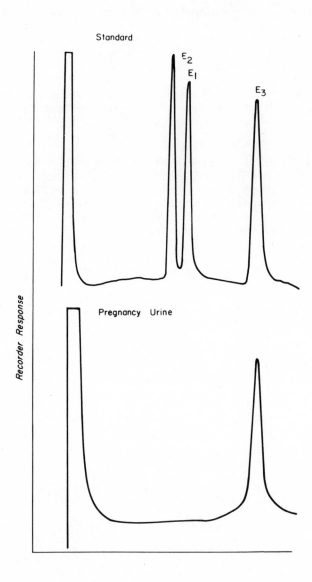

Figure 10.2   Chromatogram showing estriol from urine of a
pregnant woman.   The upper chromatogram shows the separation of
reference estriol on the same column.

twice with 30 cm^3 of ethyl acetate. Discard the aqueous (lower) portion, and combine and evaporate the ethyl acetate extracts under reduced pressure, without washing.

To the residue, add 30 cm^3 of dilute HCl (15 cm^3 plus 85 cm^3 of water) and reflux the mixture for 20 min in a heating mantle, timing from the start of boiling, cool in an ice bath, and successively extract with two 40 cm^3 portions of diethyl ether. Discard the aqueous phase. Combine the ether extracts and wash once with 10 cm^3 of NaHCO$_3$ solution (80 g/liter). Extract the ether three times with 1N NaOH, 20-, 15-, and 15-cm^3 portions, and discard the ether. Neutralize the alkaline extract to pH 2-4 by adding 10 cm^3 of 5N HCl, and extract the mixture twice with ether, 40 cm^3 each time. Discard the aqueous portion. Wash the ether once with 10 cm^3 of water and then evaporate it. Transfer the residue to a 1.5 cm^3 conical stoppered centrifuge tube and evaporate the solvent. To each such tube add 20 µl of acetone. Inject a 2 - 4-µl aliquot of this into the gas chromatograph. Calculate concentration by comparing with standard curves and correcting for dilution.

Amniotic fluid contains estriol in a variety of forms: notably free, sulfate, and glucuronide. Therefore the determination is for total estriol content. After extraction, the procedure is the same as just described for the urinary estriol.

It should be noted that the determinations of estriol described herein did not involve derivatization. This is due to the relatively high concentration of the steroid in the urine. Because concentration is lower the method is not applicable to normal urines or urine from postmenopausal women.

The use of mixed phase columns in the gas chromatographic determination of estriol is based on earlier observations in our laboratory that detector response was greater from mixed phase columns than it was from single phase packings. This is illustrated in Figure 10.3.

The method was subjected to a comparison between two different laboratories. Table 10.1 shows the results obtained and indicated that there was no significant difference between the values obtained in the two laboratories. The use of this method as opposed to some of the others reported in the literature deserves comment. Generally, most investigators describe derivatization as a necessary step in obtaining usable chromatograms. However, derivative formation is not always a reliable method, but appears necessary in biological samples with low steroid levels to increase sensitivity as well as selectivity.

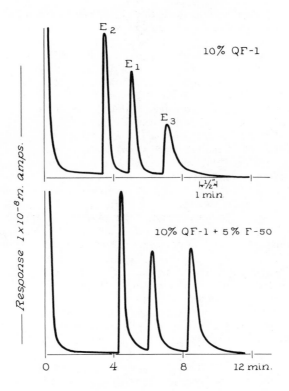

Figure 10.3   Effect of mixed phase packing on detector
response to estriol. Peaks in lower section correspond in order
and identity of those in upper section.

10.2.2 Urinary Pregnanediol

Pregnanediol, 5-β-pregnane-3-α,20-α-diol, is a major metabolite
of progesterone excreted in the urine. The cyclical fluctuations
in the urinary excretion of pregnanediol observed in normal ovu-
lating women reflect changes in the secretion by the ovaries.
The clinical value of pregnanediol determination is found in gyn-
ecological disorders, in cases of abnormal pregnancies, and as
evidence of actively secreting corpus luteum and ovulation during
the normal menstrual cycle.
     The traditional method for the determination of urinary preg-
nanediol has generally been that of Klopper et al. (9).

TABLE 10.1   COMPARISON OF ESTRIOL VALUES OBTAINED IN TWO DIFFER-
ENT LABORATORIES[1]

Collection Interval	Lab	No. Cases	Mean ± SD	p
7 am - 3 pm	Temple	11	3.94 ± 1.08	0.5-0.6
	Penn	11	3.53 ± 1.34	
3 pm -11 pm	Temple	11	2.92 ± 1.34	0.8-0.9
	Penn	11	3.05 ± 1.55	
11pm - 7 am	Temple	11	2.46 ± 1.40	0.5-0.6
	Penn	11	2.83 ± 1.82	
Total 24 hr	Temple	33	3.11 ± 1.57	0.9-1.0

[1] In mg/8 hr.

This method involves acid hydrolysis of the urine followed by extraction with toluene, oxidative purification with permanganate,
adsorption chromatography, acetylation, column chromatography on
alumina, and finally quantification by colorimetric determination
of the sulfuric acid chromogen. It is now generally accepted
that this method is not as sensitive or as specific as the newer
methods of GC.

Cooper et al. (10) and Cox (11) were the first to report the
gas chromatographic analysis of urinary pregnanediol. These
methods normally involved hydrolysis, extraction, and the chromatographic analysis of the free pregnanediol, the acetate derivative, the trimethylsilyl ether derivative, or the bis-heptafluor-
obutyrate derivative.

GC offers two advantages when compared to the other procedures. The first is the increase in speed of the analysis and the
second is the reduced number of manipulative procedures, which
usually allows a greater recovery of the substance being determined. As can be seen from the preceding discussion, three questions have to be answered when employing a pregnanediol procedure: (a) Which method of hydrolysis is to be used, acid or enzymes? (b) How much cleanup is necessary to achieve the desired
results?   (c) Is a derivative necessary for specificity and sensitivity? The literature indicates pros and cons for all three
questions, which will be discussed here.

The primary form of pregnanediol excreted into the urine is
that of the conjugate, and the major conjugate present is the
glucosiduronate. Very little work has been done on the gas
chromatographic separation of glucosiduronates, either directly

or as derivatives. Until such methods are perfected, it will be necessary to hydrolyze the pregnanediol conjugate before it can be determined by GC. Curtius and Muller (12) have compared 10 different methods of hydrolysis for 17-ketosteroids and progesterone metabolite conjugates in urine. This is recommended reading for anyone commencing work in the steroid field. The method they prefer, which gave the greatest yields, is hydrolysis with β-glucuronidase as follows:

Fifty cubic centimeters of urine are adjusted to pH 6.2 with 10 cm^3 of phosphate buffer (pH 6.2), and 5000 units of β-glucuronidase are added. The mixture is placed in a 37° incubator for 24 hr. Three drops of chloroform are added to prevent bacterial decomposition.

The hydrolysate is extracted three times with 50 cm^3 of ether. The pooled ether is then extracted twice with 20 cm^3 of 0.1N sodium hydroxide and five times with 30 cm^3 of water until neutral. Sodium hydroxide and water washes are discarded. The ether is dried over sodium sulfate and evaporated. These extracts are purified by chromatography on an alumina column.

Other researchers have used very similar enzyme hydrolysis and extraction methods followed by GC without further purification. This is preferred when possible because of the time saved and the decreased chance for sample loss due to manipulation. Many of these procedures are nearly identical, and it is often just a matter of choosing the one which is preferred.

Acid may be used to hydrolyze urinary pregnanediol conjugates in a reproducible manner with approximately 90% recovery. The overall advantage to this method is the amount of time saved. Acid hydrolysis normally takes about 15 min; enzyme hydrolysis takes 24 hr. Since many steroid molecules can be broken by acid hydrolysis, enzyme hydrolysis is the preferred method.

HYDROLYSIS AND EXTRACTION (FROM METCALF (13). Urine is acidified to pH 2-3 and stored at 4°C. Samples of pregnancy urine with suspected high levels are diluted before assay. Five cubic centimeters of filtered urine is hydrolyzed for 12.5 min at 100°C, with 1 cm^3 of 6N HCl. Five cubic-centimeter water blanks and standards (0.2 cm^3 of 100 μg/cm^3 pregnanediol in ethanol + 5 cm^3 water) are treated similarly. The hydrolyzate is cooled and shaken vigorously for not less than 1 min with 4 cm^3 of chloroform containing progesterone as the internal standard, and the aqueous phase is discarded. The organic phase is washed with 2 cm^3 of 1N NaOH and the aqueous phase is again discarded. Three cubic centimeters of the chloroform is transferred to a conical-tipped tube and evaporated to dryness. The residue from this is dissolved in 50 μl of chloroform, and 1-4 μl is injected into the gas chromatograph.

GAS CHROMATOGRAPHY (METCALF {13}). Silanized glass columns 2 feet long, packed with 2% XE-60 on 80/100 mesh Gas Chrom Q, operating at 215°C, are used. The flame ionization detector is set at 230°C and the injection port at 230°C.

For the internal standard, progesterone concentrations are chosen so that peaks obtained are comparable to those of pregnanediol. Concentrations of 8 $\mu g/cm^3$ are convenient for urine from women in the luteal phase of the cycle or pregnant women. Other urines would show lower levels of about 4 $\mu g/cm^3$.

CALCULATION OF RESULTS. The ratio of the two peaks is measured after triangulation of the areas R_A. The concentration of pregnanediol is calculated from that of the standard by the expression:

$$\frac{R_A \text{ (sample)}}{R_A \text{ (standard)}} \times 4 \text{ mg}/\ell \qquad (10.2)$$

RECOVERY AND REPRODUCIBILITY. Reproducibility was determined for both urine samples and standards. Quadruplicate pregnanediol standards (4 mg/ℓ) were analyzed on eight different occasions. The standard deviation (SD) of all the results from their group means was 0.07 mg/ℓ. The SD from the results of duplicate assays on urine samples was 0.12 mg/ℓ in the 1-10 mg/ℓ range (30 samples with a mean pregnanediol concentration of 3.88 mg/ℓ). This data implies that pregnanediol concentrations in urine of below 0.2 mg/ℓ cannot be analyzed with acceptable precision by this method.

This procedure uses no purification methods, and the steroids are determined as underivatized compounds.

Derivative formation prior to chromatographic analysis has been used successfully. An unidentified component of urine was found which had a retention time very close to that of pregnanediol and which could not be separated from it by thin layer chromatography. The trimethylsilyl ether derivatives and the trifluoroacetate derivatives of the two compounds would not provide resolution; only the acetate derivatives could be separated.

The most commonly used stationary phase for the gas chromatographic separation is 3% XE-60 on Gas Chrom Q operated at about 220°C. An inert, all-glass injection port and column must be used for any steroid analysis.

Figure 10.4 shows the chromatogram obtained using the procedure described from the pregnanediol extract of normal female urine, and that obtained from the urine of a woman with hypopituit-
arism. Note the lack of the pregnanediol peak.

10.2.3 Urinary Pregnanetriol

Pregnanetriol is formed by the reduction of 17-α-hydroxyprogest-
erone, which is formed by hydroxylation at the 17- α-position of
progesterone. Progesterone is produced by the corpus luteum of
the ovary, the placenta, and the adrenal gland. Under normal
circumstances, very little pregnanetriol appears to accumulate;
rather it is secreted as the glucosiduronate in the urine.

17-α-Hydroxyprogesterone is further hydroxylated at positions
11 and 21 in the adrenal gland to form cortisol. Under unusual
conditions, such as adrenal hyperplasia, the gland is unable to
produce enough enzyme for C-21 hydroxylation. It has been postulated (14) that C-21 hydroxylation is necessary before hydroxylation at C-11. This would result in excess amounts of 17-α-
hydroxyprogesterone in the adrenal, which would be converted to
pregnanetriol and result in increased urinary levels. The increased urinary levels can be measured and associated with these
metabolic defects.

The methods used for urinary pregnanediol can also be applied
to urinary pregnanetriol, except that acid hydrolysis should not
be used because it will cause decomposition of the pregnanetriol.
Only enzymatic methods should be employed to release the free
steroid from its conjugate. Purification of the extracted hydrolysate is either done by thin layer chromatography or by column
chromatography prior to forming the trimethylsilyl ether (TMSI)
derivative for GC. A paper by Garmendia et al. (15) contains
considerable information including values for pregnanetriol and
neutral 17-ketosteroids of urine from patients with various diseases as well as that of normals.

The following procedure was developed for urinary pregnanediol, pregnanetriol, and 17-ketosteroids in urine. It is similar
to the one presented earlier for pregnanediol (12). Hydrolysis
is complete because sulfatase for the cleavage of sulfates is included with the β-glucuronidase.

HYDROLYSIS AND EXTRACTION.   A 30-cm^3 aliquot of a 24-hr urine
sample is adjusted to pH 4.6 with glacial acetic acid and buffered with 1 cm^3 of a pH 4.6 acetate buffer. Four thousand units

of β-glucuronidase and 2000 units of sulfatase per cm^3 of urine are added. This is incubated at $37^{\circ}C$ for 36-40 hr.

The hydrolysate is extracted with 45 cm^3 of ether, which is then washed twice with 20 cm^3 of lN NaOH and twice with 20 cm^3 of distilled water. The aqueous wash is discarded. The ether extract is dried over NaOH pellets and filtered into a tube or flask for evaporation.

PREPARATION OF TRIMETHYLSILYL ETHER (TMSI) DERIVATIVES OF SAMPLE AND INTERNAL STANDARD. The dry residue from the extraction is dissolved in 0.5 cm^3 of tetrahydrofuran containing 20 μg of cholestanol (internal standard) (5-β-cholestan-3-α-ol). A quantity of 0.15 cm^3 of hexamethyldisilizane(HMDS) and 0.05 cm^3 of trimethylchlorosilane (TMCS) are added, mixed, and allowed to remain overnight in a stoppered tube at room temperature. The reagents are evaporated under dry nitrogen in a warm water bath, and the residue dissolved in 0.2 cm^3 of tetrahydrofuran. This solution is transferred to a microcentrifuge tube and centrifuged for 5 min. The supernatant is transferred to a microtube prepared by sealing the end of a Pasteur pipet. The sample volume is reduced to 10-15 μl by evaporation under dry nitrogen. This sample tube is placed in a 10-cm^3 vial with a screw cap, which contains 0.5 cm^3 of tetrahydrofuran, to inhibit evaporation, for storage.

GAS CHROMATOGRAPHIC ANALYSIS. A 6-foot long, 3-mm i.d. column is used, composed of silanized glass, and packed with 3% XE-60 on Gas Chrom P or Q (100-120 mesh) which has been conditioned for 24 hr at $240^{\circ}C$. A hydrogen flame detector with helium carrier gas is employed.

Operating parameters are: column, $218^{\circ}C$; detector, $230^{\circ}C$; injection port, $235^{\circ}C$. Input gas pressures and flowrates: helium, 30 cm^3/min; hydrogen, 50 cm^3/min; and air, 200 cm^3/min. Sample size, 1-2 μl.

The unknown peaks are identified by comparing their retention times with those of their respective standard steroids, as the TMSI derivatives. The average retention time for the last compound, 11-β-hydroxyetiocholanolone, is 60 min and for the internal standard 5-β-cholestan-3-α-ol, 30 min (43).

CALCULATION OF RESULTS. Peak area measurement was done by triangulation. The equation below would be used for the calculation of the amount of steroid in 20 cm^3 urine:

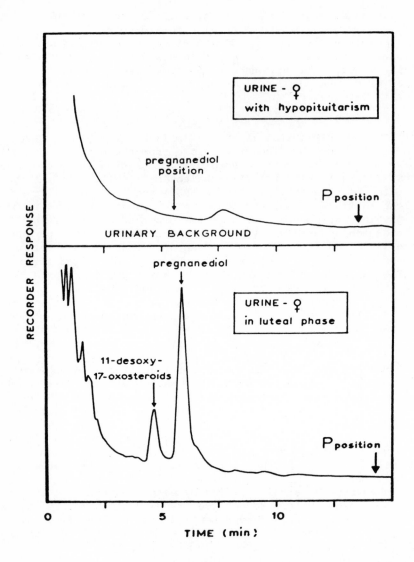

Figure 10.4   Gas chromatograms showing the absence of
pregnanediol in urine collected from a woman with panhypopituit-
arism, and its presence in urine collected from a normal woman in
the luteal phase of the menstrual cycle (liquid phase, XE-60).
P-position = retention time for progesterone.

$$\mu g/20 \text{ cm urine} = \frac{\text{peak area of steroid in sample}}{\text{peak area of internal standard}} \text{ x } 20 \text{ x } \frac{1}{RR} \quad (10.3)$$

where 20 = number of micrograms of internal standard added, and
RR (relative response) is the ratio of the peak area of each
steroid standard to the peak area of the same amount of internal
standard.

RECOVERY AND REPRODUCIBILITY. Duplicates of 20 cm^3 of pooled
urine are subjected to the procedure on different days of the
investigation. The coefficient of variation for the different
steroids varied between 3.2 and 9.8% on different days, indicating satisfactory reproducibility.

Figure 10.5 shows the separation of the TMSI derivatives of a
standard steroid mixture on an XE-60 column using this procedure.
Table 10.2 lists the abbreviations used on these chromatograms.
Figure 10.6 is the chromatogram obtained from the separation of
the TMS derivatives of the steroids extracted from normal urine
according to the above procedure.

Urinary levels in normal females range from 0.28 - 183 mg/24
hr with a mean of 1.06 mg/24 hr.

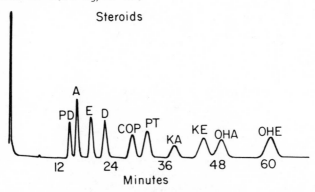

Figure 10.5   Gas chromatogram of the TMSI derivatives of
standard steroid mixture on a 3% XE-60 column.

TABLE 10.2   LIST OF ABBREVIATIONS FOR STEROIDS IN FIGURE 10.5

Pregnanediol	PD
Pregnanetriol	PT
Androsterone	A
Etiocholanolone; 3α-hydroxy-5α-androstan-17-one	E
Dehydroepiandrosterone; 3α-hydroxy-5β-androstan-17-one	D
11-Keto-androsterone; 3β-hydroxyandrost-5-en-17-one	KA
11-Keto-etiocholanolone; 3α-hydroxy-5α-androstan-11,17-dione	KE
11β-Hydroxyandrosterone; 3α-hydroxy-5β-androstane-11,17-dione	OHA
11β-Hydroxyetiocholanolone; 3α,11α-dihydroxy-5α-androstan-17-one	OHE

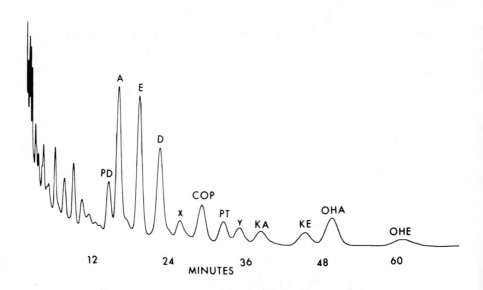

Figure 10.6   Gas chromatogram of TMSI derivatives of the
steroids from the urine of a normal adult.

10.2.4. Urinary Cholesterol

Cholesterol and cholesterol sulfate have been isolated and identified in the urine of normal and pregnant women. Increased interest is certain to arise concerning the measurement of these

and their relationship to various clinical disorders. It has
been shown (15) that the levels of urinary cholesterol are different according to different pathological conditions. These same
researchers note a lack of correlation between the urinary and
the total plasma cholesterol in some pathological conditions.
This suggests that the unconjugated urinary cholesterol may reflect an endogenous pool.

A method was developed for the analysis of urinary cholesterol
by GC which permits the investigation of urinary cholesterol during the menstrual cycle, in pathological conditions, and during
certain functional tests, such as ACTH stimulation. The method
is presented here.

EXTRACTION. A 24-hr urine sample is collected; using boric acid
as (8 g/l) preservative. A 10-cm^3 aliquot is placed into a 100-
cm^3 centrifuge tube with 10 cm^3 of distilled water. The pH is
adjusted to 4.5 using 5N H_2SO_4, and 2 cm^3 of 1M sodium acetate
buffer of pH 4.5 is added. Then 50 cm^3 of redistilled chloroform
is added and the tube shaken gently on a mechanical shaker for 10
min. After centrifugation at 1500 rpm for 5 min, the aqueous
layer is discarded. The chloroform extract is washed with 15 cm^3
of 0.1N NaOH by shaking vigorously, centrifuged, and the aqueous
phase discarded. Repeat the wash with 15 cm^3 of distilled water.
Add about 7 g of anhydrous Na_2SO_4 to the chloroform extract.
Agitate, and filter through Pyrex glass wool into a 125-cm^3
Erlenmeyer flask. Wash the tube with chloroform, filter, and add
to the flask. Do not use a plastic wash bottle for chloroform,
because contaminants may be extracted which will interfere. Evaporate the chloroform and quantitatively transfer the residue
into a 13-cm^3 centrifuge tube containing 10 or 20 µg of the 5-α-
cholestane internal standard using additional chloroform. Evaporate using nitrogen.

ACETYLATION. Acetylation of the residue may be carried out using
1 cm^3 of acetic anhydride and incubating at $100^\circ C$ for 1 hr, or
acetyl chloride at $60^\circ C$ for 15 min in a capped tube. When the
acetyl chloride is used, the residue in the tube is rinsed down
with 0.5 cm^3 of benzene and 0.5 cm^3 of the acetyl chloride is
added. Stopper and heat at $60^\circ C$ for 15 min. After either reaction is complete, the reaction mixture is evaporated to dryness
under dry N_2 in a water bath at $40^\circ C$. The walls are rinsed down
with 0.5 cm^3 of chloroform and evaporated to dryness. The residue is dissolved in 0.1 or 0.2 cm^3 of CS_2 for gas-liquid chromatographic analysis.

STANDARDS. A standard solution of 5-α-cholestane, 20 µg/cm^3, is made up in absolute ethyl alcohol. This serves as the internal standard, which is placed in the sample tubes as described above. The chromatographic standard is a mixture of 5α-cholestane and cholesterol acetate (0.5 µg/µl of each, in CS$_2$).

GAS CHROMATOGRAPHIC ANALYSIS. The 6-foot glass column is packed with 1.5% UC-W98 on 100/120 mesh Gas Chrom Q. Column temperature is 250°C. (If a 12-foot column is employed, 270°C is used.)

CALCULATION OF RESULTS. Peak heights are measured in millimeters and the following equation is used to calculate the amount of cholesterol in the unknown:

$$R_{unk} \times R_{std} \times \frac{IS}{A} \times TV \times 0.902 = \frac{mg\ cholesterol}{24\ hr} \qquad (10.4)$$

where $R_{unk} = \dfrac{\text{Peak height cholesterol acetate in unknown}}{\text{Peak height 5-α-cholestane in unknown}}$

$R_{std} = \dfrac{\text{Peak height 5-α-cholestane in GLC standard}}{\text{Peak height cholesterol acetate in GLC standard}}$

IS = µg 5-α-cholestane added to unknown
TV = total urine volume in liters

$0.902 = \dfrac{\text{mol. wt. cholesterol}}{\text{mol. wt. cholesterol acetate}}$

REPRODUCIBILITY AND RECOVERY. Three different urine extracts, containing different cholesterol levels, were each analyzed 10 times. The maximum percent error ranged from 0.39 to 1.25% and the standard deviation ranged from 0.785 to 12.368. The average recovery in 13 experiments was 99.5% with a range to 98.44 - 100.93%.

The repeatability (precision) of the technique was determined by analyzing 139 urine samples in duplicate. Values ranged from 0.06 to 13.86 mg/24 hr. The mean and standard deviation were 1.597 ± 0.022 mg/24 hr.

Normal cholesterol values are: children, 0.21 - 0.68 mg/24 hr; males, 0.44 - 1.24 mg/24 hr; and females, 0.41 - 1.57 mg/24 hr. The values included that of urine from pathological states.

Figure 10.7 shows the chromatograms obtained from a male urine sample carried through this procedure and the corresponding

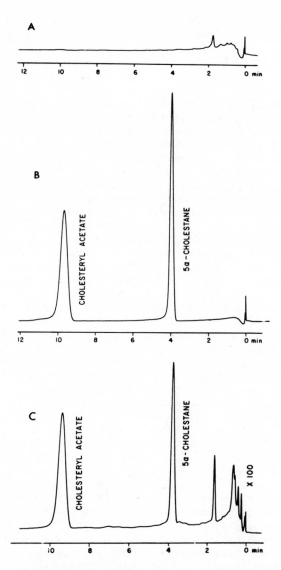

Figure 10.7.   GLC parameters: 6 ft x 0.75 in. i.d. glass
column, 1.5% UC-W98 on 100/120 mesh Gas Chrom Q at 250°C, N_2
carrier gas at 50 psi, H_2 flame ionization detector. (A) Method
blank. (B) GLC standard: 1 μg each of 5 alpha-cholestane and
cholesteryl acetate. (C) Urinary extract from a 44-year-old normal male; "free" cholesterol: 0.76 mg/24 hr.

standard mixture.

10.2.5 Catecholamines and Metabolites in Urine

The catecholamines are a group of hormones secreted by the adrenal medulla. The major urinary metabolite of norepinephrine and epinephrine is vanillylmandelic acid (VMA). Urinary levels of VMA are considerably higher than those of total catecholamine. From the standpoint of laboratory methodology, VMA estimation is preferable to total catecholamine estimation, although it is not a simple procedure. VMA has been shown to be elevated in some patients who had phenochromocytoma and normal urinary catecholamines, even though patients with neuroblastoma have a normal VMA level and elevated catecholamine levels.

Metanephrines represent metabolites of the catecholamines; urinary levels are greater than total catecholamines but less than those of VMA. In tumors, variations in the metabolic pathways can cause an increase in the metanephrines alone.

Since dopamine is present in sympathetic nervous tissue as a precursor of norepinephrine, and it has a separate metabolic pathway that yields homovanillic acid (HVA), tumors such as neuroblastomas may cause elevations of the urinary dopamine and its metabolite HVA. In some cases these elevations have been observed with normal VMA, total catecholamine, and metanephrine. Urinary HVA is usually normal in patients with phenochromocytoma. Increased HVA is found in special fluids of patients with Parkinson's disease treated with L-dopa.

Gas chromatography permits the determination of these metabolites simultaneously, and in conjunction with some clinical background, can make diagnosis much easier. Colorimetric methods lack adequate selectivity for differentiation.

There are a number of gas chromatographic procedures for the determination of the catecholamines. Methods generally accepted are making the trimethylsilyl ether derivative of the alcoholic group after using diazomethane to prepare the methyl ester and the heptafluorobutyryl derivative. The ethyl esters have been used for the separation of HVA and isoHVA and mass spectrometry applied to study their characterization.

The following method is from that reported by Karoum et al. (16,17).

URINE EXTRACTION. Neonatal urine (15 cm^3 if the 24-hr volume was greater than 100 cm^3 or 10 cm^3 if less), normal adult urine (10 cm^3), or urine of patients with suspected catecholamine secreting tumors (5 cm^3) was adjusted to pH 1 with 6N HCl. The urine was saturated with NaCl and extracted twice with 20 cm^3 of ethyl-acetate.

Urines (10 cm^3) are first hydrolyzed with sulfatase-glucuron-idase preparations prior to extraction and assay of phenolic alcohols.

Aliquots (15 cm^3 , 20 cm^3) of each ethyl acetate extract are combined and evaporated to dryness under reduced pressure at 40°C. An internal standard, equivalent to 0.05 mg of homovanill-ic acid (HVA) and 0.1 mg of 4-hydroxy-3 methoxymandelic acid (VMA), was added to a duplicate specimen and included with each batch of samples when quantification was to be carried out. Re-coveries of HVA and VMA from urine were all greater than 90%. This is a common recovery in clinical analyses.

PREPARATION OF THE METHYL ESTER-TRIMETHYLSILYL ETHER. The stand-ard acid (0.5 mg) or the residue obtained from the urine extrac-tion was dissolved in 0.5 cm^3 of methanol. Ethereal diazomethane (1 cm^3) was then added and allowed to stand for 1 min. Ethereal diazomethane was prepared by the reaction of p-tolylsulphonyl-methyl nitrosamide with an alcoholic solution of KOH.

After evaporation of the alcohol, the trimethylsilyl ether was prepared by dissolving the residue or the standard acid (after methylation) in 0.2 cm^3 of dioxane, 0.2 cm^3 of hexamethyldisila-zane, and 0.1 cm^3 of trimethylchlorosilane and mixing thoroughly. The preparation in a 2-cm^3 test tube with Teflon lined cap was left for 30 min at room temperature. After centrifugation, 0.5 µl of the supernatant can be injected into the gas chromatogra-phic instrument. The derivatives are stable if kept in a de-siccator. It is recommended that the analysis be carried out the same day.

These steps are simple, and have high yield and reproduci-bility.

GAS CHROMATOGRAPHIC ANALYSIS. The gas chromatographic analysis was performed using an argon ionization detector, although flame ionization should be suitable. The column was a 7 ft x 5 mm i.d. silanized glass, packed with 10% SE-52 on 80/100 mesh silanized Celite. The column temperature was 190°C; the detector (argon ionization) was 250°C, and the voltage for ionizing was 1250 V. The columns were conditioned for 48 hr at 250°C with a slow flow of argon. The detector was disconnected during conditioning. The inlet pressure was 1500 torrs using argon gas.

QUANTITATIVE ASPECTS. Normally the presence of high quantities
of the organic acid is sufficient for diagnosis. However, quantification is sometimes required. To do this, serial amounts of
the individual acid must be carried through the derivatization,
and a calibration curve set up. The various acids show widely
different responses to the argon detector. From the standard
curves the amounts can be derived by interpolation. Calculation
on the basis of urine dilution must be included as well as the
aliquot of the final solution that was taken for injection. The
procedure for this is similar to that described for estriol in
pregnancy urine.

A linear detector response was obtained in the concentration
range 0.1 - 0.5 µg when increasing amounts of trimethylsilyl
ether/methyl ester derivatives of the standard compounds (HVA,
dehydroxyphenyl acetic acid, VMA) were assayed.

The trimethylsilyl ether of noncarboxyl phenols gives a lower
(1/10) detector response. Furthermore, the trimethylsilyl ether
methyl esters are more readily separated on both SE-30 and XE-60
columns.

DISCUSSION. Caution should be exercised when interpreting chromatograms from a complex mixture of aromatic compounds in urine.
This is especially true when excretion of the catechols is very
high. When the derivatives of the compound extracted from urine of patients with phenylketonuria are prepared decomposition
products of the large amounts of phenylpyruvic acid present give
rise to peaks which interfere with and resemble HVA. When difficulties of this type are suspected, differential extraction of
the urine can be performed. When alkaptonuria urine is extracted
with dichloromethane, homovanillic acid is extracted preferentially, leaving the spurious compounds behind.

Figure 10.8 shows the results of analysis of authentic organic
acids derivatized as described and separated on a column of 5%
SE-52 on silanized Celite (5 ft). Figure 10.9 shows an example
of a chromatogram of extracts from urine of patients with alkaptonuria.

   10.2.6 Determination of Blood Alcohol

The effects of alcohol have medical as well as medicolegal implications. The estimation of alcohol in the blood or urine is
relevant when the physician needs to know whether it is responsible for the condition of the patient. From the medicolegal
standpoint the alcohol level is relevant in cases of sudden

death, accidents while driving, and in cases when drunkenness is
the defense plea. The various factors in determining the time
after ingestion showing maximum concentration and the quantity of
alcohol are the weight of the subject, the amount and concentration of the alcohol, how it was taken, the presence or absence of
food, and the physical state of the subject concerned.

The effects of alcohol vary among individuals, and are different for the same individual at different times. The action depends mostly on the environment and the temperament of the individual and on the degree of dilution of the alcohol consumed.
The habitual drinker usually shows relatively less effect than
would be seen with an occasional drinker from the same amount of
alcohol. Drugs potentiate the effect of alcohol.

Many cases document the synergistic effect of alcohol and barbiturates as a cause of death in cases appearing to be suicide.
Alcohol itself is probably the most frequent cause of death due
to poisoning.

A gas-solid chromatographic technique using flame ionization
detection and a Porapak Q column has been used for the identification and the determination of ethanol, isopropanol, and acetone in pharmaceutical preparations. The technique involves direct injection of an aqueous dilution of the product, and therefore is simple and direct.

GC is used to produce reliable results in the determination of
alcohol in biologic samples. Cravey and Jain (18) have done extensive work on the blood alcohol method and their review covers
the current status of methodology, including enzyme and osmomet-
ric methods as well as GC. Pereira et al (19) have described the
GC-MS determination of ethanol in blood using a modification of
the procedure described by Jain (20).

Jain (20), in his introduction, notes that the direct injection of the blood sample does not give "ghost" or interfering
peaks. The column can be used for a long time. The method is
not limited to ethanol, since methanol, isopropanol, and acetone
as well as toluene, methylethylketone, and xylene (associated
with glue sniffing) can be quickly determined.

PREPARATION OF THE SAMPLE. Two 0.5-cm^3 volumes of an isobutanol
internal standard ($10mg/cm^3$ water; pipet 12.4 cm^3 of isobutanol
and dilute to 1 liter with water) are pipetted into two different 2-dram (7.4 cm^3) shell vials, one marked "known" and the
other "unknown." A 0.5-cm^3 portion of the ethanol working standard (50 mg/100 cm^3 of blood; pipet 5 cm^3 of ethanol stock solution; 12.7 cm^3 of absolute ethanol diluted to 1 liter with water
and dilute with 100 cm^3 of blood from blood bank) is transferred
to the vial marked "known." The pipet is washed three times with

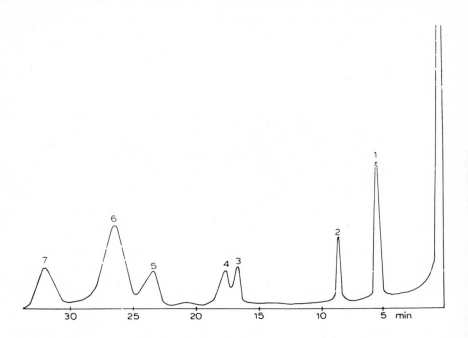

Figure 10.8.   Chromatogram of derivatives of authentic
compounds of biological importance run as a mixture on a 5-ft 5%
SE-52 column. Peak numbers correspond to the trimethylsilyl
ether/esters (TE/E) of (1) tryptophol, (2) indolylacetic acid,
(3) 5-hydroxytryptophol, (4) 5HIAA, (5) 5HIAA and the methyl ester/trimethylsilyl ethers (ME/TE, (6) 5,6-dihy-
droxyindolylacetic acid, (7) indolylpyruvic acid.

Figure 10.9.   (a) Chromatogram of trimethylsilyl ether/ester
(TE/E) derivatives prepared from an ether extract of 0.2 cm^3
phenylketonuric urine, and separated on a 7-ft 10% SE-52 column.
(b) Chromatogram prepared and run as in (a) on an aqueous extract
(10 cm^3) from a square foot of paper-tissue (napkin) previously
soaked in a dilution of 0.2 cm^3 of the phenylketonuric urine (a)
in water (5 cm^3) and then dried for two days. The peak numbers
(a,b) correspond to the TE/E derivative of (1) phenylacetic acid
(2) o-hydroxyphenylacetic acid, (3) phenyllactic acid, (4) p-
hydroxyphenylacetic acid, (5 phenylpyruvic acid.

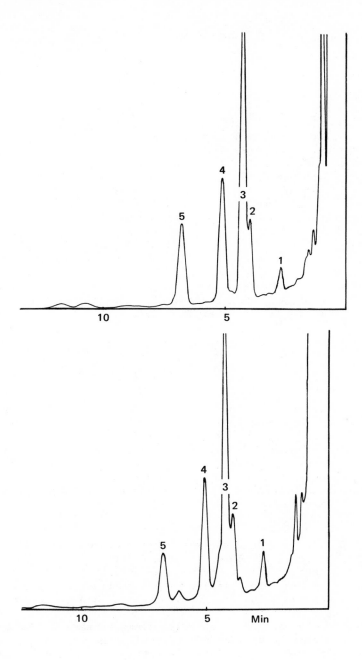

the internal standard solution. With a 10 µl Hamilton syringe,
0.5 µl of "known" and "unknown" are injected in duplicate and
the alcohol and isobutanol peak heights are measured for each.

GAS CHROMATOGRAPHIC PARAMETERS. A gas chromatograph equipped
with a hydrogen flame detector is used. Flame ionization detection allows the use of samples containing water. The column is
stainless steel (6 ft x 0.25 in. o.d.) packed with 30% Carbowax
20 M on acid-washed 60/80 mesh Chromosorb W. The column was conditioned overnight at 180°C with a flow of nitrogen carrier gas.
    Operating conditions:
    Oven temperature   100°or 130°C
    Injection temperature   160°C
    Nitrogen flow   35 cm^3/min
    Hydrogen   28 cm^3/min
    Oxygen   100 cm^3/min

CALCULATIONS. Calculations are done using peak heights of both
the alcohol and the internal standard. Use of the following formula can be employed for determinations:

$$\text{mg ethanol/100 cm}^3 \text{ of unknown} = C_S \frac{h_x}{h_s} \times \frac{R_s}{R_x}$$

where C_S = concentration of standard ethanol solution (here it is
          50 mg/100 cm^3),
      h_x = peak height of ethanol from "known,"
      h_s = peak height of internal standard (isobutanol) in
          unknown,
      R_x = peak height of ethanol from known,
      R_s = peak height of internal standard (isobutanol) in known
          sample.

DISCUSSION. Figure 10.10 shows a typical chromatogram obtained
by injecting 0.5 µl of blood-isobutanol internal standard mixture. The peaks are clearly separated and sharp. Table 10.3
gives the retention and relative retention times of some common
volatile compounds associated with drinking, diabetes, and glue
sniffing.
    The operating parameters such as column temperature or gas
flow are not critical. They may vary with the type of compound
under investigation, as well as the nature of the gas chromato-

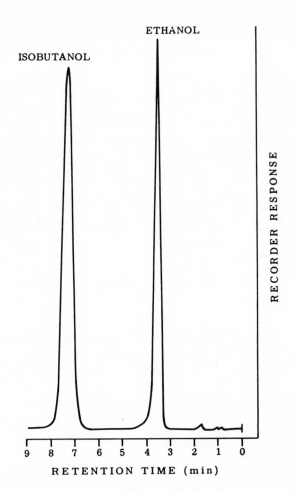

Figure 10.10.   Gas chromatogram obtained on injecting 0.5 μl of blood-isobutanol internal standard mixture at 100⁰ oven temperature.

graph in use.
    Other packings may also be used in this determination, includ-
ing Hallcomid M180L on Chromosorb W 80-100 mesh, or Porapak Q 80-
100 mesh. Porapak has different characteristics but has the advantage that it represents a packing without a coating.
    The blood that is injected into the system enters the injec-

TABLE 10.3   RETENTION DATA FOR SOME COMPOUNDS[1]

	Oven Temp. 100°C		Oven Temp. 130°C	
	Retention Time (Min)	Retention Time Relative to Ethanol	Retention Time (Min)	Retention Time Relative to Ethanol
Acetaldehyde	2.56	0.73	1.25	0.60
Acetone	2.38	0.68	1.62	0.76
Benzene	4.31	1.23	2.81	1.33
Chloroform	5.62	1.61	3.25	1.53
Ethanol	3.50	1.00	2.12	1.00
Isobutanol	7.25	2.10	3.94	1.86
Isopropanol	3.19	0.91	2.06	0.97
Methyl ethyl ketone	3.38	0.97	2.19	1.03
Propyl acetate	4.50	1.29	2.75	1.30
Toluene	6.94	1.98	4.06	1.92
Xylene	11.38	3.25	6.38	3.01

[1] 6 ft x 0.25-in. stainless steel column packed with 30% Carbowax 20 M on Chromosorb W, AW, 60/80 mesh. Flowrates: nitrogen, 35 cm^3/min, hydrogen, 28 cm^3/min; oxygen, 100 cm^3/min.

tion port and is trapped there, while the volatiles are carried by gas flow into the column. On column injection will plug the column. Thus it would be of advantage, if this method is to be used, to have a long injection port to provide for multiple injections over a long period. After 6-8 months the injection port should be disconnected and the trapped blood removed, and if necessary, it is cleaned with water. Allow the instrument to stabilize overnight before reuse.

Quantification based on peak height is satisfactory, although peak areas may be used if the peaks obtained are too wide or are not symmetrical.

10.2.7 Anesthetics in Blood

Analysis by GC is well established for quantifying the levels of commonly used anesthetic and inert gases in blood or other liquids. Most studies on methods involve only a single species. In a method for measuring the distribution of gases in the lung and blood it is necessary to measure concentrations in arterial blood and expired gas of six to eight inert gases present together. These gases range from very insoluble (sulfur hexafluoride) to very soluble (acetone). This requires a method capable of handling combinations of gases with a wide range of solubilities.

There are two separate requirements at present for measuring volatile anesthetics in blood. One is retrospect analysis of blood samples. The other is on-line monitoring of blood concentration during animal and clinical toxicity studies. Usually a method allowing analysis of samples within minutes is desired. The continued influx of new anesthetics, including the fluorinated types, will increase the demand for methodology involving direct injection GC. There are a number of methods for determination of individual series of anesthetics. Direct injection of the blood sample was used as early as 1964.

The method described as follows was chosen because it will separate nine different gases after removal of blood and can also be used for individual gases (21,22).

PREPARATION OF THE SAMPLES. The method of extraction is one of equilibration of the liquid sample with a similar volume of helium in a gas-tight ungreased glass 50-cm^3 syringe. The equilibration is performed by shaking in a waterbath at the desired temperature, usually $37^\circ C$ ($\pm 0.5^\circ C$). Following the equilibration, the gas phase is anaerobically transferred into another glass syringe which is to be used to introduce the sample into the gas chromatograph. This approach, as contrasted with extraction by negative-pressure systems, is adequate for gases of all solubilities, and thus permits analysis of acetone (very soluble) or sulfur hexafluoride (poorly soluble).

Since solubility in blood is altered by many factors, including temperature, lipid level, protein, hemoglobin, and hemocrit values, the solubilities of the gases in the samples should be measured individually. This requires only a second equilibration of the original sample with a new aliquot of helium, and from the ratio of the chromatogram peak heights of the two extractions the solubility of the gas concerned can be calculated.

THE CHROMATOGRAPHIC PARAMETERS. Samples are introduced into the gas chromatograph via a 2-cm^3 constant volume inlet loop, requiring 5 cm^3 for adequate flushing and filling. The sample was separated on a stainless steel column 6 ft long x 1/8 in diameter

packed with 80/100 mesh Porapak-T and helium was employed as
carrier gas. When a flame ionization detector was used the temperature was 160-170°C and the carrier gas flowrate was between 30
and 50 cm^3/min.

For the analysis of sulfur hexafluoride an electron capture
detector must be used. Analysis time for halothane, acetone,
ether, cyclopropane, acetylenes, ethane, methane, and sulfur
hexafluoride, under normal conditions, will average 8 min. Since
the two detectors operate under different conditions samples containing both the hydrocarbons and sulfur hexafluoride are run
through the gas chromatograph twice. By simultaneous use of separate ovens, the FID and ECD measurements can be made concurrent-
ly. Otherwise, the separate injections must be made after switching the detectors.

CALCULATIONS. To calculate the concentration of each gas in the
original liquid sample from those in the equilibrated gas, it is
necessary to measure the volume of the liquid and the gas in the
syringe following equilibration. The solubility of each gas in
the liquid must also be known. The volume of the liquid is obtained by subtracting the weight of the dry syringe from that of
the syringe plus blood, measured to the nearest 0.1 g. Then divide by the density of the blood which is measured after the
chromatography has been completed using pooled liquid from all
samples. The volume of the gas in the sample is obtained from
the difference between the total volume of gas plus liquid, read
from the syringe graduations to the nearest 0.5 cm^3 and the above
calculated volume of liquid. All the measurements of the gas
volume must be made at the temperature of the equilibration. If
V_ℓ and V_g are the volumes of the liquid and the gas in cubic centimeters, respectively, and S is the solubility of the gas in the
sample in cm^3 gas/100 cm^3 liquid then performing a mass balance,
the original pressure p_o can be expressed in terms of partial
pressure p_e:

$$P_o = P_e \left(1.0 + \frac{V_g}{V_\ell} /k \cdot S\right) \tag{10.5}$$

where K equals (the barometric pressure - saturated water vapor
pressure)/100. The value of water vapor pressure is that corresponding to the temperature of the equilibration procedure.

In many applications it is not necessary to know the absolute
gas partial pressure or concentrations. In some cases, for example, arterial, venous, or expired-to-venous concentrations are
sufficient. This avoids calibrations, removes a source of error,

and improves the accuracy.

To calculate the solubility of the gas in question, retaining P_e, S, and K from the above, define P_e* as the partial pressure of the gas after the record equilibration if V_g* and $V_\ell*$ are corresponding gas and liquid values, then performing a mass balance

$$S = \left[\frac{1}{k} \quad \frac{V_g^*}{V_\ell^*}\right] \div \left[\frac{Pe}{Pe*} - 1.0\right] \tag{10.6}$$

If the gas chromatograph is linear with respect to the peak height, the ratio Pe/Pe* reduces to the corresponding ratio of peak heights.

DISCUSSION. Figure 10.11 shows the separation of a group of gases from blood. The procedure described depends on complete equilibration between the gas and liquid in the syringe. The period to achieve equilibration depends on the gas, and can be as little as 1 min for acetone and as much as 30 min for sulfur hexafluoride. Linear calibration curves can be obtained when peak height is plotted against gas concentration for either the flame ionization or electron capture detector. The reproducibility (n = 10) for ethane gave a standard deviation of 23% of the mean concentration while it was 2.3% for halothane and 1.8% for ether.

10.2.8 Urinary Aromatic Acids

Some inborn errors of metabolism can be characterized by excessive urinary excretion of aromatic acid metabolites. These acids are distinct from the vanillyl acids discussed in a previous section. Phenylketonuria, alkaptonuria, and tyrosinosis can be diagnosed by determination of the aromatic acid metabolites. Aromatic acid profiles are characteristic of specific metabolic defects, and can be used to confirm diagnoses obtained from amino acid and other studies. Quantification of the individual aromatic acid gives information as to the fate of ingested amino acid in diseases such as phenylketonuria, where there is a block in the metabolic pathway involving the particular amino acid.

A variety of methods, including wet chemistry methods and chromatographic techniques, have been described for quantitication of this type of metabolite. Since the detection of inborn errors of metabolism is becoming a more easily performed technique, it is appropriate to include a common example of the determination of other aromatic acids such as those found in the

urine of phenylketonuria, alkaptonuria, and tyrosinosis. The
following example is taken from Hill et al. (23). A more advanced procedure involving GC-MS was recently described (24).

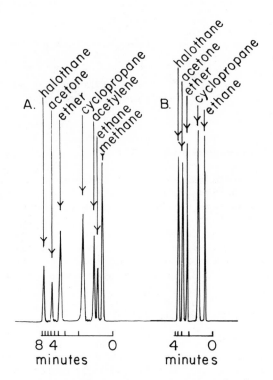

Figure 10.11.   Examples of chromatograms of mixtures of eight
(a) and six (b) gases. In each case, one gas (SF_6) is not
detected by flame ionization detector, so only seven and five
peaks are seen, respectively. Elution time for halothane, the
slowest gas, is 8 min (a, column temperature 160^0 and carrier
flow 30 cm^3/min) and 4 min (b, column temperature 170^0 and
carrier flow 50 cm^3/min). Note adequate separation of all gases
in both cases.

EXTRACTION OF URINE. One cubic centimeter of urine was acidified
by addition of two drops of concentrated HCl. After addition of
2 cm^3 of saturated NaCl solution, the urine was extracted with
3 x 3 cm^3 of ethyl acetate. Centrifugation was used to help sep-

aration.   The organic phases were evaporated under reduced pressure.

PREPARATION OF DERIVATIVES.   To each extract obtained above was
added 150 µl of N,O-bis (trimethylsilyl) acetamide (BSA).   The
internal standard was added and the volume was made up to 500 µl
with dioxane.   The reaction mixture was kept at room temperature
for 15 min in a stoppered tube.   The tube was kept sealed until
samples were to be drawn.

GAS CHROMATOGRAPHIC ANALYSIS.   The gas chromatograph was equipped
(in the example used) with a flame ionization detector.   Two
stainless steel columns were packed with 5% SE-30 and 8% SE-52
on 60/80 mesh Chromasorb W-AW HMDS.   The detector was maintained
at 240°C and helium carrier was used at a flowrate of 3.5 cm^3/min.
Column temperature was programmed.   The operator should determine
the best rate of progression of temperature with the set-up in
use.   For example, one can start at 80°C and program at 4°/min up
to 250°C.
       Five microliters of derivatized sample is injected into the
gas chromatograph.   The injection port temperature should also be
controlled.   However, the amount injected will have to be medi-
ated according to the concentration of any individual aromatic
acid present in the urine.   Some aromatic acids may be present in
amounts that may overload the column.

QUANTITATIVE ASPECTS.   Peak areas are generally used (area =
height x width at half-height).   The peak area ratio sample to
internal standard was plotted against the corresponding weight
of serial amounts of standard compound.   A standard curve must be
obtained for each individual acid because of differences in detector response.

DISCUSSION.   The concentrations of six unconjugated aromatic
acids in urine of 12 cases of phenylketonuria on normal and
phenylalanine-restricted diets are shown in Table 10.4.   The
amounts were calculated as mg/g creatinine (done on an aliquot by
established procedures).   An example of the separation of six
individual known   aromatic acids carried through the procedure is
seen in Figure 10.12.   An example of the normal profile of the
acids in a normal urine is also shown.

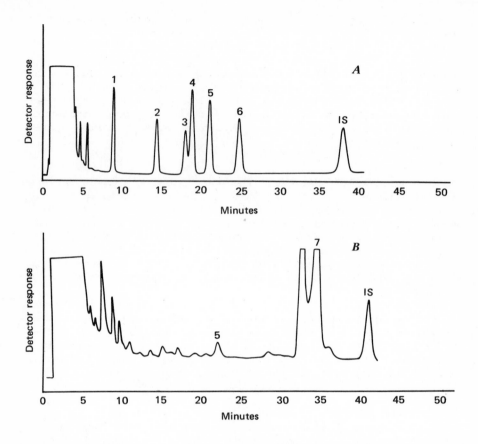

Figure 10.12.   (a) Separation of mixture of authentic aromatic
acids.   1, phenylacetic acid; 2, mandelic acid; 3, o-hydroxyphenylacetic acid; 4, beta-phenyllactic acid; 5, p-hydroxyphenyl-
acetic acid; 6, beta-phenylpyruvic acid; 7, hippuric acid; IS,
internal standard.   (b) Aromatic acid profile of normal human
urine.

10.2.9 Amino Acids in Blood

Inborn errors of amino acid metabolism, though rare, do occur,
often with lethal results.   Many of these errors are now well
defined, and clinicians know the major characteristics of the

TABLE 10.4 CONCENTRATION OF UNCONJUGATED AROMATIC ACIDS IN URINE SPECIMENS OF 12 CASES OF PHENYLKETONURIA ON NORMAL AND PHENYLALANINE-RESTRICTED DIETS

Subject	Phenylacetic Acid	Mandelic Acid	o-Hydroxy-phenylacetic Acid	Phenyllactic Acid	Phenyl-pyruvic Acid	p-Hydroxy-phenylacetic Acid
Normal diet
F. S.	213	18	83	687	515	22
P. H.	62	15	68	164	226	15
E. P.	73	32	115	532	433	30
K. J.	72	38	146	704	392	16
K. K.	60	34	161	515	592	24
E. L.	85	24	63	294	405	17
V. L.	75	39	114	635	610	37
C. A.	135	71	219	1695	1910	47
Restricted Phenylalanine Diet
D. K.	38	6	25	6.5	18	17
D. S.	33	13	38	24	96	36
L. L.	120	21	76	186	271	26
V. B.	2.0	0.6	0.2	0.1	0.1	24

533

diseases and the chemical compounds to analyze for diagnosis.
Gas chromatographic methods can aid in these diagnoses, for amino
acids in blood and urine and for organic acids in urine.

Disturbances in the metabolism of amino acids occur mainly
with the aromatic amino acids, such as phenylalanine and tyrosine. The error may be caused by the deficiency of a specific
enzyme, a defect in the enzyme itself, or by the absence of factors necessary for the proper function of the enzyme.

For example, alkaponuria is characterized by homogentisic acid
in urine; phenylketonuria, which results in mental retardation,
is characterized by quantities of phenylpyruvic acid in the urine. It is diagnosed in a suspected patient by determining the
amount of this acid in the urine and the increased levels of
phenylalanine in the plasma. Maple sugar disease is diagnosed
the presence of large amounts of the branched chain amino acids,
such as valine, leucine, and isoleucine in the blood and urine.

Many of these compounds can be determined by fluorometric or
paper chromatographic procedures, which often are time-consuming,
and the fluorometric procedure is useful for only a few of the
amino acids, each of which has to be determined by a separate
procedure. With the chromatographic procedures, more than one
compound of interest may be determined. This reduces the number
of procedures in use, and allows a profile of the compounds of
interest to be observed.

The naturally occurring amino acids are very polar, and cannot
be separated as the free compounds by GC at a temperature below
decomposition. If the polar groups in the molecule are chemically modified to produce a more volatile derivative a suitable
temperature is then possible. Weinstein (25) reviews all the
various derivatives which may be formed from amino acids and the
GC conditions necessary to separate them. In actual practice
only three derivatives are in popular use. These include the N-
heptafluorobutyryl n-propyl ester derivatives, the N-trimethyl-
silyl ether derivatives, and the n-trifluoroacetyl n-butyl ester
derivatives.

The extraction and purification procedure for amino acids in
plasma (26) will be presented here. Proteins are removed from
plasma by precipitation with picric acid, and the extract is
placed on an ion-exchange column. The amino acids are retained
on the column while the unknown substances which will interfere
with the analysis are eluted off first, and discarded. The
amino acids are then eluted, derivatized, and analyzed by GC.

SAMPLE EXTRACTION. For removal of proteins from blood plasma
with picrate, 5 cm^3 of blood plasma is placed in a 125-cm^3 Erlenmeyer flask with 25 cm^3 of 10% aqueous picric acid and stirred

5 min with a magnetic stirrer. The resulting suspension is centrifuged at 3500 rpm for 10 min and the clear supernatant containing the free amino acids and excess picrate then decanted.

PREPARATION OF CATION-EXCHANGE RESIN. Amberlite CG-120-H cation-
exchange resin is placed in a 600 cm^3 beaker; about 3 g of dry
resin (10-meq capacity) is required for each 0.9 x 15 cm column.
The resin is covered with 3N NH_4OH, swirled for 30-60 min, and
allowed to settle. The NH_4OH is decanted off, and the process
is repeated twice. The resin is then washed with doubly distilled water until it is approximately neutral.
     The resin was regenerated with 3N HCl (three times) as des-
cribed above and then washed with doubly distilled water until
neutral.
     The 0.9 x 15 cm. columns with sintered glass filters (Fischer
and Porter, Warminster, PA) are filled about halfway with the
wet resin, being careful to avoid channeling. Doubly distilled
water is added with more resin. The column is filled to 7 cm,
washed with distilled water, and the liquid is drained to the
level of the resin before the sample extract is applied.

ISOLATION OF AMINO ACIDS FROM DEPROTEINIZED SAMPLES. A 25-cm^3
aliquot of the supernatant is placed on top of the 0.9 x 7 cm
bed of Amberlite CG-120-H, prepared as in the preceding section.
The liquid is allowed to flow through the column at a rate of 1-2
cm^3/min, and five distilled water washings of 5-10 cm^3 each are
used to wash the column and discarded.
     The amino acids are eluted using 5 x 2 cm^3 of 3N NH_4OH and
5 x 5 cm^3 of distilled water, at a flow of 1-2 cm^3/min. The
combined fractions are placed in a 125 cm^3, 24/40 neck flask and
evaporated to dryness on a rotary evaporator.

PREPARATION OF DERIVATIZATION REAGENTS. Methanol is refluxed
over magnesium turnings, and methylene chloride and n-butanol are
refluxed over calcium chloride before distilling in an all-glass
system. The solvents are stored in all-glass inverted top
bottles.
   The methylation reagents are made by passing anhydrous HCl gas
through a H_2SO_4 drying tower before bubbling into the distilled
n-butanol or methanol.

DERIVATIZATION OF THE ION-EXCHANGE CLEANED PHYSIOLOGICAL FLUID.
Micro Method (2-200 µg total)

An aqueous aliquot of the amino acid eluate from the ion-exchange column, or standard, is put in a microreaction vial such as the Reacti-Vial (Pierce Chemical), and evaporated just to dryness using dry nitrogen and a 100°C sand bath. To assure removal of water, add 100 μl of CH_2Cl_2 to vial and the mixture dried. The process is then repeated. It is important not to heat the tube too long when dry, as some sample decomposition may occur.

The amino acid methyl esters are formed by adding 100 μl of the methanolic HCl to the reaction vial. A Teflon-lined cap is put on and the contents of the tube agitated ultrasonically for 15-30 sec, and allowed to esterify at ambient for 30 min with mixing every 10 min.

For interesterification to the n-butyl esters, 100 μl of n-butanol HCl is mixed ultrasonically in the reaction vial. The solution is heated for 2 hr, 30 min in the 100°C sand bath with ultrasonic mixing every 30 min. This reaction mixture is evaporated just to dryness under nitrogen as described above.

To acylate the n-butyl esters, 60 μl of CH_2Cl_2 and 20 μl of TFAA are added, then mixed with ultrasound and heated at 150°C for 5 min in an oil bath. The samples are then ready for gas-liquid chromatographic analysis.

GAS CHROMATOGRAPHIC ANALYSIS. The gas chromatographic separation is performed on 0.325% ethylene glycol adipate (EGA) on 80/100 mesh acid-washed Chromosorb G in a 1.5 m x 4 mm i.d. glass column. Carrier gas is nitrogen at a flowrate of 70 cm^3/min. The initial column temperature is set at 70°C and programmed to a final temperature of 210°C/220°C using a rate of 4°C/min. The flash evaporator and hydrogen flame detector temperatures are set at appropriately higher temperatures.

While using this procedure, it is often noted that the peak due to the trifluoroacetic acid interferes with the most volatile amino acid derivatives alanine and glycine. A useful procedure was developed to overcome this and to allow the injection of larger sample volumes. This is done with an injection port venting device called a Sol-Vent. A small precolumn of a mixed phase 2% OV-17 and 1% 210 on 100-120 mesh is placed on Gas Chrom Q in front of the EGA column normally used. As the sample injection is made, a timer-activated solenoid valve opens for a prescribed time allowing the solvent and other volatiles to vent to the atmosphere. At the end of the time period the valve closes, the temperature program is begun, and the analysis continued in the normal manner. Retention of the amino acids is based on the selection of the precolumn length and packing, the initial temperature, the venting time, and the carrier gas flowrate. This venting system allows injections of 100 μl to be made, with resulting increased sensitivity and better baseline stability.

PROCEDURE CONSIDERATIONS. The following points should be considered when using the analytical procedure just described in
order to obtain optimum results:

1. Less gas-liquid chromatographic interference occurs when protein in plasma and serum is precipitated with picric acid than
   with trichloroacetic acid or sulfosalicylic acid. Ultracent-
   rifugation or membrane techniques could be used. The cation-
   exchange columns should be washed with water until no visible
   picric acid remains.
2. After the alkaline column eluate is dried, the exclusion of
   moisture must be maintained. Use dry reagents, and careful
   technique.
3. Methionine forms insoluble salts with metals, particularly
   iron. It is important to exclude metals from the reagents and
   procedure or methionione may be lost. Magnetic stirring bars,
   if used, should be soaked in 6N HCl several hours before use.
4. Time and temperature for reactions must be carefully controlled in order to avoid decreased values for some amino acids.
   Oil baths are recommended for uniform heat transfer.
5. Procedural blanks and reagent purity checks should be performed on a regular basis.
6. Stainless steel transfer lines leading from the effluent and
   of the column to the detector should not be used because they
   will promote derivative decomposition. Glass columns should
   be used instead. Tyrosine, serine, and threonine are easily
   decomposed on hot metal surfaces.

   Normal procedures for the qualitative and quantitative evaluation of the chromatograms are employed. Table 10.5 lists normal
values for plasma amino acids as determined by this procedure.

10.2.10 Drugs of Abuse

Many abused drugs may be separated, identified, and quantified by
GC methods. These methods may be used for the broad screening
for drugs in biological samples such as blood or urine, or they
may be used as confirmatory methods to aid the identification of
a drug after it has been tentatively identified by another procedure such as ultraviolet spectrophotometry or thin layer chromatography. A general review of the many procedures for identification of drugs in forensic toxicology has been presented by

Touchstone and Dobbins. (27).

Methods for the extraction of drugs from biological fluids must be carefully chosen so as to be fairly selective for the compound type(s) being sought; otherwise any number of extraneous substances may be extracted with the drugs of interest   This often occurs no matter how selective the extraction is and it is usually advantageous to submit the extract to a purification or "cleanup" procedure to remove the co-extracted extraneous substances which may interfere with the analysis and identification of the drugs.   Such substances found in direct chloroform extracts of blood have been considered and should be borne in mind. Many procedures employ extraction with nonpolar organic solvents followed by back extraction into polar aqueous solvents or into other organic solvents after pH adjustment with concentration by evaporation for analysis by TLC or GC.   Resins, such as Rohm and Haas Amberlite XAD-2 in columns, have also been used for the purification of urine extracts prior to chromatography.

Despite the large number of stationary phases available for GC, few are widely used for drug analysis; popular phases include SE-30, OV-1, OV-17, Apiezon L/KOH, Carbowax, and some of the newer phases developed especially for drug analysis such as Supelco's SP-2250DA for barbiturates.   Moffat et al. (28) recommended a single low-polarity phase, such as SE-30 or OV-17, as the "preferred liquid phase" for the separation of 62 basic drugs.   The liquid phase diethylene glycol succinate with potassium hydroxide (DEGS/KOH) was recommended for low molecular-weight drugs.   More polar columns not only failed to separate many compounds altogether, but the retention indices of those drugs that were eluted were not greatly different.

Polar phase columns generally have a lower maximum operating temperature than nonpolar phase columns and are therefore somewhat less useful for the broad screening of extracts for basic drugs.

The total number of drugs eluted from KOH-treated phases is less than the number eluted from similar phase nontreated colums. The KOH treatment is done to reduce the adsorption of compounds by the support phase, and although it does this well, it also prevents the elution of phenolic base drugs such as morphine because of conversion to nonvolatile potassium phenolates.   Many higher molecular weight alkaloids become thermolabile under such alkaline conditions and do not elute.   Non-KOH treated phases have less tendency to cause such decomposition.

Derivatization techniques are often employed in drug analysis to extend the limits of detection sensitivity and to aid in identification.   Often the chromatographic properties of the derivative are different from those of the drug itself and may aid in the separation as well as provide another parameter for identifi-

TABLE 10.5   AMINO ACID NORMAL VALUES IN BLOOD PLASMA

Amino Acid	Average Value (mg/100 cm^3 Plasma), 5 cm^3 Aliquot
Alanine	2.6
Valine	2.3
Glycine	1.9
Isoleucine	0.7
Leucine	1.5
Proline	3.3
Threonine	1.8
Serine	1.4
Methionine	0.2
Hydroxyproline	0.1
Phenylalanine	0.8
Aspartic acid	0.6
Glutamic acid	4.5
Tyrosine	0.9
Lysine	2.5
Histidine	0.7
Arginine	1.4

cation. The fact that a drug forms a derivative with a particular reagent indicates the functional groups that may be present,
and this provides additional information. Cimbura and Kofoed
(29) have reviewed some of the derivatization techniques used in
forensic toxicoloty for gas chromatographic analysis.

Derivatives may be formed prior to analysis in a tube or reaction vial, or some derivatives may be formed directly on the
gas chromatographic column at the start of the analysis. Chapter
12 of this volume deals more extensively with this topic and the
reader is referred to the appropriate section.

As in the preceding section, techniques will be given for the
gas chromatographic analysis of the major classes of abused drugs
as isolated from biological material.

AMPHETAMINES. Amphetamines, used in medicine as central nervous
system (CNS) stimulants, are among the most commonly abused
drugs. Amphetamine itself is chemically di-α-methylphenethyl-
amine, the most common commercial name being Benzedrine, nick-
named "Bennie." Other commonly encountered amphetamines in urine

include p-methoxyamphetamine, N-methylamphetamine (methamphetamine), N,α-dimethylamphetamine (mephentermine), and α-methyl-
amphetamine (normephentermine).

Fales and Pisano (30) were the first to describe the gas
chromatographic analysis of amphetamine. They used 4% SE-30 in a
silanized glass column and an argon ionization detector for the
separation of β-phenethylamine, amphetamine, norephedrine, eph
drine, histamine, tyramine, dimethyltriptamine, tryptamine, 5-
methoxytryptamine, benzylamine, N-methylbenzylamine, octopamine,
synephrine, normetanephrine, serotonin, N-acetyltryptamine, and
melatonin. The method was also used for the analysis of amphetamine extracted from urine.

Many gas chromatographic methods for amphetamine have been developed since then, most of which include a recommended internal
standard such as N,N-dimethylaniline, phenethylamine, and diphenylamine. These methods also include the formation of derivatives such as the acetylated amines and the ketone derivative of
amphetamine as a further check on identity.

The following procedure for the extraction and analysis of
amphetamine is from Setchell and Cooper (31), employing diphenylamine as the internal standard.

EXTRACTION. To 10 cm^3 of urine in a glass-stoppered tube is added 0.5 cm^3 of diphenylamine (in ethanol) internal standard (10
μg). This solution is acidified with 0.4 cm^3 of 6N HCl and acidic substances extracted by shaking with 5 cm^3 of diethyl ether,
which is then discarded. The aqueous phase is made alkaline with
1 cm^3 of 20% KOH and extracted with 5 cm^3 of ether on a rotary
mixer. The ether is evaporated to dryness on a water bath in a
10-cm^3 centrifuge tube. The extract is not allowed to evaporate
in the hot water bath any longer than necessary. The residue is
dissolved in 50 μl of ether immediately before injection into the
gas chromatograph. Five microliters of the dissolved extract is
injected.

PREPARATION OF STANDARDS. For the amphetamine standard, 126 mg
of amphetamine hydrochloride are dissolved in 100 cm^{3}of water.
This is equivalent to 1 mg/cm^3 of amphetamine. To 5 cm^3 of the
amphetamine solution is added 0.5 cm^{3}of 20% KOH. After extraction with 5 cm^3 of diethyl ether, this is diluted 1:10 with
ether, giving a 0.1 mg/cm^3 solution of amphetamine.

The internal standard solution is made by dissolving 100 mg of
diphenylamine in 100 cm^{3}of ethanol and diluting an aliquot 1:50
with ethanol to obtain a final concentration of 20 μg/cm^3.

GAS CHROMATOGRAPHY. A 6-foot long silanized glass column, packed with 5% SE-30, is connected to a flame ionization detector. Nitrogen is the carrier gas. Operation conditions were: column temperature 140°, injection port 190°, detector temperature 200° C, nitrogen flow 45 cm^3/min, air flow through detector 500 cm^3/min, and hydrogen flow through detector 33 cm^3/min.

After injection of 5 µl of the dissolved extract, with the detector amplifier set at 1 x 10^{-10}A, the separation is allowed to proceed 20 min when the diphenylamine internal standard should be eluted. The retention times of any compounds eluted are measured relative to the internal standard and compared to the relative retention time to the internal standard as pure amphetamine.

PREPARATION OF KETONE DERIVATIVE. For additional evidence of the presence of amphetamine, a ketone derivative is prepared by adding 0.5 cm^3 of acetone to the urine extract and evaporating at 60° to a volume of about 50 µl. Some unreacted amphetamine remains. A 5-µl aliquot of the mixture is chromatographed and the relative retention times calculated. The retention time of diphenylamine will not change, as it does not form a ketone derivative.

A urine sample giving rise to peaks under both of the above sets of conditions may be strongly suspected of containing amphetamine. Figure 10.13 shows the presence of amphetamine from a urine sample obtained four hours after the oral ingestion of 5 mg of amphetamine sulfate. Peaks indicate amphetamine (A), diphenylamine (I) and nicotine (N) which is normally found in the urine of smokers. Figure 10.14 is a chromatogram obtained after the preparation of the ketone derivative of amphetamine on the same urine extract as in Figure 10.13. The presence of amphetamine is confirmed by the additional peak of the ketone derivative (K). Some unreacted amphetamine (A) is present as well as the diphenylamine (I).

QUANTIFICATION. The addition of the internal standard (10 µg) to the urine before extraction is necessary for the quantification procedures. Known amounts of amphetamine and diphenylamine in the range 0-5 µg are chromatographed and calibration curves are plotted for peak area (y axis) versus concentration (x axis).

The detector response is greater for diphenylamine than for an equivalent amount of amphetamine. Over the 0-5.0 µg range, the ratio expressed as a percentage averaged 83%. Quantification is linear over the range used.

The following formula is used to calculate the amount of amphetamine present in a urine sample (µg/100 cm^3):

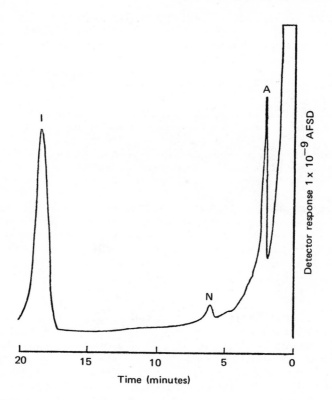

Figure 10.13.   Chromatogram obtained after the analysis of a
urine specimen obtained 4 hr after the oral administration of
5 mg of amphetamine sulphate. Gas chromatography was on 5% SE-30
with a column temperature 140°. Peaks indicate amphetamine (A),
the internal standard diphenylamine (I) and the presence of
nicotine (N).

$$\frac{\text{Area of amphetamine peak}}{\text{Area of internal standard peak}} \times \begin{array}{c}\text{Amount of internal}\\ \text{standard added}\\ \text{initially (10 } \mu\text{g)}\end{array} \times \begin{array}{c}\text{Response}\\ \text{ratio}\end{array} \times 10$$

$$(10.7)$$

RECOVERY AND REPRODUCIBILITY. The chromatographic conditions
are chosen such that the retention time for the amphetamine would
be suitable for rapid identification and quantification.

Table 10.6 lists the retention times. Methylamphetamine is recovered by this extraction procedure and the retention time for
it is given in the table.

TABLE 10.6   RELATIVE MEAN RETENTION TIMES. COLUMN; 5% SE-30 at
                            140°C

Substance	Relative Mean Retention Time	Standard Deviation	Number of Tests
Amphetamine	0.14	0.0019	20
Methylamphetamine	0.16	0.0032	20
Ketone derivative of amphetamine	0.20	0.004	20
Nicotine	0.33	0.0032	22
Diphenylamine	1.00	---	--

(approx. 18 min.)

To determine recovery, amounts of amphetamine and diphenylamine in the range 0-20 µg are added to normal urine and then extracted and assayed according to the procedure. Quantification
is done and the mean percent recovery for amphetamine was 82% and
for diphenylamine it was 83%.
     Determination of precision was made by adding 1 µg of amphe-
tamine to twenty 10-cm^3 aliquots of a sample of urine. The mean
amount of amphetamine detected was 0.98 µg ± 0.09 standard deviations.
     The sensitivity is such that a discernible peak is produced by
0.01 µg of amphetamine in 1 cm^3 of urine. Forty-eight hours after the ingestion of 5 mg of amphetamine sulfate, it is still
possible to detect the amphetamine in the urine.

BARBITURATES. Barbiturates are 5,5'-disubstituted derivatives of
barbituric acid (malonyl urea). Barbituric acid does not have
hypnotic properties, but its derivatives do, and are widely used
in medicine as hypnotic-sedatives. Because of their widespread
use, the intentional and unintentional abuse of these drugs is
most frequent, and most larger hospitals and medical centers have
clinical screening and analytical procedures established to determine such compounds in the body fluids of patients.
     Commonly encountered barbiturates include amobarbital, buta-

barbital, pentobarbital, secobarbital, and phenobarbital. Barbital has been used as an internal standard in the analysis of the
these drugs. Other sedatives commonly encountered include
glutethimide, meprobamate, diphenylhydantoin, and primidone
(primaclone, 2-desoxyphenobarbital).

Spectrophotometric methods have traditionally been used for
the analysis of barbiturates, but these procedures are time consuming, do not identify the barbiturate in question, and it is
difficult to quantify accurately unless this identity is known.
For these reasons, the development of thin layer and gas chromatographic methods for barbiturate analysis has been occurring.
Jain and Cravey (32) have reviewed all the different types of
methods being used for the identification of barbiturates from
biological specimens.

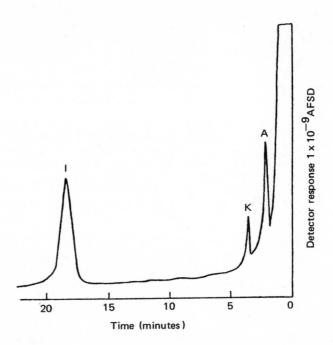

Figure 10.14.   A chromatogram obtained after the preparation
of the ketone derivative of amphetamine on the same urine extract
as in Figure 1. Confirmation of the presence of amphetamine is
indicated by the additional peak for the ketone derivative (K).
Some unconverted amphetamine (A) and the internal standard
diphenylamine (1) are also indicated.

As with most other drugs, no single extraction method will suffice for removing any barbiturate from any biological sample, and recovery will depend on technique. Generally, multiple solvent extraction under different pH conditions has been used.

A wide variety of solvents have been utilized for the extractions. Other procedures have been used, particularly with urine in drug screening programs. Included are drug adsorption on cation-exchange resin paper, XAD-2 nonionic resin, and charcoal.

The GC of barbiturates has made significant progress since Janak first tried 800°C pyrolysis in 1960 (33). The newer liquid phases have reduced problems with tailing and adsorption generally associated with the gas chromatographic analysis of the barbiturates. These phases allow the separation of the barbiturates without derivatization, at concentrations of 0.1 µg/µl.

Methylated derivatives of the barbiturates have been preferred by many workers for chromatographic separation. Barbiturates can be methylated in the flash heater of a chromatograph by injecting a methanol solution of their tetramethylammonium salts. Trimethylanilinium hydroxide has also been used, with the advantage that lower injection port temperatures may be used.

Street (34) developed an "on column" derivative formation technique using N,O-bis (trimethylsilyl) acetamide (BSA). A few microliters of BSA are drawn into a 10-µl syringe followed by several microliters of extract to be chromatographed. The whole is injected onto the column with the resultant formation of and separation of the derivatives. This procedure is recommended for the formation of methyl and trimethylsilyl derivatives in qualitative determinations, and the separation of nonderivatized barbiturates for quantification procedures.

A number of methods have been reported recently for the gas chromatographic separation of nonderivatized "free" barbiturates. The method (35) below is applicable to the qualitative and quantitative analysis of barbiturates, glutethimide and Dilantin from serum. It is a simple, rapid but accurate procedure using only 3 cm^3 of serum. Dilantin could be separated on the same column as the barbiturates after it had been converted to its methylated derivative.

REAGENT PREPARATION. Phosphate buffer of pH 3.5, a saturated solution of NaH_2PO_4 in distilled water, and tetramethylammonium hydroxide (TMAH) pentahydrate (Sigma) (48% by weight in methanol) are required.

Methyl laurate, used as the internal standard, was obtained from Applied Science Laboratories, State College, PA.

All solvents are distilled.

SAMPLE EXTRACTION.

1. Sedatives. To 3.0 cm^3 of suspect serum or normal serum containing a known amount of standard, 2.5 cm^3 of pH 3.5 phosphate buffer is added. This solution is extracted with 18 cm^3 of dichloromethane and the organic layer is filtered into a screw cap tube and evaporated under nitrogen until about 0.5 cm remains. This is quantitatively transferred into a 1.5-cm^3 Reacti-vial (Pierce Chemical Co.) and evaporated to dryness under nitrogen. The residue is dissolved in 50 µl of internal standard solution and after capping, the vial is heated at 50°C for 15 min. A 0.5- to 0.7-µl aliquot is injected into the column.

2. Dilantin. To 3 cm^3 of serum, 0.2 cm^3 of conc. HCl is added; the pH is checked to make certain that it is close to 1. The serum is extracted with 15 cm^3 of chloroform-internal standard solution and the organic phase is filtered into a screw-top tube and evaporated to dryness under nitrogen. The residue is taken up in 0.6 cm^3 of conc. HCl and extracted with 1.2 cm^3 of chloroform. The organic layer is filtered into a 1.5 cm^3 Reacti-vial and evaporated to dryness under nitrogen. The residue is dissolved in 50 µl of the 48% TMAH. A 0.5- to 0.7-µl aliquot is injected into the column.

PREPARATION OF STANDARDS. Barbiturate sodium salts in aqueous solution and glutethimide in 80% ethanol are used as stock standards. The sodium salt of Dilantin in NaOH (0.2 mol/ℓ) is used as the stock standard.

For routine analyses working standards are prepared in ion-free serum obtained from Clinton Laboratories and treated as the unknown. Sedative standards I, II, and III contain 10 mg/l, 30 mg/l and 50 mg/l, respectively, of each of the following: butabarbital, secobarbital, phenobarbital, allobarbital, amobarbital, heptabarbital, pentabarbital, and glutethimide in 10 mg/l, 20 mg/l, and 40 mg/l, respectively. Dilantin standards I, II, and III contain the drug in 10 mg/l, 20 mg/l, and 30 mg/l levels, respectively.

The internal standard for sedative analysis is 10 µl of methyl laurate in 1 cm^3 pyridine. The internal standard for Dilantin analyses is 1 mg of 5-(p-methyl, phenyl)-5-phenylhydantoin in 100 cm^3 chloroform.

GAS CHROMATOGRAPHY. Six-foot columns are packed with a mixed

phase of 6% SE-30 and 4% XE-60 on 80/100 mesh Chromosorb W-AW
(acid washed). The column is conditioned for 5 hr at 260° with
20 cm³/min helium flow. Table 10.7 lists the operating conditions for the gas chromatograph.

TABLE 10.7   OPERATING CONDITIONS FOR THE ANALYSIS OF BARBITURATES
             GLUTETHIMIDE AND DILANTIN

Condition	Barbiturates and Glutethimide	Dilantin
Helium flow (cm³/min)	60	60
Column	(1) 12°C linear rise from 220 to 240°C (2) 8 min isothermal 240°C	Isothermal 250°C
Injection port	265°C	350°C
Manifold	265°C	280°C

QUANTIFICATION. Calibration curves are obtained by plotting the
ratio of peak area of the drug to internal standard against the
corresponding drug concentration. A linear relationship is obtained for all drugs in the concentration range of 5-100 mg/ℓ.
The limit of detection varied from 3-4 mg/ℓ depending on the
barbiturate and 5 mg/ℓ for glutethimide and Dilantin.

RECOVERY AND REPRODUCIBILITY. Recovery of the individual barbiturates and glutethimide added to serum in concentrations of 10,
20, 50, and 100 mg/ℓ varied from 95 - 102%. Recovery of Dilantin
added to serum in concentrations of 5, 10, 30, and 50 mg/ℓ is
90/98%. Retention time for Dilantin is 4.8 min with the conditions described above.

DISCUSSION. By extracting the drugs on the basis of their pK
values, interference from coextraction of other drugs is decreased. Three commonly used drugs are extracted by this procedure, but because of the polarity of the column, their retention times did not interfere. These are listed in Table 10.8
with the barbiturates and glutethimide.

   Salicylate did not elute from the column under the conditions
used. Barbiturates did not interfere in the Dilantin analysis

because of the transmethylation reaction and the different, higher column temperature employed for the Dilantin. Total analysis
time for sedatives and Dilantin is 1.5 hr.

   Additional applications of GC to the analysis of drugs in biological material include tetrahydrocannabinol, the active constituent of marijuana (36), morphine (37), methadone (38), propoxyphene (Darvon), and norpropoxyphene (39) meperidine (Demerol)
(40), nortriptyline and desmethylnortriptyline (41), and trimethadione and dimethadione (42). Many of these are abused and
gas chromatographic methods would be of great value to the clinical laboratory concerned with screening for drugs and overdose
and post-morten patients.

TABLE 10.8   RETENTION TIMES AND LIMITS OF DETECTION[1]

Drug	Retention Time Min	Relative Retention Time (t_D/t_{rn}) [2]	Limit of Detection (µg)	Limit of Detection in Serum (mg/ℓ)
Methyl laurate (internal standard)	0.67	--	--	--
Allo barbital	2.20	3.29	0.09	3.0
Butabarbital	2.44	3.64	0.12	4.0
Amobarbital	2.68	4.0	0.12	4.0
Pentabarbital	2.80	4.2	0.12	4.0
Secobarbital	3.16	4.7	0.12	4.0
Phenobarbital	6.84	10.2	0.12	4.0
Glutethimide	3.36	5.0	0.15	5.0
Hexabarbital	2.84	4.25	0.12	4.0
Heptabarbital	7.20	10.8	0.10	3.5
Chlorpromazine	6.60	--	--	--
Librium	2.07	--	--	--
Valium	9.20	--	--	--
Dilantin[3]	4.0	--		5.0

[1]See Table 10.7 for operating conditions.
[2]t_D = retention time drug (min); t_{rn} = retention time methyl
laurate (min).
[3]Different column conditions.

10.3   CONCLUSIONS

In the future the continued use of, and further development of,
sophisticated methodology for high-sensitivity gas chromatographic determination of compounds of interest in clinical chemistry
will be necessary. The work in profiles of metabolites in human
urine by Horning et al. (43) and of Zlatkis et al. (44), and the
direct determination of steroid sulfates at high sensitivity in
biological fluids by Touchstone and Dobbins (45) are some examples of progress being made in this area. These methods present a broad view characterization of a sample which is difficult to obtain by other highly sensitive methods such as radioimmunoassay (RIA), for example, which measures only single components. Determination of many biologically important compounds
will prove nearly impossible by RIA, and dependency on GLC for
such methods will continue.

A renewed interest in glass capillary columns for gas chromatographic analysis is being shown. Generally such columns have
some advantages when working with biological and clinical
samples. These include the use of small sample size, and high
resolution often with increased sensitivity. Some applications
include biological amines, urinary steroids, and gas chroma-
tography-mass spectrometry.

Mass spectrometry in conjunction with gas chromatography will
become increasingly used in biological-clinical applications.
The use of the two instruments in combination allows the optimum
utilization of the separating ability of the gas chromatograph
with the unequivocal identification possibilities of the mass
spectrometer. Most compounds that are amenable to this technique,
particularly the abused drugs, can be identified on the basis of
the mass spectrum alone, at microgram and submicrogram levels.
Quantification by mass spectrometry is possible in the range
10^{-12} to 10^{-15}g.

Finkle et al. (46) have established a GC/MS reference data
system for the identification of drugs of abuse. The data include phenethylamine derivatives, opiate and synthetic narcotics,
barbiturates, and urinary metabolites. These data have been established for use with the gas chromatographic retention time index previously developed.

Two indices are used to present the data, each in the form of
a unique digital code that permits either manual or computer
search for identification purposes. A coded drug name index is
supplemented by a base peak index designed to aid in identification of unkonwn spectra.

Eldjarn et al. (47) reviewed GC/MS in routine and research
clinical chemistry. Based on their 10 years of experience, four

areas in which GC/MS has provided valuable information are indicated:

1. Identification of noxious or toxic compounds and their quantification in serum and urine from emergency cases.
2. The diagnosis of inborn errors of metabolism, particularly in
   newborns.
3. The routine analyses of substances occurring in minute amounts,
   such as steroids.
4. Evaluation of pronounced metabolic disturbances, often to sub-
   stantiate clinical diagnosis.

Two prerequisites are necessary for inclusion of GC/MS in
clinical chemistry: The laboratories involved must consider the
above problems as part of their responsibility, and the presently
available apparatus should be improved, including simplification
of operation, and cost. The development of libraries of mass
spectra, such as presently found at the National Institutes of
Health in Bethesda, Maryland, will also be necessary for widespread application.

REFERENCES

1. W. J. A. VandenHeuvel and E. C. Horning, Biochem. Biophys.
   Res. Commun. 3, 356 (1960).
2. M. J. Smith, Bull. Johns Hopkins Hosp. 41, 62 (1972).
3. F. Spielman, M. A. Goldberger, and R. T. Frank, J. Am. Med.
   Assoc., 101, 266 (1933).
4. J. C. Touchstone, J. Gas Chromatog. 2, 170 (1964).
5. F. Wolff, R. Claus, G. Skoulios, H. Frenkel, P. Leissner,
   P. Muller, A. R. Shick, and B. Keller, Nouv. Presse Med., 3,
   301 (1974).
6. J. C. Touchstone, T. Murawec, D. K. O'Leary, L. Luha, and
   G. Mantell, Clin. Chem., 17, 119 (1971).
7. J. C. Touchstone, T. Murawec, and R. J. Bolognese, Clin.
   Chem., 18, 129 (1972).
8. H. H. Bancroft, in Introduction to Biostatistics. Harper and
   Row, New York, 1956, p. 172.
9. A. Klopper, E. A. Michie, and J. B. Brown, J. Endocrinol.,
   12, 209 (1955).
10. J. A. Cooper, J. P. Abbott, B. K. Rosengreen, and W. R.
    Claggett, Am. J. Clin. Pathol. 38, 388 (1962).
11. R. I. Cox, J. Chromatog. 12, 242 (1963).
12. H. C. Curtius and M. Muller, J. Chromatog., 30, 410 (1967).
13. M. G. Metcalf, Clin. Biochem., 6, 307 (1973).

14. H. H. Wotiz and S. J. Clark, Gas Chromatography in the Analysis of Steroid Hormones, Plenum Press, New York, 1966, p. 131.

15. F. Garmendia, G. L. Nocolis, and J. L. Gabrilove, Steroids, 18, 113 (1971).

16. F. Karoun, C. R.J. Ruthven, and M. Sander, Clin. Chim. Acta, 20, 427 (1968).

17. F. Karoum, C. O. Anali, C. R. J. Ruthven, and M. Sandler, Clin. Chim. Acta, 24, 341 (1969).

18. R. H. Cravey, and N. C. Jain, J. Chromatogr. Sci., 12, 209 (1974).

19. W. E. Pereria, R. E. Summons, T. C. Rindfleisch, and A. M. Duffield, Clin. Chim. Acta, 51, 109 (1974).

20. N. C. Jain, Clin. Chem., 17, 82 (1971).

21. P. D. Wagner, H. A. Saltzman, and J. B. West, J. Appl. Physiol., 36, 588 (1974).

22. P. D. Wagner, P. F. Nauman, and R. B. Lavaruso, J. Appl. Physiol., 36, 600 (1974).

23. A. Hill, E. H. Hoag, and W. A. Zaleski, Clin. Chim. Acta, 37, 455 (1972).

24. R. J. Follitt, Clin. Chim. Acta, 55, 317 (1974).

25. B. Weinstein, in Methods of Biochemical Analysis, V. 14, D. Glick (Ed.), Interscience, New York, 1966, p. 203.

26. R. W. Zumwalt, K. Kuo, and C. W. Gehrke, J. Chromatog., 55, 267 (1971).

27. J. C. Touchstone and M. F. Dobbins, in Laboratory Medicine, Vol. 3, G. V. Race (Ed.), Harper and Row, Hagerstown, 1975, Chapter 15.

28. A. C. Moffat, A. H. Stead, and K. W. Smalldon, J. Chromatog., 90, 19 (1974).

29. G. Cimbura and J. Kofoed, J. Chromatogr. Sci., 12, 261 (1974).

30. H. M. Fales and J. Pisano, Anal. Biochem., 3, 337 (1962).

31. K. D. R. Setchell and J. D. H. Cooper, Clin. Chim. Acta, 35, 67 (1971).

32. N. C. Jain and R. H. Cravey, J. Chromatogr. Sci., 12, 228 (197 (1974).

33. J. Janak, Nature, 185, 684 (1960).

34. H. V. Street, J. Chromatog., 41, 358 (1969).

35. S. K. Levy and T. Schwartz, Clin. Chim. Acta, 54, 19 (1974).

36. E. R. Garrett and C. A. Hunt, J. Pharm. Sci., 62, 1211 (1973).

37. J. E. Wallace, J. D. Biggs, and K. Blum, Clin. Chim. Acta, 36, 85 (1972).

38. C. E. Inturrisi and K. Verebely, J. Chromatog., 65, 361 (1972).

39. K. Verebely and C. E. Inturrisi, J. Chromatog., 75, 195 (1973).

40. T. J. Goehl and C. Davison, J. Pharm. Sci., 62, 907 (1973).
41. O. Borga, L. Palmer, A. Linnarson, and B. Holmstedt, Anal.
    Lett., 4, 837 (1971).
42. H. E. Booker and B. Darcey, Clin. Chem., 17, 607 (1971).
43. E. C. Horning, M. G. Horning, J. Szafranek, P. van Hout,
    A. L. German, J. P. Thenot, and C. D. Pfaffenberger, J.
    Chromatog., 91, 367 (1974).
44. A. Zlatkis, W. Bertsch, D. A. Bafus, and H. M. Liebich, J.
    Chromatog., 91, 379 (1974).
45. J. C. Touchstone and M. F. Dobbins, J. Steroid Biochem., 6,
    1389 (1975).
46. B. S. Finkle, D. M. Taylor, and E. J. Bonelli, J. Chromatog.
    Sci., 10, 312 (1972).
47. L. Eldjarn, E. Jellum, and O. Stokke, J. Chromatog., 91,
    353 (1974).

CHAPTER 11

Physicochemical Measurements by Gas Chromatography

MARY A. KAISER

University of Georgia

11.1	SPECIFIC SURFACE AREA	554
11.2	COMPLEX FORMATION	558
11.3	THERMODYNAMICS.	562
	11.3.1 Thermodynamic Studies by Gas–Liquid	
	Chromatography	563
	Activity Coefficient	565
	Enthalpy, Free Energy, and Entropy of	
	Solution	568
	Enthalpy and Entropy of Vaporization	571
	Henry's Law Constant	573
	Equilibrium Constants	574
	11.3.2 Studies by Gas–Solid Chromatography.	575
	Distribution Constant	575
	Enthalpy, Entropy, and Free Energy of	
	Adsorption	576
	11.3.3 Concurrent Solution and Adsorption.	578
11.4	CHEMICAL KINETICS	578
11.5	VIRIAL COEFFICIENTS	579
	11.5.1 Second Virial Coefficients	580
	11.5.2 Gas–Solid Virial Coefficients	582

11.6   TRANSPORT PROPERTIES 583
11.7   MISCELLANEOUS PHYSICOCHEMICAL MEASUREMENTS BY 585
        GAS CHROMATOGRAPHY
11.8   PRECISION AND ACCURACY 585
REFERENCES 586

The place of gas chromatography (GC) in chemical analysis has been well established. Recent developments in theory and improvements in technique have made it possible to apply GC to a variety of physicochemical measurements. The advantages of GC over other techniques lie in its accuracy, convenience, specificity, versatility, speed, and ability to use only small quantities of sample. Thus, in recent years, many reviews and hundreds of papers emphasizing nonanalytical applications have appeared.

11.1   SPECIFIC SURFACE AREA

Often it is necessary to know the quantity of surface of a solid which is available for a certain application. This fundamental property of the solid is called the specific surface area and is generally given as the number of m^2/g or m^2/cm^3 of solid. The specific surface area is one of the first quantities that must be known if any physicochemical interpretation of a material's behavior as an adsorbent is to be possible.

The most well-known method for determining the specific surface area of powders is based on a theory of multimolecular adsorption of gases developed by Brunauer, Emmett, and Teller (BET) (1). The BET method involves the determination of the quantity of a gas which, when adsorbed on the surface of the solid, would completely cover the solid with a monolayer of the gas.

The classical method involves admitting a known quantity of gas to the sample chamber, which is usually maintained near the condensation point of the gas. Adsorption of the gas on the surface of the solid occurs, decreasing the pressure in the chamber until the adsorbed gas is in equilibrium with the free gas phase. The volume of gas adsorbed is determined by subtracting the volume of gas required to fill the free space (dead space) at equilibrium pressure from the volume of gas admitted. The dead space is obtained by precalibration of the chamber volume or by repeating the determination with a sample of negligible adsorption. The specific surface area (S), in m^2/g, is given by the following equation:

$$S = \frac{V_m \; \sigma \; N}{V_o \; W} \qquad (11.1)$$

where V_m = gas uptake corrected to STP,
    σ = average area occupied by the adsorbed gas molecule in
      Angstrom's squared,
    N = Avagadro's number (6.02×10^{23}),
    V_o = ideal gas volume $(22,410 \; cm^3)$,
    W = mass of the sample (g).
V_m is found by determining the volume of gas adsorbed (reduced to
STP) at a number of equilibrium relative pressures by use of the
BET equation:

$$\frac{P}{V_a(P_o - P)} = \frac{1}{V_m C} + \frac{C-1}{V_m C} \; \frac{P}{P_o} \qquad (11.2)$$

where V_a = volume of gas adsorbed, reduced to STP,
    P = equilibrium pressure of adsorbate,
    P_o = saturation pressure of adsorbate at the temperature of
      adsorption,
    C = a constant related to the energy of adsorption.
A plot of $P/(V_a(P_o - P))$ versus P/P_o gives a straight line with an
intercept $1/V_m C$ and a slope of $(C-1)/V_m C$. From this plot C and
$V_m C$ are obtained; thus V_m can be calculated.

    The determination of specific surface area using gas chroma-
tographic equipment was first demonstrated by Nelson and
Eggertsen (N/E) (2). In the N/E method, a quantity of sample is
placed in an outgassing furnace with helium flow for 1 hr. This
procedure, called outgassing or degassing, removes any volatile
components from the surface of the sample (although it may also
alter the surface). Then the sample is cooled to room temperature, purged with nitrogen, weighed, and connected to the surface
area apparatus. The total nitrogen-helium flowrate is set to
approximately 50 cm^3/min by choosing one of three possible nitrogen flowrates. After the gas composition of the system is shown
to be constant as indicated by a stable baseline on the recorder
chart from the signal of the thermal conductivity detector, the
sample tube is immersed in a liquid nitrogen bath. Adsorption of
the nitrogen on the sample occurs, causing an adsorption peak to
be recorded. After equilibrium is obtained, the liquid nitrogen
is removed to produce the desorption peak. To determine the
quantity of nitrogen adsorbed, the area of the desorption peak is
calculated and compared to the peak areas from known quantities
of nitrogen which were calculated in the absence of sample.

Since tailing effects were often observed with adsorption peaks,
they were not used in the calculation of surface area (Figure
11.1). A comparison of the specific surface area of several
materials obtained by the classical BET method and the N/E method
is given in Table 11.1

TABLE 11.1   COMPARISON OF SURFACE AREAS USING BET AND N/E
METHODS[1]

Sample	BET (m^2/g)	N/E (m^2/g)
Furnace Black	24	25.7
Silica-alumina cracking catalyst, used	103	101
Silica-alumina cracking catalyst, new	438	455
Alumina	237	231
Firebrick	3.1	3.4

[1](Ref. 2) Reprinted with permission from Anal. Chem., _30_, 1387
(1958). Copyright by the American Chemical Society.

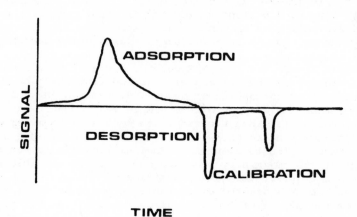

Figure 11.1   Chromatogram from a surface area measurement.

From the quantity of nitrogen adsorbed at the three nitrogen
flowrates which, in turn, correspond to three relative pressures,
a plot of the BET equation is obtained. The surface area is then
calculated according to Equation 11.1.

Improvements on the N/E system extended the range of the
method to specific surface areas as low as 0.04 m^2/g by utilizing
a stopped flow technique during desorption.

Kuge and Yoshikawa (3) related a change in the gas chromatographic peak shape to the beginning of multilayer adsorption on
the surface of the solid. For small adsorbate volumes, the peak
shape is symmetrical. As the adsorbate volume is increased, a
sharp front, diffuse tail, and a defect at the front of the peak
top is observed (Figure 11.2). It then acquires a diffuse front
and sharp tail. This point corresponds to the B point of the
BET Type II adsorption isotherm at which the relative surface
area may be calculated.

Computer-aided frontal analysis chromatography can be used to
determine surface areas and adsorption isotherms (4). Both permanent gases and adsorbates that are liquids above room temperature could be used in this system.

Hydrocarbons are often used as adsorbates in specific surface
area determinations by GC, especially with flame ionization detectors (FIDs).

The adjusted retention time of a probe molecule, such as n-
octane, can be used as a means of determining the surface area of
certain polymers. Different weight samples are placed in a chromatographic column onto which the probe is injected. A plot of
adjusted elution time versus sample weight is fairly linear with
a slope proportional to the specific surface area.

In the determination of the specific surface area of catalysts, emphasis is placed on the determination of the catalytically active surface area rather than the total surface area.
The most common techniques utilize the chemisorption of hydrogen,
oxygen, or carbon monoxide on the surface of the sample. The
volume of a gas produced in a catalytic reaction may also be used
to calculate catalytically active surface area.

Attention has also been given to possible sources of error in
specific surface area analysis. Lobenstein (5) has noted the
problems associated with the lack of additivity in BET theory as
applied to mixtures and the "interaction" between two or more
components of dental materials. He has also derived equations to
estimate the extent of the "interaction" as well as to correct
for it to give a more realistic surface area value.

Dollimore and Heal (6) have shown that in calculating a surface area from adsorption data, it is possible to calculate an
apparent surface area which is approximately one-fourth the actual surface area. This error occurs when the adsorbate molecule

fits snugly in the pore of the sample, touching the walls, thus
covering a cylindrical area of $4\pi r^2$ rather than the area
$2(3^{\frac{1}{2}})r^2$. Thus, the actual area is $2\pi(3^{\frac{1}{2}})$ times the apparent
area. As the pore size becomes larger with respect to the size
of the adsorbate molecule, the factor decreases quickly to one.

Certain precautions are necessary in the determination of specific surface area, such as obtaining a representative sample,
choosing the appropriate technique, and selecting standard materials for specific surface area analysis. The analyst should
choose the technique and conditions of analysis that most closely
approximate the intended use of the sample. Furthermore, precision studies should be carried out to see how sample handling
affects the surface area.

11.2   COMPLEX FORMATION

Metal ions can act as electron-pair acceptors, reacting with
electron donors to form coordination compounds or complexes. The
electron donor species, called the ligand, must have at least one
pair of unshared electrons with which to form the bond. Chelates
are a special class of coordination compound which results from
the reaction of the metal ion with a ligand that contains two or
more donor groups.

Complexes often form in steps, with one ligand being added in
each step:

$$M + L \rightleftharpoons ML$$

$$ML + L \rightleftharpoons ML_2 \quad \text{etc.}$$

The stepwise constants written for the equilibria above are called formation or stability constants and may be written:

$$K_1 = \frac{\{ML\}}{\{ML\}\{L\}}$$

$$K_2 = \frac{\{ML_2\}}{\{ML\}\{L\}}$$

As the complex becomes more stable, the formation constant becomes larger. The reciprocal of the formation constant is called
the instability constant.

Purnell (7) surveyed numerous gas chromatographic approaches
to the study of complex equilibria and developed some generalized

retention theories for each. He established a classification
system which greatly simplified the approach to complexation reactions. The Purnell classification system is summarized in
Table 11.2

TABLE 11.2   PURNELL CLASSIFICATION SYSTEM FOR COMPLEXES[1]

Class	Type i	Type ii	Type iii
A	AX_n	XA_m	A_mX_n
B	SX_n	XS_p	S_pX_n
C	X polimerizes in solution	X depolimerizes in solution	
D	$SA_{m, m+1}$, etc.	$AS_{p, p+1}$, etc.	

A = additive, X = solute, S = solvent, n≥1, m≥1, p≥1.

Formation constants of complexes between aromatic electron
donors and di-n-propyltetrachlorophthalate in an inert solvent
(class A(ii)) have been measured. The complex formation constant
is obtained from the following equation:

If a 1:1 complex formation is assumed,

$$A(G) \underset{}{\overset{K_R^0}{\rightleftharpoons}} A(S) \tag{11.3}$$

$$A(S) + D(S) \underset{}{\overset{K_1}{\rightleftharpoons}} AD(S)$$

where A = a volatile acceptor solute,
      D   = a nonvolatile electron donor,
      I   = a nonvolatile inert solvent in which D is dissolved,
      S   = solvent phase (D + I),
      G   = gas phase,
      K_R^0 = the distribution constant (partition coefficient) of
            uncomplexed A between S and G,
      K_1 = the formation constant of AD in solution.
If AD(S) and A(S) are considered to approach infinite dilution(a
reasonable assumption in a gas chromatographic system), then the
activities of each approach the concentrations as the concentrations approach zero, and

$$K_1 = \frac{C_{AD}}{(C_A \alpha_D)} \tag{11.5}$$

where C = concentration and α = activity. The apparent gas chromatographic distribution constant (partition coefficient), assuming an ideal solution, is given by the following equation:

$$(K_R)_S = K_R^o \ (1 + K_1 C_D) \tag{11.6}$$

where $(K_R)_S$ = the apparent distribution constant (partition coefficient). $(K_R)_S$ is obtained from the following equation:

$$V_N = (K_R)_S \ V_S \tag{11.7}$$

where V_N = corrected net maximum retention volume and V_S = total volume of the solvent phase.

This method requires the use of several columns containing different concentrations of donor. A plot of $(K_R)_S$ versus C_D is constructed from which K_R^o, the intercept, and K_1, from the slope divided by K_R^o are obtained.

Martire and Riedl and coworkers (8,9) developed a method of determination of complex constants which is less time consuming than the Cadogen-Purnell method, although it makes additional assumptions. They have demonstrated that the specific retention volume of A is related to the association constant by the following equation:

$$(V_g^o)_D = \frac{273 \ R}{\gamma_A^D \ p_A^o M_D} \ (1 + K_1 \alpha_D) \tag{11.8}$$

where (V_g^o) = the specific retention volume of A in a system containing pure D,

$\quad$ R $\quad$ = ideal gas constant,

$\quad$ γ_A^D $\quad$ = infinite dilution activity coefficient of uncomplexed A in D,

$\quad$ p_A^o $\quad$ = saturated vapor pressure of A,

$\quad$ M_D $\quad$ = molecular weight of D.

In the Martire-Riedl method, only two columns are used, one containing D and the other containing a reference liquid phase (R) of approximately the same molecular size, shape, and polarizability as D (class B(ii)).

Utilizing a noncomplexing solute N on R and D, the following equations may be obtained:

$$(V_g^o)_R = 273R/\gamma_A^R \; p_A^o \; M_R \qquad (11.9)$$

$$(\overline{V}_g^o)_D = 273R/\gamma_N^D \; p_N^o \; M_D \qquad 11.10)$$

$$(\overline{V}_g^o)_R = 273R/\gamma_N^R \; p_N^o \; M_R \qquad (11.11)$$

where the $\overline{V}_g^o$ refer to the noncomplexing solute and V_g^o refers to the electron acceptor solute. If R and D have the same approximate molecular size, shape, and polarizability, then

$$\gamma_N^R / \gamma_N^D \sim \gamma_A^R / \gamma_A^D \qquad (11.12)$$

From Equations 11.8 and 11.12,

$$\frac{(V_g^o)_D \; (\overline{V}_g^o)_R}{(V_g^o)_R \; (\overline{V}_g^o)_D} = K_1 \alpha_D + 1 \qquad (11.13)$$

where $\alpha_D = \gamma_D \, c_D$.

Liao, Martire, and Sheridan (10) used the methods of Cadogan-Purnell and Martire-Riedl with three electron donor systems. The association constants obtained were in excellent agreement. (Table 11.3).

Equilibria may be expressed in terms of activity, instead of concentrations. The expression that relates the relative retention volume to the association constant is given by the following equation:

$$\frac{\left[(V_A/V_{A*})_{S+B} \; (V_{A*}/V_A)_{S+C} - 1\right]}{x_B} = K_x \left[\frac{f_A f_B}{f_{AB}}\right]_{S+B} \qquad (11.14)$$

where V = retention volume, corrected for gas holdup at the mean
          column temperature and pressure,
      A = complex-forming solute,
      A*= solute that is closely related to A but does not form a
          complex,
      S = solvent,
      B = complex-forming compound, dissolved in S,
      C = compound that is closely related to B but does not form
          a complex,
      X = mole fraction,
      K_X= association constant (mole fraction scale),
      γ = activity coefficient (mole fraction scale).

A theoretical study to compare the methods of gas-liquid chromatography and spectrometry in the measurement of weak complexation constants is given by Eon and Guiochon (11). They show
that both chromatography and spectrometry should lead to the same
results when necessary corrections are made and the solvent is
properly chosen. Discrepancy in the data from comparison of
these methods usually can be attributed to misinterpretation of
the chromatographic measurements.

TABLE 11.3   COMPARISON OF CADOGAN—PURNELL AND MARTIRE-REIDL
             METHODS OF ASSOCIATION CONSTANT DETERMINATION[1]

| Electron Donor | Electron Acceptor | Association Constants (1/mole) at 40°C | | |
		C-P Method	M-R Method	
Di-n-octylamine	$CHCl_3$	0.405±0.19[2]	0.403[3]	0.392[4]
Di-n-octylamine	CH_2Cl_2	0.179±0.014	0.187	0.181
Di-n-octylamine	CH_2Br_2	0.222±0.004	0.219	0.224

[1](Ref. 10) Reprinted with permission from Anal. Chem., 45, 2087
(1973). Copyright by the American Chemical Society.
[2](Ref. 88),
[3](Ref. 88),
[4](Ref. 112).

11.3   THERMODYNAMICS

Thermodynamics is that branch of physical science dealing with the transfer of heat and the appearance or disappearance of work relating to chemical and physical processes. With knowledge of some thermodynamic values associated with a particular reaction, one can tell whether a reaction can occur and often can describe what factors contribute to the overall change. Thermodynamics does not describe how fast a reaction will occur or the character of the process which the reaction may undergo.

11.3.1 Thermodynamic Studies by Gas-Liquid Chromatography

Gas-liquid chromatography is a logical technique to choose for the study of the interaction of a volatile species (in a mobile phase) with a nonvolatile species (stationary phase).

Generally, the distribution constant (partition coefficient) as a function of temperature is the first measurement obtained in such a study since other constants may be derived from these data. The distribution constant, K_D, is the ratio of a component in a single definite form in the stationary phase per unit volume to its concentration in the mobile phase per unit volume at equilibrium.

$$K_D = \frac{\text{Concentration component in stationary phase}}{\text{Concentration component in mobile phase}}$$

Both concentrations should be calculated per unit volume of the phase. The term "distribution constant" is recommended in preference to the term "partition coefficient" which has been used with the same definition.

The value for K_D is high when most of the component is retained in the stationary (liquid) phase. The distribution constant is temperature dependent, so the column temperature must be stated when reporting K_D data. Generally, the distribution constant is halved for each $30^\circ C$ increase in temperature.

K_D may be related to the retention volume, the mobile phase volume, and the stationary phase volume. The total moles of substance injected onto the column, n, is divided between the stationary phase and the mobile phase.

$$n = n_S + n_M \qquad (11.15)$$

where n_S = the number of moles in the stationary phase and n_M = the number of moles in the mobile phase. The distribution constant, K_D, may be defined as

$$K_D = \frac{n_S/V_S}{n_M/V_M} \tag{11.16}$$

where V_S = volume of the stationary phase and V_M = the mobile phase volume between the point of application of the sample and the point of detection (holdup volume). The velocity of the substance, U_S, is defined by the following equation:

$$U_S = U_U \, f \tag{11.17}$$

where U_U = velocity of an unretained substance and f = the fraction of the time spent by the substance in the mobile phase. Since f is equivalent to n_M/n,

$$U_S = U_U \, (n_M/n) \tag{11.18}$$

Substituting,

$$n_M/n = n_M/(n_M + n_S) \tag{11.19}$$

$$n_S = n_M K_D V_S/V_M \tag{11.20}$$

$$n_M/n = n_M/(n_M + n_M(K_D V_S)/V_M) \tag{11.21}$$

$$n_M/n = V_M/(V_M + K_D V_S) \tag{11.22}$$

The velocity of an unretained component, U_U, and the velocity of the substance, U_S, are related to the flowrate of the carrier gas, F_C, and the column length, L, by the following expression:

$$U_U = F_C/(V_M/L) \tag{11.23}$$

$$U_S = \left(\frac{F_C}{V_M/L} \right) \left(\frac{V_M}{V_M + K_D V_S} \right) = \frac{F_C L}{V_M + K_D V_S} \tag{11.24}$$

The velocity of the substance is also defined by the following expression:

$$U_S = F_C L / V_R \qquad (11.25)$$

where V_R = the retention volume of the substance. Therefore,

$$U_S = F_C L / V = F_C L / (V_M + K_D V_S) \qquad (11.26)$$

$$K_D = (V_R - V_M) / V_S \qquad (11.27)$$

$$K_D = V_R^o / V_S \qquad (11.28)$$

where V_R^o = adjusted retention volume, which may be obtained from the chromatogram, and V_S = volume of the liquid phase.

Since a finite pressure drop exists over the length of the column, the adjusted retention volume is often multiplied by the pressure gradient correction factor, j, to give a net retention volume, V_N.

$$j = \frac{3}{2} \frac{(p_i/p_o)^2 - 1}{(p_i/p_o)^3 - 1} \qquad (11.29)$$

where p_i = pressure of the carrier gas at the inlet and p_o = pressure of the carrier gas at the outlet.

$$V_N = j \, V_R^o \qquad (11.30)$$

The specific retention volume, V_g, the net retention volume per unit mass may be given by

$$V_g = \left(\frac{(V_R - V_M')}{W} \right) \left(\frac{273}{T} \right) j \qquad (11.31)$$

where W = mass of the solvent and T = column temperature.

ACTIVITY COEFFICIENT. The description of the behavior of an ideal solution, based on experimental vapor pressure data, is given by Raoult's law,

$$p = y \, p^o \qquad (11.32)$$

where p = vapor pressure above a solution in which the mole fraction of the substance is y, and p^0= vapor pressure of pure liquid
at the same temperature. Since ideal solutions are rare, the
activity coefficient, γ, may be added to the equation to account
for nonideality. Raoult's law may be rewritten

$$p = \gamma y\ f^0 \tag{11.33}$$

where y = the mole fraction of solute in the mobile phase and f^0=
the fugacity of the pure solute. For very dilute nonideal solu-
tions, γ is constant and equal to γ^∞, the activity coefficient at
infinite dilution.

$$p = \gamma^\infty\ yp^0 \tag{11.34}$$

In Raoult's law at infinite dilution, the reference state is the
pure liquid state. The activity coefficient at infinite dilution
accounts for deviation from pure solute-solute interactions in
solution. When positive deviations from Raoult's law occur, the
partial pressure of the substance above the solution is greater
than it is in an ideal solution, $\gamma^\infty > 1$, and the chromatographic
retention time is decreased.

The activity coefficient at infinite dilution may be obtained
directly from gas chromatographic data.

$$\gamma^\infty = RT/V_g M_p 0 \tag{11.35}$$

where R = ideal gas constant,
      T = temperature (^{0}K),
      V_g= specific retention volume,
      M = molecular weight of the solvent,
      p^0= vapor pressure of the pure liquid.
(See Table 11.4). In the derivation of Equation 11.35, the ideal
gas law was used. For high-pressure GC and situations in which
carrier gases which deviate from ideality are used, this equation
may not be valid.

The substituent constant of the Hammett equation has been related successfully to the logarithm of the activity coefficient
ratio at infinite dilution for a series of meta and para isomers
of phenol. Hammett stated that a free energy relationship should
exist between the equilibrium or rate behavior of a benzene derivative and a series of corresponding meta and para monosubstitut-
ed benzene derivatives. The Hammett equation may be written

$$\log(K'_x/K'_o) = \sigma\rho \tag{11.36}$$

where K'_x nd K'_o = the rate (or equilibrium) constants for a re-
                      action of the substituted and unsubstituted
                      benzene derivative, respectively,
    σ = the Hammett substitutent constant,
    ρ = the Hammett reaction constant.
ρ may be used to establish a selectivety scale for stationary
liquid phases in gas-liquid chromatography through the following
relationship:

$$\log(\gamma_O^\infty / \gamma_X^\infty) = \sigma\rho + b \tag{11.37}$$

where b is a constant. The chromatographic substituent constant,
σ_c, is obtained from the following equation:

$$\sigma_c = 0.09 + 0.621 \log (\gamma_O^\infty/\gamma_X^\infty) \tag{11.38}$$

These relationships were shown to be capable of predicting the
types of liquid phases necessary for optimum resolution.

TABLE 11.4   SAMPLE ACTIVITY COEFFICIENT CALCULATION

V_R = 600 cm^3	W = 2 g	Mol. wt. = 114 g
V_m = 100 cm^3	T = 29°C	p^o = 140 torrs

V_g = $(V_R-V_m)/W$ $T(^oK)$ = 273 + $T(^oC)$
    = (600 cm^3 - 100 cm^3)/2g   = 273 + 27
    = 250 cm^3/g   = 300°K

p^o(atm) = p^o(torr)/760   R = 82.0573 cm^3-atm/^{o}K-mole
        = 140 torr/760
        = 0.184 atm

$$\gamma^\infty = \frac{RT}{V_g M_p{}^o}$$

$$= \frac{(82.0573 \text{ cm}^3\text{-atm}/^oK\text{-mole})(300^oK)}{(250 \text{ cm}^3/g)\ (114 \text{ g/mole})\ (0.184 \text{ atm})}$$

$$= 4.69$$

ENTHALPY, FREE ENERGY, AND ENTROPY OF SOLUTION. Measurement of
entropy, enthalpy, or free energy of solution have been reported
for hundreds of compounds. Enthalpy, or heat content, defines
the quantity of heat absorbed at constant pressure when only
mechanical work (PΔV work) is done. Free energy (Gibbs free
energy) is used to define equilibrium. A decrease in Gibbs free
energy attending an isothermal, constant pressure process is an
upper limit to all work, other than P-V work. A chemical reaction whose net change in free energy is less than zero can
occur spontaneously. A zero change in free energy is a condition
for equilibrium. Entropy is a measure of the degradation or disorder in a system.

Enthalpy, free energy, and entropy at constant temperature
often are related by the expression:

$$\Delta G = \Delta H - T\Delta S \tag{11.39}$$

where G = free energy (kcal/mole),
H = enthalpy (cal),
T = temperature (^{0}K),
S = entropy (cal/^{0}K-mole).

In most common chemical reactions, one or more of the reactants is in solution. Thus, an easy method to determine thermodynamic quantities of solution is desirable. Enthalpy of solution (heat of solution) is defined as the change in the quantity
of heat which occurs due to a combination of a particular solute
(gas, liquid, or solid) with a specified amount of solvent to
form a solution. If the solution consists of two liquids, the
enthalpy change upon mixing the separate liquids is called the
heat of mixing. When additional solvent is added to the solution
to form a solution of lower solute concentration, the heat effect
is called the heat of dilution. The definitions of free energy
of solution, entropy of solution, and so on follow the pattern of
definitions above.

Since ideal conditions simplify calculations, an ideal gas
at 1 atm pressure in the gas phase which is infinitely dilute in
solution will be utilized. Then the total standard partial molar
Gibbs free energy of solution (chemical potential), $\Delta \bar{G}_T^0$, can be
directly related to K_D, the distribution constant, by the expression

$$\Delta \bar{G}_T^0 = -RT \ln K_D \tag{11.40}$$

If the chromatographic system is at low pressure and small sample
sizes are used, ideal conditions are approximated. (For an ideal

gas at unit pressure, the Gibbs free energy, the enthalpy, and
entropy are defined as having their "standard" values.)

Another expression for the total standard partial molar Gibbs
free energy is obtained from Henry's law. Henry's law may be
written

$$p = yK_H \tag{11.41}$$

where K_H = Henry's law constant,
     y = mole fraction of solute,
     p = partial pressure of the solute.
If $\gamma\infty p^0 = K_H$ (Equations 11.34 and 11.41), then Equation 11.40 may
be rewritten:

$$\Delta\overline{G}^0_T = -RT \ln\gamma\infty p^0 \tag{11.42}$$

In Equation 11.42, note that two definitions of ideal solution
behavior are used. $\Delta\overline{G}^0_T$ is based on an ideal inifnite dilution
system and $\gamma\infty$ is based on an ideal pure liquid solute system. As
long as the differences are noted and understood, the use of
these two definitions should not cause any problems.

A plot of log V_g versus $1/T$ for a compound is linear and that
the slope is proportional to the heat of solution. Two equations
to determine heat of solution, ΔH_T may be used.

$$\Delta H_T = -2.30 R (S_V - T^2 \, d\log p_S/dT) \tag{11.43}$$

$$\Delta H_T = -R\{2.30 S_V + (T^2/V_S) \, dV_S/dT\} \tag{11,44}$$

where S_V = slope of the plot log V_g vs $1/T$,
     dV_S/dT = cubical expansion coefficient of the liquid phase,
     R = gas constant (1.987 kcal/^{0}K-mole).

The cubical expansion coefficient generally may be assumed to be
approximately 1×10^{-3} cm^3/^{0}K.

The standard partial molar Gibbs free energy of solution is
related to the enthalpy and entropy functions at the column temperature T by the expression

$$\Delta\overline{G}^0_T = \Delta\overline{H}^0_T - T\Delta\overline{S}^0_T \tag{11.45}$$

where $\Delta\overline{H}^0_T$ = standard partial molar enthalpy and $\Delta\overline{S}^0_T$ = standard partial molar entropy. The total standard partial molar Gibbs free energy, entropy, and enthalpy of solution may be divided into two parts. The first part is that due to ideal solution behavior; the second part is the difference or excess portion.

$$\Delta \overline{G}_T^0 = \Delta \overline{G}_i^0 + \Delta \overline{G}_e^0 \tag{11.46}$$

$$\Delta \overline{H}_T^0 = \Delta \overline{H}_i^0 + \Delta \overline{H}_e^0 \tag{11.47}$$

$$\Delta \overline{S}_T^0 = \Delta \overline{S}_i^0 + \Delta \overline{S}_e^0 \tag{11.48}$$

where the "i" terms refer to the "ideal" contribution and the "e" terms refer to the "excess" contributions to the total free energy. $\Delta \overline{G}_i^0$ is equal to the molar free energy of vaporization (reference state is the pure liquid solute). $\Delta \overline{H}_i^0$ and $\Delta \overline{S}_i^0$ can be obtained from temperature studies of the partial free energy term.

The partial molar excess Gibbs free energy and partial molar excess enthalpy of mixing are defined by the following equations:

$$\Delta \overline{G}_e^0 = RT \ln \gamma^\infty \tag{11.49}$$

$$\Delta \overline{H}_e^0 = R \frac{d(\ln \gamma^\infty)}{d(1/T)} \tag{11.50}$$

The partial molar excess entropy is obtained from the following equation, after $\Delta \overline{H}_e^0$ and $\Delta \overline{G}_e^0$ are determined.

$$\Delta \overline{S}_e^0 = \frac{\Delta \overline{H}_e^0 - \Delta \overline{G}_e^0}{T} \tag{11.51}$$

From the combination of Equations 11.42, 11.46, and 11.49, an expression for $\Delta \overline{G}_i^0$ may be obtained.

$$\Delta \overline{G}_i^0 = RT \ln p^0 \tag{11.52}$$

Determinations of the differences in free energies of two components, $\Delta(\Delta G^0)$, can be obtained readily from the chromatogram.

$$\Delta G_2^O - \Delta G_1^O = \Delta(\Delta G^O) = -RT \ln\alpha \qquad (11.53)$$

$$\alpha = \frac{t_{R_2} - t}{t_{Rl} - t} \qquad (11.54)$$

where ΔG^O = standard molar Gibbs free energy of solution,
    α = relative retention,
    t_{R2}= retention time of component 2,
    t_{Rl}= retention time of component 1,
    t = retention time of a nonretained component.
$\Delta(\Delta G^O)$ is a measure of relative retention and is unaffected by flowrate, percent loading, or solvent molecular weight.

In addition to its use as a measure of relative retention, $\Delta(\Delta G^O)$ may also be used to study isotopic and isomeric interactions.

ENTHALPY AND ENTROPY OF VAPORIZATION.   The first accurate thermodynamic quantities obtained using GC were enthalpies of vaporization.   In 1960 Mackle, Mayrick, and Rooney, (12) measured the heats of vaporization of 2-thiobutane, 3-thiobutane, and 4-thioheptane by using gas-liquid chromatography coupled with a bypass sampling system.   In this system the sample is placed in a sample tube, connected to the apparatus, and cooled with dry ice. Then the sample is warmed to the specified temperature and equilibrated in one arm of the apparatus.   The liquid is then isolated from the system and carrier gas allowed to pass through the arm, sweeping the vapors onto the gas chromatographic column. The greater the vapor pressure, the greater the peak height on the chromatogram.   A plot of the logarithm of the peak height versus the reciprocal of the temperature is made from which ΔH_v is obtained from the slope of the line.

Generally, an approximate heat of vaporization can be extracted from the slope of a plot of net retention volume versus the reciprocal of the column temperature.   Table 11.5 lists some compounds whose vapor pressures have been determined by GC.

The entropy of vaporization, ΔS_v, may be obtained from the following equations:

$$\Delta S_v = \Delta S_m - \Delta S_s \qquad (11.55)$$

where ΔS_m = entropy of mixing and ΔS_s = entropy of solution.

For a pure vapor (standard state),

$$\Delta S_m = -R \ln X \qquad (11.56)$$

where X = mole fraction of solute in the stationary phase. The entropy of solution is determined in the manner shown above.

TABLE 11.5 VAPOR PRESSURES OF COMPOUNDS DETERMINED BY GC

Compounds	Reference
Inorganic chlorides and oxychlorides	12
Propanol, butanol, 2-butanone, 3-butanone, n-heptane, p-dioxane	13
Benzene, toluene, butyl acetate	14
Metal carbonyls	15
Fatty acids, fatty esters, fatty alcohols, chloro-alkanes	16
Alcohols	17
Perfumes	18
Hydrogen saturated with methanol	19, 20

The partial molar excess Gibbs free energy of the methylene group can be used as a means of characterizing stationary phases. From rearrangement of Equation 11.46, an expression for the standard partial molar Gibbs free energy can be obtained.

$$\Delta \overline{G}_e^0 = \Delta \overline{G}_T^0 - \Delta \overline{G}_i^0$$

For an ideal solution, the total standard partial molar excess Gibbs free energy, $\Delta \overline{G}_T^0$, equals the ideal contribution, $\Delta \overline{G}_i^0$, and $\Delta \overline{G}_e^0 = 0$. In other words, there is no interaction occurring between the solute and the stationary phase. When a nonpolar molecule such as an alkane is chosen as the solute, $\Delta \overline{G}_e^0$ will increase as the polarity of the stationary phase increases. Therefore, $\Delta \overline{G}_e^0$ may be useful in determining the extent of the polarity of the stationary phase.

Of course $\Delta \overline{G}_e^0$ will not be the same value for every alkane, especially as the polarity of the stationary phase increases. So it becomes necessary to define a unit of the alkane, the methylene group, on which to base comparisons. Thus, the partial molar excess Gibbs free energy of the methylene group, $\Delta \overline{G}_{e(CH_2)}^0$, is defined by

$$\Delta \overline{G}^{o}_{e\,(CH_2)} = -RT \frac{d \ln V_g p^{o}}{dn} \tag{11.57}$$

where R = gas constant,
       T = column temperature (^{o}K),
       V_g = specific retention volume,
       p^{o} = saturated vapor pressure of pure solute at T,
       n = number of methylene groups in the molecule of solute.

An alternate method of characterization of stationary phases does not require knowledge of the saturated vapor pressure. Instead, the differential molar enthalpy of evaporation from solution, ΔH^{s}_{evp}, is calculated from the specific retention volume V_g at various temperatures.

$$\ln V_g = \frac{\Delta H^{s}_{evp}}{RT} + constant \tag{11.58}$$

The differential molar enthalpy of evaporization of the methylene group from solution, $\Delta H^{s}_{evp\,(CH_2)}$, is obtained using the linear relationship between the specific retention volume and the number of carbon atoms in a homologous series of solutes. The slope of the line equals the change of the specific retention volume per methylene group. If this slope versus the reciprocal of the column temperature is plotted, $\Delta H^{s}_{evp\,(CH_2)}$ is obtained.

The differential enthalpy from solution for -OH, -O-, $>$C=O, -CHO, -OC(O)H, and -OC(O)CH$_3$ may also be obtained. The value of ΔH^{s}_{evp} for a particular functional group is a measure of the strength of interaction between the solute and the stationary phase. Thus, knowledge of ΔH^{s}_{evp} for a mixture of solutes can help in choosing the liquid phase to obtain optimum separation.

HENRY'S LAW CONSTANT. The Henry's law constant, K_H, is obtained from rearrangement of Equation 11.41.

$$K_H = p/y \tag{11.59}$$

where p = vapor pressure of the solute and y = mole fraction of the solute. At infinite dilution, the state which is approached by gas-liquid chromatography, K_H at infinite dilution is written

$$K_H(\text{infinite dilution}) = RT/MV_g \tag{11.60}$$

where M = molecular weight of the stationary phase and V_g = specific retention volume.

EQUILIBRIUM CONSTANTS. For a reaction at equilibrium, the rate of forward reaction equals the rate of backward reaction.

$$aA + bB \rightleftharpoons cC + dD$$

K, the equilibrium constant, is defined by the expression

$$K = \frac{(C)^c (D)^d}{(A)^a (B)^b} \tag{11.61}$$

where ( ) indicates concentration, A and B are the reactants, C and D are the products, and a,b,c, and d are the number of molecules that occur in a balanced equation. By convention, the concentration terms for the products are placed in the numerator. K is approximately constant at a specified temperature so that if the concentrations of starting materials and K are known, it is possible to predict the concentrations of products.

   K can often be calculated from the relationships of thermodynamics, if the Gibbs free energy for the chemical reaction can be obtained. For a gaseous reaction in which the equilibrium constant K_p is determined from equilibrium pressures of each component, the following expressions may be written:

$$K_p = \frac{P_C^c \, P_D^d}{P_A^a \, P_B^b} \tag{11.62}$$

$$\Delta G^o = -RT \, \ln K_p \tag{11.63}$$

For systems in which the concentrations are known in moles/liter or mole fractions,

$$K_p = K_c \, (RT)^{\Delta ng} \tag{11.64}$$

$$K_x = K_p \, P^{-\Delta ng} \tag{11.65}$$

where K_p = equilibrium constant in terms of pressure,

K_C = equilibrium constant in terms of concentrations (moles/l),

K_X = equilibrium constant in terms of mole fractions,

Δn_g= number of moles of products less that of reactants.

When the equilibrium constant is expressed in terms of activities, the constant K_a is obtained which is valid for all gases and solutions.

$$K_a = \frac{a_C^c \; a_D^d}{a_A^a \; a_B^b} \qquad (11.66)$$

$$\overline{\Delta G} = -RT \ln K_a \qquad (11.67)$$

The equilibrium constant can be derived from free energy considerations of gas chromatographic data. It may also be obtained directly from the measurement of all concentrations at the specified temperature by selection of a column that can separate each component and by a quantitative method of detection.

11.3.2   Thermodynamic Studies by Gas-Solid Chromatography

In the early years of GC, more consideration was given to partition (GLC) than to adsorption (GSC) systems. For GLC, the mechanism of retention was well understood, all of the mathematics were derived, and the chromatographic peak shapes were symmetrical. At that time, GSC had been utilized only for the separation of permanent gases. In recent years much has been accomplished in the determination of thermodynamic parameters in GSC separations. Part of the reason for the upsurge of interest was due to the desire to predict sample separations at any temperature, since most GSC data was reported at only one temperature.

DISTRIBUTION CONSTANT. For elution GSC the distribution constant K_D can be obtained from the adjusted retention volume V_R', and the surface area of the adsorbent, $A(m^2)$.

$$\frac{V_R'}{A} = V_g = K_D = C_S/C_g \quad (cm^3/m^2) \qquad (11.68)$$

where V_g = specific retention volume at the column temperature T,
C_S = sorbate equilibrium surface concentration (moles/m^2),
C_g = sorbate equilibrium gas phase concentration (moles/
cm^3).

If C_S is expressed in moles/cm^2, the distribution constant is given by K_D'.

$$K_D' = 10^{-4} K_D \tag{11.69}$$

ENTHALPY, ENTROPY, AND FREE ENERGY OF ADSORPTION. The free energy of adsorption, $\Delta G_{ads}'$, can be obtained from the distribution constant by the expression

$$\Delta G_{ads}' = -RT \ln K_D = -RT \ln V_g = \Delta H_{ads}' - T\Delta S_{ads}' \tag{11.70}$$

where R $\quad$ = gas constant,
T $\quad$ = absolute temperature,
$\Delta H_{ads}'$= enthalpy of adsorption,
$\Delta S_{ads}'$= entropy of adsorption.

Standard state free energies (ΔG_{ads}^o) and entropies (ΔS_{ads}^o) may also be determined from GSC retention data if ideal conditions are assumed. For the adsorbate behaving as an ideal gas in the mobile phase, the standard state is defined as a partial pressure of 1 atm. The adsorbed standard state is defined as a two-dimensional perfect gas at 1 atm where the mean distance between adsorbed molecules is the same as in the three dimensional gas phase standard state. Thus, the sorbate equilibrium surface concentration C_S becomes 4.07 x 10^{-9}/T (moles/cm^2) and the gas phase sorbate concentration becomes 4.07 x 10^{-9}/TK_D'.

The ideal gas equation may be written

$$P_{eq} = C_g RT \tag{11.71}$$

Combining the calculated value of C_g with the ideal gas equation gives the expression for P_{eq}, the equilibrium partial pressure of sorbate vapor.

$$P_{eq} = (4.07 \times 10^{-9} R)/K_D' \tag{11.72}$$

The standard differential molar Gibbs free energy (ΔG^o) for transfer of 1 mole of vapor at 1 atm to P_{eq}, the equilibrium vapor pressure, is given by

$$\Delta G^{o} = RT \ln(P_{eq}/1) = RT \ln(4.07 \times 10^{-9} R/K_D') \qquad (11.73)$$

Since $K_D' = 10^{-4} K_D$ (Equation 11.69),

$$\Delta G^{o} = RT \ln(4.07 \times 10^{-5} R/K_D) = RT \ln(4.07 \times 10^{-5} R/V_g) \qquad (11.74)$$

After combining terms and substituting the values for the constants,

$$\Delta G^{o} = \Delta G_{ads}' - 11.33 \ T \qquad (11.75)$$

For the specific retention volume, V_g, in cm^3/m^2 and $\Delta G_{ads}'$ in calories, an equation relating V_g and $\Delta G_{ads}'$ as a function of T may be written.

$$\log V_g = (-\Delta G_{ads}^{o}/4.58) - 2.48 = -\Delta G_{ads}'/4.58T \qquad (11.76)$$

The heat of adsorption may be calculated readily from retention data by using the following expression:

$$\log V_g = -\Delta H_{ads}/2.3RT + \text{constant} \qquad (11.77)$$

Plotting $\log V_g$ versus $1/T$ gives a straight line with the slope $-\Delta H_{ads}/2.3R$.

Heats of adsorption are due to three fundamental processes, physical adsorption, reversible chemisorption, and irreversible chemisorption. Physical adsorption can be attributed to van der Waals forces (dispersion forces). Reversible and irreversible chemisorption are due to bond formation between the adsorbent and the adsorbate. Generally heats of adsorption less than 10-15 kcal/mole are considered due to physical adsorption alone, although some physical adsorption processes may exceed these values. Any chemisorption process generally involves all three processes so that the heat of adsorption value reflects the sum of the average contribution of each. The interaction of transition metals with unsaturated compounds is attributed by most authors, at least in part, to chemisorption.

11.3.3 Concurrent Solution and Adsorption

The possibility of mixed solution and adsorption phenomena contributing to chromatographic retention must be considered. Several factors are responsible for retention in GC, including bulk liquid partition, liquid interfacial adsorption, and solid support adsorption. One or all of these factors may play a major role depending upon the experimental parameters chosen (e.g., temperature, percent liquid phase, nature of "inert" support, solute.

The net retention volume is affected by temperature according to the following equation:

$$\ln V_N = -(\Delta H_S/RT) + \text{constant} \qquad (11.78)$$

where H_S is the heat of sorption. Furthermore, the entropy associated with the principle mechanism (adsorption or solution) can be computed with a precision of the same order of magnitude as the heat of sorption.

11.4 CHEMICAL KINETICS

Chemical kinetics is that branch of chemistry that describes the progress of a chemical reaction. The most common description of the progress is given by the term "rate of reaction," a positive quantity which expresses how the concentration of a reactant or product changes with time. For the reaction A $\longrightarrow$ P, the rate may be expressed as the appearance of the product per unit time or the disappearance of reactant per unit time.

$$\frac{\Delta(P)}{\Delta t} = -\frac{\Delta(A)}{\Delta t} = \text{rate} \qquad (11.79)$$

Gas chromatography can be a versatile tool in studying many reactions, especially in multicomponent systems, process reaction studies, or catalytic reactions. Samples can be taken from a reaction mixture at different time intervals, chromatographed, and the rate calculated from changes in concentration.

The chromatographic column has been used as a reactor to study the kinetics of the dissociation. The reactant is introduced as a pulse at the head of the column and is continuously converted

to product and separated as it travels through the column. The
apparent rate constant (k_{app}) is a function of the rate of the
liquid (stationary) phase reaction (k_1), the rate of the gas
(mobile) phase reaction (k_g), the residence time in the gas phase
(t_g), and the residence time in the liquid phase (t_1).

$$k_{app} = k_1 + (t_g/t_1)\ k_g \qquad (11.80)$$

A mathematical statement of the dependence of the rate on the
concentration of reactants is called the rate equation, for example, rate = k (A)(B), where k = the rate constant and (A) and
(B) represent the concentrations of reactants. From the rate
equation one can frequently extract information on the mechanism
(i.e., the exact path followed to convert reactants to products).
Furthermore, it has been shown that activation energies may be
derived from gas chromatographic data. (Activation energy is an
empirical constant with units of energy which can be visualized
as the quantity of energy needed before a reaction may begin.)
A plot of log k versus 1/T is linear and the slope equals
$-E_a/2.3R$ where E_a = the activation energy and R = the gas const-
and (1.99 cal./$^\circ$K-mole) (Figure 11.3). A useful expression relating the rate constants k_2 and k_1, at two different tempera-
turess T_2 and T_1, is given below.

$$\log\ (k_2/k_1) = \left(\frac{-E_a}{2.30R}\right)\left(\frac{T_2 - T_1}{T_2 T_1}\right) \qquad (11.81)$$

Thus, if E_a and k_1 at T_1 are known, a rate constant for any temperature may be calculated.

11.5 VIRIAL COEFFICIENTS

The mathematical relationship between pressure, volume, temperature, and number of moles of a gas at equilibrium is given by its
equation of state. The most well-known equation of state is the
ideal gas law, PV=RT, where P = the pressure of the gas, V = its
molar volume (V/n), n = the number of moles of gas, R = the ideal
gas constant, and T = the temperature of the gas. Many modifications of the ideal gas equation of state have been proposed so
that the equation can fit P-V-T data of real gases. One of these
equations is called the virial equation of state which accounts
for nonideality by utilizing a power series in ρ, the density.

$$\frac{PV}{RT} = 1 + B\rho + C\rho^2 + D\rho^3 + \cdots \qquad (11.82)$$

where $\rho = 1/V$ and B, C, D, etc. are called the second, third, fourth, etc. virial coefficients, respectively. The values of the virial coefficients are a function of temperature and the particular gas under consideration, but are independent of density and pressure.

11.5.1 Second Virial Coefficient

The virial equation of state is especially important since its coefficients represent the nonideality resulting from interactions between two molecules. The second coefficient represents interactions between two molecules.Thus,a link is formed between the bulk properties of the gas (P,V,T) and the individual forces between molecules.

For a multicomponent mixture, a virial coefficient is needed to account for each possible interaction. The second virial coefficients for a two component mixture are B_{11}, B_{12}, and B_{22} where B_{11} represents the interaction between two molecules of component 1, B_{12} represents the interaction between a molecule of 1 and a molecule of 2, and B_{22} represents interaction between two molecules of 2. A tabulation of some compounds whose virial coefficients have been measured by GC is given in Table 11.6.

TABLE 11.6   SECOND VIRIAL COEFFICIENTS MEASUREMENTS BY GC

Compounds	Reference
H_2, N_2, O_2, CO_2, hydrocarbons	21
Hydrocarbons	22-24
Hydrocarbons, permanent gases	25
Benzene-gas mixtures	26
Benzene-CO_2, benzene-N_2	27
Higher hydrocarbons and their derivatives	28
Benzene-N_2, cyclohexane, n-hexane, di-isodecyl-phthalate	29

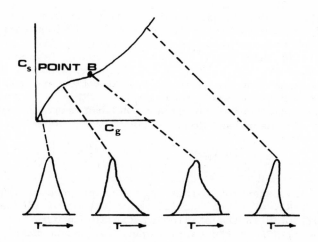

Figure 11.2   BET adsorption isotherm and corresponding peak
shapes.   Reproduced by permission of the Chemical Society of
Japan.

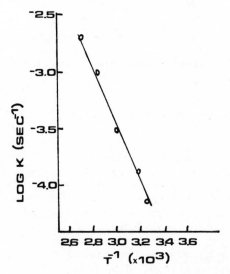

Figure 11.3   Plot of log k versus T^{-1} from which E_a may be
obtained.   Slope = $-E_a/2.3R$ = $(-2.2 - (-3.7)/(2.8 - 3.3(10^{-3}))$;
E_a = 13.73 kcal/mole.

The method of determination of virial coefficients by GC consists of measuring the retention volumes at various carrier gas
pressures and extrapolating to zero pressure. Three procedures
of extrapolation have been suggested, although they do not give
the same results. The method of Cruickshank, Windsor, and Young
(30), which takes into account carrier gas flowrate and local
pressure, seems to be most promising.

The least complicated equation to determine B_{12} is given by
Cruickshank, Windsor, and Young (31).

$$\ln V_N = \ln V_N^o + \beta p_o J_3^4 \qquad (11.83)$$

where V_N = net retention volume, p_o = outlet pressure, β =
$(2B_{12} - v_1^\infty)/RT$, v_1^∞ = the partial molar volume of the sample at
infinite dilution in the stationary (liquid) phase, (1) refers
to the sample, (2) to the carrier gas, (3) to the stationary
liquid, and J_3^4 is a function of the column inlet and outlet
pressures, p_i and p_o.

$$J_n^m = \frac{n}{m} \frac{(p_i/p_o)^m - 1}{(p_i/p_o)^n - 1} \qquad (11.84)$$

Thus, B_{12} may be obtained from the slope of the plot $\ln V_N$ versus
$p_o J_3^4$ (Figure 11.4).

11.5.2 Gas-Solid Virial Coefficients

The determination of gas-solid virial coefficients can be a useful technique to explain the interaction between an adsorbed gas
and a solid surface. The terms are defined so that the number of
adsorbate molecules interacting can be readily ascertained. For
example, the second order gas-solid interaction involves one adsorbate molecule and the solid surface; the third order gas-solid
interaction involves two adsorbate molecules and the surface, and
so on. The number of adsorbed molecules under consideration is
expanded in a power series with respect to the density of the adsorbed phase.

Few determinations of gas-solid virial coefficients have been
made. Halsey and coworkers (32,33) used the temperature dependence of the first gas-solid virial coefficient to calculate the
potential energy curve for a single molecule in the presence of a
solid. Hanlan and Freeman (34) showed this coefficient may be

obtained from frontal analysis gas chromatographic data.
Rudzinski and coworkers (35,36) first used the second and third
gas-solid virial coefficients obtained from GC data to estimate
surface areas. The surface area of silica gel determined using
virial expansion data was greater than that obtained using the
BET method (Section 11.1). The discrepancy was explained by noting the BET method does not take the lateral interactions into
account. These interactions have an effect of decreasing the
effective area of the adsorbent, thus making the calculated BET
area less than it should be (Table 11.7).

11.6   TRANSPORT PROPERTIES

The Van Deemter equation (Chapter 2) shows how diffusion and
transfer processes affected HETP. Applications of GC to the
measurement of transport properties of gases have shown to be

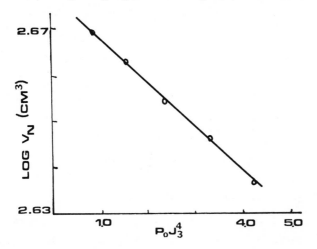

Figure 11.4.   Plot of log V_N versus $p_o J_3^4$ from which B_{12} may be
obtained. β = slope = $(2.68-2.64)/(1.0-5.0)_\infty$ = -0.01 atm^{-1}; β=
$(2B_{12} - v_1)/RT$ = -0.01 atm^{-1}; B_{12}= $\beta RT/2 + v_1/2$ = $\{(-0.01$ atm$^{-1})$
$(82.1$ cm^3-atm/deg.K-mole$)(296^\circ K)\}/2 + 40.0$ cm^3/mole/2; B_{12} =
$-121.5 + 20.0 = -101.5$.

TABLE 11.7   SPECIFIC SURFACE AREA FROM BET AND THIRD VIRIAL
             COEFFICIENT MEASUREMENTS

Specific Surface Area of Schuchardt's Silica Gel (m^2/g)

Adsorbent	BET	Virial Expansion	$T(^{\circ}K)$
CCl_4	13.1	62.6	359.8
Cyclohexane	11.9	58.6	359.8
CCl_4	13.9	64.0	374.5
Cyclohexane	18.3	71.5	374.5
CCl_4	9.9	49.2	410.5
Cyclohexane	13.2	64.7	410.5
CCl_4	13.6	66.8	426.2
Cyclohexane	12.4	58.7	426.2

especially useful.

Unpacked tubes within a heating chromatograph may be used to
determine mutual gas diffusion coefficients. For low flowrates
(v) the gas diffusion coefficient (D_g) can be related to the
height equivalent to a theoretical plate (h) by the following expression:

$$D_g = v \ (h/2) \tag{11.85}$$

h, may be calculated from the following equation:

$$h = (L_1-L_s) \ (\tau_1-\tau_s)/(t_1-t_s) \tag{11.86}$$

where l = data from the long tube,
       s = data from the short tube,
       L = length of the diffusion tube,
       τ = standard deviation (w/4 where w = peak width at the
           baseline),
       t = time.
For a typical case, He-H_2, (carrier gas velocity 3.39-8.45 $cm^3/$
min., L_1 = 14.27 m, L_s = 1.041 m), the mutual gas diffusion coefficient equaled 1.34.

An improvement of this method involves the injection of a
sharp band of gas onto the column and elution at a controlled
velocity. When the band is about halfway down the column, the
flow is stopped and band spreading occurs by diffusion only.

Then the band is eluted and the gas diffusion is obtained from one of the following equations:

$$\frac{d\delta_t^2}{dt} = \frac{2D_g}{v^2} \quad \text{(empty tube)} \tag{11.87}$$

$$\frac{d\delta_t^2}{dt} = \frac{2\gamma_2 D_g}{v^2} \quad \text{(packed tube)} \tag{11.88}$$

where δ^2 = variance,
    γ = a constant approximately equal to one,
    t = time,
    v = gas velocity.

A plot of δ_t^2 versus residence time is linear with a slope of $2D_g/v^2$ (empty tube) or $2\gamma D_g/v^2$ (packed tube).

11.7   MISCELLANEOUS PHYSICOCHEMICAL MEASUREMENTS BY GAS CHROMATOGRAPHY

Many physical and chemical properties of substances have been determined by GC. Characterization of complex materials such as asphalts, polymers, and catalysts have been performed. Carbon number and the placement of hydroxyl groups in alcohols as well as vapor pressures of numerous compounds (Table 11.5) have been determined by GC.

Very little work has been done on the measurement of many of these properties since most of them can be measured more conveniently by other techniques. They are listed in this section to testify to the versatility of GC. (Table 11.8).

11.8   PRECISION AND ACCURACY

The precision and accuracy of a physicochemical measurement by GC

TABLE 11.8   PHYSICAL PROPERTIES DETERMINED BY GAS CHROMATOGRAPHY

Property	Reference
Freezing point	37
Boiling point	38-40
Molecular weight	41-47
Solubility	48-56
Ionization potential	57
Electron affinity	58
Hydrogen bond energies	59-61
Gas-solid interaction potentials	62
Wetting phenonena	63
Surface tension	64
Viscosities	65
Degree of order of liquid crystals	66
Transitions between smectic and nematic phases	67

relies on the ability of the instruments to control and measure
all parameters relating to the required chromatographic data.
The temperature must be known and controlled to $\pm 0.05^{\circ}C$ and, in
some cases, the mass of the stationary phase must be known to
$\pm 0.2\%$.   Goedert and Guiochon (68) in 1973 commented that most
commercial GC equipment is unsuitable for thermodynamic measurements.

Unfortunately it is not easy to assess what parameters play
the greatest role in maintaining acceptable precision and accuracy.   One must consider the assumptions inherent in the theory as
well as the chemical, mechanical, and instrumental parameters,
Generally, gas chromatographic methods agree within 1-5% of other
methods.   The speed and simplicity of the gas chromatographic
method continue to make it very attractive for physicochemical
measurements.

REFERENCES

1. S. Brunauer, P. Emmett, and E. Teller, J. Am. Chem. Soc., 60,
   309 (1938).
2. F. M. Nelson and F. T. Eggertsen, Anal. Chem., 30, 1387
   (1958).
3. Y. Kuge and Y. Yoshikawa, Bull. Chem. Soc. Jap., 38, 948
   (1965).
4. M. F. Burke and D. G. Ackerman, Anal. Chem., 43, 573 (1971).

5. W. V. Loebenstein, J. Biomed. Mater. Res., 9, 35 (1975).
6. D. Dollimore and G. Heal, Nature, 208, 1092 (1965).
7. J. H. Purnell, "Gas-chromatographic Study of Chemical Equilibria," in A. B. Littlewood (Ed.), Gas Chromatography 1966, Institute of Petroleum, London, 1967, p. 3.
8. J. P. Sheridan, D. E. Martire, and Y. B. Tewari, J. Am. Chem. Soc., 94, 3294 (1972).
9. D. E. Martire and P. Riedl, J. Phys. Chem., 72, 3478 (1968).
10. H. Liao, D. E. Martire, and J. P. Sheridan, Anal. Chem., 45, 2087 (1973).
11. C. Eon and G. Guiochon, Anal. Chem., 46 1393 (1974).
12. S. Se, J. Bleumer, and G. Rijnders, Sep. Sci., 1, 41 (1966).
13. G. Geisler and R. Jannash, Z. Physik. Chem., 233, 42 (1966).
14. A. Franck, H. Orth, D. Bidlingmaier, and R. Nussbaum, Chem.-Ztg., Chem. App., 95, 219 (1971); through Chem. Abstr., 75, 41223s (1971).
15. C. Pommier and G. Guiochon, J. Chromatogr. Sci., 8, 486 (1970).
16. A. Rose and V. Schrodt, J. Chem. Eng. Data, 8, 9 (1963).
17. F. Ratkovics, Acat. Chim. Acad. Sci. Hung., 49, 57 (1966); through G. C. Abstr., 442 (1967).
18. S. A. Voitkevich, M. M. Shchedrina, N. P. Soloveva, and T. A. Rudol'fi, Maslo-Zhir, Prom., 37, 27 (1971); through Chem. Abstr., 76, 251 (1972).
19. F. Ratkovics, Acta. Chim. Acad. Sci. Hung., 49, 71 (1966); through G. C.Abstr., 443 (1967).
20. F. Ratkovics, Magyar Kem. Folyoirat, 72, 186 (1966); through Chem. Abstr., 65, 1450 (1966).
21. D. H. Desty, A. Goldup, G. Luckhurst, and W. Swanton, in M. van Swaay (Ed.), Gas Chromatography 1962, Butterworths, London.
22. L.Chekalov and K. Porter, Chem. Eng. Sci., 22, 897 (1967).
23. E. M. Dantzler, C. M. Knobler, and M. L. Windsor, J. Chromatog., 32, 433 (1968).
24. R. L. Pecsok and M. L. Windsor, Anal. Chem., 40, 1238 (1968).
25. P. Chovin, Bull. Soc. Chim. Fr., 1964, 1800; through G. C. Abstr., 965 (1966).
26. C. R. Coan and A. D. King, J. Chromatog. 44, 429 (1969).
27. A. J. B. Cruickshank, B. W. Gainey, C. P. Hicks, T. M. Letcher, R. W. Moody, and C. L. Young, Trans. Faraday Soc., 65, 105 (1969).
28. M. Vigdergauz and V. Semkin, Zh. Fiz. Khim., 45, 931 (1971).
29. B. K. Kaul, A. P. Kudchadker, and D. Devaprabhakava, Trans. Faraday Soc., 69, 1821 (1973).
30. A. J. B. Cruickshank, M. L. Windsor, and C. L. Young, Proc. Roy. Soc. (London), A295, 271 (1966).
31. A. J. B. Cruickshank, M. L. Windsor, and C. L. Young, Proc.

Roy. Soc. (London), A295, 259 (1966).

32. J. R. Sams, G. Constabaris, and G. D. Halsey, J. Chem. Phys., 36, 1334 (1962).

33. W. A. Steele and G. D. Halsey, J. Phys. Chem., (1955).

34. J. F. Hanlan and M. P. Freeman, Can. J. Chem., 37, 1575 (1959).

35. W. Rudzinski, Z. Suprynowic , and J. Rayss, J. Chromatog., 66, 1 (1972).

36. W. Rudzinski, A. Waksmundzki, Z. Suprynowica, and J. Rayss, J. Chromatog., 72, 221 (1972).

37. E. T. Rangel, M. S. Thesis, Rice University, Houston, 1956.

38. D. E. Willis, Anal. Chem. 39, 1324 (1967).

39. K. Otozai, Z. Anal. Chem., 268, 257 (1974).

40. R. Bach, E. Dotsch, H. A. Friedrichs, and L. Mary, Chromatographia, 4, 561 (1971).

41. M. J. Cohen and R. F. Wemlund, Ind. Res., 17, 58 (1975).

42. R. S. Swingle, J. Chromatogr. Sci., 12, 1 (1974).

43. J. D. Burger, J. Gas Chromatogr., 6, 177 (1968).

44. S. C. Bevan, T. A. Gough, and S. Thorburn, J. Chromatog., 44, 241 (1969).

45. C. Phillips and P. Timms, J. Chromatog. 5, 131 (1961).

46. E. Kirar and J. K. Gillham, Anal. Chem., 47, 983 (1975).

47. B. M. Mitzner, G. Hild, and J. Gates Clarke, Jr., Anal. Chem., 47, 1880 (1975).

48. G. A. Kurkchi, S. F. Shakhova, and L. E. Sergeeva, Zh. Fiz. Khim., 46, 2302 (1972); through Chem. Abstr., 78, 8456j (1973).

49. H. Massaldi and C. King, J. Chem. Eng. Data, 18, 393 (1973).

50. T. Sugiyama, T. Takeuchi, and Y. Suzuki, J. Chromatog. 105, 265 (1975).

51. R. L. Brown and S. P. Wasik, J. Res. Nat. Bur. Stand., Sect. A., 78, 453 (1974).

52. K. Konno, M. Sakiyama, T. Suzuki, and A. Kitahara, Nippon Kagaku Kaishi, 1974, 2211; through Chem. Abstr., 82, 90700y (1975).

53. F. Franks, Nature, 210, 87 (1966).

54. D. Carter and G. Esterson, J. Chem., Eng. Data,18, 166 (1973).

55. K. Aulif and V. Mak, Org. Geokhim., 1970, 168; through Chem. Abstr., 76, 34802j (1972).

56. T. J. Roseman and S. H. Yalkowsky, J. Pharm. Sci., 62, 1680 (1973).

57. R. J. Laub and R. L. Pecsok, Anal. Chem., 46, 1214 (1974).

58. W. E. Wentworth and R. S. Becker, J. Am. Chem. Soc., 84, 4263 (1962).

59. L. Belyakova, A. Kiselev, and N. Kovaleva, Anal. Chem., 36, 1517 (1964).

60. A. Apelblat, J. Inorg. Nucl. Chem., 31, 483 (1969).
61. A. Iogansen, G. Kurkchi, and O. Levina, Gas Chromatogr. Int. Symp. Anal. Instrum. Div. Instrum. Soc. Am., 6, 35 (1966); through Chem. Abstr., 70, 109218y (1969).
62. W. Hammers and C. Delingny, J. Polym. Sci., Polym. Phys., Ed., 12, 2065 (1974).
63. K. Bartle and M. Novotny, J. Chromatog., 94, 35 (1974).
64. D. E. Martire, R. Pecsok, and J. H. Purnell, Trans. Faraday Soc., 61, 2496 (1965).
65. N. Dyson and A. B. Littlewood, Trans. Faraday Soc., 63, 1895 (1967).
66. L. Chow and D. Martire, Mol. Cryst. Liquid Cryst., 14, 293 (1971); through Chem. Abstr., 75, 122954h (1971).
67. H. Kelker, Ber. Der Bunsen Gesellshaft, 67, 698 (1963).
68. M. Geodert and G. Guiochon, Anal. Chem., 45, 1180 (1973).

CHAPTER 12

Drug Analysis Using
Gas Chromatography

EUGENE J. MCGONIGLE

Merck, Sharp and Dohme Research Laboratories

12.1   INTRODUCTION 592
       12.1.1 Types of Analyses 593
       12.1.2 Why GC or Why Not? 594
       12.1.3 Laws with which the Analyst Should be Familiar .594
       12.1.4 Safety Precautions 596
12.2   METHOD DEVELOPMENT TECHNIQUES 597
       12.2.1 Raw Material Studies 597
       12.2.2 Selection of Internal Standard (Quantitative) . .598
       12.2.3 Selection of Resolution Standard 598
       12.2.4 Linearity 599
       12.2.5 Reproducibility 600
       12.2.6 Drug Analog Behavior 601
       12.2.7 Stress Conditions 602
12.3   SAMPLE PREPARATION .603
       12.3.1 Legitimate Formulations 604
              Formulation Types and Typical Ingredients 605
       12.3.2 Illicit Formulations 606
       12.3.3 Sample Preparation with Legitimate and Illicit .606
              Formulations
       12.3.4 Biological Samples 609

591

      12.3.5 Summary .610
12.4   DERIVATIZATION TECHNIQUES 611
      12.4.1 Silylation 611
      12.4.2 Acetylation 613
      12.4.3 Methylation or Esterification 614
      12.4.4 Summary 615
12.5   APPLICATIONS 616
      12.5.1 General Applications 617
      12.5.2 Nonaddicting Analgesics and Antipyretics 618
      12.5.3 Barbiturates and Related Amides 618
      12.5.4 Organic Nitrogenous Bases 619
      12.5.5 Hormonal Drugs 628
      12.5.6 Antibiotics 629
      12.5.7 Vitamins .630
      12.5.8 Resolution of Enantiomers by GLC 632
REFERENCES 633

12.1 INTRODUCTION

The application of any method of analysis for one or more components is successful only insofar as the analyst can cope with
the medium in which the compound is dissolved or dispersed. This
is particularly true in drug analyses where the medium might
either be a tablet, capsule, or ointment, in the case of a pharmaceutical preparation or formulation, or blood, urine, or tissue
in the case of a biological specimen. The problems facing the
analyst, therefore, in developing a gas chromatographic method
are not all that unique, and, with a few exceptions, the rationale for method development in drug analyses by gas chromatography (GC) is similar to many other methods. It is the application of this rationale, with special emphasis on GC, which is the
topic of this chapter. Of necessity, the discussion will be
limited in scope as the intention is only to familiarize the novice with the technique. Scores of articles are added to the literature weekly, rendering a thorough treatment of GC applications
in drug analyses in one chapter a virtual impossibility. Fortunately, there is a wealth of information in the literature on
the analysis of most drugs so that the analyst need only apply a
described, studied system (column, temperature, and flow characteristics) to his specific problem.
    By law, all drugs must meet certain minimum specifications
which certify acceptable purity for human consumption. Most manufacturers of licit preparations, or products prepared in accord-
ance with regulations defined by law and enforced by the Food
and Drug Administration (FDA), in the United States, will adopt

more rigid specifications to assure themselves a quality product.
Even so, the analyst should not equate purity grades, such as
United States Pharmacopeia (USP) Certified, or Authentic Specimen
British Pharmacopeia (BP) with chemical purity. In many instances, these compounds are of the highest quality, but the analyst
should always establish acceptability of a drug species for purposes of a control, or a standard, by using some technique in
addition to GC. Detector response is not the same for all molecules, and trace impurities may contribute significant interference, particularly if sensitive detectors (or selective) such
as electron capture (see Chapter 5) are employed.

12.1.1 Types of Analyses

The kinds of analyses encountered in the laboratory are selected
according to the nature of the sample and the purpose of the
analysis. The following flow chart illustrates common areas
that require analytical data:

Bulk Drug

1. Purity data
2. Data to indicate optimum yield
    during synthesis
3. Data to evaluate charcteristics
    of derivatives

Formulated by: Pharmacist, licit or illicit
manufacturer

Product

Quality	Safety	Identification
1. Stability	1. Metabolism	1. Forensic
2. Uniformity	studies. (tissue)	2. Clinical diag-
(dose to dose)	residues)	nosis post
3. Dissolution rate	2. Pharmacokinetics	mortem
4. Identification	(rate of biologi-	
to insure against	cal transfer to	
contamination	produce drug	
5. Data to compare	effect)	
with a competitor		

Obviously, each of these categories can be subdivided further.
For example, toxicity studies may require analyses of tissue and
sera using several species of animals, as well as humans. It is

very rare to find an analyst equally proficient in all areas even
though he may possess astute capability in applying GC in any one
area. The reason lies not just in differences in the nature of
the sample, but, also in the whole philosophy of recognizing and
defining the analytical problem, and interpreting the significance of the data.

12.1.2 Why GC or Why Not?

Virtually all drug analyses require both identification and quantitation. In this regard, GC offers major advantages as well as
some noteworthy disadvantages. The high sensitivity of GC is an
important factor, particularly in drug metabolism or toxicity
studies, where only trace quantities of drug are available.
Some elemental detectors (see Chapter 5) can yield accurate
measurements in the picogram range. On the other hand, sensitivity can be a disadvantage in the presence of interfering materials, even though only a few percent of a contaminating material is present. High-resolution GC columns allow excellent selectivity in the presence of structurally similar analogs or degradates. Inherent, is the ability to adjust parameters such as
carrier gas flow and column temperature, enabling the analyst to
optimize selectivity. But, despite high-resolution capability,
retention volumes are a poor indication of component identification since column packings vary in efficiency from day to day,
and from column to column. Reproducibility of columns is somewhat "artistic" depending on the materials employed and the
packing technique (see Chapter 3). The analyst must protect himself against reproducibility problems by including standards that
will serve as a check on resolution. This topic is discussed in
a later section. To further complicate the problem of identification, thermal stress on the molecular species as it passes into and through the chromatographic system sometimes results in an
altered or degraded structure, so that isolation and subsequent
independent analysis using a splitter can be confusing. Mass
spectroscopy has aided tremendously with this problem as fragmentation patterns can confirm identity, even though the species
may be degraded. However, many laboratories cannot afford this
luxury, so the analyst should be aware of his limitations.

12.1.3 Laws with which the Analyst Should be Familiar

If the analyst is employed in a pharmaceutical laboratory,

or a laboratory associated with some regulatory agency, the legal
aspects of handling drugs, even for legitimate purpose of analysis, will be included in the employee orientation. However,
individuals who are part of a consulting laboratory, or some
other area not dealing exclusively with drugs, should familiarize
themselves with laws which control the procurement, storage, use,
and disposal of drugs. Obviously, the treatment here is cursory
and individuals should secure a complete listing of regulations
as applicable to his or her purposes from the local Drug Enforcement Agency (DEA). In the United States, there are two Federal
laws of primary concern although individual states can, and have,
supplemented these. The first is the Controlled Substances Act
which regulates the handling of drugs of abuse. There are five
(5) schedules, or lists, which classify drugs according to medical use and degree of abuse:

Schedule I	– Drugs have no accepted medical use in the United States but a high rate of abuse, e.g., heroin, LSD.
Schedule II	– Drugs have an accepted medical use in the United States and a high rate of abuse, e.g., morphine, codeine, amphetamines.
Schedule III	– Drugs have an accepted medical use, but a lower potential for abuse than Schedule I or II, e.g., lysergic acid, derivatives of barbituric acid.
Schedule IV	– Drugs have an accepted medical use and are generally the long-acting barbiturates, hypnotics and minor tranquilizers, e.g., meprobamate, phenobarbital.
Schedule V	– Any of the above usually in solution at low concentration (of controlled substance) which also contains noncontrolled ingredients in sufficient quantity to attribute to the mixture valuable medicinal qualities other than those possessed by the controlled substance alone, e.g., cough syrup.

As one might expect, accountability is more stringent for substances classified in Schedules I and II. These compounds can be
procured only from licensed sources, and amounts are recorded.
After a certain period of time, the quantities of drug used,
which have been recorded and witnessed in the laboratory notebook
are tabulated and compared with amount remaining (unused), and
the original amount issued. The balance is usually returned, or
destroyed in the presence of an authorized witness. Dilute standard solutions can be purchased, in limited quantities, usually

without difficulty. Not all drugs are controlled substances because not all drugs are subject to the kinds of abuse which require controls. However, the balance of all drugs intended for
human or animal consumption are covered by the Food, Drug, and
Cosmetic Act. To procure and use drugs covered by this act, requires licensing, but issuance is based primarily on an established, justifiable purpose. Most legitimate laboratories performing drug analyses would qualify. Accountability for amounts
used are not required, although a good analyst should keep accurate records as a matter of professional policy.

12.1.4 Safety Precautions

Finally, it is imperative to say a few words about caution in
working with drugs in the laboratory. Many people experience
dermatological effects when exposed to drug substances. The
effects of inhaling dusts can be even more dramatic...even dangerous. Weighing standards, preparing synthetics for recovery
studies, or handling solution concentrates, all require that the
analyst be alert to potential dangers. Weighing bulk drug for
use as a standard or in synthetic mixtures requires particular
caution, since the drug is undiluted, and may be present in
amounts representing 1, 10, or 10^5 times the recommended dose.
The analyst should familiarize himself with the properties of any
drug before attempting to work with bulk quantities. These include: therapeutic effect, side effects, reported allergic reactions, or incompatibility with any medical incompacity the analyst may possess. Two good rules are as follows: (a) Use gloves
whenever possible when handling the pure solid; surgical gloves
are excellent for this purpose. (b) Undiluted drug should be
transferred to weighing bottles prior to analytical weighings.
The transfer should be completed in a hood, and quantities should
be no more than two times the anticipated amount required.
    Solutions are somewhat less dangerous but this fact can lead
to careless handling due to a false sense of security. Precautions against internal consumption are obvious, but even more
insidious is the potential danger of skin absorption. Many drugs
are amine hydrochlorides. In alkaline solution, the free base has
increased lipophillic character, so the rate of skin absorption
is increased.
    A third consideration is the cumulative effect that some drugs
may have on sensitive individuals. These would produce toxic
effects over a period of time which may be irreversible.
    These points are not mentioned to unnecessarily frighten any
individual who may be working with drugs. Rather, it is just a

reminder that these compounds are not innocuous, they are of value because they are known to produce a biological effect (in controlled amounts). All drug substances should be treated as toxic and experience should not lead to carelessness.

12.2   METHOD DEVELOPMENT TECHNIQUES

12.2.1 Raw Material Studies

The first step in the development of a gas chromatographic procedure is to establish the performance characteristics of the drug in the absence of any matrix. Of principal importance is the selection of some portion that may serve as a standard. As was pointed out earlier, the analyst must be careful in his or her judgment of a chemically pure standard. Purity should be established by independent techniques which will reinforce confidence in structural integrity before any GC is attempted. Phase solubility and thin layer chromatography are two very commonly employed techniques which will verify purity. Not only do these techniques indicate initial suitability of a standard, but they also help to monitor changes in standard purity with time which could render the standard no longer acceptable. Of the two, phase solubility is the most reliable from the standpoint of accuracy and the fact that it can be employed in most laboratories. On the other hand, when phase solubility data indicate differences in bulk purity, it is good practice to establish whether the change is purity or stability related. Phase solubility is not readily employed in this capacity; however, thin layer chromatography can provide excellent supplemental data. There are additional techniques, but these two can provide considerable information with relative ease and low cost. In capable hands, data is reliable and acceptable to regulatory agencies (provided appropriate method justification has been included).

Assuming no derivatization is necessary, the analyst may now proceed to determine which column will best suit his objectives (see Chapter 3). Normally, with the state of the art what it is, this decision will be based on the results of a careful search of the literature. Many "experts" presume to judge intuitively the best column(s) to achieve a separation. To be sure, experience will allow many good judgments, but no amount of experience can replace a thorough literature search. In the applications section, we will point out some successfully employed drug/ column systems and it will become apparent how few liquid phases are really necessary for most analyses.

    After the column has been selected, the analyst should now
determine an acceptable range of linearity suitable for the col-
umn/detector capacity. No internal standard is required at this
point, although some care should be taken to duplicate injections. An ideal range would be a concentration that is compatible with sample availability and linear over a range C/100-c-
10C, where C is the proposed working concentration. Linearity
should be confirmed at more than one temperature. Peak height or
area can be employed depending on the detector. If possible, the
solvent employed should be that which will serve as the extraction medium in the sample preparation. This is important to
avoid possible drug/solvent interactions which will mask true
response characteristics. Many amines are extracted by liquid/
liquid partitioning from an aqueous, alkaline medium using chloroform or methylene chloride; high-temperature interactions are
possible between the amine and the solvent which will produce
"ghost" peaks.

12.2.2 Selection of Internal Standard (quantitative)

The analyst is now ready to select an internal standard for
quantitating the data (assuming quantitation is the objective,
not estimation). The purpose of the internal standard is twofold. The first, and most obvious function is to allow for the
uncertainty of a syringe/septum injection. A standard and sample
containing the same concentration of internal standard will compensate for a reproducibility problem in filling and emptying a
syringe and most septum leaks (but not all). The second function
of the internal standard is to aid in assuring the analyst that
during the analysis, chromatographic changes do not occur which
result in different partitioning characteristics. Of course,
such problems may be due to columm bleeding or degradation. An
obvious manifestation would be a change in retention volume and
subsequent peak height. For best control, the internal standard
should be structurally similar to the sample compound so that
both have similar volatility (and hence partitioning). Even with
good column performance, some septum leaks can go uncorrected if
the internal standard has a volatility (vapor pressure) significantly different than the sample, such that more (or less) of the
standard or sample escapes through the septum. Hence, volatility
differences should be minimal.

12.2.3 Selection of Resolution Standard

Too often, the analyst omits a very important step in method development. A standard should be selected which will assure that the desired separation is reproducible on a day-to-day, month-to-month basis. The internal standard (quantitative) discussed above is not satisfactory for this purpose because it is not one of the compounds that is a "potential" compound to be resolved. Consider, if you will, a situation wherein a method has been developed which will resolve a drug from a given metabolite. This metabolite is a product indicative of some toxic reaction within the body. Let us assume that the drug produces this reaction and, hence, the metabolite occurs only rarely in some people. Obviously, the internal standard (quantitative) must be resolved from the drug and the metabolite. However, because the toxic reaction occurs rarely, in most analyses, the metabolite will not be present. Could the analyst assume that because the retention volumes of the internal standard (quantitative), and the drug standard, are about the same as they were the last time the analysis was done, that the metabolite would still be resolved, if present? The answer is apparent: The analyst must include a metabolite "standard" to reaffirm selectivity and sensitivity. In other words, the analyst should select a "resolution standard" that closely resembles the compound to be resolved (in the analysis) from the sample. For example, in a gas chromatographic analysis for dexamethasone acetate wherein dexamethasone alcohol is a potential degradate, prednisolone acetate could serve as a quantitative internal standard while desamethasone alcohol (or prednisolone alcohol) could be the resolution standard. The former quantitates the drug from injection to injection while the latter is only employed occasionally (perhaps only once) to establish day-to-day selectivity characteristics. Fortunately, in drug analysis, a variety of structural analogs are usually available, so that availability of appropriate standards is ordinarily not a problem. A calculated resolution range is a good factor to include in method specifications under the heading "performance characteristics." Let both standards determine whether the chromatographic procedure is performing acceptably. The analyst must remember that in drug analysis, his method invariably will be employed by someone else, either in a regulatory agency or in a control laboratory. It is rare that drug analyses are carried out personally, on a "one-time" basis. Accordingly, delays are minimized when justification of a method is clear. There is no substitute for including demonstrable resolution characteristics with method description and justification.

12.2.4 Linearity

Some discussion of linearity was included under Section 12.2.1,
Raw Material Studies. The purpose there was to show that the
column/detector system was compatible with the desired drug concentration range. This compatibility must be reaffirmed in the
presence of any known compound that will possibly be chromatographed, parcicularly the internal standard.   The reason is to
be sure that the presence of the resolved spécies will not affect
the partitioning of the drug in the chromatographic system during
the separation. This is usually a rapid test and, under normal
circumstances, need only be carried out once. If the analyst
sees evidence of phenomena indicative of irreversible adsorption
of a drug, or the formation of a thermal degradate or drug analog,
then the linearity check may have to be repeated more frequently.
Aside from the appearance of foreign peaks, symptoms of such problems would be changes in retention volume and/or peak height (or
area) magnitude with successive injections. The analyst should
be cautious, even though the linearity response passes through
origin, since it may take time to build up an interfering molecular "coating" on the stationary phase.   Interactions like these
are the major reason that all drug analyses should be carried out
using glass columns and on-column injections whenever possible.
Many times, supports are silylated to minimize active adsorptive
sites. However, it should be pointed out that these phenomena
are not all that common, and if symptoms such as those described
above are encountered, the novice should revert back to basic
troubleshooting techniques first, and be sure that all fittings
are tight and that something has not occurred to alter flow/tem-
perature equilibrium.   There have been many theories of sophisticated drug/substrate interaction ultimately attributable to a
leaky septum or a faulty flow regulator.

    12.2.5 Reproducibility

There are many criteria for reproducibility depending on the
journals one reads, or the professional societies to which one
belongs. In developing a gas chromatographic method, no two prepared columns are identical. No two instruments project the same
temperature gradients from injection port to detector. In short,
the only certain thing is that the chromatographic system, if developed properly, should exhibit the same resolution and quantitation characteristics over a period of several days. In evaluating his method, the chemist should shut down the column oven,
and remove the column. Injection port and detector heaters may
be left on. The next day, the unit should be again set up and
equilibrated. Equilibration time and instrument parameters

should be noted. After a week, the chemist should observe if any
parameters (temperatures or flow) required an adjustment in the
day-to-day settings, which were consistently positive (or negative), to maintain resolution characteristics. If the system is
reproducible, then daily adjustments should be random during the
week, and the range of each parameter, observed over the period,
should be the specified setting included in the method procedure,
not just the average value. If the analyst intends to publish
the procedure, then another column should be prepared (ideally
by someone else) and the procedure repeated. Retention volumes,
resolution, and linearity responses should be recorded as daily
evaluation criteria.

12.2.6 Drug Analog Behavior

At this point, the analyst has established the basic chromatographic system and is ready to optimize conditions to provide the
best resolution. There are those who feel that the studies described in this section as well as the following section should be
carried out earlier, since the column may not perform the desired
separation, and the chemist will have to start over again. There
are arguments for both approaches and, ultimately, it probably
does not matter which is done first as long as the procedure is
followed to completion.

   If the application is for a biological system, wherein the objective is to isolate or quantitate a specific drug and its metabolite which is known, then the chemist can proceed to work with
the compounds directly. In most cases, however, the chemist must
be prepared to resolve the unexpected compound, which is not all
that difficult, depending on the equipment available. The possibility that a thermal environment, such as GC, can initiate a reaction involving a labile drug or metabolite can be distressing,
to say the least. Sample preparation usually includes some isolation technique which restricts classes of compounds. The need
for sample cleanup will depend on the nature of the sample (see
Section 12.3). The point here is that there will be some sample
preparation as the first step in the analysis which will provide
an initial "cleanup" of the sample.

   A good initial approach is to study the behavior of known synthetic precursors, drug analogs or degradates. In our previous
example, dexamethasone acetate, one can readily purchase compounds such as cortisone, cortisone acetate, dexamethasone alcohol, dexamethasone-17-keto analog, prednisolone alcohol, and so
on. Their chromatographic responses, relative to the compound of
interest, will indicate which structural modifications are detec-

table in the system. It should not be implied that these compounds form in any quantity at all; the sole objective is to test
resolution characteristics. If possible, the resolution characteristics should be included in the justification of the method,
when presented to anyone for reproduction. The novice should remember that no matter how carefully he or she has developed the
method to determine drug X in the presence of drug Y, and drug Z,
the question will most likely be asked, "What about compound M or
drug Q?" These possibilities may not seem realistic to the analyst, but could be of paramount importance to a regulatory official, toxicologist, or pharmacologist. Many times, the compounds
in question are not available, but a knowledge of the selectivity
of the method can reassure those concerned that should compound
M or drug Q be present, they would be resolved.

12.2.7 Stress Conditions

There is much debate among groups involved in drug analysis as to
the value of stress studies in method development. Stress studies can be grouped as follows:

1. Thermal: (A) High temperature--approach or exceed the melting
   point of solids: autoclaving solutions. (B) Low temperature-
   usually below $100^{\circ}C$ and may be conducted in a controlled humidity environment.
2. Oxidative: usually solutions that are treated with permanganate, peroxide, or air (bubbling).
3. Hydrolysis: acid or base.
4. Light: fluorescent, sunlight, intense ultraviolet.

The difficulty is that these conditions may initiate reactions
that would never normally occur because potential energy barriers
required for reaction initiation would not be exceeded under less
stringent conditions. An obvious exception is the analogy between thermal stress studies, and stress encountered within the
chromatographic system itself. To be sure, this is a major disadvantage in the application of GC in drug analysis, wherein
structures can be thermally altered, sometimes with relative
ease. Nevertheless, stress studies provide affirmative information which instills confidence (or the lack of it) in a given
chromatographic separation by demonstrating selectivity under ad-
verse conditions. For example, it is generally known that catecholamines undergo rapid and extensive oxidative degradation including cyclization and polymerization. Solutions of catecholamines in mild alkali, or even water, can be degraded as

evidenced in part by extensive discoloration on standing with
time or, more rapidly, by bubbling air through them. Even though
a method of analysis may be designed to monitor a single catecholamine, information should be gathered that demonstrates the
resolution and quantitative performance of the analytical system,
in the presence of stress-induced structural analogs. In such
studies, the analyst should not concern himself if resolution is
not complete, since some interference is quite likely. An independent method of analysis such as thin-layer or high-speed
liquid chromatography should be used to corroborate the results.
A successful study involving samples wherein loss of intact drug
is in the range of about 25-50% might include the following:

1. No peak distortion of the drug moiety.
2. No change in retention value for the intact drug moiety.
3. The appearance of several peaks resolved from the intact drug
   moiety.
4. Quantitative correlation within 10-20% between corroborating
   methods (with some similarity in selectivity characteristics,
   e.g., TLC, LC).

Ideally, more than one stress condition should be employed, and
two different time intervals should be used to determine if any
loss is linear.

12.3   SAMPLE PREPARATION

The preliminary step in any qualitative or quantitative analysis
is sample preparation. The objective is usually to isolate the
compound(s) of interest from substances which will cause interference of any of the following types:

A. Interference that decreases sensitivity by contributing background signal in the same region as the signal response of the
   compound of interest.

B. Interference that decreases accuracy by contributing an
   interfering signal masked by the signal of the compound of
   interest.

C. Chemical interactions that result in apparent loss of the
   compound of interest.

D. Interactions with the packing which changes the resolution

characteristics or efficiency (N). Active sites can be altered reversibly or irreversibly.

E. Interference which decreases sensitivity by contaminating the
   detector.

To compound the problem, interference may not always be reproducible, or may worsen with time (viz., D and E). Because of the
rather unpredictable qualities of interfering materials, whenever
possible, the analyst should avoid the use of sample blanks to
compensate for interference. Instead, efforts should be directed
to optimize a "sample cleanup" procedure which will eliminate as
much as possible the contaminating materials. This is relatively
easy to accomplish in most instances when the sample is a drug
formulation. However, when the objective involves a biological
specimen such as might be encountered in a metabolism or toxicological study, use of a blank is imperative. In recent years,
concerns about environmental impact of drugs have spurred interests in analyses of soil and water resources, including effluents. It is obvious that the problem of sample preparation becomes more significant as sensitivity requirements increase and
knowledge or familiarity with the sample matrix decreases. One
could discuss sample matrices and their analytical treatment as
the subject of a treatise. Our objective here is only to introduce the novice to some of the more commonly encountered excipients and how they can be handled as part of a chromatographic
procedure.

12.3.1 Legitimate Formulations

The broadcast category, about which we know most, comes under the
heading of legitimate formulations. These are drug dosage forms
which are manufactured ethically by a government-sanctioned or
licensed company, or by a trained, licensed pharmacist. The intended purpose of these products is therapeutic or diagnostic use
by patients themselves (over-the-counter products) or through
physicians (prescription items). Analyses of these products are
usually conducted by chemists who are interested in stability,
purity, uniformity, or other tests of quality, and are conducted
within the manufacturing area itself. In addition, agencies
within the Federal Government such as the Food and Drug Administration or the United States Pharmacopeia, or consulting laboratories may have similar interests for a variety of reasons. The
following includes some examples of dosage forms and typical excipients. A more complete description of formulation ingredients and the "hows and whys" of their use may be found in

Remington's Pharmaceutical Sciences (1).

FORMULATION TYPES AND TYPICAL INGREDIENTS.

A. Tablets: Compressed powders or granules which form a predetermined shape and weight.   Tablets may be coated or uncoated. Excipients include:
         a. binding agent: corn starch, methyl cellulose,
            glyceryl monostearate (sustained release)
         b. bulk for varying dose: lactose, microcrystalline
            cellulose, or dicalcium phosphate.
         c. lubricant: magnesium stearate, talc
         d. coating agents: to mask unpleasant tastes or
            preserve stability (sugar, cellulose derivatives)

B. Capsules: No compression employed. Powders are supplied by
   weight to a gelatin container of fixed volume. Excipients
   include materials similar to tablet ingredients.

C. Parenteral or ophthalmic solutions: Solutions used for injection (parenteral) or application to the eye (ophthalmic). Excipients include:
         a. solvent: ethanol, glycerin, water, propylene glycol
         b. buffer: phosphates, citrates, etc.
         c. surfactants: sodium lauryl sulfate, "Tween"
         d. preservatives (antibacterial or bacterostatic):
            sorbic acid, benzalkonium chloride, phenol, paraben,
            benzyl-alcohol, benzoic acid
         e. antioxidants: sodium bisulfite, ascorbylpalmitate
            ascorbic acid
         f. additives for isotonicity or isoosmoticity: sodium
            chloride, dextrose, mannitol

D. Oral suspensions: Additives are very similar to C, only flavorings are included along with suspending agents such as the
   gums, tragacanth, acacia, and bentonite.

E. Syrups: Additives are very similar to C, only flavorings are
   included along with high concentrations of sugar (or artific-
   ial sweetner) to increase viscosity.

F. Elixirs: Additives are very similar to C, only flavorings and
   alcohol are included.

G. Creams or ointments: Semisolid preparations intended for top-
   ical application. Excipients include petrolatum, isopropyl

myristate, polyethylene glycols, hydroxypropyl cellulose, propylene glycols, vegetable oils, silicones, alcohols, and varying degrees of water.

H. <u>Suppositories</u>: Solid preparations which melt at body temperature delivering medication for at-site treatment or for absorption at that point (usually rectal, vaginal, or urethral). Excipients include cocoa butter, waxy fatty acids, and derivatives, polyethylene glycol, theobroma oil, as well as many ingredients found in G.

12.3.2 Illicit Formulations

A second category comes under the heading, illicit, when the dosage form is manufactured without any regard for regulatory control. Such products can come under two general types:

1. Preparations that are a counterfeit of a legitimate product, intended to be clandestinely substituted and used as the legitimate product.

2. Preparations that are designed to make a controlled substance (Schedules I-V, inc.) palatable or consumable, for purposes of abuse.

Obviously, very little is known about illicit preparations since the major specification is minimal cost and manufacturing convenience. Heroin, for example, has been diluted with quinine, procaine, lactose, and many other excipients. To be safe, the chemist should regard such samples as completely unknown and not assume that the excipients will be innocuous to the chromatographic system. One important precaution is to anticipate contamination and impurities. The former may not be present uniformly per sample weight while the latter will likely be uniform.

12.3.3 Sample Preparation with Legitimate and Illicit Formulations

These preparations have one thing in common which can be used advantageously in sample preparation. The body should be able to distinguish the drug from the matrix and extract it. With this in mind, the analyst can decide on an initial approach to sample preparation by first considering the dosage form. Solid dosage

forms, viz., tablets or capsules, are ingested into the stomach.
Some are absorbed at that point; others are carried into the in-
testinal tract and released there. Therefore, a reasonable first
approach might be to perform a liquid/solid extraction of the
ground up tablets (or capsule contents) using 0.1N hydrochloric
acid or pH 7-8 phosphate buffer. Sometimes a preliminary treatment with small amounts of methanol or ethanol may be helpful to
prevent lumping and subsequent drug entrapment. Usually, mechanical agitation is adequate although ultrasonic vibration is preferred. Precautions include solution stability (hydrolysis,
photolysis, etc.), adsorption of drug from solution onto an insoluble excipient, drug/dye interactions in solution, and entrapment of the drug in insoluble granules. Most excipients in solid
dosage forms offer few problems in GC from an interference standpoint after liquid/solid extraction. However, it is not uncommon
to encounter suspended particulate material, even after centrifugation (2000-3000 rpm). Since filtering should generally be
avoided in drug analyses because of possible adsorption phenomena, liquid/liquid partitioning is employed to preclude syringe
clogging (or worse) due to barely discernable particles. Parent-
erals, ophthalmics, syrups, and elixirs are solutions that may be
diluted as necessary with an appropriate solvent. Suspensions
present a similar problem as discussed previously, not from the
drug itself, but, more often, from the suspending agent which may
display little solubility in any of the more commonly employed
solvents. Many excipients in these matrices are distributed in
both phases after liquid/liquid partitioning, viz., phenol or
benzyl alcohol. In addition, many present their own stability
problems, viz., benzyl alcohol can oxidize to benzaldehyde. To
avoid this, multiple partitioning systems are employed. For example, some secondary and tertiary amine hydrochlorides can be
extracted into chloroform or methylene chloride as an ion pair
using sodium chloride and sulfuric acid. The amine/chloride ion
pair can be extracted into sulfuric acid from the organic phase
leaving behind most interfering species. It is not at all uncommon to use more than one partitioning system. Some surfact-
ants can cause extreme difficulties in liquid/liquid partitioning
systems by reducing the surface tension at the interface. The
result is emulsions. These can be most troublesome to the analyst, not only in affecting the ability to recover the organic
phase, but in changing the distribution coefficient of the drug
between the two phases. Emulsions can sometimes be avoided by
using gentle agitation. If unavoidable, a few drops of ether or
alcohol can sometimes be effective in breaking an emulsion. However, these techniques are sometimes difficult to reproduce, let
alone to describe to others for them to reproduce. Whenever one
encounters emulsions, it is probably best to admit defeat and try

another system such as ion exchange chromatography or column
liquid/liquid partitioning chromatography.

Many ointments, creams, and suppositories are water insoluble.
Some are alcohol soluble. Suppositories can be dissolved after
melting, if not directly soluble. Liquid-liquid partitioning is
almost always employed since most excipients exhibit some solubility in most solvents. The more common organic solvents used
in liquid-liquid partitioning are chloroform, diethyl ether,
hydrocarbons, ethyl acetate. As before, column chromatography is
employed when simpler liquid-liquid partitioning is inadequate.
Most of the difficulties encountered with these formulations are
of two kinds. The first difficulty is, as before, emulsions, and
are handled in the same fashion as was with solutions. The second problem inherent with ointments and creams is an interference
problem enhanced by the fact that the ratio of excipients/active
drug is usually quite high. Purity of excipients are not as
closely controlled as is the purity of the drug, and, hence,
trace impurities that are not uniformly distributed from batch
to batch can cause erratic results, depending on the nature of
the interference.

It is virtually impossible to predict purity or stability
characteristics of a given drug-matrix system when sensitive
chromatographic detectors are employed. Whenever possible, it is
best to study the matrix (in the absence of drug) under stress
conditions. When this is not possible, the method of additions
is sometimes effective. Using this procedure, more than one concentration of standard is added to the drug-matrix system. The
result of chromatographic analysis should be a straight line with
a given intercept. The experiment is repeated using a stressed
formulation and the resulting plot of detector response versus
standard concentration should be superimposable with the first
plot, or if the intercept may be less, due to some instability of
the drug under the stress conditions, then the slopes should be
equal. However, changes in the matrix should not produce any
conditions that would alter chromatographic response of the drug.
If possible, some independent technique other than GC (such as
TLC) should be included to corroborate intercept values (drug
assay before addition of standard). As a general rule, samples
of unknown origin or composition should not be examined by GC as
part of a preliminary study, even though the drug component might
be known. Not only does the analyst risk ruining the chromatographic system by contamination, but, in addition, valuable
sample may be consumed for a fruitless test yielding erroneous
data. It is best to study an unknown drug composition by a
simpler technique which is nondestructive. Since most drugs display some ultraviolet absorption characteristics, TLC (using
phosphor quench) offers excellent selectivity while preserving

sample. TLC is also employed very often to confirm the efficiency of an extraction/partitioning technique.

12.3.4 Biological Samples

Applications in biological systems were taken up in a previous
chapter. However, a few points can be included here with reference to sample preparation and method development. Unlike sample
matrices considered so far, in biological systems such as those
involved in metabolism or toxicology studies, formulation excipients are usually not a problem. What is of significance is
the ability to distinguish a real metabolite from a degradate
that may be structurally related to a metabolite. Even more
problematic, is the fact that individuals vary in their sera or
tissue composition and concentration of various compounds. For
this reason, accurate GC methods are required, since the accuracy
of the method of analysis must be assumed as compared to the normal variation in assay values in individual studies. It is a
poor idea to attempt to study the metabolic path of a drug by GC
alone, this is more effectively determined by scintillation
counting. Once, the metabolic path is established, GC can be
used to quantitate the drug-metabolite in various media. For
these reasons, method parameters are more rigidly defined. GC/
mass spectroscopy has been a tremendous aid in this area, when
identification of foreign peaks is of paramount importance.
  Very simple liquid-liquid partitioning systems have been employed to isolate many drugs from biological media. Alkaloids,
antihistamines, barbiturates, and tranquilizers from blood, can
be extracted from urine, stomach juices, and tissue, using acetone: diethyl ether, 1:1, and aqueous phases of varying pH (3-6).
Typical drugs include caffeine, cocaine, atropine, codeine, heroin, morphine, quinine, doxylamine, chlorpheniramine, diphenyl-
pyraline, amobarbital, pentobarbitol, secobarbitol, and pheno-
barbitol (2-5).
  Blood can be a particular problem since it contains proteins,
in addition to quantities of most any compound which has been
ingested. Proteins are most often denatured (if necessary) using
EDTA or tungstic acid. Even then, direct injection can produce
interference from the resulting denatured fragments which contain
free amino acid or peptide groups. Again, liquid-liquid partitioning is effective in avoiding difficulties, since many interfering materials will remain preferentially in a mildly acidic or
alkaline medium. If electron capture detection is used, however,
cleanup will doubtless need more rigorous evaluation. Despite
the fact that blanks are employed frequently, contamination of

the chromatographic system may be irreversible. Amphetamines,
antihistamines, and phenothiazines may be partitioned from pH 9
buffer into ethylene chloride/isoamyl alcohol (10%) (6). All the
barbiturates, methylprylon, glutethimide, meprobamate, propoxyphene, methaqualone, primidone, diphenylhydantoin, methadone,
codeine, morphine, amphetamine, and methamphetamine can be extracted from urine and sera buffered at pH 4.9 or 8.3 using
chloroform (7). Finkle et al. (8) presented similar partitioning
systems for isolating up to 600 drugs, poisons, and human metabolites from blood, tissue, and sera but cautioned against spurious results that might occur using filter paper and plastics.
Wallace (9) similarly discussed isolation of drugs from biological specimens by liquid-liquid partitioning. The scheme was
based primarily on acidic-basic aprotic properties (as are all
partitioning systems).

Gravity-fed columns have been employed using charcoal, alumina,
and silica. Nonionic resin columns, ion exchange paper (Reeve
Angel SA-2), and disposable ion exchange resins (Amberlite XAD-2)
have also been used to isolate drug species from urine (10,11).
Catecholaminoacids and metabolites in brain tissue have been isolated using alumina columns after homogenization with perchloric
acid. Alumina is selective for the catechol group (12).

When employing these techniques, the novice is cautioned
against contamination, and sample loss due to irreversible adsorption, particularly if low concentrations are involved.

12.3.5 Summary

As a summary of observations on sample preparation of all types,
some generalizations can be reiterated. Whenever possible,
blank samples spiked with standards should be employed and recovery studies should include assays of each extract, to monitor
possible drug-substrate interaction. One technique effective in
this regard, is to examine the aqueous phase by TLC after all the
"free" drug has been extracted. Standard calibration curves
should be compared to similar plots in the presence of sample
matrix constituents whenever possible. GC should never be employed as the sole criterion in any analytical evaluation.

While gravity-fed columns including adsorption or ion exchange
are recommended when extensive sample cleanup is required, the
analyst should be cautious about contamination and adsorptive
effects contributing to sample loss. Liquid-liquid partitioning
is the preferred technique. When concentrating solvents, care
should be taken to insure against loss of volatile extracts.
Plastic equipment or fibrous filters should be avoided.

Finally, the literature can be searched exclusively for the determination of the drug in a particular medium, even though another analytical technique might have been employed for quantitation or estimation.

12.4   DERIVATIZATION TECHNIQUES

Many drugs have a relatively high molecular weight. In addition, they contain relatively polar substituent groups which contribute to interactions with the solid support. Structural configurations, in many instances, do not afford good thermal stability. In short, many drugs are not readily suited to gas chromatographic analysis. Fortunately, derivatives can be prepared that result in a thermally stable, volatile compound which, for the most part, exhibits minimal tailing (tailing is usually suggestive of interaction of the compound with the packing support). Sometimes, depending on the detector, increased sensitivity can be achieved by virtue of the response of a particular functional group on the derivatizing agent. Some common derivatives employed are described in the following section. As before, the treatment is not intended to be all encompassing. However, most literature references include complete derivatization procedures, since the success or failure in the application of a given GC method may lie in the ability to complete the formation of some derivative.

12.4.1 Silylation

The formation of silyl ethers is applicable to most pharmaceutical preparations and is employed in most laboratories. Typical reagents are listed here:

1. <u>Trimethylsilyl Derivatives (TMS)</u>
A. Hexamethyldisilizane (HMDS)-$(CH_3)_3SiNHSi(CH_3)_3$
B. Trimethylchlorosilane (TMCS)-$(CH_3)_3SiCl$ (often used as a catalyst)
C. Bis(trimethylsilyl)acetamide (BSA)-$CH_3C(OSi(CH_3) = NSi(CH_3)_3$
D. Bis(trimethylsilyl)trifluoroacetamide (BSTFA)-$CF_3C(OSi(CH_3) = NSi(CH_3)_3$
2. <u>Dimethylsilyl Derivatives (DMS)</u>
A. Dimethylchlorosilane (DMCS or DMMCS)-$(CH_3)_2HSiCl$
B. Tetramethyldisilazane (TMDS)-$(CH_3)_2HSiNHSi(CH_3)_2H$
C. Bis(dimethylsilyl)acetamide (BDSA)-$CH_3C(OSiH(CH_3)_2)=HSiH(CH_3)_2$

D. Chloromethyldimethylchlorosilane (CMDMCS)-$(CH_2Cl)(CH_3)_2SiCl$

In each case the reagent is used to replace an active proton, rendering the compound less reactive. The result is a decreased tendency to thermally degrade and hydrogen bond. Compounds derivatized successfully include sugars, phenols, alcohols, amines, thiols, steroids, amino acids, and ordinary carboxylic acids. Sweeley et al. (13) have described a general reaction procedure but manufacturers' literature is available which describes a variety of techniques depending on the drug and the analysis objective (14,15). Trimethylsilyl ether (TMS) derivatives are generally higher boiling than their dimethylsilyl (DMS) counterparts. Many drugs can be separated at lower temperatures as DMS derivatives; other times, TMS derivatives must be employed for better resolution. The choice is really a matter of what experience dictates is adequate. The technique usually involves adding excess reagent to about 10 mg of drug (or less) in a stoppered vial. Many times, room temperature is adequate, but reaction temperatures in excess of $100^\circ C$ are not uncommon. Solvents employed are pyridine or dimethylformamide, although others may be employed including DMSO or THF. At the end of the reaction period, the mixture can be injected directly or the excess reagent can be evaporated (usually with nitrogen) and the residue dissolved in a more suitable solvent, although the former procedure is more common. When developing a method, several time intervals and reaction temperatures should be evaluated to optimize yield. As might be expected, TMS derivatives are somewhat more thermally stable than DMS derivatives. Reaction solvent is sometimes an important consideration. For example, whereas pyridine is the most common medium, DMF has been recommended many times for certain steroids (17-keto variety). Moisture should be avoided as foreign peaks are sometimes observed due to hydrolysis of reagents and derivatives; DMS derivatives are somewhat less stable in this regard. However, in the absence of moisture, TMS and DMS derivatives are stable for many days. Metal injection ports should be avoided as these can catalyze any thermal instability. Frequent use of silyl derivatives can build up reagent products and by-products in the detector such as SiO_2, introducing noise and decreasing sensitivity.

Chlorinated silyl derivatives are mainly used for derivatizing compounds to optimize sensitivity when using electron capture detectors. The same precautions, advantages, and disadvantages apply as for TMS or DMS derivatives.

Some compounds are more difficult to derivatize than others, because of variations in the reactivity of the replaceable proton. In the case of the 17-keto steroids, substituting a solvent is adequate but such is not always the case. More powerful or

more reactive silylating agents are available such as BSA or
BSTFA. Both form essentially the same derivatives but the latter
is more volatile. Since reactivity is greater, moisture is a
greater problem than with the less reactive TMS and DMS reagents.
Also, excipients, particularly acidic or polyhydroxylated compounds, will derivatize (ointments, creams, suppositories, sus-
tained release products). A particularly useful application is
in the area of amino acid analysis (16-20). Reactions can be
carried out in acetonitrile or pyridine.

12.4.2 Acetylation

As is silylation, acetylation is used to replace an active proton rendering a compound less polar. Common candidates are alcohols,phenols, amino acids, and primary and secondary amines.
Amino acids are usually esterified to the methyl or butyl ester
prior to acetylation (16,21). Typical solvents are pyridine,
dimethylformamide, or even benzene. Reagents consist of acid
anhydrides such as acetic anhydride or trifluoroacetic anhydride.
To take maximum advantage of sensitive electron capture detection
heptafluorobutyric anhydride has been used. A typical reaction
may be carried out at room temperature for a few hours, but more
often, temperatures ranging from 50 to 150°C are employed with
a reaction time of 15 min to 2 hr. Excess reagent can be evaporated using a stream of nitrogen. Dopamine, epinephrine and norepinephrine can be determined as the trifluoroacetate derivatives
according to the following reaction scheme (22).

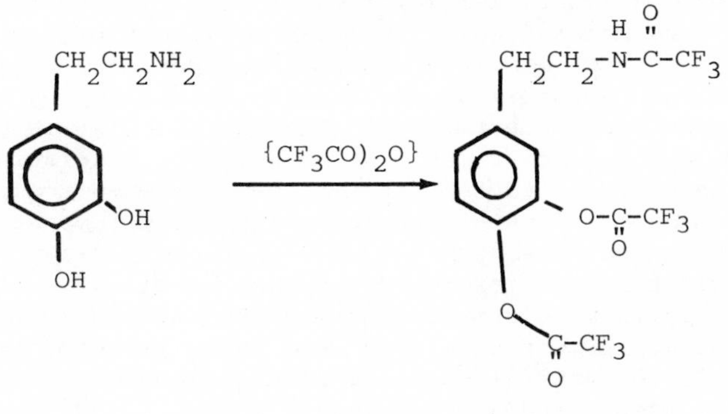

Reaction time is about 1 hr at room temperature. Moisture presents a stability problem and should be avoided; however, some
hydrolysis products are often encountered. Excess reagent should

be driven off completely, as free acid (hydrolysis products) can
ruin a column.

12.4.3 Methylation or Esterification

Carboxylic acid analogs, principally long-chain fatty acids and
barbiturates can be derivatized using HCl in methanol or diazomethane. Several techniques can be employed. In all cases,
completeness of reactions as well as side reactions should be
carefully evaluated. Rogozinski (23) dissolved 500 mg of sample
in 10 cm^3 of absolute methanol. One cubic centimeter of sulfuric
acid (conc.) was added followed by diethyl ether and water. The
ether extract was evaporated after extraction to remove excessive acid and the derivative was redissolved and chromatographed.
Amino acids can be esterified by reacting them with butanol in
3N HCl at 150°C for 15 min (20). Excess reagent is evaporated
under nitrogen(dry) at 50 C (the amino acids can then react with
BSTFA to form silyl derivatives). Dimethyl sulfite in methanolic HCl has been used to methylate amino acid hydrochlorides.
Reaction time is about 30 min at reflux temperatures (21).
    Another approach to methylation is to use diazomethane (24).
This material is somewhat less reactive, so minimal side reactions are observed although spurious peaks have been reported
in methylating long-chain fatty acids when freshly distilled reagent was not used (25). One preparation of diazomethane is to
add 3 cm^3 of 40% KOH to 10 cm^3 of ether precooled to 5°C. With
the temperature maintained at 5°C, 1 g of finely powdered nitrosomethylurea is added in increments over a period of 1-2 min.
The ether layer (yellow) can be decanted into a separatory funnel
funnel; it contains about 0.28 g of CH_2N_2. Excess water can be
removed by decanting over KOH pellets. Several variations have
been described including substituting N-methyl-N-nitroso-p-tolu-
ene sulfonamide for nitrosomethylurea (26). An important thing
to remember is that the process can be extremely dangerous. The
reagent(s) is toxic. Diazomethane is both toxic and potentially
explosive. The novice should seek out the best conditions suitable to his environment, equipment and analysis objective. One
precaution is to avoid all contact with diazomethane solution and
vapor (carcinogenic and destructive to tissue). Rubber gloves
should be worn and preparations and derivatizations should be
carried out in a hood. Only enough reagent as will be used in a
given day should be prepared; any remainder should be destroyed
by adding (dropwise) glacial acetic acid.

The disappearance of the yellow color is a good indication of complete degradation. If storage is necessary, ground glass stoppers should be avoided and the reagent should be stored in a freezer. Aside from these rather distasteful properties, diazomethane is a good methylating agent. Significant advantages include rapid, low-temperature reactivity, virtually instantaneous. The ethereal solution (cold) is added to the drug residue dropwise until the yellow color persists. Excess reagent can be evaporated (in the hood). Good descriptions of the application of the technique can be found in the literature which describe the use of diazomethane to derivatize pyruvic and lactic acids (27). Reaction is clean, since nitrogen is the other product (in addition to the derivative). Although ether is the more common solvent, methanol has been employed to derivatize amino acids (21). Diazomethane has been introduced as a gas into ethereal solutions of fatty acids containing 10% methanol (28). Caution against this technique is warranted, however, from the viewpoint of safety. Dropwise addition of ethereal solutions is usually adequate. As with other derivatives described, method development should include evaluation of side reactions and completeness, especially if the compound formulation contains acidic excipients.

A brief comment should be made concerning the technique of "on-column" methylation. This is a technique wherein a methylating agent is injected along with the drug into the gas chromatograph whose temperatures initiate and complete derivatization. Trimethylanilinium hydroxide is the most commonly employed reagent and has been used to form the 1,3-dimethyl derivatives of barbiturates which do not form stable TMS derivatives (29). Pierce Chemical Co. (14) supplies the reagent under the name Methelute. Their literature describes applications for various sedatives, barbiturates, xanthine bases, phenolic alkaloids and Dilantin. Supelco Inc. also supplies the reagent for barbiturate determinations (30). Care should be taken to insure the column is stable to traces of methanol. Also, chromatographic temperatures must be high enough to insure adequate rates of reaction.

Boron trifluoride and boron trichloride are employed to catalyze methylating reactions between alcohols and fatty acids. (31, 32). Because of the high toxicity of these materials, and the availability of acceptable alternatives, these reagents are not recommended for the novice and will not be discussee further here.

12.4.4 Summary

Derivatization is an important technique to the analytical chemist in GC analyses of drugs. In developing a method, the novice should remember three important factors:

1. Be sure to determine the completeness of reaction. Use more than one reaction temperature and time.

2. Beware of possible side reactions with additional functional groups on the drug moiety itself or with incompletely removed matrix constituents.

3. Determine the thermal stability of the derivative in terms of on-column partitioning time.

Finally, remember to consider the overall analytical objective first, then the derivatization/chromatography. Especially important is an awareness that the derivatization process or reaction conditions may possibly generate or destroy the very species required to be resolved. For example, consider the formulation that contains sodium benzoate as a preservative. Also present are various excipients, many of which are esters. The analytical objective is to determine intact sodium benzoate after a period of several weeks to ascertain the remaining effectiveness of sodium benzoate. A reasonable first approach might be to isolate free sodium benzoate as benzoic acid by extracting an acidic solution/suspension of the formulation with some organic solvent like chloroform. Benzoic acid could then be chromatographed as the methyl ester. But, in fact, transesterification in the formulation during the storage period may be a possible mode of degradation, generating various esters of benzoic acid. How likely that this would happen is of no significance at this point, since that was not the proposed question. Acidic environment may convert esters representative of instability back to benzoic acid, falsely indicating that the formulation was stable; alkali may do the same thing. Derivatization to form the methyl (or other) ester for chromatography could generate a derivative identical to a degradate. The answer to the problem lies in considering first the objective, then the chromatography. In this example, anion exchange chromatography would isolate only ionized carboxyl groups, allowing the esters to pass through the column. After appropriate rinsing, benzoic acid could be recovered from the resin using dilute acid. Then the methyl ester could be formed, and the sample chromatographed.

12.5   APPLICATIONS

The following discussion is a presentation of some demonstrated
applications of GC in drug analyses. The objective is to point
out examples of drug, derivative, column combinations which have
met with success. As was pointed out earlier, a thorough treatment is neither appropriate nor feasible. These examples were
selected as representative from a series of randomly encountered
references under the general heading, "drug analyses by gas
chromatography." The basis of selection was the depth to which
the author(s) pursued the description of their procedures. Note,
also, the relatively few liquid phases needed to undertake many
different chromatographic analyses.

12.5.1 General Applications

Several authors have discussed different classes of drugs or
written reviews surveying the technique as applied in general.
Finkle et al.(8) reported retention data for almost 600 drugs,
poisons, and metabolites. Included are chromatographic conditions, sample preparation, and derivatization techniques. Some
typical columns employed were 2.5% SE-30 on Chromosorb G, 5%
Hallcomid M-18 and 0.5% Carbowax 600 on Teflon 6, 10% Carbowax
6000 and 5% KOH on Chromosorb W. Sample media were blood, tissue
and urine specimens. Kazyak and Knoblock (33) in a review of
GC as applied to analytical toxicology reported the chromatographic behavior of 59 drugs on 1% SE-30 on Anakrom ABS using a
Strontium 90 argon ionization detector (see Table 12.1).
Samples were chromatographed as the free acid or base. More recently, Kazyak and Permisohn (33) described the use of retention
indices as a means of compound identification in the gas chromatographic analyses of over 150 drugs, insecticides, and alkaloids using QF-1, OV-1, and OV-17 as liquid phases. Moffat (34)
reported retention indices for 480 drug compounds using 2-3% SE-
30 as the liquid phsse. The support was of minor significance
although acid-washed, silanized Chromosorb G was more commonly
used. GC separations of heroin and related "street drugs" were
discussed in a report issued by Supelco Co. (35). Drugs included
were heroin, quinine, codeine, procaine, cocaine, methadone,
methapyrilene, and procaine. Columns included 3 or 5% SP-2401
and 3% SP-2100 on Supelcon AW-DMCS, or 3% SP-2250 on Supelcoport.
Column temperatures ranged from 230 - 250°C. Applied Sciences
Laboratories, Inc. has available a bulletin which lists over 500
references on GC applications, most of which are drugs (36).
Adams and Slavin (37) reported on automated GLC system for drugs
of abuse including barbiturates, tranquilizers, amphetamines and
narcotics. Alha and Korte (38) analyzed 34 toxic drugs in post

mortem tissues using 3% Hi-Eff BP on Gas Chrom Q, 2.5% SE-30 on
Chromosorb G, 5% Carbowax 2M + 5% KOH on Chromosorb, 10% OV-17 on
Gas Chrom Q. Kern et al. (39) have presented an excellent review
for the novice including various applications and techniques for
the determination of different classes of drugs and pharmaceuticals by GC.

12.5.2 Nonaddicting Analgesics and Antipyretics

Kaneo et al. (40) determined levels of aminopyrine in plasma on
1.5% OV-17 or 3% OV-101 on Chromosorb W. Column temperatures
were 200-210°C. Amsel and Davison (41) separated aspirin and
acetaminophen using OV-17 on Gas Chrom Q at 190°C. Thomas and
Coldwell (42) used TMS derivatives of phenacetin and paracetamol
(acetaminophen) to effect a GC separation on 3% OV-1 on Gas Chrom
Q. Column temperature was 160°C. Pellerin and Mancheron (43)
described a general technique for determining analgesics using
GLC and TLC. Drugs included amidopyrine, acetanilide, methyl
acetanilide, phenacetin, antipyrine, and caffeine. Prescott and
Adjepon-Yamoah (44) determined antipyrine in plasma samples using
0.5% SE-30 and 0.5% Carbowax columns. Thomas and Solomonraj (45)
formed TMS derivatives of aspirin and various salicylates using
BSTFA as the derivatizing reagent. The analysis was carried out
on 5% OV-17 on Gas Chrom Q at 150°C. Bruce et al. (46) used GLC
to study the bioavailability of phenylbutazone. Heptafluoro-
butyrl derivatives were prepared after isolation from plasma and
urine samples. The separation was effected on 3% OV-210 on Gas
Chrom Q. Ali (47) described similar separations in various
pharmaceutical preparations. Crippen and Freimuth (48) were able
to determine aspirin and free salicylic acid in pharmaceutical
preparations and tissue residues by forming the methyl esters using anhydrous methanol and BF_3 at reflux temperatures 10-15 min.
The column consisted of 30% Carbowax 20M on Chromosorb W.
Nikelly (49) performed a similar separation without the need for
derivatives. The column was 0.25% Carbowax 20M/0.4% isophthalic
acid on acid-washed microbeads. Included in Kern's work (39) was
a description of a separation of aspirin and salicylic acid as
the TMS derivatives on 3% OV-17 on Chromosorb W-HP at 130 C°as
well as the separation of phenacetin, antipyrine, and aminopyrine
without derivatization on SE-30, 3% on VarAport 30 at 180°C.

12.5.3 Barbiturates and Related Amides

Cimbura and Kofoed (50), in their review of GLC techniques applied to forensic toxicology discussed the use of on-column
methylation to derivatize diphenylhydantoin. Trimethylanilinium
hydroxide in methanol served as the derivatizing reagent. The
analysis was carried out on 3% SE-30 as shown in the chromatogram
in Figure 12.1. Note both the residual reagent as well as the
appearance of underivatized drug. The separation of barbital,
amobarbital, pentobarbital, secobarbital, glutethimide, and
phenobarbital can be achieved on 10% OV-17 on Gas Chrom Q deactivated with 0.2% H_3PO_4 (51). Column temperature was 240°C. Various barbiturates, glutethimide, and meprobamate have been separated on a special grade of 3% SP-2250 on Supelcoport (52).
Glass wool column plugs were treated with phosphoric acid to insure reproducible peaks. Column temperatures ranged between
215 and 230°C. Barrett (53) used on-column methylation in determinations of Dilantin. Derivatizing reagents were trimethyl-
ammonium hydroxide in methanol. Separations were accomplished on
3% OV-17 on Gas Chrom Q at column temperature of 250°C. Flanagan
and Withers (54) determined butobarbital, amylobarbital, hexa-
barbital, and glutethamide on 4% cyclohexanedimethanol succinate
(CDMS) on Diatomite CQ. Column temperature was 225°C. Pentobarbital, quinalbarbital, and methaqualone could also be determined in plasma samples. No derivatization was required.
Skinner et al. (55) used GC/MS in their analyses of barbiturates.
Samples were methylated (1,3-dimethyl derivatives) on-column using trimethylanilinium hydroxide. The analysis was conducted on
3% OV-1 on Gas Chrom Q. Temperature was programmed from 120 to
220°C at 10°C/min.

12.5.4 Organic Nitrogenous Bases

Cimbura and Kofoed (50),mentioned earlier, used GLC to separate
amphetamine and methamphetamine after acetylation with acetic
anhydride in methanol. Derivatives were extracted using diethyl
ether and chromatographed on columns of either 3% OV-17, OV-1,
or SE-30. Column temperature was 160°C. They also reported the
chromatographic determination of acetylated morphine on 3% SE-30,
OV-1, or OV-17 at temperatures of 220°C. Cruickshank et al.(21)
separated 21 amino acids as their trifluoroacetylated methyl
esters. The column was 5% neopentyl glycol succinate on Gas
Chrom P. Column temperatures were both isothermal and programm-
ed: 65°C for 20 min at 1.5°C/min; then 2°C/min until 42.5 min;
then 4°C/min until 60 min; then isothermal until about 75 min
(see Figure 12.2). Chang et al. (19), used BSA/pyridine to form
the TMS derivatives of levodopa, methyldopa, tyrosine,

TABLE 12.1 PERFORMANCE CHARACTERISTICS OF A 1% SE 30 COLUMN[1]

	115	130	150	165	180	200	210	225	250
Temperature (°C)	115	130	150	165	180	200	210	225	250
Inlet pressure (psig)	19	20	21	21	26	33	36	39	42
Flowrate (mL/min)	50	50	60	60	65	65	70	80	80
Compound				Retention Time (min)					
Ethchlorvynol (Placidyl)	1.4	1.2							
Phenethylamine	2.1	1.7							
Amphetamine	2.4	1.9							
Desoxyephedrine (Methamphetamine)	3.0	2.1	1.3						
Nicotine	6.7	4.1	2.2						
Ephedrine	7.5	4.2	2.3	1.2					
Ethynycyclohexyl carbamate (Valmid)	9.5	4.9	2.5						
Warfarin				1.7	1.2	0.8			
Methyprylon (Noludar)		10.8		2.9	1.8				
Barbital			4.3	2.8	1.6				
Probarbital (Ipral)			5.6	3.5	1.9				
Acetophenetidin (Phenacetin)			8.6	5.7	2.6				
Meperidine (Demerol)			9.7	5.9	3.0	1.6			
Amobarbital (Amytal)			10.9	6.0	3.0				
Hydroxyphenamate (Listica)			11.3	6.5	3.2	1.6			
Pentobarbital			13.2	6.5	3.2				
Pheniramine (Trimeton)			13.0	7.5	3.6				
Caffeine			14.7	8.1	3.8				
Secobarbital (Seconal)			16.8	8.2	3.8				
Diphenhydramine (Benadryl)			16.9	9.3	4.2	1.9			
Glutethimide (Doriden)				9.1	4.2	2.0	1.0		
Meprobamate (Miltown)				10.4	4.4	1.8			
Lidocaine (Xylocaine)			18.0	9.9	4.5	2.1			
Prominal (Mebaral)			23.1	11.2	5.2				

Antipyrine	20.8	11.8	5.3	2.3	1.9
Aminopyrine		12.0	5.7	2.6	
Tripelennamine (Pyribenzamine)		14.7	6.7	3.0	
Methapyrilene (Histadyl)		14.8	6.8	3.0	
Chlorpheniramine (Chlor-Trimeton)		16.3	6.8	3.0	
Phenobarbital		19.2	7.8	3.3	
Procaine		18.6	7.9	3.4	
Methadone (Dolophin)		12.1	4.9	3.3	
Bromodiphenhydramine (Ambodryl)		12.3	4.7	3.3	2.0
Propoxyphene (Darvon)		13.8	5.5	4.0	
Atropine		14.0	5.6	4.0	2.3
Thonzylamine (Anahist)		14.7	5.7	4.0	2.3
Chlorcyclizine (Perazil)		15.5	6.3	4.4	2.4 (1.3)
α-Cyclohexyl-α-phenyl-1-piperidine-propanol (Artane)				4.3	
Tetracaine (Pontocaine)		16.4	6.5	4.3	
Promazine (Sparine)		17.2	6.4	4.2	
Librium			5.5	3.0	1.6
Scopolamine			5.6	3.0	1.6
DDT (Dichlorodiphenyltrichloroethane)			5.7	3.1	1.6
Antazoline (Antistine)	26.0	8.1	5.5		1.6
Codeine	20.3	8.5	5.5		1.8
Ethylmorphine (Dionin)		8.6	6.0		2.0
Diphenylhydantoin (Dilantin)		9.2	6.8		1.9
Chlorpromazine (Thorazine)		10.7	7.4		2.2
Morphine		9.9	4.7		2.3
Cinchonine		13.6	6.2		2.8
Cinchonidine		14.6	6.4		2.8
Diacetylmorphine (Heroin)		14.5	6.9		3.0
Chloroquin		14.7	6.9		2.6
Dibucaine (Nupercaine)		20.9	9.1		3.1
Quinidine		11.0	4.4		
Quinine		11.2	4.4		
Anileridine (Leritine)	32.2	13.9	4.4		
Meclizine (Bonamine)		20.3	6.2		
Strychnine		23.7	8.7		

[1]Reprinted from ref. (32), p. 1450, through the courtesy of the American Chemical Society, Washington, D. C.

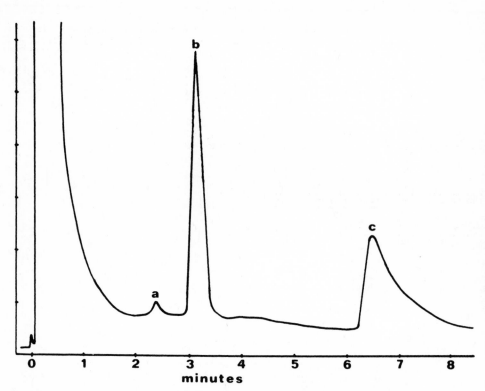

Figure 12.1.   Chromatogram of methylated diphenylhydantoin
using on-column methylation: (a) residual trimethylaniline; (b)
methylated diphenylhydantoin (attenuated); (c) unreacted
diphenylhydantoin.   Column: 3% SE-30 at 215°C. reprinted from
ref. (50), p. 263, through the courtesy of The Journal of
Chromatographic Science, Niles, IL.

3-methoxytryrosine, methyldopate, 3-aminotyrosine, and 2,4,6-tri-
hydroxyphenylalanine in pharmaceutical preparations.   The column
was OV-1(2%)/OV-17(6%) on Chromosorb G.   Column temperature was
225°C (see Figure 12.3).   Street (56) cited the chromatographic
conditions for determining 29 alkaloids in biological media.
Columns were modified SE-30 on Chromosorb W; chromatographic temperatures ranged from 130 - 310°C, depending on the compound(s).
Fish and Wilson (57) determined morphine and nalorphine as their
TMS derivatives using BSA.   The column was 2.9% OV-17 on Chromosorb W-AW-DMCS at 205°C.   Cocaine was determined, underivatized
in the same system at 185°C.   Parker et al. (58) described the
retention/temperature characteristics for 41 alkaloids using

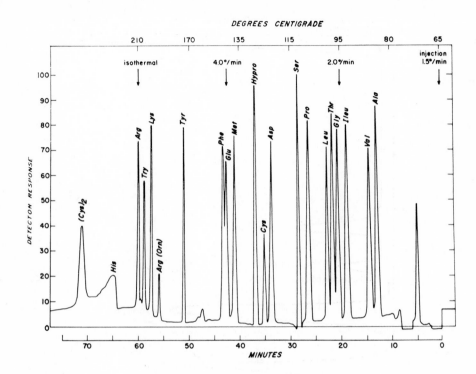

Figure 12.2.   Separation of a mixture of 20 trifluoroacetylated amino acid methyl esters.   Reprinted from ref. (21), p. 1195,
through the courtesy of the American Chemical Society, Washington,
D. C.

5% SE-30 on Chromosorb W at 250°C (see Figure 12.4 and Table
12.2).   De Silva and Puglisi (59) used GLC to study the metabolic
path of medazepam in Niobrium and diazepam in Valium.   Determinations were conducted on 3% OV-17 on Gas Chrom Q at 230°C.   Similarly, Inturrisi and Verebely (60) studied methadone and two
metabolites in plasma and urine samples using 3% SE-30 on Gas
Chrom Q.   Column temperatures were 200°C for plasma and 180°C for
urine samples, respectively.   Viala and Cano (61), used GLC to
determine diazepam (Valium), chlordiazepoxide (Librium), oxazepam
(Saresta) and dipotassium chlorazepate (Tranxene), nitrazepam
(Mogadon), medazepam (Nobrium) in blood and urine.   Maruyama and
Takemori (62) chromatographed norepinephrine and dopamine as the
respective TMS derivatives on 3% OV-17 at 170°C.   Townley and

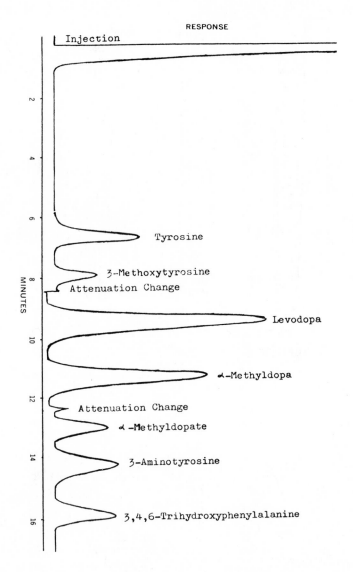

Figure 12.3.   Chromatogram of the trimethylsilyl derivatives
of a synthetic mixture of levodopa and other analogs. Reprinted
from ref. (19), p. 1339, through the courtesy of the American
Pharmaceutical Association, Washington, D. C.

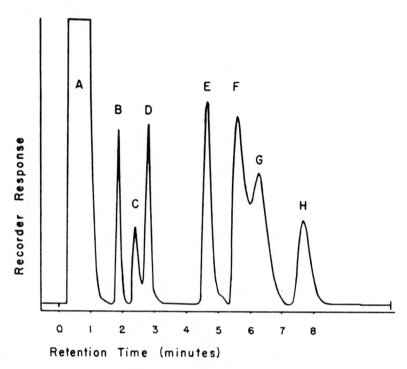

Figure 12.4.   Separation of a mixture of alkaloids on a 5%
SE-30 Column at 250°C using nitrogen as the carrier gas.
Reprinted from ref. (58), p. 357, through the courtesy of the
American Chemical Society, Washington, D. C.

Parez (63) used temperature-programmed GLC to determine chlorpheniramine. The column consisted of glass beads coated with
Ucon oil. Temperatures were programmed. Garrett and Tsau (64)
used GLC/TLC to study the stability of tetrahydrocannibinols in
aqueous media as a function of pH. TLC was used in aiding sample
cleanup. Samples were chromatographed on 3% OV-225 on Gas Chrom
Q at 240°C. Recently, Supelco (65) described the determination
of cannabinoids in marijuana using 3% SP-2250 on Supelcoport.
Column temperature was 235°C. Compounds resolved included:
cannabidiol, cannabichromene, cannabigerol Δ6-THC, Δ1-THC, and
cannabinol. Jask et al. (66) studied hydralazine blood levels.
The drug was not extractable with organic solvents, but reaction
with acidic sodium nitrite formed the tetrazolo (5,1-α) phthala-
zine derivative in plasma which was extracted with benzene.

TABLE 12.2 RETENTION DATA ON SOME ALKALOIDS AND RELATED COMPOUNDS ON A 5% SE-30 COLUMN.[1]

Compound (free base)	M.I., p.[2]	Retention Data — Chromatography Temperatures (°C)								
		190 RT[3]	210 RT	210 r.RT	230 RT	230 r.RT	250 RT	250 r.RT	270 RT	270 r.RT
Aconitine	15	[4]			10.6	1.08	5.1	1.06	3.0	1.00
Apomorphine	94		21.3	1.38	12.3	1.28	5.1	1.06	3.2	1.07
Atropine (d-hyos-cyamine)	111		9.2	0.60	5.7	0.59	3.0	0.63	2.1	0.70
Berberine	143				3.9	0.41	2.1	0.44		
Betaine	147		24.0	1.56	11.3	1.18	5.2	1.08	3.1	1.03
Caffeine	187	6.0	3.4	0.22	2.1	0.22	1.2	0.25		
Cinchonine (d-cincho-nine)	262				13.9	1.45	7.9	1.65	5.3	1.77
Cinchonidine (l-cin-chonine)	261				13.9	1.45	7.9	1.65	5.2	1.73
Cocaine	273		9.2	0.60	5.5	0.57	2.8	0.58	1.9	0.63
Codeine	275		15.4	1.00	9.6	1.00	4.8	1.00	3.0	1.00
Colchicine	277				11.0	1.15	5.2	1.08	3.0	1.00
Cotarnine	291		3.1	0.20	2.2	0.23	1.2	0.25		
Diacetylmorphine	333		32.0	2.08	13.3	1.39	6.3	1.31	3.9	1.30
Dibucaine	342				20.3	2.12	9.4	1.96	5.5	1.83
Dihydrocodeinone	360				11.3	1.18	5.4	1.12		
Dihydrohydroxy-codeinone	361				1.8	0.19	1.1	0.23		
Dihydromorphinone	362		25.0	1.62	14.1	1.47	6.1	1.27	1.2	0.40
Emetine	401				11.6	1.21	5.1	1.06		
Ethylmorphine	433		19.5	1.27	11.6	1.21	5.0	1.04	3.2	1.07
Gelsemine	474				27.0	2.81	10.0	2.08	6.1	2.04
Honatropine	522		7.3	0.47	4.2	0.44	2.4	0.50	1.6	0.53

626

Compound	Page[2]		RT[3]	r.RT	RT[3]	r.RT	RT[3]	r.RT	RT[3]	r.RT
Hydrastine	528				11.3	1.18	5.2	1.08	3.2	1.07
l-Hyoscyamine	547		9.4	0.61	5.7	0.59	2.9	0.60	2.1	0.70
Meperidine	646				1.7	0.18	1.1	0.23		
Mescaline	657		2.3	0.15	2.0	0.21	1.0	0.21		
d,l-Methadone	662				4.7	0.49				
Morphine	691		17.0	1.10	11.6	1.21	5.8	1.19	4.2	1.40
N-Allylnormorphine	701				17.8	1.85	8.0	1.67	5.0	1.67
Nicotine	719	5.0	2.6	0.17	0.8	0.08				
Papaverine	771				35.6	3.71	13.2	2.75	7.8	2.60
Pilocarpine	818				7.0	0.73	2.3	0.48	1.7	0.57
Piperine	823				28.5	2.97	13.6	2.84	7.8	2.60
Procaine	854		6.1	0.40	3.5	0.36	2.0	0.42	1.4	0.47
Quinine	888				27.0	2.81	13.0	2.70	8.2	2.74
Scopolamine	925		13.6	0.88	8.1	0.84	3.8	0.79	2.6	0.87
Sparteine	971		2.8	0.18	1.9	0.20	1.6	0.33		
Strychnine	986						26.0	5.40		
Thebaine	1028		25.6	1.66	13.8	1.44	6.0	1.25	3.8	1.27
Theobromine	1029		24.3	1.58	10.4	1.08	5.0	1.04	3.2	1.06
d-Tubocurarine	1075				11.2	1.17	5.0	1.04		
Yohimbine	1111				3.6	0.38	1.9	0.40		

[1] Reprinted from Ref. (58), p. 358, through the courtesy of the American Chemical Society, Washington, D.C. [2] Page references to Merck Index (10). [3] RT retention time (min); r.RT retention time relative to codeine. [4] Compound injected but no response observed.

Assay was completed by chromatographic analysis of the extract on
3% OV-225 on Chromosorb W-HP at 220°C.

12.5.5 Hormonal Drugs

Olson used GLC to determine urinary estriol as the TMS derivative
on 2% XE-60 or 2% OV-1 on Gas Chrom Q at 210°C (67). Edwards
(68) studied some 5-unsaturated steroids, 17-oxosteroids, cortisol metabolites, and pregnanediols. The acetate esters were
chromatographed on 3% OV-1 at 265°C. Noujaim and Jeffery (69)
studied 41 different steroids using various chromatographic systems including GLC, TLC, and paper. Baily et al. (70) and
Brooks et al. (71) as part of a monograph on the subject of steroids described many analytical schemes employing GC for the determination of steroids including separation techniques from biological media. Various derivatization schemes included TMS and
the methyloxime/TMS treatments of 17-hydroxy-corticosteroids.
VandenHeuvel and Kuron (72) studied the effects of column diameter and sample size in the determination of some derivatized
and underivatized steroids. Columns were 1.8% OV-17 and 2.0% SE-
30 on acid-washed and silanized Gas Chrom P. Column temperatures
ranged from 240 to 250°C. The data indicated that smaller diameter columns were more efficient when sample sizes were less
than 30 mcg (4-mm columns) but when sample sizes were larger,
50-500 mcg, larger-diameter columns were more efficient (11-mm
columns). Fales and Luukkainen (73) studied the effectiveness of
the O-methyloxime derivatives of the keto position in various
steroids in characterizations of individual drugs. The derivatives could be formed in the presence of excess methoxyamine in
pyridine at room temperature overnight. Excess reagent could be
evaporated under a stream of nitrogen and the derivatized residue redissolved in organic solvents like benzene. Depending on
the position of the functional group, reactivity varied. Nuclear
magnetic resonance and mass spectroscopy were used to study the
structure of derivatives chromatographed on 1% SE-30 or 1% BGS
columns on Gas Chrom P. Column temperatures ranged from 220 to
240°C. The data indicated a definite loss in polarity upon derivatization. Feher and Bodrogi (74) studied the relationship
between different steroid structural configurations and chromatographic characteristics. Columns consisted of 3% SE-30 or 3%
QF-1 on Diatomite CQ at 220°C. VandenHeuvel (75) presented a review of different derivatization schemes and their effectiveness
in steroid analysis using GC/MS. Kaiser et al. (76) determined
the level of medroxyprogesterone acetate in plasma as hepta-
fluorobutyrl derivatives on 10% OV-17 on Gas Chrom Q. Column

temperature was 250°C. Accuracy was reported to be 10% at a concentration of 5-20 ng/cm^3 . Recently, VandenHeuvel (77) discussed the qualitative and quantitative effects of classical reactions of steroids by GLC including derivatization techniques such as Wolff-Kishner removal of keto groups and reduction with NaBH$_4$, TMS, and so on.

12.5.6 Antibiotics

Cohen and Brennan (78) reported determinations of the antineoplastic, 5-fluorouracil as the TMS derivative on 3% OV-1 on Gas Chrom Q. Column temperature was 110°C. Other antitumor agents, cyclophosphamide, and isophosphamide were determined by derivatizing as the trifluoroacetyl analogs and chromatographing the samples at 200°C on a column of SE-30 on Chromosorb Q (79). Margosis (80) determined chloramphenicol as the TMS derivative using any of the following columns: OV-1, OV-101, or SE-30. Chromatographic temperatures ranged from 190 to 240°C, depending on the column. He reported results to be more precise, accurate and specific than the microbiological procedures. Tesnick et al. (81) also used GLC for determining chloramphenicol in serum toxicity studies. The sample was derivatized as the TMS analog and chromatographed on 3% QF-1, 2% SE-30 or 2% XE-60 columns at 225°C. Maximum sensitivity was derivatized as the TMS analog and chromatographed on 3% QF-1, 2% SE-30 or 2% XE-60 columns at 225°C. Maximum sensitivity was obtained using an electron capture detector. Shaw (82) also used the TMS derivatives of chloramphenicol and 18 structurally related compounds in studying drug bisynthesis by Streptymyces Venezuelae. Samples were chromatographed on 1.5% SE-30 on Gas Chrom Q pretreated with dimethyldichlorosilane (DMCS). Column temperature was programmed from 110°C at 6.4°C/min for 12 min then 2.3°C/min up to 260°C. Brodasky and Sun (83) used GLC to study the ratio of unhydrolyzed to hydrolyzed clindamycin-2-palmitate in human sera. The drug and degraded species were isolated from pH 8 buffer using benzene and derivatized as the TMS analogs. The sample was then chromatographed on a column consisting of 1.5% lexan and 1.5% polysulfone on Gas Chrom Q at 295°C. Brown also studied clindamycin by GLC (84) using the trifluoroacetyl derivative on OV-17 on Gas Chrom Q or SE-54 on Diatoport S. Burchfield et al. (85) determined 4,4'-diaminodiphenylsulfone and 4-acetamidophenyl-4'-aminophenylsulfone used in the treatment of leprosy by first converting the drugs to their iodo derivative and then chromatographing them on 3% Poly-A-103 on Gas Chrom Q at 285°C. This technique was employed to take advantage of an electron capture detector.

Otherwise, samples could be reduced with lithium aluminum hydride and chromatographed on 3% SE-30 at 240°C. Baker et al. (86) used GLC to study the metabolism of 1-methyl-5-nitro-2-(2'-pyrimidyl)-imidazole. Samples were acetylated and chromatographed on 5% SE-30 isothermally at 200°C for 6 min then programmed at 10°C/min for 5 min and 250°C for 5 min. Evrard et al. (87) reported the separation of penicillin and penicillanic acid as the methyl esters derivatized using diazomethane/ether.. Samples were chromatographed on 0.4% SE-52 on Gas Chrom P(silanized). Column temperature was 240°C. A strontium detector was employed. Figure 12.5 shows the separation of benzylpenicillin, phenoxymethyl-penicillin, 3,4-dichloro-α-methoxybenzylpenicillin (D&L isomers). Probenecid is not an antibiotic but, sometimes, it is employed to block the excretion of some antibiotics, thus maintaining a blood level of antibiotic for an extended period of time. Conway and Melethil (88) extracted probenecid and metabolites from urine and derivatized the acids as propyl esters using diazo-propane. Samples were then chromatographed on 10% OV-1 at 250°C. Kern et al. (39), mentioned earlier, discussed the chromatographic separation of various antimicrobials including sulfonamides (prontosil, sulfanilamide, sulfonamide derivatives), antibiotics (penicillin G, penicillin V, para-amino benzoic acid, chlor-amphenicol, para-aminosalicylic acid, pyrazine analogs of nico-tinamide). They reported tetracyclines as very unstable and not amenable to GC. Columns were, for the most part, 3% OV-17 on Chrom W-HP at temperatures ranging from 235 to 260°C depending on the separation. Nicotinamide was chromatographed on Porapak Q at 240°C; no derivative was employed. Pyrazinamide was determined using 5% Versamide 900 on Chrom W-KOH washed at 180°C.

12.5.7 Vitamins

Dwivedi and Arnold (89) determined the thiazole moiety in thiamine by derivatizing first to form the TMS analog. Samples were chromatographed on 3% OV-17 on Chromosorb G (DMCS treated) at 110°C. Vitamin D_2 (calciferol) and its analog Vitamin D_3 have been the subject of many applications in GC analyses. The compounds have been derivatized as the TMS analog and chromatographed on OV-17 (90) and SE-30 (91) and on 3% silicone on Celite after treatment with antimony trichloride (92). Several references to vitamine E (α-tocopherol) are included in the review by Kern et al. (39) on GLC determinations of pharmaceuticals and drugs. The acetate was determined on 3% OV-17 on Chrom W-HP at 280°C. Vecci and Kaiser (93) determined Vitamin C as the TMS derivative on 3% SE-30 or 10% XE-60 on Anachrom ABS. Column

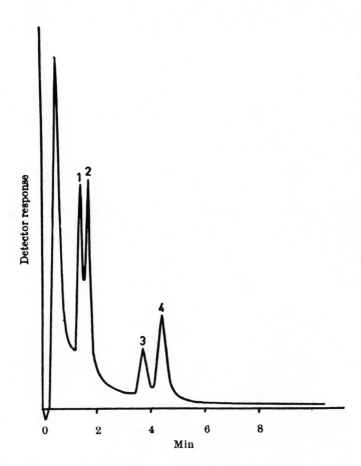

Figure 12.5.   Chromatogram showing the separation of the
methyl esters of benzylpenicillin (1), phenoxyethylpenicillin
(2), 3,4-dichloro-α-methoxybenzylpenicillin, L isomer (3) and
D isomer (4) on 0.4% SE-52.   Reprinted from ref. (87), p. 1125,
through the courtesy of MacMillan Journals Ltd., Washington, D.C.

temperatures were 170°C using SE-30 and 135°C using XE-60.   GLC
determinations of Vitamin B_6 was also described in the review by
Kern et al (39).   The derivatives were used and chromatography
was carried out using 3% SE-30 on Varaport 30 at 180°C.

12.5.8   Resolution of Enantiomers by GLC

Many drugs are active only insofar as they are biologically
available as one or the other optical isomer. It is of interest,
many times, to know if one isomer has converted to another within
the body during metabolism or if conversion has occurred in a
dosage form representative of instability involving a mechanism
that includes racemization. GLC has not been used extensively to
approach this problem largely because of the difficulty in obtaining routinely reproducible results. On the other hand, not
many techniques are available which will routinely provide the
data. Gil-Av and Feibush (94,95) have employed the technique of
packing GC columns with optically active stationary phases with
the idea of separating optically dissimilar species. They were
successful in separating some primary amines and amino acids.
Trifluoroacetyl derivatives of amines were resolved on acids.
Trifluoroacetyl derivatives of amines were resolved on trifluoro-
acetylated-L-valyl-L-valine isopropyl ester coated in capillary
columns. Better efficiency was obtained when a ureide, which
was formed by the condensation of phosgene with L-Valine isopropyl ester, was coated on glass capillaries of about 70 meters
in length and 0.25 mm i.d. Purity of the reagent (column coating) will affect resolution but not the relative peak areas.
The work was also repeated using 5% N-TFA-L-valyl-L-valine
cyclohexyl ester coated on Chromosorb W. Resolution was obtained for the (d,l) isomers of alanine chromatographed as the trifluoroacetyl, t-butyl esters. Column temperature was 100°C.
Koenig and Nicholson (96) used the optically active stationary
phases, N-TFA-L-phenylalanyl-L-phenylalanine cyclohexyl ester
and N-TFA-L-phenylalanyl-L-aspartic acid, bis (cyclohexyl) ester.
(d,l) isomers of TFA-aminoacid isopropylesters were separated on
a 39-meter capillary column at 110°C or programmed from 130°C to
165°C at 1°C/min. Pollock and Kawauchi (97) formed diastereomers
of aspartic acid, tryptophan, serine, hydroxyproline, tyrosine,
and cysteine. The diastereomer was formed using (+)-2-butanol
and (+)-2,2-dimethyl-3-butanol for esterifying the carboxyl
group, while various derivatizing reagents served to block other
reactive functional groups. The amino acid-alcohol diastereomers
were separated on capillary columns of Carbowax 20M and DEGS.
Column length was 150 feet and 0.02 in. i.d. Column temperatures
ranged from 140 - 190°C. Completeness of derivatization was determined to be between 85 - 100% depending on the amino acid.
Structures were verified by mass spectroscopy. Analysis time ran
as high as 50 min for two isomers, and as low as 10 min. Murano
(98) described the chromatographic characteristics of 19 assymm-
etric amines derivatized as amide diastereomers using N-tri-

fluoroacetyl-L-prolyl chloride. Stationary phases consisted of
3% SE-30, 2% QF-1, 2% DEGS, 3% OV-17, 5% XE-60, 2% Carbowax 20M,
2% LAC-2R-446 on acid-washed, silanized Chromosorb W. Column
temperatures ranged from 160 to 220°C depending on the amine
diastereomer and the column. Mass spectroscopy was used to characterize all derivatives. Brooks and Gilbert (99) used GLC to
study the metabolic breakdown of ibuprofen. In urine, residues
were found to be enriched with the (+) enantiomer. Ibuprofen,
(±)-2-(4-isobutylphenyl) propionic acid, was treated with (+)-α-
phenylethylamine in ether. Derivatives were chromatographed on
5-meter columns packed with 1% SE-30 or 1% OV-17 on Gas Chrom Q.
Column temperature ranged from 220 to 245°C depending on the
column. The gas chromatograph was equipped with a flame ionization detector. Nitrogen served as the carrier gas. Van Giessen
and Kaiser (100) performed a similar separation of enantiomers
of ibuprofen using l-α-methylbenzylamine to form the diastereomer. The analysis was carried out on 3% OV-17 on Gas Chrom Q,
1.5 meters in length. Column temperature was maintained at 220°C.
Helium served as the carrier gas. Linearity was observed over
a range of 0-80 µg/ml of drug. Matrices employed were urine or
plasma. Tosolini et al. (101) also employed l-α-methylbenzylamine to determine plasma levels of enantiomers of d,l-α-{4-(1-
oxo-2-isoindolinyl)phenyl} propionic acid. Sensitivity was 0.3
µg/ml plasma. The diastereomers were resolved on a 2-meter column consisting of 1% OV-7 coated on acid-washed, silanized
Chromosorb W. Nitrogen served as the carrier gas.

REFERENCES

1. Remington's Pharmaceutical Sciences, 15th edit., Mack Publishing Co., Easton, PA, 1975.
2. N. Jain and P. Kirk, Microchem. J., 12, 249 (1967).
3. N. Jain and P. Kirk, Microchem. J., 12, 256 (1967).
4. N. Jain and P. Kirk, Microchem. J., 12, 242 (1967).
5. N. Jain and P. Kirk, Microchem. J., 12, 229 (1967).
6. H. Street and C. McMartin, Nature, 199, 456 (1963).
7. J. Shipe and J. Savoy, Ann. Clin. Lab. Sci., 5, 57 (1975).
8. B. Finkle, E. Cherry, and D. Taylor, J. Chromatogr. Sci.,
   9, 393 (1971).
9. J. Wallace, K. Blum, and J. Singh, Clin. Toxicol., 7, 477
   (1974).
10.K. Kaistha, J. Pharm. Sci., 61, 655 (1972).
11.Applied Science Laboratories Technical Bulletin No. 10,
   Applied Science Laboratories, Inc., State College, PA.
12.V. Lotti and C. Porter, J. Pharm. Exp. Ther., 172, 406 (1970).

13. C. Sweeley, R. Bentley, M. Makita, and W. Wells, J. Am. Chem. Soc., 85, 2497 (1963).
14. Handbook of Silylation-GPA-3A, Pierce Chemical Co., Rockville IL (1972).
15. Applied Science Technical Bulletin No.11B, Applied Science Laboratories, Inc., State College, PA.
16. R. Zumwalt, D. Roach, and C. Gehrke, J. Chromatog., 53, 171 (1970).
17. C. Gehrke and K. Leimer, J. Chromatog., 53, 201 (1970).
18. C. Gehrke and K. Leimer, J. Chromatog., 57, 219 (1971).
19. B. Chang, B. Grabowski, and W. Haney, J. Pharm. Sci., 62, 1337 (1973).
20. J. Hardy and S. Kerrin, Anal. Chem., 44, 1497 (1972).
21. P. Cruickshank and J. Sheehan, Anal. Chem., 36, 1191 (1964).
22. D. Clarke, S. Wilk, S. Gielow, and M. Franklin, J. Gas Chromatog., 5, 307 (1967).
23. M. Rogozinski, J. Gas Chromatog., 2, 136 (1964).
24. A. Blatt (Ed.), Organic Synthesis, Collective, Vol. 2, John Wiley and Sons, London, 1943, 0. 165.
25. W. Morrison, Chem. Ind., 38, 1534 (1961).
26. T. DeBoer and H. Backer, Rec. Trav. Shim., 73, 229 (1954).
27. F. Estes and R. Bachmann, Anal. Chem., 38, 1178 (1966).
28. H. Schlenk and J. Gellerman, Anal. Chem., 32, 1412 (1960).
29. E. Brochmann-Hanssen and T. Oke, J. Pharm. Sci., 58, 370 (1969).
30. Supelco Technical Bulletin 721 B, Supelco, Inc., Bellefonte, PA.
31. Applied Science Laboratories Technical Bulletin No. 17, Applied Science Laboratories, Inc. State College, PA.
32. L. Kazyak and E. Knoblock, Anal. Chem., 35, 1448 (1963).
33. L. Kazyak and R. Permisohn, J. Forensic Sci., 15, 346 (1970).
34. A. Moffat, J. Chromatog., 113, 69, (1975).
35. Supelco Technical Bulletin 734, Supelco Corp., Bellefonte, PA.
36. Applied Science Laboratories Technical Bulletin No. 3, Applied Science Laboratories, Inc. State College, PA.
37. R. Adams and W. Slavin, Clin. Chem., 17, 660 (1971).
38. A. Alha and T. Korte, Ann. Med. Exp. Biol. Fenn., 50, 175 (1972).
39. H. Kern, P. Schilling, and S. Muller, Pharmaceuticals and Drugs, Varian Aerograph, Walnut Creek, CA.
40. Y. Kaneo, T. Goromaku, and S. Iguchi, Yakugaku Zasshi, 93, 258 (1973); Chem. Abstr., 79, 27022b.
41. L. Amsel and C. Davison, J. Pharm., Sci., 61, 1474 (1972).
42. B. Thomas and B. Coldwell, J. Pharm. Pharmacol. 24, 243 (1972).
43. F. Pelleren and D.Mancheran, Sixth International Symposium on

Chromatography and Electrophoresis, Belgian Soc. Pharm. Sci., Ann Arbor Sci. Publ., Ann Arbor, MI, 1971, p.501.

44. L. Prescott, K. Adjepon-Yamoah, and E. Roberts, J. Pharm. Pharmacol., 25, 205 (1973).
45. B. Thomas, G. Solomonraj, and B. Coldwell, J. Pharm. Pharmacol., 25, 201 (1973).
46. R. Bruce, W. Maynard and L. Dunning, J. Pharm. Sci., 63, 446 (1974).
47. S. Ali, Chromatographia, 8, 33 (1975).
48. R. Crippen and H. Freimuth, Anal. Chem., 36, 273 (1964).
49. J. Nikelly, Anal. Chem., 36, 2248 (1964).
50. G. Cimbura and J. Kofoed, J. Chromatog. Sci., 12, 261 (1974).
51. Retention Times, Vol. 1, (No. 2), Tracor Inst., Austin, Texas, (1974).
52. Chromatography Lipids, VI (No.3), Supelco Corp., Bellefonte, PA., (1972).
53. M. Barrett, Clin. Chem., Newsletter, 3, 16 (1971).
54. R. Flanagan and G. Withers, J. Clin. Pathol., 25, 899 (1972).
55. R. Skinner, E. Gallaher, and D. Predmore, Anal. Chem., 45, 574 (1973).
56. H. Street, J. Chromatog., 29, 68 (1967).
57. F. Fish and W. Wilson, J. Chromatog., 40, 164 (1969).
58. K. Parker, C. Fontan, and P. Kirk, Anal. Chem., 35, 356 (1963).
59. J. de Silva and C. Puglisi, Anal. Chem., 42, 1725 (1970).
60. C. Inturrisi and K. Verebely, J. Chromatog., 65, 361 (1972).
61. A. Viala and J. Cano, Mrs. A. Angeletti-Philippe, J. Euro. Toxicol., 3, 109 (1971); Chem. Abstr., 75, 107697h.
62. Y. Maruyama and A. Takemori, Anal. Biochem., 49, 240 (1972).
63. E. Townley, I. Parez, and P. Kabasakalian, Anal. Chem., 42, 1759 (1970).
64. E. Garrett and J. Tsau, J. Pharm. Sci., 63, 1563 (1974).
65. Chromatography Lipids, IX (No. 1), Supelco Corp., Bellefonte, PA (1975).
66. D. Jask, S. Brechbuhler, P. Dege, P. Zbinden, and W. Riess, J. Chromatog., 115, 87 (1975).
67. A. Olson, Clin. Chem., Newsletter, 2, (1), (1970).
68. R. Edwards, Acta Endocrinol., 80 (Suppl. 199), 253 (1975).
69. A. Noujaim and D. Jeffery, Can. J. Pharm. Sci., 5, 26 (1970).
70. E. Baily, D. Murphy, and H. West, in Gas Chromatographic Determination of Hormonal Steroids, F. Polvani, M. Surace, and M. Luise (Eds.), Academic Press, New York, 1967, p. 105.
71. C. Brooks, E. Chambay, W. Gardiner, and E. Horning, Gas Chromatographic Determination of Hormonal Steroids, F. Polvani, M. Surace, and M. Luise (Eds.), Academic Press, New York, 1967, p. 35.

72. W. VandenHeuvel and G. Kuron, J. Chrom. Sci., 7, 651 (1969).
73. H. Fales and T. Luukkainen, Anal. Chem., 37, 955 (1965).
74. T. Feher and L. Bodrogi, J. Chromatog., 71, 17 (1972).
75. W. VandenHeuvel, Sixth International Symposium on Chromato-
    graphy and Electrophoresis, Belgian Soc. Pharm. Sci., Ann
    Arbor Sci. Pub., Ann Arbor, Mich., 1971, p. 14.
76. D. Kaiser, R. Carlson, and K. Kirton, J. Pharm. Sci., 63,
    420 (1974).
77. W. VandenHeuvel, J. Chromatog., 103, 113 (1975).
78. J. Cohen and P. Brennan, J. Pharm. Sci., 62, 572 (1973).
79. C. Pantarotto, A. Bossi, G. Belvedere, A. Martini, M. Donelli
    and A. Frigerio, J. Pharm. Sci., 63, 1554 (1974).
80. M. Margosis, J. Pharm. Sci., 63 435 (1974).
81. G. Resnick, D. Corbin, and D. Sandberg, Anal. Chem., 38, 582
    (1966).
82. P. Shaw, Anal. Chem., 35, 1580 (1963).
83. T. Brodasky and F. Sun, J. Pharm., Sci., 63, 360 (1974).
84. L. Brown, J. Pharm. Sci., 63, 1597 (1974).
85. H. Burchfield, E. Storrs, R. Wheeler, V. Bhat, and L. Green,
    Anal. Chem., 45, 916 (1973).
86. K. Baker, M. Coerezza, L. Del Corona, A. Frigerio, G.
    Massaroli, and G. Sekules, J. Pharm. Sci., 63, 293 (1974).
87. E. Evrard, M. Claesen, and H. VanderHaeghe, Nature, 201,
    1124 (1964).
88. W. Conway and S. Melethil, J. Chromatog., 115, 222 (1975).-
89. B. Dwivedi and R. Arnold, J. Food Sci., 37, 889 (1972).
90. T. Kobayashi and A. Adachi, J. Nutr. Sci., Vitaminol., 19,
    311 (1973).
91. J. Schlack, J. Assoc. Official. Anal. Chemists, 57, 1346
    (1974).
92. T. Murray, K. Day, and E. Kodicek, Proceedings of the Bio-
    chemical Society, 450th Meeting at Dundee, June 18, 1965,
    Biochemical J., 96, 29 (1965).
93. M. Vecchi and K. Kaiser, J. Chromatog., 26, 22 (1967).
94. E. Gil-Av and B. Feibush, Tetrahedron Lett.,35, 3345 (1967).
95. B. Feibush and E. Gil-Av, J. Gas Chromatog., 5, 257 (1967).
96. W. Koenig and G. Nicholson, Anal. Chem., 47, 951 (1975).
97. G. Pollock and A. Kawauchi, Anal. Chem., 40, 1356 (1968).
98. A. Murano, Agr. Biol. Chem., 37, 981 (1973).
99. C. Brooks and M. Gilbert, J. Chromatog., 99, 541 (1974).
100. G. Vangiessen and D. Kaiser, J. Pharm. Sci., 64, 799 (1975).
101. G. Tosolini, E. Moro, A. Forgione, M. Ranghieri, and V.
    Mandelli, J. Pharm. Sci., 63, 1073 (1974).

Index

A

Acetylation,613
Acetate derivative,507,509
n-Acetyl derivatives,540
n-Acetyl tryptamine,540
Acid hydrolysis,499,508
Acid,phosphatidic,464
Acids, amino,
 optically active,472
Acids,fatty
 methyl esters,453-460
Activation energy,579
Active surface area,557
Activity coefficient,565
 measurement of,566
 sample calculation,567
Adjusted retention volume,21
Adsorbent,21,153
Adsorbents,
 carbon,125
 molecular sieves,124
 porous polymers,124
 silica gel,124
Adsorption,10-12,47-49,74,82,
     599,603,610
 chemical,48,
 Freundlich equation,11,
 Langmuir equation,12,
 physical,48,

 thermodynamics\of,154
Adsorption chromatography,21
Adsorption column,21
Air or gas sampling,374,375
 "grab" samples,375
 quiescent air sampling,375
 stack sampling,374
Air peak,21
Alanine(optical isomers),632
Alcohol,127,520
 blood,520
 separation,127
Alcohols,613,615
Alditols,484
Aldonitrile acetates,484
Alkali flame ionization detector,
     269
 flowrates,270
 linearity of response,274
 minimum detectable level,274
 response factors,273
Alkaloids,610,617,628
Alkaptonuria,530
Alumina,612
Amberlite cation resin,535
Amides,619
Amine separation,128,549
Amines,597,598,607,612,613,632

Index

Amino acid derivatives,467-473,
        475
  n-butyl esters,467,470,471
  carbonyl bis(N-L-valine iso-
      propyl ester),472
  N-O-perfluoropropionyl deriva-
      tives,468
  phenylthiohydantoids,475
  n-propyl esters,467
  silyl derivatives,473
  stationary phases,468
  TFA-D-2-butyl esters,472
  TFA-L-hydroxyprolyl,472
  TFA-L-prolyl,472
  N-trifluoroacetyl derivatives,
      467-473
Amino,acid,sequence,473
Amino acids,619,632,612-615
  blood,532
  optically active,472
  protein,534
  stationary phases,468,471
Amobarbital,543
Amphetamines,539,610,617,619
Analgesics,618
Analysis time,22,137
Analytical chemistry,44
Anesthetics,blood,526
Anomerization,481,482
Antibiotics,629
Antihistamines,609
Antipyretics,618
Argon detector,265
Aromatic acids,urinary,529
Attenuator,22
Automatic sampling devices,317-
        321
  consideration for selection,317
  capsule type for solids,317
  headspace type,321
  syringe type for liquids,318
  valves for gases,321
  valves for liquids,318
  for process applications,
        318

B

Background current,256
Ball and disc integrator,354
  precision of,354
  role of recorder with,354
Band,22
Band area,22,28
Band dispersion,72
Band spreading in GSC,152
Barbiturates,537,609,614,615-618
Barbituric acid,543
Baseline,22
Baseline correction,354-356
  electronic integrators and,354
  mechanical integrators and,354
Benzylamine,540
Bed volume,22
Beilstein test,274
Beroza's "p" value,160,387
BET method,554
β,β-oxydipropionitrile,90
Beta-ray argon detector,255
Beta-ray ionization detector,25!
BF_3/BCl_3 in methylation,615,618
Bis(trimethylsilyl)acetamide
      (BSA),531,445
Blood estrogens,499
Boiling point,586
Brunel mass detector,284
Bulk property detector,228
Butylation,614
Butabarbital,543
Butane-1,4-diolsuccinate,486

C

Calibration curves,184
Caffeine,609,618
Cannabinoids,625
Capacity factor,22,55
Capillary column,22,145
Carbohydrate derivatives,451,
        477-479,484-488
  acetates, aldonitrile,484
  acetyl esters,478,484

Index

butane boronic esters,484-488
  ethers,methyl,477,478,484,487
  methyl ethers,   478,484,487
  oligosaccharides,476
  trifluoroacetyl esters,478,484
  trimethylsilyl ethers,451,479,
    484
Carbohydrates,stationary phases,
    482,484,486,487
Carbon skeleton,162
Carbon supports,125
Carboxylic acids,612
Carrier gas,22,74,136,291-304,
    336
  contamination,136
  density of,293
  diffusion coefficients of, and
    column performance,291
  inert, need for,291
  purity of,291
        effect of, in TPGC,293
        effect of, on baseline,293
  regulation and control of,292
  thermal conductivity of,336
  viscosity of,291
Catecholamines,518,602,609
Cellulases,478
Chemiluminescence,268
Chloromethyldimethylchlorosilane
    (CCMDMCS),611
Cholestane,511
Cholesterol acetate,516
Cholesterol,plasma,515
  plasma,515
  urinary,515
Choline,phosphatidyl,464
Chromatogram,22
Chromatograph,22
Chromatographic separation,
  components of,42
Catalytic,reaction studies,557
Chromatography,2-9,13-18,22,23,
    25,28,39-112,461,465,466,
    476
  applications of,16-18

classification of methods,4
  definition of,5,22
  development of,2-4
  displacement,7,8,23
  elution,8,9,25
  frontal,6,7
  gas-liquid,25
  gas-solid,25
  history of,2-4
  ideal,13-16
  liquid,465,466,476
  nonideal,13-16
  partition,28
  theory,39-112
  thin layer (TLC),461
  types of,43
  uses of,4
Chromothermography,328
Clinical analysis,495-552
Colorimetry,497,507
Column,22
  bleed,23
  conditioning,144
  cross-sectional area of,88
  deterioration,135
  effect of carrier gas on per-
    formance of,291
  efficiency of,24,61,63,85,92,
    117,291
  inlet splitters for,346
  length of,95
  material,23
  performance characteristics,599
  preparation,143
  selection,137,139
  support,600,611
  volume,23
Column bleed,322,342
  ability to compensate in FID,
    342
  ability to compensate in TCD,
    342
  compensation of, in TPGC,342
  excessive,322
Column,capillary

Index

inlet systems (See inlet system, capillary columns), 304-311
regulation of carrier, 229
sample size for, 311
Column operation,
 evaluation of, 90
Column ovens, 321-334
 for TPGC, 326
 requirements of, 321
 size and configurations of, 332
  effect on column, 333
Column selection, 113-149
Column temperature, 322, 594
 constancy of, 322
 control of, 322
  full power on-off type, 323
  power proportioning for, 323
  proportional integral, 324
  using microprocessor for, 324
  sensors used for, 324
  use of powerstatt for, 324
  use of vapor bath for, 323
 effect on retention values, 322
 selection of, 322
Columns. See adsorbents, packings, supports, stationary phases, tubing.
 capillary, 22
 coiled, 333
 open tubular, 28, 29, 77
 packed, 77
 U-tube, glass, 333
Complexes, 558
 Purnell classification, 558
 theory, 558
Component, 22
Computers. See Integrators.
Computing/reporting integrators

 and systems, 356, 357
 analog-to-digital conversion in, 357
 auto-calculation capabilities, 357
 capabilities of, 357
 costs of, 357
 multi-channel, 357
 precision of integration, 357
 role of microprocessor in, 357
 single-channel, 357
Concentration distribution ratio, 23
Conditioning columns, 144
Continuous stream analysis, 317, 321
Controlled Substances Act, 595
Corrected retention volume, 23
Coulometric detector, 278
Coulometry, 256
Countercurrent extraction, 59
Cross-section detector, 266

D

2,4-D (2,4-dichlorophenoxy acetic acid), 389
Dalton's Law, 46
Darcy's law, 79
Data reduction and handling, 290, 352-361
 computer systems, 356
 computing/reporting integrators, 356
 digital reporting chromatograph, 359
 electronic integrators, 354
 manual means, 352
Deadbanding, 218
Dead volume, 23, 26
Debye induction forces, 89
DEGS, 486
Derivatization, 385, 505, 611-616
Desmethylnortriptyline, 549
Detection, 18, 19, 23

Index

cumulative,19
differential,18
instantaneous,18
integral,19
types of,18
Detector,18,23,26,217-222,225,
    228,254,265,274,497,502
  argon ionization,265,497,502
  bulk property,228
  differential,23
  drift,218
  electron capture,254, 497
  flame ionization,274,497,502
  nitrogen(N),497
  phosphorus(P),497
  specific ion,497
  integral,26
  linear dynamic range,225
  minimum detectable level,219
  response factor,221
  sensitivity,220
  signal-to-noise ratio,217
  specific,222
  universal,222
  volume,23
  wander,218
Detector/recorder deadbanding,
    218
Detector types,213-288
  alkali flame ionization,269
  Brunel mass,284
  conductivity,280
  coulometric,278i
  cross-section,266
  dielectric constant,284
  flame ionization,244
  flame photometric,266
  flow impedance,285
  for preparative GC,103
  gas density balance
  gas pressure,285
  gas volume,285
  helium,265
  hydrogen flame temperature,284
  infrared,286
  ionization,27

  mass spectrometer,285
  microwave plasma,275
  miscellaneous,283
  piezoelectric sorption,282
  radioactivity,282
  reaction coulometer,280
  semi-conductive thin-film,283
  thermal conductivity,228
  thermionic,269
  titration,284
  ultrasonic,281
  ultraviolet,286
Detectors,335-348
  amplification of output,339,
      348
  dual,342-348
      as compensation for column
      bleed,342
      in parallel (dual channel
      operation),346
      in series,345
      independent operation,348
      thermal conductivity de-
      tector,344
  effect of carrier gas and flow-
      rate,336
  effect of sample type and
      amount,335
  electron capture.(See Electron
      capture detectors).
  flame ionization. (See Flame
      ionization detectors).
  function of,335
  thermal conductivity. (See
      Thermal conductivity de-
      tectors).
  selective,335,348
  temperature control,322,336,338
Diabetic urine,499
Diatomaceous earth,81
Diatomite,81
    pink,81
    white,81
Dielectric constant detector,284
Diffusion,49-52,61,65,66,68-72,
    74,75

Index

coefficient,49,75,584
eddy,61,66,68,69
free,51
in mobile phase,65
into particles,65
Knudsen,51
local non-equilibrium,66,68-70
longitudinal,50,52,61,65,71,72
molecular,51,71,74
ordinary,66,68,69
surface,51
Volmer,51
Diffusion tubes. See Standards.
Diffusion coefficient,583-585
  measurement of,584
  theory,583-585
Diffusivity,117
Digital chromatograph,359
Dilantin,546
Dilute samples,injection
    problems of,305
Dimethadione,549
N,N-Dimethyl aniline,540
Dimethyltrytamine,540
Dipeptidylaminopeptidases,475
Diphenylamine,540
Diphenylhydantoin,544
Disc integrator. (See Peak size
    measurement).
Dispersion forces,48
Displacement development,7
Distillation technique,52-53
Distribution coefficient,24,54-
    55,62,161,389
Distribution constant,10,24,54,
    58,559,560,564
Distribution ratio,59,388
DMCS,82
Double column chromatography,
    315
Dopamine,518
Drift,218
Drug analysis,591-636
Drug enforcement agency,595
Dynamic gas standards. See

Standards.

E

Effective carbon number,253
Effective number of plates,92
Effective plate height,93
Effective theoretical plate
    number,24
Efficiency,column,24,61,63,85,
    92,114-117
Electrolytic conductivity de-
    tector,280
Electrometers,252,339-342
    digital type,342
    auto-ranging,342
    early types,340
    controls on,340
    ranging errors and trans-
    ents,340
    reduction of,340
    transistor feedback type,
    342
Electron capture detector,254-
    257,259,261-264
    background current,256
    cell design,256
    concentric cylinder,256
    constant voltage,259
    coulometry,256
    linearity,264
    minimum detectable level,
    264
    photometric,257
    pin-cap,256
    plane parallel,256
    pulsed constant frequency,
    259
    pulsed variable frequency,
    259
    response factors,261-263
    solute switching and syn-
    chronous demodulation,259
    standing current,256
    voltage,259

Index

Electron capture detectors,335, 336,339
 applications of,335
 carrier for,339
 control of,339
 effect of impurities in,339
 overloading of,336
 temperature effect on sensi-
     tivity of,339
 temperature limitations of
     various sources for,339
 selectivity of,335
 sensitivity of,335
Electronic integrators,354
 accuracy and precision of,354
 adjustable peak detection
     limits,354
 baseline correction in,354
Elemental analysis,163
Eluent,25
Elution,25
Elution development,8
Enthalpy,57,568-576,
 of adsorption,576
 of solution,568
 of vaporization,571
Entropy,47,568-576
 of adsorption,576
 of solution,568
 of vaporization,571
Enzyme hydrolysis,508
Enzymes,461,464,475,478
 cellulases,478
 pancreatic lipase,461
 phospholipase A_2,464
 phospholipase A_1,464
 phospholipase C,464
 phospholipase D,464
 proteolytic dipeptidylamino-
     peptidases,475
 proteases,475
 trypsin,475
Epinephrine,518,539
Equation,
 BET,555
 Hammett,566

virial,of state,580
Equilibration(of columns),600
Equilibrium constant,558,574,575
Errors in quantitative analysis,
     177,178,186,187,189,202-
     204,206,208-210
 peak size measurements,177,210
 sample size,186,204,206,208
 sampling techniques,203
 standards,187,189,202
 system errors,178,209
Esters,451-46-,478,484-488
 acetyl,478,484
 butane boronic,484-488
 methyl,451-460
 trifluoroacetyl,478,484
Estriol,urinary,499,500,503
 in amniotic fluid,500,503
Estrone,502
Ethanolamine,phosphatidyl,464
Evaporation technique,52
Excess thermodynamic functions,
     570,
 enthalpy,570
 entropy,570
 free energy,570
External standardization. See
     Standardization techniques.
Extraction,160

F

Fatty acids,614
Fatty acid methyl ester,station-
     ary phases,454,455,458
FDA(Food and Drug Administra-
     tion),592,604
Fick's law of diffusion,49,50
Filament element,25
Film thickness,73
Flame chemistry,246
Flame emission detector,274
Flame ionization detector,244,
     248,250,252
 design,248

effective carbon number,248
electrometer,252
flow,249
ignitor,252
ion-collection,250
jet-248
polarity,252
response factors,252
thermal control,250
triboelectricity,251
Flame ionization detectors,335-
    348
carrier gas for,337
effect on linearity of,337
effect on noise of,337
effect on response of,337
sensitivity to changes in
    flow of,337
flame gases,337
control of,337
optimization of flow of,337
Flame photometric detector,
    266-269
air pollution,269
fuel analysis,269
gas flow,268
minimum detectable level,269
operation,267
response,268
response factors,268
Flow,25,74,77,80,297-300
controller,25,297-300
laminar,77,80
plug,77
programming,25
rate of,25,74,77,80
turbulent,77,80
Flow impedance detector,285
Flow meters,80,87,300-304
rotameters for,302
soap-film type,300
    accuracy of,300
thermal mass type,303
    range and resolution of,
    303

Flowrate,control of,294-297
by fixed restrictor,296
by flow controller,294
by needle valve,294
by pressure regulator,294
for isothermal column operation,
    294
for programmed temperature
    operation,294
Flowrate,linear,27,72
Fluid velocity,79
Food,Drug and Cosmetic Act,596
Food analysis,449-493
Fourier Transform infrared
    spectroscopy,286
Free energy,47,568-577
of adsorption,576,577
of solution,568-571
Freezing point,586
Freundlich equation,11
Frontal analysis,6
Fronting,25
Fuel analysis,229
Functional group identification,
    163

G

Gas, carrier. See Carrier Gas.
Gas chromatographs,290,359
basic components of,290
number in use today,290
processor based,290,359
Gas chromatography,16-21,30,34-
    36,60,128,498,549,550
abstracts of,128
advantages,19-21
applications of,16-18
definitions,21-33
instruments,36
limitations,19-21
literature of,34-36,128
mass spectrometry(GC/MS),498,
    549,550

Index

nomenclature,21-33
pyrolysis,16,30
Gas density balance,88,276
Gas-liquid chromatography,25
Gas pressure detector,285
Gas sampling,321
Gas sampling devices,375,376
  glass containers,375
  plastic bags,376
  Teflon,376
  Mylar,376
  Tedlar,376
  trapping systems,in series,376
Gas sampling valve. See Sample
    introduction.
Gas-solid chromatography,25,
    103-109
  adsorbent properties,104
  adsorbents used,107
  adsorption of gases,105
Gas standards. See Standards.
Gas-tight syringes. See
    Sample introduction.
Gas volume detector,285
Gases,permanent,126
Gaussian curve,equation for,50
Geiger counter,283
Geometrical isomers,286
Ghost peaks,305
  elimination of,305
  septum related,305
Gibbs free energy. See Free
    energy.
Glass columns,498
Glass wool,135,144
Glucosiduronate,507
Glucuronic acid,499
Glucuronidase,499,508,511
Glutethimide,544

H

Hagen-Poiseuille equation,146
Halogen derivatization,265
Halothane,527

Hammett equation,566
Heart cutting,315
Heat,of adsorption. See Enthal-
    py of adsorption.
Heat,of solution. See Enthalpy
    of solution.
Heat,of vaporization,571-573
Height equivalent to an effect-
    ive theoretical plate,26
Height equivalent to a theoret-
    ical plate,26,53,62,64,71,
    74
Helium detector,265
Henry's law,46,573
Herbicide analysis,280
Herbicides,389
Heroin,606,609,617
HETP,26,51,60,62,69,72,117
Heptafluorobutyrate derivative,
    507,518,534
Heptafluorobutyryl n-propyl
    esters,534
Hexamethyldisilizane(HMDS),511,
    519,611
Histamine,540
History of chromatography,2-4
HMDS,83
Hold-up volume,26
Homogentisic acid,534
Homovanillic acid(HVA),518
Hormones,628
Hot-wire detector,228
Hydrogen bond energies,586
Hydrogen flame temperature de-
    tector,284
Hydrolysis of proteins,465,475
17 α-hydroxyprogesterone,510

I

Ignitor,252
Illicit formulation,606-69
Indium tube. See Sample intro-
    duction.

Index

Infrared detectors,286
Injection point,26
Injection port,26,498
Injection temperature,26
Injector volume,26
Inlet systems for capillary
   columns,311
Instrumentation,289-363
Integral detector,26
Integrator,26
Integrators,422,424-448
 advantages and disadvantages,
   422
 description of sizes and
   classes,424-426
 planning and economics,426-
   441
 troubleshooting,442-448
Interaction potential,gas-
   solid,586
Internal normalization. See
   Standardization techniques.
Internal standard,27,499,598
Internal standard selection,
   201
Internal standardization. See
   Standardization techniques.
Interstitial fraction,27
Interstitial velocity of
   carrier gas,27
Interstitial volume,27
Ion exchange,466,476
Ionization potential measurements,586
Isolation and concentration
   techniques,380,381
 concentration of liquid or
   solid samples,381
 concentration of water-
   based samples,381
 evaporation,380
 extraction,380
 trapped materials,381
   removal and transfer,381
   porous polymers,381

   Charcoal,381
   Chromosorb,101,102,etc.
   381
   Porapak,381
   XAD-1,2,etc.,381
 trapping systems,380
   packed,380
   alumina,380
   charcoal,380
   liquid-coated solid supports,380
   porous polymers,380
   silica gel,380
Isotherms,10-12,60
 adsorption,10-12
 concave,10-12
 convex,10-12
 linear,10-12,62
 nonlinear,10

J

"j" factor, 565

K

Katharometer,27
Keesom orientation forces,89
17-Ketosteroids,508,510
Kieselguhr. See Diatomite.
Knudsen diffusion,51
Kovat's retention index,93,155

L

Langmuir equation,12
LAC-4R-886,486,489
Leaks,136
Lecithin,464
Linear dynamic range,225
Linear flowrate,27
Linearity,598-600
Lipase,pancreatic enzyme,461
Lipids,451-465

Index

Lipids, stationary phases,456
Liquid chromatography,382,465,
    466,476
Liquid crystals,586
 degree of order,586
 transitions,586
Liquid liquid extraction,53-59
 efficiency in,58
 equilibrium of,54-58
Liquid phase,27,88,89,342
 bleeding of,342
 choice of,88
 see stationary phase
 viscosity of,89
 wetting ability of,89
Liquid sampling valve. See
    Sample introduction.
Liquid standards. See
    Standards.
Liquid syringes. See Sample
    introduction.
Literature sources,128
London dispersion forces,89

M

Malonyl urea,543
Marijuana,549
Marker,28
Mass chromatograph,278
 carrier gas,278
Mass chromatography,158
Mass distribution ratio,28
Mass spectrometer,285
Mass transfer,61
 resistance to,61,71-75
Mean interstitial velocity of
    carrier gas,28
Melatonin,540
Memory peaks. See Ghost peaks.
Meperidine,549
Meprobamate,544
Mesh size of supports,84,86,
    118,123
Metal columns,498

Metanephrine,518
Methadone,549
Methionine,537
Method development(for drugs),
    597-603
Methods, chromatographic,4-6
p-Methoxyamphetamine,540
5-Methoxytryptamine,540
N-Methylamphetamine,540
Methylation,614
N-Methylbenzylamine,540
Methylthiohydantoins,475
Microwave plasma detector,275
Minimum detectable level,219
Miscellaneous detectors,283
Mobile phase,22,28,68,85,291-
    300
 velocity of,68
Model methods,397
 sources of,397
 AOAC,397
 EPA,397
 NIOSH,397
Molecular sieves,124
Molecular sieve traps,292
 for removing impurities in
    carrier,292
 reconditioning of,292
Molecular weight,586
Molecular weight chromatography,
    158
Morphine,459
Moving phase,22,28
Mutarotation,481

Mc

McReynolds constants,130,131,157

N

Narcotics,617
Net retention volume,28
Nickel,63,257
Nicotine,541

Index

Nonvolatile solutes in water,
  determination of,398-403
  analytical procedure,399
  column preparation,402
  sample processing,402
  Snyder column,403
  Soxhlet extractor,399
Norepinephrine,518,539
Normal error curve equation,
      63
Normetanephrine,540
Norpropoxyphene,549
Nortriptyline,549
Number of plates,63,65,92

                O

Octopamine,540
On-column methylation,615,618
Open tubular columns,22,28,
      145-148
  column performance,147
  column permeability,146,147
  description and capabilities,
      145
  modifications of,147
  performance index,148
  resolution of,148
  specific permeability co-
      efficient,146
Optically active amino acids,
      472
Organic nitrogenous bases,
      619-628
O-Ring memory effects,304
Oxygen,
  effect on column,136
  removal from carrier gas,136

                P

"p" value,160,387
Packing material,28
Packings,column,
  preparation of,138

recommended,139
Particle size,65,75,83,84
Partition chromatography,28
Partition coefficient,10,161
Peak,28,29,61,503
  area,28,503
  base,28
  height,29
  maximum,29'
  resolution,29
  sharpness of,61
  width,29
  width at half-height,29
Peak area measurement. See peak
      size measurement.
Peak height,factors influencing,
      168
Peak height measurements. See
      peak size measurement.
Peak size measurement,167-179
  comparison of techniques,177,
      179
  cut and weigh,172
  disc integrator,172
  electronic integrators,175
  error of manual methods,177
  height and width at half-
      height,170
  peak height,167
  planimeter,172
  survey of methods used,176
  triangulation,171
Peak width, factors influencing,
      168
Pentobarbital,544
Peptides,465,466,473,475
Performance index,29
Permeability coefficient,79
Permeation tubes. See Standards.
Pesticides,265,269
Pesticide analysis,157,270,280
Phase. See Stationary phases.
Phase equilibria,43-47
Phase ratio,29,55
Phase solubility analysis,597

Phases,stationary. See
    Stationary phases.
Phenethylamine,540
Phenochromocytoma,518
Phenolic alcohols,519
Phenobarbital,544
Phenylalanine,531
Phenylketonuria,520,530,531
Phosphatidic acid,464
Phosphatidyl choline,464
Phosphatidyl ethanolamine,464
Phosphatidyl serine,464
Phospholipase A_1 enzyme,464
Phospholipase A_2 enzyme,464
Phospholipase C enzyme,464
Phospholipase D enzyme,464
Phospholipids,464
 lecithin,464
 phosphatidyl choline,464
 phosphatidic acid,464
 phosphatidyl ethanolamine,464
 phosphatidyl serine,464
Physicochemical measurements,
    553-589
Piezoelectric sorption detect-
    or,282
Planimeter. See Peak size
    measurement.
Plate height,73
Plates,theoretical,64,116
Plate theory,14,61-65
PLOT columns,29,146
Polarity,128
Polarity,liquid phases,90
Polarity scale,90
Polychlorinated biphenyls,265
Porapak,124
Porous polymers,124
Post-column,166
Potentiometric recorder,29
Pre-column,161
Pregnanediol,urinary,506
Pregnanetriol,urinary,510
Pre-heater,87
Preparative gas chromatography,

    138,142
 essential components of,142
Pressure devices,87
Pressure gradient correction
    factor,29,565
Primidone,544
Printer/plotter,348
 as a stand-alone recorder,352
 inkless plotting,352
 recorder needs,fulfilling,352
Process analyzers. See continu-
    ous stream analysis.
Progesterone,508,510
Programmed temperature gas
    chromatography(PTGC),33,
    243,326,332
Proportional counter,283
Propoxyphene,549
Proteases,475
Protein sequence,473
Proteins,hydrolysis of,465,475
PTGC. See Programmed tempera-
    ture gas chromatography.
Pyrogram,17,30
Pyrolysis,16,30,162
Pyrometer,30

Q

Qualitative analysis,153-166
 by carbon skeleton determin-
    ation,154
 by controlled pyrolysis,162
 by extractions,160
 by functional group analysis,
    163
 from Beroza's "p" value,160
 from Kovat's indices,93,155
 from log retention versus
    carbon number plots,154
 from mass chromatography,158
 from multiple column retention
    data,157
 from retention data,153
 logic of,166

Index

of pesticides,161
using elemental analysis,163
using GC and other instru-
    mental methods,164
using pre-columns,161
using selective detectors,
    158
using water-air equilibria,
    161
Quantitative analysis,166
Quantitative analysis errors.
    See Errors.
Quantitative analysis in trace
    analysis,
standardization,390
    adsorption on column
    packings,391
    degradation on column
    packings,391
    internal standard,
    390,391
    peak tailing,391

                R

Radioactivity detectors,282
Radioimmunoassay,497,549
Random walk theory,66-71
Raoult's law,44-46,565-566
Rate theory,14,65-81
Rates of reaction,578
Reaction coulometer,280
Reaction rates,578
Recorders,348-352
Recorder span,32
Regulators,pressure,292-297
  for carrier flow control,
    297
Relative retention,30
Relative volatility. See
    separation factor.
Reproducibility,600
Required plate numbers,30,94
Resolution,30,93,601-603
Response factors,181,199,221

Retention behavior,
  in PTGC,326
Retention index,30,129,130
Retention time,31
Retention volume,21,23,28,31,32
  adjusted,21
  corrected,23
  net,28
  specific,32
Reynold's number,77,80
Rohrschneider constants,130

                S

Sample,31
Sample injector,31
Sample introduction,198,204,206-
    208,305,315,321
  gas sampling valve,207,321
  gas-tight syringes,206
  gases,206,207
  indium tube,208
  liquid sampling valve,208
  liquid syringes,204
  liquids,204,305
  solids,198,208,315
  with capillary columns. See
    Inlets for capillary col-
    umns.
Sample ports,305
Sampling techniques,203
Scintillation counter,283
SCOT,32
SCOT columns,75,146
Screen openings,84
Secobarbital,544
Selectivity,column,114,119,130
Semiconductive Thin-film detect-
    or,283
Sensitivity,220
Separation,31
  phase concentration,effect of,
    132
Separation factor,31,59,89,136
Separation number,31,136

Index

Separation techniques,40-60
  classification of,41
  processes of,41
Separation temperature,32,133
Septum-less injectors,309
Septums,309
Sequence,amino acid,473
Sequence,protein,473
Serine,phosphatidyl,464
Serotonin,540
Sieve size,84
Signal-to-noise ratio,217
Silanization,82,498,611-613
Silica gel,124
Silicone stationary phases,
    conditioning,145
Silylation,82,498,611-613
Solid sampling,315
Solid support,32,81-85
Solubility measurements,586
Solute,32
Solution coating,142
Solvent,32
Solvent efficiency,32
Sorbates,149-151
  classification of,149-151
  retention of,151
Sources of analysis errors. See
    Errors in quantitative
    analysis.
Specific detector,222
Specific retention volume,32
Specific surface area,32,554
Spectrophotometric detector,286
Splitter,32
Squalane,90
Standard deviation,51.67
Standard solutions,stability of,
    397
Standard stability. See
    Standards.
Standardization techniques,
  external standardization,184
  internal normalization,181
  internal standardization,199

Standards,180,187,189,190,191,
    193,194,196-198,395
  diffusion tube technique,193
  dynamic,190
  gas,187,190,194
  liquid,197,395
  make or buy?,180
  multiple component,189,198
  multiple dilution,191
  permeation tube technique,194
  purity of,180
  stability,180,189,198
  static,187
  vapor,189,191,196
  vapor pressure technique,191
Standing current,256
Static gas standards. See
    Standards.
Stationary phase,32,88,127,498
Stationary phase,
  choice of,88,127
  concentration of,132,138
  effect on efficiency,117
Stationary phase fraction,32
Stationary phase volume,32
Stationary phases
  amino acids,468,471
    Apiezon M,468,471
    CD-430,471
    OV-3,471
    OV-7,471
    OV-25,471
    OV-61,471
    OV-210,468,471
    OV-225,471
    SE-54,471
    SP-2401,471
  amino acid derivatives,468
    Carbowax 20M,468
    Dexsil-400 GC,468
    EGA,468
    Lexan,468
    MS-200,468
    QF-1,468
    SE-30,468

Index

Silar SCP,468
Tabsorb,468
Ucon-75H90,468
XE-60,468
carbohydrates,482,484,487
Carbowax 20M,487
ECN55-M,484
EGS,482
OV-17,484,487
QF-1,487
SE-30,487
SE-52,481
XE-60,484,487
fatty acid methyl esters,458
DEGS,458
EGA,458
EGS,454,455
OV-275,458
Silar 10 C,458
SP-22-PS,458
XF-1150,454
lipids,456
300 GC,456
400 GC,456
410 GC,456
238-149-99,456
BbS,456
DEGA,456
DEGS,456
EGA,456
EGS,456
EGSS-X,456
EGSS-Y,456
NGA,456
NGS,456
NG Sebacate,456
OV-275,456
Silar,5CP,456
Silar 7CP,456
Silar 9CP,456
Silar,10C,456
SP222,456
SP2300,456
SP2310,456
SP2330,456

SP2340,456
XE-60,456
XF-1150,456
triglycerides,453
JXR,453
OV-1,453
OV-17,453
SE-30,453
SE-52,453
Stefan-Boltzman equation,232
Sterchmal. See Diatomite.
Steroids,497,612,628
Substrate,32
Sulfatase,511
Sulfur hexafluoride,527,528
Supports,solid
deactivation,121
efficiency,effect on,117
porous polymers,124
silanization,121
size,mesh,117,118
Teflon,123
types,120,122
Surface area,32,554-558
of catalysts,557
theory,554-558
Surface diffusion,51
Surface tension,measurement of, 586
Synephrine,540
Syringes. See sample introduc- tion.

T

Tail reducers,123
Tailing,33,74,82,115,119,611
reduction of,121,123,132,138
Teflon supports,123
Temperature control,
of column oven,321
of detectors,338
of injection ports,305
Temperature
effect on column efficiency,119

Index

initial and final,26
injection,26
programming,33,243,326-332
separation,32
Temperature limit,
 of column,123
Temperature ramps,328
 ballistic,328
 linear,328
 multi-linear,328
Tetrahydrocannabinol,548
Tetramethylammonium hydroxide,
 545
Theoretical plate number,33,64
Theory of chromatography,39-
 112
Thermal conductivity,33
 cell,33,87
Thermal conductivity detector,
 228,277,335-348
 carrier gas,243
 DPS number,234
 geometry factor,229
 minimum detectable level,240
 operating hints,241
 response factors,238
 temperature programming,243
 typical conditions,231
 variations,237
Thermionic detector,269
Thermistor bead element,33
Thermodynamics,562-578
 activity coefficient,565
 by GLC
        enthalpy of solution,568
        enthalpy of vaporiza-
        tion,571
        entropy of solution,568
        entropy of vaporiza-
        tion,571
        equilibrium constants,
        558,574
        free energy of solution,
        568-571
        heat of solution,568

        heat of vaporization,571-
        573
        Henry's law constant,573
 by GSC
        distribution constant,575
        enthalpy of adsorption,576
        entropy of adsorption,576
        free energy of adsorption,
        576
        heat of adsorption,576
 excess functions,570
Thin layer chromatography(TLC),
 461,608,619,628
Thiols,611
Titration detector,284
TLC (thin layer chromatography),
 461
Trace analysis,
 accuracy requirements,369
 analytical ethics,416-417
 atmospheric pressure sampling,
 414
 cleanup methods,pesticide
 recovery,382
 "column chromatography,"385
 concentration range (definition),369
 in forensic science,367
 GC/mass spectrometry,389
 of gases,265
 general scheme for,409
 in industrial products,367
 in medical and pharmaceutical
 research,367
 methylation,389
 in pollution analysis,367
 qualitative GC analysis,386
 quantitative analysis,GC,390-
 395
 sampling and sampling devices,
 372
 "single ion monitoring,"389
 spiking,411
 statistical variations,416-417
 syringe,gas-tight,416

Index

thin-layer chromatography,385
water analysis, use of macro-
    reticular resins,400-401
Tranquilizers,609,618
Transpost properties,583-585
Trapping,165
Trichloroethylene,527
Trifluoroacetate derivatives,
  509
n-Trifluoroacetyl n-butyl
    esters,534
Triglycerides,460-464
 stationary phases,453
Trimethadione,549
Trimethylanilinium hydroxide,
  545
Trimethylchlorosilane,511,519
Trimethylsilyl ether deriva-
    tives,507,509,510,511
    519,534,540
Tritium H^3,257
Troubleshooting,134
True adsorbent volume,33
Trypsin,475
Tubing,treatment of,134,143
Tubular columns. See capillary
    colums .
Types of chromatography,1-37
Tyramine,540
Tyrosinosis,530

U

Ultrasonic detector,281
Universal detector,222

V

van Deemter equation,71-77,117
van der Waals forces,48
Vanillyl mandelic acid (VMA),
  518
van't Hoff equation,48
Vapor Standards. See Standards.
Variance,51,67

Vented septum injectors,310
 effect on baseline,310
Virial coefficients,579-583
 gas-solid,582
 second,580
Viscosity,measurement of,586
Vitamins,630
Volatile organic solutes in
    water,determinatinn of,
    404-409
 chlorinated hydrocarbons,406
 equilibration technique,404
 aromatic hydrocarbons,406
Volmer diffusion,51
Wander,218
Water,
 effect on column,135
 removal from carrier gas,135
Water sampling,
 automatic sampling,373
 composite sample,374
 "grab" samples,373
 ponds,lakes,marshes,374
 preservatives,373
 pressurized water system ,372
 waste or natural streams,373
WCOT columns,33,146
Weight stationary liquid phase,
  33
Wetting phenomena,586
Wool, glass,135,143

XYZ

XF-1150,454
Zone,13,33
 broadening of,13,62,67,69,70,
  74,76
 thickness of,62
 velocity of,69